Lecture Notes in Computer Science 8610

Dominik Ślęzak Gerald Schaefer
Son T. Vuong Yoo-Sung Kim (Eds.)

Active Media Technology

10th International Conference, AMT 2014
Warsaw, Poland, August 11-14, 2014
Proceedings

Volume Editors

Dominik Ślęzak
University of Warsaw, Poland
and Infobright Inc., Poland
E-mail: slezak@mimuw.edu.pl

Gerald Schaefer
Loughborough University, UK
E-mail: gerald.schaefer@ieee.org

Son T. Vuong
University of British Columbia
Vancouver, BC, Canada
E-mail: vuong@cs.ubc.ca

Yoo-Sung Kim
Inha University, Incheon, Korea
E-mail: yskim@inha.ac.kr

ISSN 0302-9743 e-ISSN 1611-3349
ISBN 978-3-319-09911-8 e-ISBN 978-3-319-09912-5
DOI 10.1007/978-3-319-09912-5
Springer Cham Heidelberg New York Dordrecht London

Library of Congress Control Number: 2014945241

LNCS Sublibrary: SL 3 – Information Systems and Application, incl. Internet/Web and HCI

Typesetting: Camera-ready by author, data conversion by Scientific Publishing Services, Chennai, India

Printed on acid-free paper

Springer is part of Springer Science+Business Media (www.springer.com)

Preface

This volume contains the papers selected for presentation at the 2014 International Conference on Active Media Technology (AMT 2014), held as part of the 2014 Web Intelligence Congress (WIC 2014) at the University of Warsaw, Poland, during August 11–14, 2014. The conference was organized jointly by the Web Intelligence Consortium, the University of Warsaw, the Polish Mathematical Society, and Warsaw University of Technology.

The series of Active Media Technology conferences began in Hong Kong in 2001. Since then, AMT has been held in China and Japan (multiple times in both countries), as well as in Australia and Canada. In 2014, this series of events visited Europe for the first time ever.

AMT 2014 received 86 paper submissions, in the areas of foundations of active media, ubiquitous intelligence and ubiquitous services, big data management for active media, data mining for active media, social networks and social media, interactive systems and intelligent interfaces, as well as active media engineering and applications. After a rigorous evaluation process, 27 papers were selected as regular contributions, giving an acceptance rate of 31.4%, and are grouped into the first five sections of this volume.

The last four sections of this volume contain 20 papers selected for oral presentations in AMT 2014 special sessions and workshops. We would like to thank all special session and workshop organizers, as well as all authors who contributed their research results to this volume.

The congress provided a very exciting program with a number of keynote talks, regular and special sessions, workshops, tutorials, panel discussions, and social programs. We greatly appreciate our keynote speakers: Andrew Chi-Chih Yao, Karl Friston, Henryk Skarżyński, Stefan Decker, Robert Kowalski, Sadaaki Miyamoto, Yi Pan, John F. Sowa, and Andrzej Szałas. We would also like to acknowledge all tutorial and panel speakers for preparing high-quality lectures and conducting inspiring research discussions.

A program of this kind would not have been possible without the dedication of Marcin Szczuka and the whole Local Organizing Committee, the administrative support by Juzhen Dong, the strategic guidance by Andrzej Skowron, Ning Zhong, and Jiming Liu, as well as all other chairs, Program Committee members, and external reviewers.

We would also like to thank our institutional patrons: the Ministry of Science and Higher Education of the Republic of Poland and the Warsaw Center of Mathematics and Computer Science.

We wish to express our gratitude to our sponsors: Dituel (Main Sponsor), Google (Gold Sponsor), Human Brain Project, IOS Press (Web25 Event Sponsor), Core Technology, and Gemius. Without their generous support we would never have been able to make this conference such a success.

We are also grateful to the Committee on Informatics of the Polish Academy of Sciences, the Polish Artificial Intelligence Society, the World Wide Web Consortium (Web25 Anniversary initiative), the European Brain Council (Year of the Brain in Europe initiative), and the IEEE CIS ETTC Task Force on Brain Informatics for help with publicizing the congress.

Finally, we wish to acknowledge Andrzej Janusz and Marcin Możejko for taking over the job of putting this volume together. We extend our highest appreciation to Springer's LNCS/LNAI team for their generous support. In particular, we thank Alfred Hofmann, Anna Kramer, Ingrid Beyer, and Leonie Kunz for their help in coordinating the publication of this volume.

August 2014

Dominik Ślęzak
Gerald Schaefer
Son T. Vuong
Yoo-Sung Kim

Conference Organization

General Congress Chair

Andrzej Skowron	University of Warsaw, Poland

General Program Chair

Dominik Ślęzak	University of Warsaw and Infobright Inc., Poland

AMT 2014 Program Co-chairs

Gerald Schaefer	Loughborough University, UK
Son T. Vuong	UBC, Canada and NTTU, Vietnam
Yoo-Sung Kim	Inha University, Korea

Workshop Chairs

Lipika Dey	TATA Consultancy Services, India
Adam Krasuski	Main School of Fire Service, Poland
Marek Reformat	University of Alberta, Canada

Tutorial Chairs

Hiranmay Ghosh	TATA Consultancy Services, India
Radosław Katarzyniak	Wrocław University of Technology, Poland
Christina Schweikert	St. John's University, USA

Publicity Chairs

Shinichi Motomura	Maebashi Institute of Technology, Japan
Piotr S. Szczepaniak	Technical University of Łódź, Poland
JingTao Yao	University of Regina, Canada

Publication Chairs

Marcin Dziubiński	University of Warsaw, Poland
Andrzej Janusz	University of Warsaw, Poland

Financial Chair

Juzhen Dong	Web Intelligence Consortium, Japan

Local Organizing Committee

Marcin Szczuka	University of Warsaw, Poland
Łukasz Sosnowski	Dituel, Poland
Henryk Rybiński	Warsaw University of Technology, Poland
Andrzej Jankowski	Warsaw University of Technology, Poland

Steering Committee Chairs/WIC Directors

Ning Zhong	Maebashi Institute of Technology, Japan
Jiming Liu	Hong Kong Baptist University, Hong Kong, SAR China

Steering Committee/WIC Technical Committee

Jeffrey Bradshaw	UWF/Institute for Human and Machine Cognition, USA
Nick Cercone	York University, Canada
Dieter Fensel	STI/University of Innsbruck, Austria
Georg Gottlob	Oxford University, UK
Lakhmi Jain	University of South Australia, Australia
Jianchang Mao	Microsoft, USA
Jianhua Ma	Hosei University, Japan
Pierre Morizet-Mahoudeaux	Compiegne University of Technology, France
Hiroshi Motoda	Osaka University, Japan
Toyoaki Nishida	Kyoto University, Japan
Vijay Raghavan	University of Louisiana at Lafayette, USA
Andrzej Skowron	University of Warsaw, Poland
Jinglong Wu	Okayama University, Japan
Xindong Wu	University of Vermont, USA
Yiyu Yao	University of Regina, Canada

AMT 2014 Program Committee

Arun Agarwal	University of Hyderabad, India
Md. Atiqur Rahman Ahad	University of Dhaka, Bangladesh
Aijun An	York University, Canada
Pradeep Atrey	University of Winnipeg, Canada
Dominik Batorski	University of Warsaw, Poland

Gloria Bordogna	National Research Council, Italy
Tibor Bosse	Vrije Universiteit Amsterdam, The Netherlands
M. Emre Celebi	Louisiana State University at Shreveport, USA
Shampa Chakraverty	Netaji Subhas Institute of Technology, India
Chien-Chung Chan	University of Akron, USA
Carl K. Chang	Iowa State University, USA
Ruay-Shiung Chang	National Dong Hwa University, Taiwan
Santanu Chaudhury	IIT Delhi, India
Min Chen	Huazhong University of Science and Technology, China
Yiqiang Chen	Institute of Computing Technology, CAS, China
Ray Cheung	City University of Hong Kong, Hong Kong, SAR China
William Cheung	Hong Kong Baptist University, Hong Kong, SAR China
Yiu-ming Cheung	Hong Kong Baptist University, Hong Kong, SAR China
Chin-Wan Chung	KAIST, Korea
Martine De Cock	Ghent University, Belgium
Diane Cook	Washington State University, USA
Alfredo Cuzzocrea	DEIS-Unical, Italy
Bogusław Cyganek	AGH University of Science and Technology, Poland
Matthias Dehmer	University for Health Sciences, Medical Informatics and Technology, Austria
Jitender Deogun	University of Nebraska-Lincoln, USA
Scott E. Fahlman	Carnegie Mellon University, USA
Yuhong Feng	Shenzhen University, China
Xiaoying (Sharon) Gao	Victoria University of Wellington, New Zealand
Osvaldo Gervasi	University of Perugia, Italy
Ali Akbar Ghorbani	University of New Brunswick, Canada
Hiranmay Ghosh	TATA Consultancy Services, India
Adrian Giurca	Brandenburg University of Technology at Cottbus, Germany
Marco Gori	University of Siena, Italy
William Grosky	University of Michigan, USA
Venkat Gudivada	Marshall University, USA
Bin Guo	Northwestern Polytechnical University, China
Hakim Hacid	SideTrade, France
Aboul Ella Hassanien	University of Cairo, Egypt
Christopher J. Henry	University of Winnipeg, Canada
Daryl Hepting	University of Regina, Canada
Enrique Herrera-Viedma	University of Granada, Spain
Wilko Heuten	OFFIS Institute for Technology, Germany
Masahito Hirakawa	Shimane University, Japan

Mark Hoogendoorn	Vrije Universiteit Amsterdam, The Netherlands
Simo Hostikka	Aalto University School of Engineering, Finland
Ching-Hsien Hsu	Chung Hua University, Taiwan
D. Frank Hsu	Fordham University, USA
Bin Hu	Lanzhou University, China
Hiroki Inoue	Kyoto University, Japan
Hiroshi Ishikawa	Kagawa University, Japan
Andrzej Jankowski	Warsaw University of Technology, Poland
Hanmin Jung	KAIST, Korea
Yoshitsugu Kakemoto	JSOL Inc., Japan
Byeong Ho Kang	University of Tasmania, Australia
Brigitte Kerhervé	Université du Québec à Montréal, Canada
Haklae Kim	Samsung Electronics Inc., Korea
Myungchul Kim	KAIST, Korea
Mario Koeppen	Kyushu Institute of Technology, Japan
Bożena Kostek	Gdańsk University of Technology, Poland
Gang Kou	University of Electronic Science and Technology, China
Adam Krasuski	Main School of Fire Service, Poland
Noriyuki Kushiro	Kyushu Institute of Technology, Japan
Halina Kwaśnicka	Wrocław University of Technology, Poland
Juan Li	North Dakota State University, USA
Kuan-Ching Li	Providence University, Taiwan
Wen-bin Li	Shijiazhuang University of Economics, China
Kathy Liszka	University of Akron, USA
Jiming Liu	Hong Kong Baptist University, Hong Kong, SAR China
Xiaohui Liu	Brunel University, UK
Lucelene Lopes	Pontifical Catholic University of Rio Grande do Sul, Brazil
Sanglu Lu	Nanjing University, China
Marco Luetzenberger	DAI-Labor, TU-Berlin, Germany
Xiong Luo	University of Science and Technology Beijing, China
Ilias Maglogiannis	University of Piraeus, Greece
Azizah Abd Manaf	UTM, Malaysia
Wenji Mao	Institute of Automation, CAS, China
Jesus Martinez-del-Rincon	Queen's University of Belfast, UK
Stan Matwin	Dalhousie University, Canada
Ernestina Menasalvas	Universidad Politécnica de Madrid, Spain
Marcin Michalak	Silesian University of Technology, Poland
Pabitra Mitra	IIT Delhi, India
Sabah Mohammed	Lakehead University, Canada
Katarzyna Musiał	King's College London, UK

Omar Nibouche	University of Ulster, UK
Maria do Carmo Nicoletti	Federal University of Sao Carlos, Brazil
Toyoaki Nishida	Kyoto University, Japan
Cyrus F. Nourani	TU-Berlin-DAI, Germany and SFU, Canada
Mariusz Nowostawski	Gjovik University College, Norway
Kouzou Ohara	Aoyama Gakuin University, Japan
Yukio Ohsawa	University of Tokyo, Japan
Yoshihiro Okada	Kyushu University, Japan
Yoshiaki Okubo	Hokkaido University, Japan
V. Kamakshi Prasad	JNTU Hyderabad, India
Sheela Ramanna	University of Winnipeg, Canada
Felix Ramos	Research and Advanced Studies Center, Mexico
Phill Kyu Rhee	Inha University, Korea
Witold Rudnicki	University of Warsaw, Poland
Jose Hiroki Saito	Federal University of Sao Carlos, Brazil
Maria J. Santofimia	Universidad de Castilla-La Mancha, Spain
Eugene Santos	University of Conneticut, USA
Franciszek Seredyński	Cardinal Stefan Wyszyński University, Poland
Denis Shestakov	Aalto University, Finland
Władysław Skarbek	Warsaw University of Technology, Poland
Andrzej Skowron	University of Warsaw, Poland
Jacek Sroka	University of Warsaw, Poland
Ya Su	University of Science and Technology Beijing, China
Tsuyoshi Sugibuchi	Internet Memory Research, France
Kazunari Sugiyama	National University of Singapore, Singapore
Yuqing Sun	Shandong University, China
Piotr Synak	Infobright Inc., Poland
Marcin Szczuka	University of Warsaw, Poland
Piotr Szczuko	Gdańsk University of Technology, Poland
Yasufumi Takama	Tokyo Metropolitan University, Japan
Domenico Talia	University of Calabria, Italy
Xijin Tang	Academy of Mathematics and Systems Sciences, CAS, China
Xiaohui Tao	University of Southern Queensland, Australia
Takao Terano	Tokyo Institute of Technology, Japan
Juan Velásquez	University of Chile, Chile
Quoc Bao Vo	Swinburne University of Technology, Australia
Le Anh Vu	NTTU, Vietnam
Guoyin Wang	Chongqing University of Posts and Telecommunications, China
Haipeng Wang	Northwestern Polytechnical University, China
Huanjing Wang	Western Kentucky University, USA

Rajeev Wankar	University of Hyderabad, India
Alicja Wieczorkowska	Polish-Japanese Institute of Information Technology, Poland
Szymon Wilk	Poznań University of Technology, Poland
Bogdan Wiszniewski	Gdańsk University of Technology, Poland
Yue Xu	Queensland University of Technology, Australia
Jian Yang	Beijing University of Technology, China
Neil Y. Yen	University of Aizu, Japan
Tetsuya Yoshida	Nara Women's University, Japan
Zhiwen Yu	Northwestern Polytechnical University, China
Guoqing Zhang	University of Windsor, Canada
Hao Lan Zhang	Zhejiang University, China
Shichao Zhang	University of Technology, Australia
Zili Zhang	Southwest University, China
Ning Zhong	Maebashi Institute of Technology, Japan
Huiyu Zhou	Queen's University Belfast, UK
Zhangbing Zhou	Institut Telecom & Management SudParis, France
Tingshao Zhu	Graduate University of CAS, China

AMT 2014 External Reviewers

Timo Korhonen
Karol Kreński
Juan Carlos López
Michał Meina
Marcin Mirończuk
Mark Neerincx
Krzysztof Rykaczewski
Fariba Sadri
Xavier del Toro
Piotr Wiśniewski
David Villa
Félix Villanueva

Table of Contents

Regular Contributions

Active Media and Ubiquitous Learning Foundations

Big Data Management and Mining for Active Media

Active Media Engineering

Security and Privacy in Ubiquitous Environment

Social Networks and Social Media

Special Sessions

Human Aspects in Cyber-Physical Systems

Facial Image Analysis with Applications

Workshops

Human Aspects in Ambient Intelligence

Sensing, Understanding, and Modeling for Smart City

A Semi-supervised Learning Algorithm for Web Information Extraction with Tolerance Rough Sets*

Cenker Sengoz and Sheela Ramanna

Department of Applied Computer Science, University of Winnipeg, Winnipeg, Manitoba R3B 2E9 Canada
sengoz-c@webmail.uwinnipeg.ca, s.ramanna@uwinnipeg.ca

Abstract. In this paper, we propose a semi-supervised learning algorithm (TPL) to extract categorical noun phrase instances from unstructured web pages based on the tolerance rough sets model (TRSM). TRSM has been successfully employed for document representation, retrieval and classification tasks. However, instead of the vector-space model, our model uses noun phrases which are described in terms of sets of co-occurring contextual patterns. The categorical information that we employ is derived from the Never Ending Language Learner System (NELL) [3]. The performance of the TPL algorithm is compared with the Coupled Bayesian Sets (CBS) algorithm. Experimental results show that TPL is able to achieve comparable performance with CBS in terms of precision.

Keywords: Semi-supervised learning, tolerance approximation spaces, tolerance rough sets model, NELL, Coupled Bayesian Sets.

1 Introduction

In this paper, we propose a semi-supervised learning algorithm to extract categorical noun phrase instances from a given corpus (unstructured web pages) based on the tolerance rough sets model (TRSM). TRSM has been successfully employed for document representation, retrieval and classification tasks [9,8,17]. The categorical information that we employ is derived from the Never Ending Language Learner System (NELL) [3] which operates in a semi-supervised mode continuously extracts and structurizes relevant information from the Internet to grow and maintain a knowledge base of facts. The facts in question are represented by two means: category instances e.g. *City*(Winnipeg) and semantically related pairs e.g. *City-In*(Winnipeg, Canada). The core component of NELL,

* This research has been supported by the Natural Sciences and Engineering Research Council of Canada (NSERC) Discovery grants 194376. We are very grateful to Prof. Estevam R. Hruschka Jr. and to Saurabh Verma for the NELL dataset and for discussions regarding the NELL project. Special Thanks to Prof. James F. Peters for helpful suggestions.

D. Ślęzak et al. (Eds.): AMT 2014, LNCS 8610, pp. 1–10, 2014.

Coupled Pattern Learner (CPL) [2], is a free text extractor and it uses what are called *contextual patterns* to detect and extract *noun phrases* instances for the knowledge base.

Semi-supervised approaches using a small number of labeled examples together with many unlabeled examples are often unreliable as they frequently produce an internally consistent, but nevertheless, incorrect set of extractions [16]. While such semi-supervised learning methods are promising, they might exhibit low accuracy, mainly, because the limited number of initial labeled examples tends to be insufficient to reliably constrain the learning process, thus, creating concept drift problems [5]. Verma et. al [16] solve this problem by simultaneously learning independent classifiers in a new approach named Coupled Bayesian Sets (CBS) algorithm. CBS outperforms CPL after 10 iterations rendering it a good possible alternative for the CPL component of NELL. In this paper, we compare our semi-supervised algorithm based on TRSM to the Coupled Bayesian Sets (CBS) algorithm proposed by Verma and Hruschka [16].

The contribution of the paper is a Tolerant Pattern Learner (TPL) algorithm which uses noun phrase-contextual pattern co-occurrence statistics from a given corpus (NELL dataset) to extract credible noun phrase instances for predefined categories. TPL is based on the TRSM model and is a semi-supervised learning algorithm for learning contextual patterns. Overall, TPL is able to demonstrate comparable performance with CBS in terms of precision. The paper is organized as follows: In Section 2, we discuss the research closely related to our problem domain. Section 3 describes our proposed tolerant pattern learner (TPL) algorithm. Sections 4 and 5 yield experiments and discussion on TPL, respectively. We conclude the paper in section 6.

2 Related Work

2.1 Semi-supervised Learning with CBS and NELL

The ontology for NELL is initialized by a limited number of labeled examples for every category/relation [3,2]. The categories and relations are predefined by the ontology. Those examples bootstrap the system and are used to label more instances, which will then be used to label more, leading to an ever-growing knowledge base.

Coupled Pattern Learner (CPL) [2] uses *contextual patterns* to detect and extract *noun phrase* instances for the knowledge base. Those patterns are sequences of words e.g. "sports such as *arg*" or "*arg1* is the president of *arg2*" providing a context for a noun phrase argument. The main idea is, noun phrases that are likely to belong to a particular category/relation are also likely to co-occur frequently with the patterns associated to that category. The vice versa also holds. Accordingly, the co-occurrence information between noun phrases and contexts is what CPL relies on for learning.

Inspired by NELL, Verma and Hruschka proposed the Coupled Bayesian Sets (CBS) [16] algorithm to fulfill the same task: extracting noun phrases to populate category instances. Likewise, it follows a semi-supervised approach and it makes use of the co-occurrence statistics between noun phrases and contextual patterns. CBS is based on the Bayesian Sets Algorithm [7]. Provided with an ontology defining categories and a small number of seed examples along with a large corpus yielding the co-occurrence information between phrases and patterns, CBS calculates a probabilistic score by using those co-occurrence statistics for every candidate of a given category; and the top ranked ones are promoted as trusted instances for that category. The promoted instances are then used as seeds in the subsequent iterations. The algorithm also exploits the mutual exclusion relations across categories to provide further evidence for its decisions. An in-depth discussion on Bayesian Sets and Coupled Bayesian Sets can be found in [16]. Other web information extraction systems such as KNOWITALL [6] is beyond the scope of this paper.

2.2 Tolerance Rough Sets Model (TRSM) in Document Retrieval

TRSM is based on the rough set theory proposed by Z. Pawlak [11]. Rough Set Theory involves partitioning a universe U of objects into indiscernible objects i.e. *equivalence classes* via an equivalence relation $R \subseteq U \times U$ and then using those classes to define three regions for a given concept $X \subseteq U$: *positive*, *negative* and *boundary* regions. Because an equivalence relation has to be reflexive ($xRx \ \forall x \in U$), symmetric ($xRy \iff yRx \ \forall x, y \in U$) and transitive ($xRy \wedge yRz \implies xRz \ \forall x, y, z \in U$) it is deemed too restrictive for text-based information retrieval tasks [9]. Hence a TRSM model that employs a tolerance relation instead of an equivalence relation is favored where transitivity property is not required [14,12] thus permitting overlapping classes, which is desired for text-based partitioning. The earliest known work on the application of the TRSM as a document representation model was proposed by Kawasaki, Nguyen and Ho [9] where a hierarchical clustering algorithm for retrieval that exploits semantics relatedness between documents is used. A thorough investigation of application of TRSM to web mining can be found in [10]. Specifically the authors use TRSM to represent web documents and use a form of k-means clustering for retrieval. More recently a novel method, called a lexicon-based document representation was presented in [17], where terms occurring in a document are mapped to those that occur in the lexicon which results in a compact representation in a lower dimensional space. Shi et al [13] present a combination of SVM and Naive Bayes ensemble classifier in a semi-supervised learning mode for text classification based on the tolerance model.

We now recall some formal definitions that were presented in [14]. Due to space limitation, we defer a discussion on the background and history of the Tolerance model and refer the readers to [14]. A tolerance approximation space is defined by the $A = (U, I, \tau, \nu, P)$ where

- **Universe** U is the universe of objects.
- **Uncertainty function** $I : U \to \mathbb{P}(U)$ where $\mathbb{P}(U)$ is the power set of U. It defines the tolerance class of an object. It also defines the tolerance relation $x\tau y \iff y \in I(x)$. It can be any function that is reflexive and symmetric.

- **Vague inclusion function** $\nu : \mathbb{P}(U) \times \mathbb{P}(U) \rightarrow [0, 1]$ measures the degree of inclusion between two sets. Typically, it is defined $\nu(X, Y) = \frac{|X \cap Y|}{|X|}$ for TRSM and it can be any function that is monotone with respect to the second argument: $Y \subseteq Z \implies \nu(X, Y) \leq \nu(X, Z)$ for $X, Y, Z \subseteq U$
- **Structurality function** $P : I(U) \rightarrow \{0, 1\}$ where $I(U) = \{I(x) : x \in U\}$ allows additional binary conditions to be defined over the tolerance classes. It is defined $P = 1$ for TRSM.

The lower and upper approximations of set X can be defined as:

$$L_A(X) = \{x \in U : P(I(X)) = 1 \wedge \nu(I(x), X) = 1\} \tag{1}$$

$$U_A(X) = \{x \in U : P(I(X)) = 1 \wedge \nu(I(x), X) > 0\} \tag{2}$$

The main idea is to use approximation to create an enriched document representation. Given a set of documents $D = \{d_1, d_2, ..., d_M\}$ and a set of terms $T = \{t_1, t_2, ..., t_N\}$, each document is represented as a weight vector of terms. The tolerance relation is defined over the set of terms and the terms are considered tolerant if and only if they co-occur in θ documents or more: $I_\theta(t_i) = \{t_j : f_D(t_i, t_j) \geq \theta\} \cup \{t_i\}$. By using the partitions created by the tolerance relation I_θ, the upper approximation of the documents are calculated as $U_A(d_i) = \{t_j \in U : \nu(I(t_j), d_i) > 0\}$. It covers the terms d_i has and all of their tolerants. Ultimately, the term weights are assigned by a weighting scheme based on term frequency-inverse document frequency and through this scheme, non-co-occurring boundary terms $ti \in U_A(d_i) \backslash d_i$ are also assigned nonzero weights, creating the enriched representation. At the end, the documents are classified/clustered by a form of similarity of the vectors.

3 Tolerant Pattern Learner - TPL

In this section, we present our proposed Tolerant Pattern Learner (TPL) in terms of the formal model and the semi-supervised learning algorithm which uses noun phrase-contextual pattern co-occurrence statistics from a given corpus to extract credible noun phrase instances for predefined categories.

3.1 Model

Let $\mathcal{N} = \{n_1, n_2, .., n_M\}$ be the set of noun phrases and $\mathcal{C} = \{c_1, c_2, ..., c_P\}$ be the set of contextual patterns in our corpus. Furthermore, let $C : \mathcal{N} \rightarrow \mathbb{P}(\mathcal{C})$ denote the co-occurring contexts $\forall n_i \in \mathcal{N}$ and let $N : \mathcal{C} \rightarrow \mathbb{P}(\mathcal{N})$ denote the co-occurring nouns $\forall c_i \in \mathcal{C}$. We can define a tolerance space $A = (U, I, \tau, \nu, P)$ where universe $U = \mathcal{C}$ and the tolerance classes are defined by the uncertainty function I_θ in terms of contextual overlaps:

$$I_\theta(c_i) = \{c_j : \omega(N(c_i), N(c_j)) \geq \theta\} \tag{3}$$

Here, θ is the tolerance threshold parameter and ω is the overlap index function which is the Sørensen-Dice index [15]:

$$\omega(A,B) = \frac{2|A \cap B|}{|A| + |B|} \tag{4}$$

where ω is reflexive and symmetric so it is applicable to our problem. We further define $\nu(X,Y) = \frac{|X \cap Y|}{|X|}$ and $P(I(c_i)) = 1$. The resulting tolerance classes shall be used to approximate the noun phrases n_i as the target concept:

$$U_A(n_i) = \{c_j \in C : \nu(I_\theta(c_j), C(n_i)) > 0\} \tag{5}$$

$$L_A(n_i) = \{c_j \in C : \nu(I_\theta(c_j), C(n_i)) = 1\} \tag{6}$$

In a similar manner that TRSM uses a tolerance approximation space to create an enriched representation for the documents, TPL uses it to create an enriched representation for the promoted (labeled) noun phrases. Unlike TRSM, TPL does not use vector-space model; noun phrases are described in terms of sets of co-occurring contextual patterns, rather than vectors. Accordingly, every promoted noun phrase n_i is associated to three descriptive sets: $C(n_i)$, $U_A(n_i)$ and $L_A(n_i)$ These descriptors are used to calculate a micro-score for a candidate instance n_j, by the trusted instance n_i:

$$micro(n_i, n_j) = \omega(C(n_i), C(n_j))\alpha + \omega(U_A(n_i), C(n_j))\beta + \omega(L_A(n_i), C(n_j))\gamma \tag{7}$$

Once again, the overlap index function ω is used. α , β , γ are the contributing factors for the scoring components and they can be configured for the application domain.

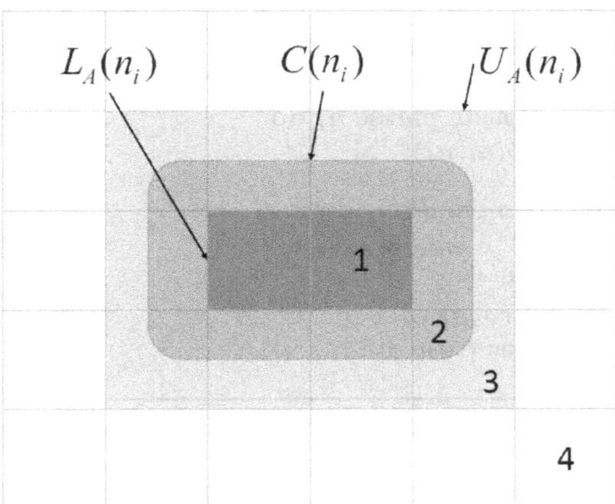

Fig. 1. Four zones of recognition for contexts emerging from approximations of n_i

The notion behind this approach can be elaborated via Fig. 1. For a trusted noun phrase n_i, its descriptors $L_A(n_i)$, $C(n_i)$, $U_A(n_i)$ partition the universe of contexts into four recognition zones. For micro-score calculation, a candidate's contexts falling in zone 1 of the trusted's are covered by all three descriptors and thus will make high contribution to its score. Contexts in zone 2 are covered by $C(n_i)$ and $U_A(n_i)$ so they will make medium contribution. Zone 3 contexts are only covered by $U_A(n_i)$ and they will make low contribution. Contexts in zone 4 will not contribute.

3.2 TPL Algorithm

The input for TPL is an ontology which is formed by a set of categories (e.g. Profession) as well as a fistful of seed instances per category (e.g. engineer, lawyer, graphic designer). In addition, another input is a large co-occurrence matrix representing the noun phrases and the contextual patterns the world wide web yields. The output of TPL are sets of trusted noun phrase instances, classified by their respective categories from the ontology.

A trusted instance serves as a "proxy" for its category. In other words, every category is represented in terms of its trusted noun phrases. Candidate nouns are ranked by their similarity to these proxies, therefore, to the category itself. Seed examples are the initial proxies.

Algorithm 1. Tolerant Pattern Learner

Input : An ontology O defining categories and a small set of seed examples; a large corpus U

Output: Trusted instances for each category

- **for** $r = 1 \to \infty$ **do**
  - **for** each category *cat* **do**
    - **for** each new trusted noun phrase n_i of *cat* **do**
      - Calculate the approximations $U_R(n_i)$ and $L_R(n_i)$;
      - **for** each candidate noun phrase n_j **do**
        - Calculate $micro(n_i, n_j)$;
    - **for** each candidate noun phrase n_j **do**
      - $macro_{cat}(n_j) = \sum_{\forall n_i \in cat} micro(n_i, n_j)$;
    - Rank instances by $macro_{cat}/|cat|$;
    - Promote top instances as trusted;

TPL uses a score-based ranking whereas the scoring mechanism employs tolerance approximation spaces. For the category *cat*, we can maintain an accumulated macro-score for the candidate n_j

$$macro_{cat}(n_j) = \sum_{\forall n_i \in Trusted_{cat}}^{n} micro(n_i, n_j) \tag{8}$$

After calculating it for every candidate of $4cat$, we rank the candidates by their macro-scores normalized by the number of trusted instances of *cat*. Ultimately, we promote the top new candidates as trusted. Like CBS and CPL, TPL is an iterative algorithm which is designated to run indefinitely. At every iteration, it learns new trusted instances and it uses its growing knowledge to make more comprehensive judgments in the subsequent iterations.

Algorithm 1 summarizes the overall flow. Semi-supervised learning is used in the context of finding the trusted instances. Initial category seeds are labeled and appointed by the user, forming the supervised step. At the end of each round, TPL marks the top-scoring new candidates as trusted examples for the category, forming the unsupervised step.

4 Experiments

TPL algorithm was implemented in Matlab. Our dataset was an excerption from Andy Carlson's all-pairs data corpus [4] harvested from ClueWeb [1], a collection of over one billion web pages. We used a subset of noun phrases used in the CBS experiments that was provided to us. For each noun phrase, we appended its most frequent 100 contexts to our dataset leading to 930,426 contexts. For feasibility reasons, we filtered out the ones which had low cross-noun phrase frequency i.e. $|N(c_i)| < 10$. Our resulting dataset consisted of 68,919 noun phrase instances and 59,325 contextual patterns. It was stored in form of a matrix M_{ij} with each cell corresponding to the co-occurrence of noun phrase j against contextual pattern i.

Throughout our experiments, we tried to follow the same conventions as the CBS [16] experiments. We employed the following 11 categories as the input ontology: *Company, Disease, KitchenItem, Person, PhysicsTerm, Plant, Profession, Sociopolitics, Website, Vegetable, Sport.* These are indeed the same categories as the ones used in [16]. We initialized each category with 5-8 seed instances and we let it run for 10 iterations. In every iteration, we promoted the top 5 new noun

Table 1. Precision@30 of TPL for all categories, by iteration (%)

Iteration	**1**	**2**	**3**	**4**	**5**	**6**	**7**	**8**	**9**	**10**
Company	100	100	100	100	100	100	100	100	100	100
Disease	100	100	100	100	100	100	100	100	100	100
KitchenItem	100	100	100	100	100	100	100	100	100	100
Person	100	100	100	100	100	100	100	100	100	100
PhysicsTerm	100	100	100	100	93	97	97	93	90	90
Plant	100	100	97	97	100	97	97	97	97	97
Profession	100	100	100	97	100	100	100	100	100	100
Sociopolitics	100	100	100	100	100	100	100	100	100	100
Website	87	90	90	90	90	90	90	90	90	90
Vegetable	77	90	90	90	93	93	87	80	73	63
Sport	100	97	97	97	97	97	97	100	100	100
Average	**96.7**	**97.9**	**97.6**	**97.3**	**97.5**	**97.6**	**97.3**	**96.4**	**95.4**	**94.5**

Table 2. TPL's top-16 ranked instances for selected categories. Incorrect instances are boldfaced.

Iteration 1			Iteration 10		
Phys'Terms	**Soc'politics**	**Vegetables**	**Phys'Terms**	**Soc'politics**	**Vegetables**
inertia	socialism	zucchini	density	humanism	zucchini
acceleration	democracy	spinach	conductivity	pluralism	cabbage
gravity	dictatorship	cucumber	intensity	federalism	kale
buoyancy	monarchy	tomato	viscosity	interna'lism	celery
velocity	independence	broccoli	permeability	nationalism	cauliflower
momentum	justice	lettuce	velocity	rationality	eggplant
magnetism	equality	celery	brightness	liberalism	carrots
resonance	pluralism	cabbage	attenuation	secularism	asparagus
curvature	interna'lism	kale	luminosity	individualism	tomatoes
electromagnet.	federalism	cauliflower	reflectance	democracy	spinach
density	secularism	asparagus	sensitivity	environ'ism	squash
elasticity	liberalism	carrots	amplitude	morality	cucumber
surface tension	hegemony	tomatoes	thickness	pragmatism	**melon**
polarization	self-determ.	**avocado**	frequency	spirituality	**chicken**
vibration	unification	eggplant	**water cont.**	regionalism	**tofu**
entropy	capitalism	carrot	**salinity**	subjectivity	**shrimp**

phrases for every category as trusted, which were to be used as seeds in subsequent rounds. Heuristically, we set the tolerance threshold $\theta = 50\%$ as it produced the most semantically accurate tolerance classes. We also set $\alpha = 0.5, \beta = 0.25, \gamma = 0.25$ to provide a balance between directly co-occurring terms and their tolerants.

In order to measure the performance, we used Precision@N, the same metric used in [16]. In any iteration, after the noun phrases are scored and ranked for a particular category, the ratio of the correct instances within the set of the top N-ranked ones is computed. The correctness of an instance is judged via human supervision, as the data is unlabeled.

It took about 2 hours and 40 minutes to run our test scenario as a single thread, on a Windows 7 machine with 2.40 GHz Intel i7 processor. Table 1 illustrates the Precision@30 results for every category per iteration and the results are promising. For most categories, the algorithm successfully maintained high precision throughout the 10 iterations without exhibiting an indication of context drift. One category which resulted in relatively poor precision was vegetable; the algorithm first drifted by confusing fruits with vegetables and then went significantly off-track by electing diaries and meat products. (See Table 2.) Nevertheless, this was a problematic category for CBS as well, so it may be regarded as a categorical anomaly.

Overall, TPL achieved comparable performance with CBS, in terms of precision. Table 3 shows the results for iterations 5 and 10. Because the experimental setups (used seeds, dataset) were not identical, a numerical comparison with CBS is not very meaningful or informative. However, the stable average precision values across iterations should be an indication of the promising nature of TPL.

Table 3. Precision@30 of TPL and CBS per category. CBS results are as seen in [16].

Categories	Iteration 5		Iteration 10	
	TPL	CBS [16]	TPL	CBS [16]
Company	100%	100%	100%	100%
Disease	100%	100%	100%	100%
KitchenItem	100%	94%	100%	94%
Person	100%	100%	100%	100%
PhysicsTerm	93%	100%	90%	100%
Plant	100%	100%	97%	100%
Profession	100%	100%	100%	87%
Sociopolitics	100%	48%	100%	34%
Sport	97%	97%	100%	100%
Website	90%	94%	90%	90%
Vegetable	93%	83%	63%	48%
Average	**97.5%**	**92%**	**94.5%**	**87%**

5 Discussion

There are several factors which play a role in TPL's performance over individual categories. First of all, some categories are harder to define and identify boundaries, as a result the measured performance might fluctuate depending on the degree of inclusion. For instance, "Sociopolitics" is a brand that has arguably no certain definition and boundary, so every concept that had an interpretation in a socio-political context was accepted (See Table 2). On the other hand, nouns corresponding to parties owning websites were rejected. Fortunately, most categories including "Person","Sport","Disease","Plant", "Company", "Profession" had rigid definitions so their outcomes were much more evident to judge. TPL is also affected by the ratio of intra-inter category similarity. "PhysicsTerm" is a good example, where the mis-extractions were chemistry terms (e.g. reactivity), or mathematical terms (e.g. curvature). We already explained that some vegetables appeared to overlap with fruits, diary, meat products because contextually, the main cluster they form is "food" and it challenged the algorithm.

As it is pointed out in [16], learning individual classifiers is a very difficult task. That is why CBS and CPL coupled the learning process by learning classifiers simultaneously. As an example, they enforced mutual exclusion constraints: evidence for an instance belonging to category A was used to reduce its chance to be assigned to B, given that A and B are mutually exclusive. Such course of an action was not taken for TPL so it can be regarded as a space for potential improvement.

6 Conclusion

We have proposed a semi-supervised learning algorithm incorporating tolerance rough sets for information extraction using contextual pattern statistics. Our experiments suggest that the approach yields promising results for the current domain. As future work, we plan to explore its capabilities over more sophisticated ontologies with more categories and over larger data sets. We also consider addressing the second half of the Nell problem: learning pairs of noun phrases for relations.

References

1. Callan, J., Hoy, M.: Clueweb09 data set (2009), http://lemurproject.org/clueweb09/
2. Carlson, A., Betteridge, J., Wang, R.C., Hruschka Jr., E.R., Mitchell, T.M.: Coupled semi-supervised learning for information extraction. In: Proceedings of the Third ACM International Conference on Web Search and Data Mining, pp. 101–110 (2010)
3. Carlson, A., Betteridge, J., Kisiel, B., Settles, B., Hruschka, E.R., Mitchell, T.M.: Toward an architecture for never-ending language learning. In: Proceedings of the Twenty-Fourth Conference on Artificial Intelligence, AAAI 2010 (2010)
4. Carlson, A.: All-pairs data set (2010)
5. Curran, J.R., Murphy, T., Scholz, B.: Minimising semantic drift with mutual exclusion bootstrapping. In: Proc. of PACLING, pp. 172–180 (2007)
6. Etzioni, O., Fader, A., Christensen, J., Soderland, S., Mausam: Open information extraction: The second generation. In: International Joint Conference on Artificial Intelligence. pp. 3–10 (2011)
7. Ghahramani, Z., Heller, K.A.: Bayesian sets. Advances in Neural Information Processing Systems 18 (2005)
8. Ho, T.B., Nguyen, N.B.: Nonhierarchical document clustering based on a tolerance rough set model. International Journal of Intelligent Systems 17, 199–212 (2002)
9. Kawasaki, S., Nguyen, N.B., Ho, T.-B.: Hierarchical document clustering based on tolerance rough set model. In: Zighed, D.A., Komorowski, J., Żytkow, J.M. (eds.) PKDD 2000. LNCS (LNAI), vol. 1910, pp. 458–463. Springer, Heidelberg (2000)
10. Ngo, C.L.: A tolerance rough set approach to clustering web search results. Master's thesis, Warsaw University (2003)
11. Pawlak, Z.: Rough sets. International Journal of Computer & Information Sciences 11(5), 341–356 (1982), http://dx.doi.org/10.1007/BF01001956
12. Peters, J., Wasilewski, P.: Tolerance spaces: Origins, theoretical aspects and applications. Information Sciences 195(1-2), 211–225 (2012)
13. Shi, L., Ma, X., Xi, L., Duan, Q., Zhao, J.: Rough set and ensemble learning based semi-supervised algorithm for text classification. Expert Syst. Appl. 38(5), 6300–6306 (2011)
14. Skowron, A., Stepaniuk, J.: Tolerance approximation spaces. Fundam. Inf. 27(2,3), 245–253 (1996), http://dl.acm.org/citation.cfm?id=2379560.2379571
15. Sørensen, T.: A Method of Establishing Groups of Equal Amplitude in Plant Sociology Based on Similarity of Species Content and Its Application to Analyses of the Vegetation on Danish Commons. Biologiske skrifter, I kommission hos E. Munksgaard (1948), http://books.google.co.in/books?id=rpS8GAAACAAJ
16. Verma, S., Hruschka Jr., E.R.: Coupled bayesian sets algorithm for semi-supervised learning and information extraction. In: Flach, P.A., De Bie, T., Cristianini, N. (eds.) ECML PKDD 2012, Part II. LNCS, vol. 7524, pp. 307–322. Springer, Heidelberg (2012)
17. Virginia, G., Nguyen, H.S.: Lexicon-based document representation. Fundam. Inform. 124(1-2), 27–46 (2013)

Identifying Influential Nodes in Complex Networks: A Multiple Attributes Fusion Method

Lu Zhong[1], Chao Gao[1,*], Zili Zhang[1,2], Ning Shi[1], and Jiajin Huang[3]

[1] College of Computer and Information Science, Southwest University, Chongqing, China
[2] School of Information Technology, Deakin University, VIC 3217, Australia
[3] International WIC Institute, Beijing University of Technology, Beijing, China
cgao@swu.edu.cn

Abstract. How to identify influential nodes is still an open hot issue in complex networks. Lots of methods (e.g., degree centrality, betweenness centrality or K-shell) are based on the topology of a network. These methods work well in scale-free networks. In order to design a universal method suitable for networks with different topologies, this paper proposes a Multiple Attribute Fusion (MAF) method through combining topological attributes and diffused attributes of a node together. Two fusion strategies have been proposed in this paper. One is based on the attribute union (FU), and the other is based on the attribute ranking (FR). Simulation results in the Susceptible-Infected (SI) model show that our proposed method gains more information propagation efficiency in different types of networks.

1 Introduction

Influential nodes in complex networks are those nodes which play important roles in the function of networks. How to identify the most influential nodes in complex networks is of theoretical significance [1,2], such as controlling rumor and disease spreading [3,4], creating new marketing tools [4], supplying electric power [5] and protecting critical regions from intended attacks [6,7,8]. The topological structure plays an important role to function and behavior of networks. According to this finding, lots of methods have been proposed to identify influential nodes in a network based on the topological properties of a node [8]. Meanwhile, identifying influential nodes is also affected by network transmission mechanism of a network [9,10,11]. Therefore, it is necessary to investigate the influence of nodes from the perspective of transmission mechanism.

Many methods identifying influential nodes from the view of the structural analysis have been proposed [4,9,12]. These methods can be categorized into two aspects: one is topology-based method (e.g., degree-based method, betweenness-based method or K-shell decomposition method) and the other is diffusion-based method (e.g., PageRank-based method and LeaderRank-based method). These methods have been used in many domains to solve different problems as shown in [4]. Because of their own computational characteristics, these methods can perform well in networks with specific topologies (e.g., scale-free structures), rather than in other types of networks (e.g., ER network or WS network) [13]. Due to the diversity of network structures in the real world, it is

* Chao Gao is the corresponding author of this paper.

D. Ślęzak et al. (Eds.): AMT 2014, LNCS 8610, pp. 11–22, 2014.

desirable for us to design a new method which can overcome the limitation of network structures.

Given a network, the ranking results of existing methods are only based on specific features of the network. Although these features reflect different topological and diffused attributes of a node in a network, the ranking results based on different methods are similar to some extent, i.e., these rankings are correlated [8]. According to this finding, this paper aims to uncover those influential nodes, which are always important no matter which method we use, through combining the advantages of existing methods together based on attribute fusion. Specifically, two strategies are used for multiple attributes fusion: one is based on the attribute union (FU) and the other is based on the attribute ranking (FR). The strategy FU gets intersectional elements in ranked lists of nodes obtained by existing methods. The other strategy FR first constructs a new directed weighted network based on the ranking results of existing methods. And then, the PageRank algorithm is used to re-rank the importance of nodes in such network. Based on the above fusion, the proposed Multiple Attribute Fusion (MAF) method can obtain adaptive ranking result and overcome topological restrictions. In order to evaluate the efficiency of the proposed method, a typical SI model is used to compare the effect of selected influential nodes (ranked by different methods) on information diffusion.

The remainder of paper is organized as follows: Section 2 introduces the related work. Section 3 formulates the proposed method. Section 4 provides some simulation results to estimate the efficiency of our method. In Section 5, we conclude our contribution.

2 Related Work

Many methods based on network topologies have been proposed to identify the influential nodes [9,14]. Generally speaking, they can be divided into two types, namely topology-based methods and diffusion-based methods.

The topology-based methods rank nodes according to attributes of nodes (e.g., node degree and betweenness). For example, degree-based method (DB) is a very simple and feasible topology-based method, but it treats the adjacent neighbors of a node as the same importance [9]. Actually, the influence of adjacent neighbors of a node are not equal, i.e., some nodes with a low degree may have a large influence [4]. Considering of this phenomenon, Kitsak [15] proposes the K-shell decomposition method (KS) to quantify the influence of a node. But this method treats nodes with the same shell value as equal and ignores their differentiation [12]. For example, for a network whose nodes are all 1-core, the K-shell decomposition method regards each node as the same influence [12,16]. Although betweenness-based method (BB) [17], a global measure, can better quantify influential nodes [8], it costs too much time due to calculating the shortest path of all nodes [4]. Besides above three common topology-based methods, Zhou et al [18] propose a node contraction method (NC). They assume that the most important nodes can enlarge the contractional degree of a network. However, this method may regard some nodes as equal in some networks with particular topological structures, such as a chain network [19]. On the other hand, although the node importance evaluation matrix method (EM) [19] overcomes the disadvantages of BB and NC, it is not suitable for directed networks.

The diffusion-based methods rank nodes by taking advantage of network-based diffusion algorithms [8], e.g., PageRank [20,21], LeaderRank [22], HITS scores [23]. They assume that a node visited by diffusion process frequently should be important in directed networks. Although PageRank-based (PR) and LeaderRank-based (LR) methods perform well in dynamical networks, they are not fit for undirected networks [4]. Compared to the computational complexity of PR and LR, HITS-based method (HITS) has a low complexity. However, its result has a deviation with the actual result if there exist abnormal links [9].

Based on above analyses, we find that existing methods only focus on specific features of networks. Therefore, these methods cannot always identify influential nodes precisely in existing networks due to the diversity of network topologies. On the other hand, some preliminary experiments show that ranking results of existing methods are overlapped partly. In order to uncover the correlation among these methods and combine these methods' diversities, a universal Multiple Attribute Fusion (MAF) method with two combined strategies is proposed. First, as for each node, the influences computed by existing methods are defined as its attributes. Then, the proposed method combines topological attributes and diffused attributes of a node together, to identify the influential nodes in existing networks.

3 Multiple Attribute Fusion Method

In this section, we formulate two combined strategies of MAF. One is based on the attribute union (FU) as shown in Sec. 3.1 and the other is based on the attribute ranking (FR) as shown in Sec. 3.2. Two strategies combine various attributes of a node together based on compound mode, in order to provide a universal method for overcoming the limitation of network topologies.

3.1 MAF Based on the Attribute Union

Node rankings assigned by existing methods are relative, i.e., parts of ranking results are coupled (as shown in Table 1), which inspires us to combine the important nodes ranked by most of existing methods. The new strategy is called Multiple Attribute Fusion method based on the attribute union (FU), to combine the important nodes ranked by most of existing methods. For a network $G = (V, E)$ with $n = |V|$ nodes and $m = |E|$ edges, the subset of V is denoted as $P = \{v_i | v_i \in V,\ i \in [1, n]\}$. The existing ranking methods, i.e., degree-based method, betweenness-based method, node contraction method, evaluation matrix method, K-shell decomposition method, PageRank-based method, HITS-based method, and LeaderRank-based method, form a set $Q = \{DB, BB, NC, EM, KS, PR, HITS, LR\}$. Specifically, J is a subset of Q, which is defined as $J = \{q_j | q_j \in Q,\ j \in [1, 8]\}$.

For a node v_i, $f_j(v_i)$ is defined as the value of node v_i computed by a method q_j. The final ranking result of FU strategy is based on Eq. (1).

$$P_k = \{v_i | v_i \in P'_{1k} \cap P'_{2k} \cap ... \cap P'_{jk}, \frac{|Q|}{2} \leq |J| \leq |Q|\} \quad (1)$$

where P_k is a set of top-k nodes selected by FU strategy. P'_{jk} is a set of top-k nodes obtained by a certain method q_j based on Eq. (2). P_{jk} in Eq. (2) is a set of nodes whose value is maximum chosen by method q_j based on Eq. (3). Here, the function $ArgMax(f_j(i))$ returns a node v_i whose value is maximal computed by the method q_j.

$$P'_{jk} = P_{j1} \cap P_{j2} \cap ... \cap P_{jk} \tag{2}$$

$$P_{jk} = \{v_i | v_i \in P \& v_i \notin P'_{jk-1}, ArgMax(f_j(v_i))\} \tag{3}$$

Specifically, the results obtained by FU strategy cannot fit our requirement sometimes. The number of influential nodes obtained by union strategy may be less or more than the number we want. In order to solve this problem, we propose a supplementary strategy as follows.

If we want to select k' nodes from a network and P_k has already chosen k ($k \neq k'$) nodes, we have to replenish or remove $k_1 = |k - k'|$ nodes. Thus, P'_k in Eq. (4) is the set containing top-k' nodes which satisfy our requirement. Here, we define P_1 is a set with $k_1 = |P_1|$ nodes that would be replenished and P_2 is a set with $k_1 = |P_2|$ nodes that would be removed, as shown in Eq. (5) and Eq. (6) respectively.

$$P_{k'} = \begin{cases} P_k \cup P_1, & k \leq k', k' = k + k_1; \\ P_k \cap \bar{P}_2, & k > k', k = k' + k_1; \end{cases} \tag{4}$$

$$P_1 = \{v_i | v_i \in P, v_i \notin P_k, v_i \in P'_{jk}\} \tag{5}$$

$$P_2 = \{v_i | v_i \in P, v_i \in P_k\} \tag{6}$$

Table 1. The top-10 nodes implemented by different methods in the Enron email network G_6

Ranking	DC	BC	CN	EM	KS	PR	HITS	LR
1	$\mathbf{v_{30}}$	$\mathbf{v_{30}}$	$\mathbf{v_{30}}$	$\mathbf{v_{137}}$	$\mathbf{v_{1}}$	$\mathbf{v_{30}}$	$\mathbf{v_{137}}$	$\mathbf{v_{30}}$
2	$\mathbf{v_{5}}$	$\mathbf{v_{110}}$	$\mathbf{v_{137}}$	v_{154}	$\mathbf{v_{5}}$	$\mathbf{v_{110}}$	v_{154}	$\mathbf{v_{5}}$
3	$\mathbf{v_{1}}$	$\mathbf{v_{137}}$	$\mathbf{v_{5}}$	v_{111}	$\mathbf{v_{15}}$	$\mathbf{v_{5}}$	v_{111}	$\mathbf{v_{1}}$
4	$\mathbf{v_{15}}$	$\mathbf{v_{1}}$	v_{154}	v_{175}	v_{21}	v_{103}	v_{175}	$\mathbf{v_{15}}$
5	$\mathbf{v_{110}}$	$\mathbf{v_{5}}$	v_{175}	v_{171}	v_{26}	$\mathbf{v_{1}}$	v_{171}	$\mathbf{v_{110}}$
6	v_{77}	v_{200}	v_{132}	v_{132}	$\mathbf{v_{28}}$	$\mathbf{v_{15}}$	v_{149}	v_{77}
7	v_{103}	v_{21}	v_{149}	v_{149}	v_{29}	$\mathbf{v_{28}}$	v_{132}	v_{103}
8	v_{51}	v_{103}	$\mathbf{v_{1}}$	v_{153}	$\mathbf{v_{30}}$	v_{51}	v_{180}	v_{51}
9	$\mathbf{v_{28}}$	$\mathbf{v_{15}}$	$\mathbf{v_{15}}$	v_{180}	v_{37}	$\mathbf{v_{100}}$	v_{112}	$\mathbf{v_{28}}$
10	v_{41}	v_{343}	v_{111}	v_{112}	v_{41}	v_{77}	v_{176}	v_{41}

Taking the Enron email network[1] which is introduced in Sec. 4.1 as an example, we find that the ranking results obtained by existing methods are correlated (as shown in Table 1), i.e., some nodes can be identified by different methods. For example, if we want to select top-10 nodes from the Enron email network, some nodes $\{v_1, v_5, v_{15}, v_{28}, v_{30}, v_{110}, v_{137}\}$ can be recognized by more than 4 methods. FU strategy takes these nodes as P_k. This means that each of these nodes should be one of the top-10 nodes

[1] http://bailando.sims.berkeley.edu/enron/enron.sql.gz

rather than other nodes from many aspects (e.g., node influence and node neighbors). But these 7 nodes cannot satisfy with our requirement of choosing 10 nodes. Thus we apply supplementary strategy to choose another three nodes in order to gain 10 important nodes in the Enron email network: $\{v_{41}, v_{51}, v_{77}\}$. In many cases, these selected nodes are the most significant nodes in a network, which will be estimated in Sec. 4. By studying this case, we illustrate that the FU strategy can obtain the most important k' nodes from a network through getting intersection elements from existing methods.

3.2 MAF Based on the Attribute Ranking

Besides directly obtaining the intersectional results from existing ranking methods, this section proposes the other compound mode, named as Multiple Attribute Fusion method based on the attribute ranking (FR). The new strategy combines ranking results obtained by existing methods based on PageRank algorithm. There are two steps for the new strategy. First, a new directed weighed network is constructed based on the ranking results of existing methods. Then, typical PageRank algorithm is used to re-rank nodes in such directed weighted network [24].

Let a network be $G = (V, E)$ with $n = |V|$ nodes, while a new directed network $G^* = (V, E^*)$ is defined as $V = \{v_1, v_2, ...v_n\}$, $E^* = \{(v_l, v_i)|\forall v_l, v_i \in V,\ f_j(v_l) < f_j(v_i)\}$. $f_j(v_i)$ is the ranking value based on the method q_j as defined in Sec. 3.1. For a node v_i, (v_l, v_i) is the directed edge from v_l to v_i. $\omega(v_l, v_i)$ represents the weight of edge from v_l to v_i and $\omega(v_l, v_i)$ doesn't equal to $\omega(v_i, v_l)$. The final ranking result of FR is from Eq. (7).

$$FR(v_i) = \frac{(1-\sigma)}{n} + \sigma \sum_{l=1}^{l'} \frac{\omega(v_l, v_i)}{\sum\limits_{r=1}^{r'} \omega(v_l, v_r)} FR(v_l) \tag{7}$$

where σ $(0 < \sigma < 1)$ is the damping coefficient. $FR(v_i)$ is the influential value of node v_i. $\sum\limits_{l=1}^{l'} \frac{\omega(v_l, v_i)}{\sum\limits_{r=1}^{r'} \omega(v_l, v_r)}$ is the sum of ratios between weight from v_l to v_i and the sum of weight from v_l to all its pointed nodes. v_l is in set $\{v_1, v_2, ..., v_{l'}\}$ and l' is the value of v_i's indegree. $\sum\limits_{r=1}^{r'} \omega(v_l, v_r)$ is the sum of weights from certain node v_l to its pointed neighbors $\{v_1, v_2, ..., v_i, ..., v_{r'}\}$ and r' is the value of v_l's outdegree. $\omega(v_l, v_i)$ represents the weight from node v_l to node v_i, denoted as Eq. (8).

$$\omega(v_l, v_i) = \sum_{j=1}^{|Q|} \sum_{l=1}^{l} (\text{sgn}(f_j(v_i) - f_j(v_l)) + 1),\ \ l \neq i \tag{8}$$

where Q is a method set defined in Sec. 3.1. For a method q_j, if $f_j(v_l) \leq f_j(v_i)$, then v_l points to v_i and weight of (v_l, v_i) is added.

Here, an example is used to illustrate how to construct G^* and how FR strategy identifies the influential nodes from G^*. Table 2 shows the ranking results of a small network G_X with 5 nodes based on different methods. According to Eq. (8), we can assign a weight to each edge. For two nodes v_l and v_i, $f_j(v_l)$ and $f_j(v_i)$ denote their ranking results based on a certain method q_j. If $f_j(v_l) \leq f_j(v_i)$, there is a directed edge from v_l to v_i. After $|Q|$ iterations, the weight between v_l and v_i is emerged. Fig. 1(a) shows the new directed weighted network G_X^* based on the small network G_X. Fig. 1(b) shows the cumulated weight of Fig. 1(a) and FR values based on Eq. (7) (σ=0.5). The larger FR value of a node is, the larger spreading ability of the node has. The node ranking obtained by FR strategy is $\{v_3, v_1, v_4, v_2, v_5\}$.

Table 2. Ranking results implemented by different methods in the small network G_X

Ranking	DC	BC	CN	EM	KS	PR	HITS	LR
1	v_3	v_3	v_3	v_3	v_1	v_3	v_3	v_3
2	v_1	v_1	v_1	v_4	v_3	v_1	v_1	v_1
3	v_2	v_4	v_4	v_1	v_4	v_4	v_5	v_2
4	v_5	v_2	v_5	v_2	v_5	v_2	v_2	v_5
5	v_4	v_5	v_2	v_5	v_2	v_5	v_4	v_4

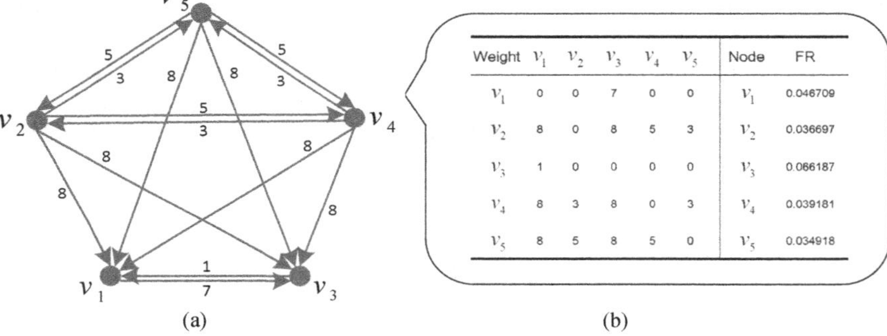

Weight	v_1	v_2	v_3	v_4	v_5	Node	FR
v_1	0	0	7	0	0	v_1	0.046709
v_2	8	0	8	5	3	v_2	0.036697
v_3	1	0	0	0	0	v_3	0.066187
v_4	8	3	8	0	3	v_4	0.039181
v_5	8	5	8	5	0	v_5	0.034918

Fig. 1. (a) The new directed weighted network G_X^* created by rankings of existing methods in a network G_X. (b) The weight between nodes accumulated by existing methods and the FR value of each node in G_X.

4 Experimental Results

This section will show the propagation results of methods mentioned above. Sec. 4.1 shows the simulation process. Networks and propagation model used in this paper are introduced in Sec. 4.2 and Sec. 4.3 respectively. Finally, Sec. 4.4 presents the simulation results.

4.1 Experimental Design

The experimental process is illustrated in Fig. 2. By using different ranking methods and our strategies, influential nodes (top-k ranked nodes) are selected. In our experiments, the most influential nodes are defined as those nodes which can gain the maximal infection coverage if viruses first infect those influential nodes. Specifically, the typical

SI model, introduced in Sec. 4.3, is used to measure the propagation efficiency of different methods. Here, the propagation efficiency means the final total number of infected nodes in a network.

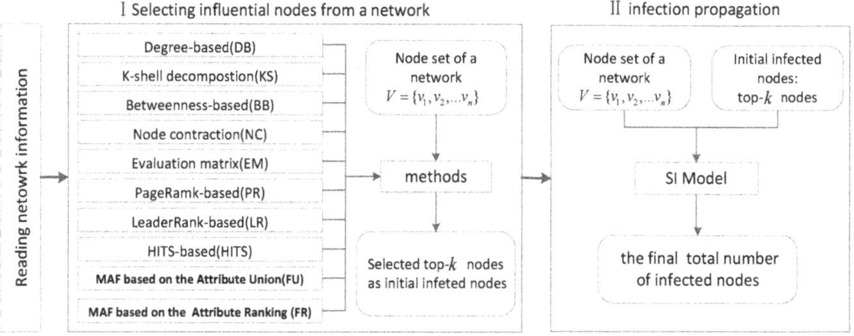

Fig. 2. The process of experiment

Table 3. Networks with different structures used in our experiments where V and E denote total number of nodes and edges in a network, respectively. α is the exponent of a power-law network. $< K >$, $< l >$, and $< C >$ are the average degree of nodes, average shortest path length and average clustering coefficient of a network.

Network	Type	V	E	α	$< K >$	$< L >$	$< C >$
Synthetic Network	G_1	3600	10000	—	5.56	5.0060	0.001322
	G_2	3600	10000	—	5.56	5.5117	0.000880
	G_3	1024	3000	2.70	5.81	3.9192	0.020657
	G_4	4000	16733	2.39	8.34	5.4634	0.064886
	G_5	4096	17084	—	8.34	5.4942	0.027597
Benchmark Network	G_6	1238	2106	2.29	3.40	4.6726	0
	G_7	1133	5451	2.83	9.62	3.6061	0.220029

4.2 Network Environments

Some synthetic and benchmark networks are used to evaluate the efficiency of our proposed method. The synthetic networks include ER (G_1) network, WS (G_2) network and BA (G_3) network generated by GLP model [25]. Meanwhile, two synthetic community-based networks are constructed: one (G_4) is composed of four communities which all follow a long-tail distribution (more information of G_4 is shown in Table 1 in [26]); the other (G_5) is an interdependent network in which four communities have different structures as shown in Fig. 3(b). The benchmark networks are Enron email network[2] (G_6) constructed by Andrew Fiore and Jeff Heer and University email network[3] (G_7) complied by the members of University Rovira i Virgili (Tarragona). The structures of used networks are shown in Table 3 and the degree distributions of networks except G_5 are shown in Fig. 4.

[2] http://bailando.sims.berkeley.edu/enron/enron.sql.gz
[3] http://deim.urv.cat/~aarenas/data/welcome.htm

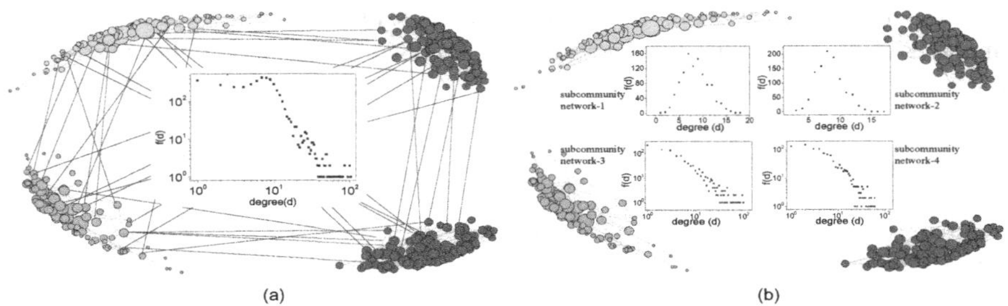

Fig. 3. (a) The degree distributions of G_5. (b) The degree distributions of communities in G_5. d represents node's degree; $f(d)$ is the distribution function of degree d.

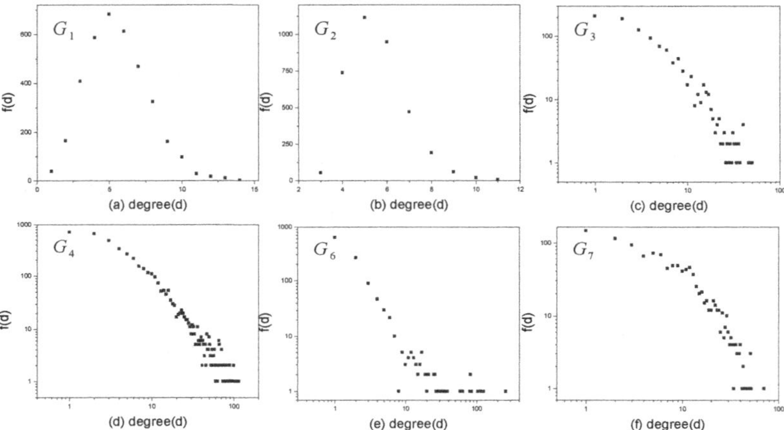

Fig. 4. The degree distributions of G_1, G_2, G_3, G_4, G_6 and G_7 respectively. d represents node's degree; $f(d)$ is the distribution function of degree d.

4.3 Simulation Model and Measurements

Here we use a SI model to evaluate the efficiency of our proposed method through examining the spreading influence of top-k ranked nodes (as the initial infected nodes) [27]. According to different states, the individuals in SI model can be divided into two types: (1) Susceptible individual, who is not infected and susceptible to diseases with probability β; (2) Infected individual, who is infected and has the ability to infect neighbors. In the beginning, we set top-k ranked nodes as the initial infected nodes. As for each time step, each infected node infects all its neighbors, and the susceptible node catches the infection with probability β. In order to wipe off the computational fluctuation, we simulate the spreading process 100 times every time step and $I(t)$ is the averaged number of infected nodes over 100 times where t represents the time step. The total number of infected nodes is denoted as $N(t) = \sum I(t)$, which evaluates the final influence of initial infected nodes. The higher $N(t)$ is, the more influence the top-k nodes are in a network.

In order to eliminate the effect of parameters of SI model and only estimate the effect of selected influential nodes on propagation, we choose a small value for β. That is because if the β is large, the propagation scope and speed are irrelevant to the role of top-k ranked nodes. Through comparing the effects of top-k nodes on the final propagation process (i.e., the infected nodes) in both synthetic and benchmark networks, we estimate the efficiency of proposed methods.

4.4 Simulation Results of All Methods

In this section, the parameters in SI model are set as following: the initial infected nodes (i.e., the top-k influential nodes) are 2% of all nodes in a network; the infected probability β=0.07.

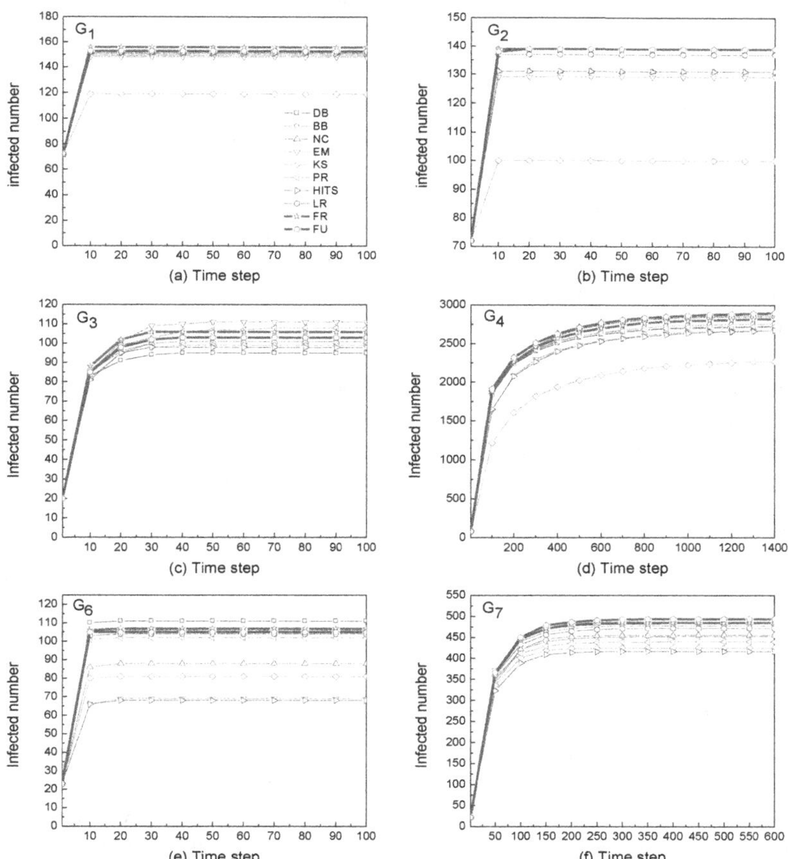

Fig. 5. Time evolution of SI model with top-k nodes selected by different methods and activated as initial infected nodes in a network. This result shows that our method is fit for different types of networks.

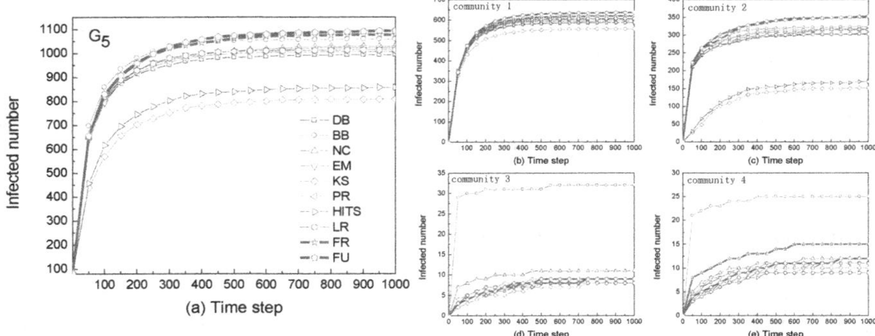

Fig. 6. (a) Time evolution of the SI model with top-k nodes by different methods on G_5. (b)(c)(d)(e) The process of propagation on each community of G_5. These results show that our method is fit for the heterogeneous network. Specially, although some methods outperform the our method in community_3 and community_4, our method perform best in the whole network.

Figure 5 compares the propagation results in both synthetic and benchmark networks. The propagation efficiency of our method with two strategies (FU and FR) in these networks is better than others. As for G_1 and G_2 networks, as shown in Figs. 5(a)(b), almost all methods (e.g., KS, DB) are not suitable for such networks. For BA networks (as shown in Figs. 5(c)-(f)), especially G_4, we find that our proposed method can always obtain a better result than others. However, the degree-based method can perform well in G_6. That is because G_6 is a special network in which all nodes are connected to a few nodes (i.e., the highest-degree nodes). As a whole, our method performs effectively in any networks.

Besides the propagation scope, our proposed method can improve the propagation speed in any networks. Taking G_7 in Fig. 5(f) as an example, our method can infect 400 nodes with 64 time steps, and existing methods BB, NM, KS, PR, HITS cost 77, 81, 97, 83 and 118 time steps, respectively. This indicates that our method is a more effective and stable method in different types of networks.

Figure 6 shows the results of another heterogeneous network G_5 which is composed of three kinds of communities (i.e., ER, WS and BA network). Figs. 6(b)-(e) show the propagation result of communities of G_5. These simulation results show that our method can also perform a better result in the heterogeneous network than others.

5 Conclusion

This paper proposed a Multiple Attribute Fusion (MAF) method to identify influential nodes in complex networks. The proposed method combines topological attributes and diffused attributes of a node and has two strategies. One is Multiple Attribute Fusion method based on the attribute union (FU), and the other is Multiple Attribute Fusion method based on the attribute ranking (FR). The SI model is used to evaluate the efficiency of our proposed method in both synthetic and benchmark networks, in which top-k infected nodes are selected by existing ranking methods. Numerical examples illustrate that our method can effectively identify influential nodes in networks with different topologies. As for the changeable and large-scale networks, our proposed strategies (FU and FR) can also work well when the fused attributes are all local attributes (e.g., degree centrality).

Acknowledgment. This project was supported by the Fundamental Research Funds for the Central Universities (No.XDJK2012B016, XDJK2012C018), SRFDP 20120182120016, the PhD fund of Southwest University (SWU111024, SWU111025) and CQ CSTC (cstc2012jjA40013, cstc2013jcyjA40022, cstc2012jjB40012), and in part by the National Training Programs of Innovation and Entrepreneurship for Undergraduates (201310635068).

References

1. Szolnoki, A., Xie, N.G., Ye, Y., Perc, M.: Evolution of emotions on networks leads to the evolution of cooperation in social dilemmas. Physical Review E 87, 042805 (2013)
2. Newman, M.E.J.: The structure and function of complex networks. Society for Industrial and Applied Mathematics 45(2), 167–256 (2003)
3. Yan, G., Zhou, T., Wang, J., Fu, Z.Q., Wang, B.H.: Epidemic spread in weighted scale-free networks. Chinese Physics Letters 22(2), 510–513 (2005)
4. Chen, D.B., Lv, L.Y., Shang, M.S., Zhang, Y.C., Zhou, T.: Identifying influential nodes in complex networks. Physica A: Statistical Mechanics and its Applications 391(4), 1777–1787 (2012)
5. Albert, R., Albert, I., L.Nakarado, G.: Structural vulnerability of the north american power grid. Physical Review E 69(2), 025103 (2004)
6. Wuellner, R.D., Roy, S., DSouza, R.M.: Resilience and rewiring of the passenger airline networks in the United States. Physical Review E 82(5), 056101 (2010)
7. Albert, R., Jeong, H., Barabasi, A.L.: Error and attack tolerance of complex networks. Nature 406(6794), 378–382 (2000)
8. Hou, B., Yao, Y.P., Liao, D.S.: Identifying all-around nodes for spreading dynamics in complex networks. Physica A: Statistical Mechanics and its Applications 391(15), 4012–4017 (2012)
9. Liu, J.G., Ren, Z.M., Guo, Q., Wang, B.H.: Node importance ranking of complex networks. Acta Physica Sinica 62(17), 178901 (2013)
10. Liu, J.G., Wu, Z.X., Wang, F.: Opinion spreading and consensus formation on square lattice. International Journal of Modern Physics C 18(07), 1087 (2007)
11. Bond, R.M., Fariss, C.J., Jones, J.J., Kramer, A.D., Marlow, C., Settle, J.J., Fowler, J.H.: A 61-million-person experiment in social influence and political mobilization. Nature 489(7415), 295–298 (2012)
12. Gao, C., Lan, X., Zhang, X.G., Deng, Y.: A bio-inspired mehtodology of identifying influential nodes in complex networks. PLoS ONE 8(6), e66732 (2013)
13. Yu, H., Liu, Z., Li, Y.J.: Key nodes in complex networks identified by multi-attribute decision-making method. Acta Physica Sinica 62(2), 20204–20204 (2013)
14. Gao, C., Wei, D.J., Hu, Y., Mahadevan, S., Deng, Y.: A modified evidential methodology of identifying influential nodes in weighted networks. Physica A: Statistical Mechanics and its Applications 392(21), 5490–5500 (2013)
15. Kitsak, M., Gallos, L.K., Havlin, S., Liljeros, F., Muchnik, L., Stanley, H.E., Makse, H.A.: Identification of influential spreaders in complex networks. Nature Physics 6(11), 888–893 (2010)

16. Chen, D.B., Gao, H., Lv, L.Y., Zhou, T.: Identifying influential nodes in large-scale directed networks: The Role of Clustering. PLoS ONE 8(10), e77455 (2013)
17. Barthlemy, M.: Betweenness centrality in large complex networks. The European Physical Journal B-Condensed Matter and Complex Systems 38(2), 163–168 (2004)
18. Tan, Y.J., Wu, J., Deng, H.Z.: Evaluation method for node importance based on node contraction in complex networks. Systems Engineering-Theory & Practice 26(11), 79–83 (2007)
19. Zhou, X., Zhang, F.M., Li, K.W., Hui, X.B., Wu, H.S.: Finding vital node by node importance evaluation matrix in complex networks. Acta Physica Sinica 61(5), 050201 (2012)
20. Page, L., Brin, S., Motwani, R., Winograd, T.: The PageRank citation ranking: bringing order to the web. Technical Report Stanford InfoLab 66 (1999)
21. Brin, S., Page, L.: The anatomy of a large-scale hypertextual web search engine. Computer Newtworks and ISDN Systems 30(1), 107–117 (1998)
22. Lv, L.Y., Zhang, Y.C., Yeung, C.H., Zhou, T.: Leaders in social networks, the delicious case. PLoS ONE 6(6), e21202 (2011)
23. Kleinberg, J.M.: Authoritative sources in a hyperlinked environment. Journal of the ACM 46(5), 604–632 (1999)
24. Zhang, K., Li, P.P., Zhu, B.P., Hu, M.Y.: Evaluation method for nod importance in directed-weighted complex networks based on PageRank. Journal of Nanjing University of Aeronautics and Astronautics 45(3), 429–434 (2013)
25. Bu, T., Towsley, D.: On distinguishing between internet power law topology generators. In: Proceedings of the Twenty-First Annual Joint Conference of the IEEE Computer and Communications Societies (INFOCOM 2002), vol. 2, pp. 638–647 (2002)
26. Gao, C., Liu, J.N., Zhong, N.: Network immunization and virus propagation in email networks: experimental evaluation and analysis. Knowledge and Information Systems 27(2), 253–279 (2011)
27. Garas, A., Argyrakis, P., Rozenblat, C., Tomassini, M., Havli, S.: Worldwide spreading of economic crisis. New Journal of Physics 12(11), 113043 (2010)

Local Regression Transfer Learning for Users' Personality Prediction

Zengda Guan[1,2], Dong Nie[1,2], Bibo Hao[1,2], Shuotian Bai[1,2], and Tingshao Zhu[1,⋆]

[1] Institute of Psychology, Chinese Academy of Sciences, Beijing 100101, China
[2] School of Computer and Control, University of Chinese Academy of Sciences, Beijing 100049, China

{guanzd,ginobilinie,haobibo}@gmail.com
baishutian10@mails.ucas.ac.cn
tszhu@psych.ac.cn

Abstract. Some research has been done to predict users' personality based on their web behaviors. They usually use supervised learning methods to model on training dataset and predict on test dataset. However, when training dataset has different distributions from test dataset, which doesn't meet independently identical distribution condition, traditional supervised learning models may perform not well on test dataset. Thus, we introduce a new regression transfer learning framework to deal with this problem, and propose two local regression instance-transfer methods. We use clustering and k-nearest-neighbor to reweight importance of each training instance to adapt to test dataset distribution, and then train a weighted risk regression model for prediction. We perform experiments on the condition that users dataset are from different genders and from different districts, and the results indicate that our methods can reduce mean square error about 30% to the most compared with non-transfer methods and be better than other transfer method in the whole.

Keywords: Local Regression Transfer Learning, Importance Reweighting, Personality Prediction.

1 Introduction

In psychology, personality reflects individual's consistent behavior patterns and interpersonal communication, and influences her/his cognitions, emotions, motivations, and behaviors [1]. It is widely accepted that personality is a driven force of users' web behaviors [2], and thus it can help people understand web users in depth. Recent research [3] showed that users personality traits could be predicted by users' web behaviors, such as micro-blogging behaviors, which is more effective comparing with self-report method, as it can be done without

⋆ Corresponding author.

D. Ślęzak et al. (Eds.): AMT 2014, LNCS 8610, pp. 23–34, 2014.

participant recruitment and less resource consumption. In this paper, we use people's microblog data for personality prediction.

It is well known that a trained model by conventional machine learning algorithm depends on the training dataset inevitably. If we apply the model to other dataset, it usually can't be assured that the distribution of new dataset is the same as the original one, thus the model may perform poorly. For example, there are various kinds of web users on internet, such as male and female, young and old, users in different districts, users of different ages, etc. If we only have one kind of labeled data, it may be problematic to apply the trained model to predict personality for another kind of unlabeled users. Existing research about web users' personality prediction doesn't pay much attention to this problem yet. To cope with the problem, we test whether two dataset follow the same distribution and employ transfer learning methods. Since personality labels are often continual values, we focus on regression transfer learning methods. In previous research, researchers mostly use the entire dataset to reweight the possibility of every instance in training dataset, which could produce big computation cost and unsatisfactory result accuracy sometimes. In this work, we propose to use a neighborhood local of each training instance to estimate the possibility instead.

In this paper, we propose a regression transfer learning framework to predict users' personality and two local regression transfer learning methods to address biased distribution. In the framework, we firstly check whether the training dataset and test dataset follow the same distribution, then introduce the regression transfer learning methods to reweight the importance of training instances and solve a weighted risk regression problem, and finally predict and test on the test dataset. Our local transfer learning framework includes two methods: clustering transfer learning method and k-NN transfer learning method.

2 The Related Work

It is usually defined as transfer learning or learning under covariate shift that people predict on test dataset with different distribution from training dataset. The keypoint for this kind of problem is to obtain the importance for each training instance according to distribution of test dataset [4] [5].

In the previous research, Huang et al. [5] proposed a kernel-mean matching (KMM) algorithm, which mapped data from both domains into a reproducing-kernel Hilbert space (RKHS) and matched the means of both dataset. Sugiyama et al. [6] proposed a Kullback-Leibler Importance Estimation Procedure (KLIEP) algorithm based on minimizing the Kullback-Leibler divergence between training and test input densities and selecting model through cross validations. Kanamori et al. [7] proposed a method called Least-Squares Importance Fitting (LSIF), which uses a Gaussian kernel model centered at the test points to compute the importance through minimizing squares importance biases. Different from ours, their method used all test instances or randomly drawn many, so sometimes it would bring more computation cost or sampling errors. Loog [8] proposed an importance weighting method using a 1-nearest neighbor classifier, but it's designed only for classification problem. Pardoe et al. [9] proposed a list of regression transfer learning algorithms based on boosting technology, some of which were developed from TrAdaBoost by Dai et al. [10]. The method needs to use some labeled test instances, and can't work without labels.

3 Local Regression Transfer Learning

3.1 The Problem

We aim to transfer the model trained under one situation to another new situation and predict users' personality. Before transfer learning, what we would do is to affirm whether datasets in the different situations follow the same distribution. Here, we apply T-test and Kolmogorov-Smirno test to perform the multi-dimension two-sample test. T-test is fit to test dataset following Gaussian distribution in each feature dimension, and Kolmogorov-Smirno can test unknown distribution. If the two datasets belong to different distributions, we employ a regression transfer learning framework to solve the problem. We propose two local regression transfer learning methods, which use clustering technology and k-NN technology to reweight the importance respectively. The methods estimate the importance of all training instances according to the test data distribution, and integrate traditional regression algorithm in a weighted risk form to train the model.

In this paper, we define the training dataset as $(x_{tr}^{(i)}, y_{tr}^{(i)})$, which has n_{tr} samples. Meanwhile, the test dataset is defined as $(x_{te}^{(i)}, y_{te}^{(i)})$, which has n_{te} samples, and y_{te} is the predicted labels. The probability distribution of training dataset and test dataset are defined as P_{tr} and P_{te}, respectively.

3.2 Method of Importance Weighting

In order to solve the problem caused by different distributions of training and test dataset, sample reweighting method are proposed. We will illustrate the method in the following similar to [5].

Generally, regression transfer method aims to reduce the expected risk in the following:

$$\min_{\theta} E_{(x_{tr}, y_{tr}) \sim P_{te}} l(x_{tr}, y_{tr}, \theta) \tag{1}$$

where $l(x_{tr}, y_{tr}, \theta)$ is the loss function, θ is the unknown parameter which decides the predicted value of x_{te} in test dataset, and $(x_{tr}, y_{tr}) \sim P_{te}$ represents the probability of (x_{tr}, y_{tr}) according to the test data distribution. Since it is difficult to directly calculate the real distribution of P_{te}, we compute the empirical risk in the below:

$$\begin{aligned} &\min_{\theta} E_{(x,y) \sim P_{tr}} \frac{P_{te}(x_{tr}, y_{tr})}{P_{tr}(x_{tr}, y_{tr})} l(x_{tr}, y_{tr}, \theta) \\ \approx &\min_{\theta} \frac{1}{n_{tr}} \sum_{i=1}^{n_{tr}} \frac{P_{te}(x_{tr}, y_{tr})}{P_{tr}(x_{tr}, y_{tr})} l(x_{tr}, y_{tr}, \theta) \end{aligned} \tag{2}$$

The predictive function over both dataset are usually assumed to be identical here, i.e. $P_{tr}(y|x) = P_{te}(y|x)$, $\frac{P_{te}(x_{tr},y_{tr})}{P_{tr}(x_{tr},y_{tr})}$ turns to $\frac{P_{te}(x_{tr})}{P_{tr}(x_{tr})}$. Thus, we can bridge the distributions of training and test dataset by adjusting $\frac{P_{te}(x_{tr})}{P_{tr}(x_{tr})}$, which reweights each instance of the training dataset.

3.3 Reweighting the Importance

Our regression transfer learning method can be divided into two stages: For the first stage, we reweight the importance of instance of training dataset by importance weighting method; On the second stage, we use training instances with the updated weight to train models and then use the model to predict on test dataset.

On the first stage, we reweight the importance of training instances by estimating the ratio $(P_{te}(x_{tr})/P_{tr}(x_{tr}))$ between each training instance in training and test distribution. Local learning, which takes use of nearest neighbors of the predicted points, has been shown to be effective in many applications [11] [12] [13]. Moreover, too many neighbor points(for example, the whole points) can result in over smooth regression functions, while too few neighbor points(for example, one nearest point) can result in regression functions with incorrectly steep extrapolations [13]. Hence, we are motivated that we can use test dataset neighbors of the trained points to learn transfer learning model, and keep quantity of neighbors either not all or not few, but in a reasonable level. Here, we propose two near-neighbors based way to compute the importance: clustering transfer method and k-NN transfer method.

For the clustering transfer method, we use clustering method to cluster the whole training and test dataset, and estimate $(P_{te}(x_{tr})/P_{tr}(x_{tr}))$ by the ratio of test instance quantity and training instance quantity in the same cluster. The idea originates from that, if training dataset and test dataset are clustered in small enough regions, all instances in the same region can be thought to have the equal probability. Then we can estimate the probability with the ratio of instance quantity in the region and the total instance quantity. Thus, we can weight the importance of each training instance by ratio of test and training instance quantity in the same cluster. The clustering importance weighting formula is as follows:

$$Weig(x_{tr}^{(i)}) = \frac{|Clus_{te}(x_{tr}^{(i)})|}{|Clus_{tr}(x_{tr}^{(i)})|} \tag{3}$$

where $Weig(x_{tr}^{(i)})$ is defined as the importance of the training instance $x_{tr}^{(i)}$, and $|Clus_{tr}(x_{tr}^{(i)})|$ and $|Clus_{te}(x_{tr}^{(i)})|$ represent the training and test instance quantity of the cluster which contains $x_{tr}^{(i)}$, respectively.

Clustering method and dataset distribution influence the risk of importance weighting. If the training and test dataset have completely different clustering division, the importance weighting method will assign every training instance 0 weight and therefore the risk will approximate to 100%. If test instances in all the clusters reflect very accurate probability, the method will assign each training instance almost real importance values and therefore the risk of reweighting will be very low. Thus, clustering transfer method may be unstable due to unstable clustering method.

For k-NN transfer method, we use the k nearest test neighbors of the training instance to compute its importance. We choose Gaussian kernel to compute similarity distance between training and test instance for the importance.

$$Weig(x_{tr}) = \sum_{i=1}^{k} exp(-\gamma||x_{tr} - x_{te}^{(i)}||_2^2) \tag{4}$$

where k represents the quantity of the nearest test neighbors for training instance x_{tr}, and γ is a coefficient which controls the relative impact of every test neighbor and $\gamma > 0$. When the quantity of population is very large compared with k, k-NN transfer method could save much computation time obviously.

Both clustering transfer and k-NN transfer methods will combine the importance and regression to build model. Additionally, in this work we just make trials to determine the unfixed parameter of the formulas, and don't regard an automatic adaptive method.

3.4 Weighted Learning Models

In this stage, we use the reweighted training dataset to train a weighted learning model for prediction on test dataset. The difference between the conventional learning model and the weighted learning model is that, for the former model every instance has a uniform importance, while for the latter model each instance has different importance.

The method is to set the risk for each instance $(x_{tr}^{(i)}, y_{tr}^{(i)})$ according to its importance, and solve the programming problem. It is as follows:

$$\min \sum_{i=1}^{n_{tr}} Weig(x_{tr}^{(i)}) \cdot l(y_{tr}^{(i)}, f(x_{tr}^{(i)})) \tag{5}$$

where $Weig(x_{tr}^{(i)})$ represents the importance of training instances $x_{tr}^{(i)}$, and $l(y_{tr}^{(i)}, f(x_{tr}^{(i)}))$ define the bias between $y_{tr}^{(i)}$ and $f(x_{tr}^{(i)})$.

Here, we take local weight regression(LWR) method and Multivariate Adaptive Regression Splines (MARS) method as examples. For LWR, we take its linear form, while polynomial form can be obtained similarly. The weighted risk form of linear LWR is as following:

$$\begin{aligned} &\min \sum_{i=1}^{n_{tr}} Weig(x_{tr}^{(i)}) \cdot K(x_{tr}^{(i)}, x_{te}) \cdot (y_{tr}^{(i)} - \alpha_{te} - \beta_{te} x_{tr}^{(i)})^2 \\ &y_{te} = \alpha_{te} + \beta_{te} x_{te} \end{aligned} \tag{6}$$

where x_{te} represents any instance in test dataset, and the formula represents the way how to compute the prediction result on instance x_{te}. In addition,

$K(x_{tr}^{(i)}, x_{te})$ is the kernel weighting function. Similar to what [14] did, we can infer to the final result.

$$\begin{aligned}
&y_{te} = (1, x_{te})(B^T W_{te} B)^{-1} B^T W_{te}\mathbf{y} \\
&where, \\
&B^T = \begin{pmatrix} 1 & 1 & \ldots & 1 \\ x_{tr}^{(1)} & x_{tr}^{(2)} & \ldots & x_{tr}^{(n_{tr})} \end{pmatrix} \\
&W_{te} = diag(K(x_{tr}^{(i)}, x_{te}))_{n_{tr}\times n_{tr}} \cdot diag(Weig(x_{tr}^{(i)}))_{n_{tr}\times n_{tr}} \\
&\mathbf{y} = (y_{tr}^{(1)}, \ldots, y_{tr}^{(n_{tr})})
\end{aligned} \tag{7}$$

where $diag(K(x_{tr}^{(i)}, x_{te}))_{n_{tr}\times n_{tr}}$ indicates to a $n_{tr} \times n_{tr}$ diagonal matrix with diagonal elements $K(x_{tr}^{(i)}, x_{te})$. $diag(Weig(x_{tr}^{(i)}))_{n_{tr}\times n_{tr}}$ is similarly defined.

We next integrate MARS method with k-NN regression transfer techniques. MARS is an adaptive stepwise method for regression [14], and its weighted risk form is as follows:

$$\begin{aligned}
&\min \sum_{i=1}^{n_{tr}} Weig(x_{tr}^{(i)}) \cdot (y_{tr}^{(i)} - f(x_{tr}^{(i)}))^2 \\
&f(x_{tr}^{(i)}) = \beta_0 + \sum_{j=1}^{m} \beta_j h_j(x_{tr}^{(i)})
\end{aligned} \tag{8}$$

where $h_j(x)$ is a constant denoted by C, or a hinge function with the form $max(0, x - C)$ or $max(0, C - x)$, or a product of two or more the above hinge functions. m is the total steps to get optimal performance, $Weig(x_{tr}^{(i)})$ is defined as the above, $f(x_{tr}^{(i)})$ denotes the prediction value of training instances, and $f(x_{te}^{(i)})$ can be defined similarly. Note that, the above formula aims to compute uniform coefficients β_j, which is different from transfer LWR method.

The second method is very simple. It's just to draw the training instances how many times they are proportional to their importance. In details, if the importance value is not integer, it should be converted to integer first, with the decimal part cut. This may be not an accurate estimation for all cases, but it's so easy to use.

3.5 Algorithm

We summarize the previous section and give the algorithm as following. The below clustering transfer method chooses K-Means method after comparing with some other cluster methods. The parameter within can be determined by many trials.

4 Experiments

Sina weibo is a popular microblog platform and has a very wide and deep influence in China. We gathered users' web behavior data from Sina weibo by

Algorithm 1. Local Regression Transfer Learning Framework

Input:

Training dataset with labels, X_{tr}, Y_{tr};
Test dataset without labels, X_{te};
Quantity of clusters, n_c;
The k value of k-NN transfer method, knn;
Intial weight $Weig_i^{(i)}$ and final weight $Weig_f^{(i)}$ of element in X_{tr};

Output:

Labels Y_{te} of X_{te};

1. For clustering transfer method, using Eq.(3) with n_c to reweight the instances of training dataset; For k-NN transfer method, using Eq.(4) with knn to reweight the of training dataset.
2. Using the reweighted training dataset integrating traditional regression algorithms, such as Eq.(7) and Eq.(8) to train the model.
3. Using the trained models to predict personality of X_{te};
4. **return** Y_{te};

API provided by Sina. We invited many users to fill self-report personality scales on internet, collected their scale answers and web behavior records with their agreement and with permission by law. In the work, we chose Big Five personality traits, which are popular and widely accepted scales in psychology. Big Five personality traits include five dimensions: agreeableness(A), conscientiousness(C), extraversion(E), neuroticism(N), and openness(O) [15]. It is generally regarded, agreeableness reflects the tendency individual behave compassionate and cooperative, whether he/she is tempered, and so on. Conscientiousness reflects self-discipline, intension for achievement, and organized behaviors. Extraversion reflects preference between sociability and solitary, energetic communication and talkative ability. Neuroticism reflects the tendency and capacity of emotional stability and control. Openness reflects the degree of curiosity for new things, creativity and imagination.

562 user data with 845 features were collected and their personality scores were computed to be as users' labels. These features included users' online time information, users' favorites, users' friends, users' comments, users' emotions and so on. For our work, we tried to extend the model built on some condition to different condition, i.e., from male data to female data, and from one district to other districts. We just selected two cases to testify that our methods can perform well and effectively, and we think they can be well in other cases.

4.1 Predicting Users' Personality between Genders

Male and female users may have different behaviors on internet so that we can't directly apply model trained on one gender to the other gender. Our task is to predict male users' personality based on female users' labeled data and male users' unlabeled data. We have a dataset with a total number 562, 347 are female and the rest are male. Since our dataset has 845 features which is more than the data size(562), we choose stepwisefit method in Matlab to reduce dimension and select the most relevant features first. We preprocess the female dataset, and obtain about 25,14,19,25 and 20 features for Big Five personality A, C, E, N and O.

We then test whether male and female dataset belong to the same distribution. T-test and Kolmogorov-Smirno test are used to measure whether both dataset are drawn from the same distribution. As a result, there are 3, 1, 2, 3 and 2 features by T-test, and 2, 0, 0, 2 and 1 features by Kolmogorov-Smirno test which have different distribution for the two datasets with probability more than 95% confidence. It is seen that there exists distribution divergence for male and female, though the divergence is not big. We then analyze how the transfer regression method can be used on the dataset.

We compared our proposed local regression transfer learning methods with non-transfer regression method, global transfer method and other transfer learning method. Our local regression transfer learning methods includes clustering transfer learning method and K-NN regression transfer learning methods. The global transfer method is also a K-NN regression transfer learning method, and it takes all data points as neighbors. We also chose a transfer learning method called Kernel Mean Matching method(KMM) [5] as a baseline method, and tuned its parameters to make it work under a level of optimization. Further more, we integrated these importance reweighting techniques with some common weighted risk learning methods. In details, we combined clustering transfer method with Support Vector Machine(SVM) and Regression Tree method (RT) which both are very common regression methods, while we combined k-NN transfer method with MARS and LWR which can be integrated naturally and simply. SVM method was from LIBSVM(`http://www.csie.ntu.edu.tw/~cjlin/libsvm/`), especially we chose the nu-SVR type. RT method is from Matlab toolbox. MARS and LWR are both open source regression software for Matlab/Octave from(`http://www.cs.rtu.lv/jekabsons/regression.html`). In our tables, GMARS addicts to global MARS which is global transfer method, and LMARS addicts to local MARS which is k-NN transfer learning method. Global LWR and local LWR represent GLWR and LLWR. In the experiments, we had each mean square error (MSE) result to be tested by 30 times, and calculated the mean.

From Table 1, we can see that regression transfer methods improve the prediction accuracy in most conditions, and obtain the highest prediction accuracy on test dataset. First, we compare clustering transfer methods with non-transfer method. For transfer SVM method, it improves on four traits, though not too much, and performs almost equally well in the other trait. For transfer Regression Tree methods, it improves much on four traits, and reduces a little in E trait. In general, the clustering transfer methods provide the higher performance than non-transfer methods in most situations. In the Table 1, transfer SVM behaves only a little better than non-transfer SVM. The reason may be that SVM only utilizes support vectors which is a small part of the whole data, and thus clustering to the whole data doesn't produce big effect.

Table 1. Local Regression Transfer Learning Results for Predicting Personality in Different-gender Dataset

Condition	A	C	E	N	O
Non-transfer SVM	25.0162	38.5171	25.6888	27.0483	29.1443
transfer SVM	24.5143	37.0168	25.1010	26.0709	29.1943
Non-transfer RT	49.5025	61.9840	44.2176	45.0610	60.9959
transfer RT	42.8706	57.3700	43.7455	45.1582	52.7108
Non-transfer MARS	34.8431	45.9335	34.0655	29.5776	32.6700
transfer KMM-MARS	26.7654	30.8683	24.0116	27.9208	28.1425
transfer GMARS	25.2125	31.5119	23.1247	27.6345	30.6127
transfer LMARS	24.3776	31.1357	23.1247	27.4160	28.2948
Non-transfer LWR	45.7402	44.3442	34.3780	31.2915	38.1603
transfer KMM-LWR	39.7646	44.3371	34.2630	31.2584	38.1433
transfer GLWR	319.3444	43.9598	163.1730	503.4256	1347.4744
transfer LLWR	38.5335	43.9598	34.0845	31.2940	38.2608

Secondly, we compare our local (k-NN) regression transfer method with the non-transfer method, global transfer method, and KMM transfer method, respectively. We choose MARS method and LWR method as a part of local and global transfer methods. For MARS method, we can see that k-NN regression transfer method(transfer LMARS) performs better than non-transfer method, especially, it reduced the MSE about 30% in trait A, C and E. For LWR methods, k-NN regression transfer method(transfer LLWR) is better than non-transfer LWR method on A, C,and E trait, and is almost as well as non-transfer method on N and O traits. Moreover, it denotes that the local regression transfer methods(k-NN regression transfer) perform better than the global transfer methods. Especially, global transfer LWR method produces very high error, while local transfer LWR method can improve very much. Compared with KMM methods, our k-NN regression transfer method performed better on more conditions, and its biggest advantage is that it doesn't need to solve a optimization problem and has a lower computation cost. It's worth noting in the table that both k-NN transfer method and KMM method don't perform obviously superior to non-transfer method in many cases for LWR method. The reason may be that transfer learning way doesn't perform well in some situations with LWR method.

We made experiments about how accuracy of prediction is effected by number of cluster in clustering transfer methods for trait A. From Fig. 1, we can see that number of cluster can affect the prediction accuracy much. The reason is that number of cluster determines the probability of training instances in test dataset distribution. In addition, it can't be seen there exists a simple and uniform corresponding relationship between number of cluster and prediction accuracy. Transfer SVM method performs stably, since it only uses the support vectors to model and reweighing all instances is not so determinant. Transfer regression tree method works better when the cluster number takes some certain value.

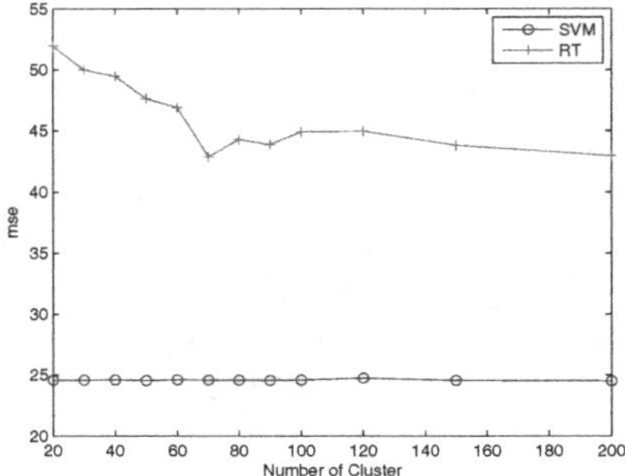

Fig. 1. Clustering Regression Transfer Learning Results by Cluster Number In A

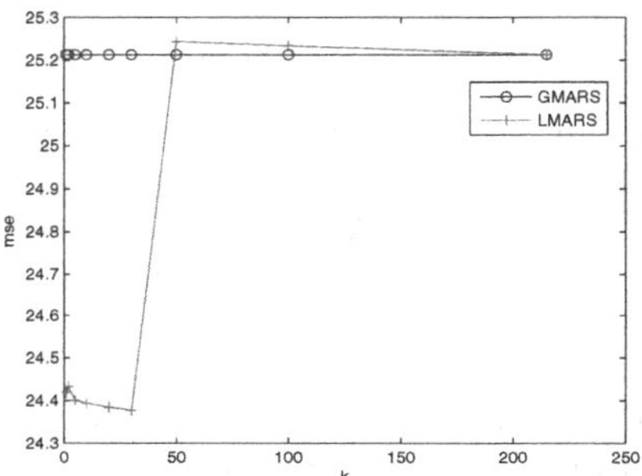

Fig. 2. The impact of quantity of nearest neighbors to the performance of k-NN transfer methods in A

Similarly, we run experiments on trait A to investigate the impact of nearest neighbor quantity to the performance of k-NN transfer methods. We take two methods GMARS and LMARS as an example. Fig. 2 shows that when k is about 30, it has the best performance, when k approximates to the size of test dataset, the performance converges to GMARS method, and when k is bigger than 50, it works a bit worse than non-transfer method. From the experiment results, it can be concluded that, when k is changed, the performance is also changed, and when k approximates to the size of test dataset, the MSE converges to GMARS situation.

4.2 Predicting Users' Personality between Districts

In personality prediction, people may also encounter using models trained in some district to predict in other district. In this experiment, we use weibo data in Guangdong province of China to train the model and predict users' personality in other districts. Firstly, we still apply stepvise to select 19, 21, 18, 22 and 20 features from total 845 features corresponding to A, C, E, N and O trait. We then use T-test and get 3, 1, 3, 3 and 2 distribution-different features, and use Kolmogorov-Smirno test and get 3, 5, 6, 9 and 2 distribution-different features, both with probability more than 95% confidence. Finally, we perform our regression transfer method on different district dataset, and compare with other methods used in the previous different-gender experiment. We analyze perfor-

Table 2. Local Regression Transfer Learning Results for Predicting Personality in Different-district Dataset

Condition	A	C	E	N	O
Non-transfer SVM	27.9419	30.6624	28.5141	27.8503	36.1007
transfer SVM	27.8795	30.8438	27.9056	27.7846	36.8981
Non-transfer RT	40.5350	52.7851	54.6466	55.1238	55.3174
transfer RT	43.2129	50.0899	51.8251	48.9670	54.7914
Non-transfer MARS	43.6764	65.0172	44.3688	47.4115	229.8742
transfer KMM-MARS	42.1194	48.9055	39.3781	47.4057	59.7330
transfer GMARS	44.7136	45.8609	43.0928	49.2114	43.1696
transfer LMARS	43.2840	42.0574	38.8104	42.9135	43.1696
Non-transfer LWR	44.5928	64.7299	50.9529	73.3051	104.4743
transfer KMM-LWR	44.3207	64.8545	50.3865	69.0167	102.6817
transfer GLWR	1110.5220	208.9482	163.7121	640.3618	1381.4780
transfer LLWR	40.5555	63.4878	51.0720	70.6810	55.8075

mances of different methods. Table 2 shows that clustering regression transfer methods work better in most cases, while k-NN transfer methods perform better than non-transfer methods and global transfer methods in most cases, and they excel KMM method in more situations.

It can be summarized from the two experiments that, clustering transfer regression methods work better than non-transfer method in most cases when they have the same settings, while k-NN transfer regression methods perform better than non-transfer, global transfer and KMM transfer methods generally under the same conditions.

5 Conclusions

In this paper, we propose a local regression transfer learning framework to predict users' personality when predicted user dataset dataset is different from training dataset. We first check whether training dataset and predicted user dataset are drawn from the same distribution, then propose two local regression transfer learning methods, including clustering and k-NN transfer methods, to reweight training instances, and finally use weighted risk regression to learn prediction model. Our experiments on different-gender and different-district datasets demonstrate that regression transfer learning is necessary, and the local regression method reduced MSE 30% under the best level than non-transfer method and it performed equally well to existing transfer learning method like KMM while it's much easier to compute. In the future, we plan to investigate how to adaptively determine an appropriate local region for each training instance and reweight its importance, so as to improve the performance of transfer learning.

Acknowledgment. The authors gratefully acknowledges the generous support from National Basic Research Program of China (2014CB744600), Strategic Priority Research Program (XDA06030800), and Key Research Program of CAS(KJZD-EW-L04).

References

1. Burger, J.: Personality, 7th edn. Thomson Higher Education, Belmont (2008)
2. Amichai-Hamburger, Y.: Internet and personality. Computers in Human Behavior 18, 1–10 (2002)
3. Li, L., Li, A., Hao, B., Guan, Z., Zhu, T.: Predicting active users personality based on micro-blogging behaviors. PLoS ONE 9(1) (2014)
4. Pan, S., Yang, Q.: A survey on transfer learning. IEEE Transactions on Knowledge and Data Engineering 99, 1041–4347 (2009)
5. Huang, J., Smola, A., Gretton, A., Borgwardt, K., Scholkopf, B.: Correcting sample selection bias by unlabeled data. In: Proc. 19th Ann. Conf. Neural Information Processing Systems (2007)
6. Sugiyama, M., Nakajima, S., Kashima, H., Buenau, P., Kawanabe, M.: Direct importance estimation with model selection and its application to covariate shift adaptation. In: Proc. 20th Ann. Conf. Neural Information Processing Systems (2008)
7. Kanamori, T., Hido, S., Sugiyama, M.: Efficient direct density ratio estimation for non-stationarity adaptation and outlier detection. Advances in Neural Information Processing Systems 20, 809–816 (2008)
8. Loog, M.: Nearest neighbor-based importance weighting. In: IEEE International Workshop on Machine Learning for Signal Processing, Santander, Spain (2012)
9. Pardoe, D., Stone, P.: Boosting for regression transfer. In: Proceedings of the 27th International Conference on Machine Learning (2010)
10. Dai, W., Yang, Q., Xue, G., Yu, Y.: Boosting for transfer learning. In: Proceedings of the 24th International Conference on Machine Learning (2007)
11. Holte, R.: Very simple classiffication rules perform well on most commonly used data sets. Mach. Learn. 11, 63–90 (1993)
12. Loader, C.: Local Regression and Likelihood. Springer, New York (1999)
13. Gupta, M., Garcia, E., Chin, E.: Adaptive local linear regression with application to printer color management. IEEE Transactions on Image Processing 17, 936–945 (2008)
14. Hastie, T., Tibshirani, R., Friedman, J.: The Elements of Statistical Learning, 2nd edn. Springer (2008)
15. Funder, D.: Personality. Annu. Rev. Psychol. 52, 197–221 (2001)

Graph Clustering Using Mutual K-Nearest Neighbors

Divya Sardana and Raj Bhatnagar

Department of EECS, University of Cincinnati,
Cincinnati, OH 45219
sardanda@mail.uc.edu, bhatnark@ucmail.uc.edu

Abstract. Most real world networks like social networks, protein-protein interaction networks, etc. can be represented as graphs which tend to include densely connected subgroups or modules. In this work, we develop a novel graph clustering algorithm called G-MKNN for clustering weighted graphs based upon a node affinity measure called 'Mutual K-Nearest neighbors' (MKNN). MKNN is calculated based upon edge weights in the graph and it helps to capture dense low variance clusters. This ensures that we not only capture clique like structures in the graph, but also other hybrid structures. Using synthetic and real world datasets, we demonstrate the effectiveness of our algorithm over other state of the art graph clustering algorithms.

1 Introduction

Community structure in a network is a subset of nodes within the graph such that connections among the nodes are stronger than the connections with the rest of the network [1]. Some examples of communities include groups of people sharing common interests in a social network and pathways in molecular biological processes. One of the many existing methods for community detection uses the graph theoretic concept of a clique to formalize this definition of dense clusters in a graph. A classic example is the maximum clique algorithm which detects fully connected subgraphs [2]. However, in many real world applications, modeling dense clusters using the concept of cliques can be too restrictive. This could lead to the missing out of many non-clique but otherwise densely connected clusters.

Many real world datasets provide useful node connectivity information embedded in the form of edge weights, for example, co-authorship networks. This information of edge weights can be used to facilitate clustering decisions to find clique and well as non- clique structures in graphs. We present here a density-based graph clustering algorithm that focuses on using the topological similarities as well as edge weights among the nodes in order to obtain cohesively connected clusters in weighted and undirected graphs. We do not make any strong assumptions about the shape of the clusters. We use a reciprocal node affinity measure called as Mutual K-nearest neighbor (MKNN) [3] as a guide towards cluster formation. MKNN defines a two way relationship between a set of nodes and has been successfully used in [3] for clustering data points. We extend an adaptation of this same concept to graphs. MKNN helps to separate the very dense cores of the graph from their sparser

D. Ślęzak et al. (Eds.): AMT 2014, LNCS 8610, pp. 35–48, 2014.

neighborhoods. Experimental results on synthetic as well as real world datasets show that this leads to the identification of dense clusters which achieve a balance between cohesive intra-cluster structure as well as homogeneous edge weights.

2 Related Work

The concept of graph clustering has been widely studied in the literature. Newmann and Girvan proposed a modularity measure to capture dense communities in a graph[1]. Most of the existing clustering algorithms based upon this concept have high computation complexities, eg., SA [4], FN [5], CNM [6]. Recently a near linear time algorithm (Louvain method) based upon the concept of greedy modularity optimization was proposed by Blondel et al. [7]. However, all modularity optimization based methods suffer from the resolution limit problem, i.e., detection of communities below a certain size is not possible by using these algorithms [8].

Molecular COmplex DEtection (MCODE) [9], proposed by Bader and Hogue is an unweighted graph clustering algorithm based upon a variant of clustering coefficient to weigh the vertices of the graph. Then, clusters are grown using seed vertices in the order defined by the vertex weights. Recently, several versions of weighted clustering coefficients have been proposed in the literature for graph clustering [10]. Van Dongen proposed a random walk based graph clustering algorithm called Markov Clustering algorithm (MCL) [11] which is based upon simulating the process of flow diffusion or random walks in a graph. MCL can be used to cluster weighted datasets as well. Liu et al. proposed a graph clustering algorithm based upon iterative merging of Maximal Cliques (CMC) for finding complexes in PPI networks [12].

3 Node Affinity Measure

We consider an undirected, weighted graph G= (V, E) where V is the set of vertices or nodes and E is the set of edges. G can be represented by a symmetric adjacency matrix A= $[a_{ij}]$. The adjacency a_{ij} between nodes i and j represents the strength of the connection or similarity between the two nodes and is a non-negative real number between 0 and 1.

Node Similarity Matrix. We define an augmented version of the adjacency matrix called as *similarity matrix* (SM) to define similarities between all the vertices which are up to four hops away. This is obtained by raising A^4. However, the definition of matrix multiplication used here is as follows.

$$(A^2)_{ij} = \begin{Bmatrix} a_{ij} & if a_{ij} \neq 0 \\ \max\limits_{k} \ (a_{ik} * a_{kj}, a_{ij}) otherwise \\ k \neq i, j & \end{Bmatrix} \quad (1)$$

We call the graph represented by SM as an augmented graph G_a. We call the edges represented by G as primary similarities. All other edges in G_a which are not in G are termed as secondary similarities. We use the term similarity from now on to point to either primary or secondary similarity.

Mutual K-Nearest Neighbors (MKNN). We define a node affinity measure called Mutual K-nearest neighbor relationship on the augmented graph G_a based upon the node similarity values. As opposed to the traditional neighborhood affinity measure KNN, MKNN defines a two way relationship between nodes. We explain the concept of MKNN using figure 1 for K=4. The four nearest neighbors for the node G are nodes A, B, E and D. However, G is not among the four nearest neighbors of nodes A, B, E and D. Nodes G and F become mutual K nearest neighbors (MKNN) of each other because all the nodes in the dense cluster in the middle refuse to have a MKNN relationship with either of them.

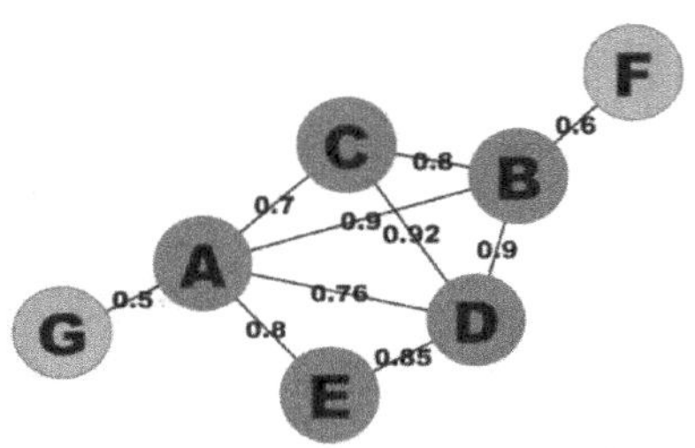

Fig. 1. Explanation of MKNN relationships

For relatively low values of K, MKNN based upon SM helps to connect vertices at approximately the same level of edge-label based density. Imposing some further constraints on the clustering algorithm to take the structural similarity between nodes into account, clustering based upon MKNN can achieve a balance between edge-label based density and structure-based density. Formally we define MKNN as follows.

Definition 1. [*Mutual K-nearest neighbor*] Two nodes i and j are MK-Nearest neighbors if either (1) in the Euclidian space between them there are fewer than K other nodes or (2) there are more than K nodes between them, but many of these nodes have found mutual K-NNs amongst themselves, and thus refuse to be mutual K-NNs with any other nodes, thereby leading to at most K-1 nodes in the intervening Euclidean space that have not found their mutual K-NNs.

MK-nearest neighbors are calculated in two steps: (1) All node pairs (v_i, v_j) in G_a, such that $sm_{ij} \neq 0$ are sorted in descending order of their similarity value. (2) MKNN relationships are then set by a top-bottom scan of the table of similarities built in (1). MKNN relationships thus formed are mutual, i.e., if v_i is an MKNN of v_j, then v_j must be an MKNN of v_i.

4 Clustering Algorithm

Our clustering framework is based upon partitioning a weighted graph G based upon both structural as well as edge weight similarity through an MKNN based node affinity measure. The algorithm begins with two initialization steps. First, it augments

G to form an augmented graph G_a as per equation 1. Second, MKNN relationships are defined amongst all the nodes of G_a. After this, the algorithm essentially works in two phases. The first phase builds candidate clusters by merging MKNN neighbors together. The second phase performs merging of relatively close candidate clusters to obtain the final clusters. We describe the two phases of clustering in detail below. We use a synthetic weighted graph, synthetic 1, with 39 vertices and 74 edges, as described in figure 2 to explain the algorithm.

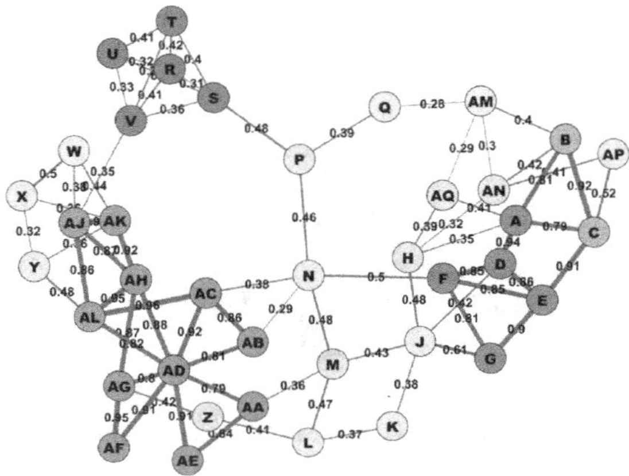

Fig. 2. A weighted synthetic dataset 1

The weights are set so as to embed three strong communities in the graph with high edge label and structural similarity. One of the communities (R, S, T, U, V) is a complete clique and the other two communities (A, B, C, D, E, F, G) and (AA, AB, AC, AD, AE, AF, AG, AH, AJ, AK, AL) have hybrid structure.

4.1 Clustering Phase 1: Preliminary Phase

The preliminary phase of clustering aims at partitioning the graph into small sized dense subgraphs. We now describe this phase in detail after defining what we mean by an *i-node*.

Definition 2. [*i-node*] We refer to vertices or nodes in the preliminary phase as individual-nodes or *i-node*. For example, in figure 2, each independent node A, B, C, etc. corresponds to an *i-node*.

All MKNN partners of an *i-node* lie at approximately the same edge similarity level from the *i-node*. Thus, it makes sense to merge these *i-nodes* together to form preliminary clusters. However, this doesn't ensure that the clusters will have good structural density as well. In order to account for this problem, we introduce a term called 'radius' to rank the *i-nodes* in a decreasing order of density surrounding them.

Definition 3. [*Radius, δ*] Given an *i-node* in_i with its similarities from all its MK-nearest neighbors being $(sm_1, sm_2, \ldots sm_p)$. Let the degrees of MKNNs be $(d_1, d_2, \ldots d_q)$. The radius of in_i is defined as:

$$\delta_{in_i} = \frac{\sum_{i=1}^{p} d_i * sm_i}{p * d_{avg}}. \quad (2)$$

Here $p <= K$ is the number of MK-nearest neighbors of in_i and d_{avg} is the average degree of all *i-nodes* in the graph. Intuitively, radius gives a measure of density surrounding an *i-node*. An *i-node* which is strongly connected to its MKNN neighbors, which in turn have a high degree will have a high value of radius. Further, such an *i-node* should be given the first chance to merge with its MKNN neighbors. We call this merging order of *i-nodes* defined by the descending order of their radius values as the *cluster initiator order*.

In figure 2, *i-node* D has the highest value of radius and is the first chosen cluster initiator to add its MKNNs A, E, F and G into one preliminary cluster, (A, D, E, F, G). Let us call this cluster as p1. The next best cluster initiator in line is *i-node* C. It's MKNN neighbors are B, E, G and A. C adds *i-node* B into its cluster. However, before adding *i-nodes* E, G and A into its cluster, it notices that these three *i-nodes* were already added by the first cluster initiator D into the preliminary cluster p1. For all such *i-nodes*, which are MKNN neighbors of two cluster initiators, we merge them with the cluster initiator where they will cause the least increase in standard deviation upon addition to the preliminary cluster. In this case, *i-nodes* E, F and G remain in p1. Thus, the cluster initiator C forms a preliminary cluster with *i-nodes* (B, C). We call this preliminary cluster as p2. It is to be noted that *i-nodes* like A, E, F, G and B, which get added by a cluster initiator are not allowed to become a cluster initiator themselves. This process continues until we exhaust all the cluster initiators. The preliminary clusters obtained for synthetic graph 1 have been shown in figure 3.

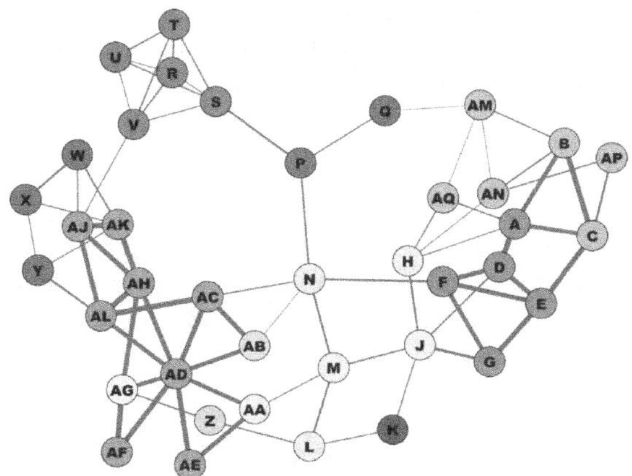

Fig. 3. Preliminary clusters obtained for Synthetic dataset1

Further, the pseudocode for this phase is given in procedure 1.

4.2 Clustering Phase 2: Merge Phase

The preliminary phase of clustering generates dense subgraphs which have strong intra-cluster similarity. However, depending upon the size of the graph, inter-cluster similarities could still be high. The merge phase (1) identifies preliminary clusters with high levels of similarity and (2) merges these highly similar subgraphs together in an iterative fashion to obtain the final clusters. Before moving on to a detailed description of the merge phase, we introduce a term called '*c-node*'.

Definition 4. [*c-node*] We refer to the clusters formed in preliminary phase as cluster-nodes or *c-nodes*. To give an example of *c-nodes*, the preliminary clusters p1 (A, D, E, F, G) and p2 (B, C) represented by blue and pink color respectively in figure 3, each represent an independent *c-node*. *c-node* p1 is comprised of five constituent *i-nodes*, i.e., A, D, E, F and G.
There are three main challenges in this phase as described below.

(1) Identifying c-nodes for merging. In order to identify the *c-nodes* which will serve as good candidates for merging, we first need to define a measure of well-connectedness or closeness among the *c-nodes*. We build a connectivity matrix defining such closeness value among the *c-nodes*. Before providing a formal definition for the connectivity matrix, we define some preliminary concepts used in its calculation.

Definition 5. [*Internal similarity of c-node (Insim)*] For a *c-node* cn_i, insim(cn_i) captures the sum of all primary similarities among all the constituent *i-nodes* of cn_i. Mathematically, insim(cn_i) is defined as:

$$insim(cn_i) = \sum_{\substack{in_i \in cn_i \\ in_j \in cn_i, \\ j \neq i}} A(in_i, in_j). \tag{3}$$

Definition 6. [*Cut of c-node (Cut)*] For a *c-node* cn_i, cut(cn_i) captures the sum of all primary similarities between the *i-nodes* belonging to cn_i and all the other *i-nodes* in the graph which may belong to other *c-nodes*. Mathematically, cut(cn_i) is defined as:

$$Cut(cn_i) \sum_{\substack{in_i \in cn_i \\ in_j \notin cn_i}} A(in_i, in_j). \tag{4}$$

Definition 7. [*Linkage between two c-nodes*] Linkage(cn_i,cn_j) between two *c-nodes*, cn_i and cn_j captures the sum of primary similarities of the connection between the *i-nodes* belonging to cn_i and the *i-nodes* belonging to cn_j. Mathematically, linkage(cn_i,cn_j) is defined as below:

$$linkage(cn_i, cn_j) = \sum_{\substack{in_i \in cn_i \\ in_j \in cn_j}} A(in_i, in_j). \tag{5}$$

Definition 8. [*Connectivity matrix (CM)*] Given the number of *c-nodes* as *n*, $CM_{|n| X |n|}$ defines a measure of similarity between all pairs of *c-nodes*. It is defined as follows:

$$CM(cn_i, cn_j) = \frac{Linkage(cn_i, cn_j)}{Cut(cn_i)} + \frac{Linkage(cn_i, cn_j)}{Cut(cn_j)}. \tag{6}$$

Intuitively, CM captures the fact that how closely the two *c-nodes* are connected to each other than to the rest of the *c-nodes* in the graph. Using CM values alone for merging would require considering every pair of *c-nodes* in the descending order of CM values as candidates for merging. In order to avoid this complexity, we take the help of MKNN relationships among c-nodes to aid in the process of merging. Using connectivity matrix (CM) values as a measure of similarity between two *c-nodes*, we define MKNN relationships among *c-nodes* using the algorithm described in phase 1.

All MKNN *c-nodes* are not good candidates for merging. Thus, before merging two MKNN *c-nodes*, we perform the following two checks to ensure that the merged clusters maintain a good level of structural similarity and edge label similarity.

- *Check for structural similarity:* When we merge two *c-nodes*, we need to make sure that the final merged cluster will also be a cohesive unit as the individual *c-nodes*. For this, we make sure that each *c-node's* participation in the merged cluster is strong. More precisely, at least 25% of the *i-nodes* belonging to a *c-node* cn_i should have primary links to a *c-node* cn_j and vice versa.
- *Check for edge-label similarity:* Before merging two c-nodes, we need to make sure that the final cluster will be homogenous as per edge label similarity. For this, we make use of the difference between average similarity of the two c-nodes and their connection. We define a term called cluster separation.

Definition 9. [*Cluster Separation*] Let cn_i and cn_j be two *c-nodes* which are MKNN neighbors of each other. Let μ_i and μ_j represent the mean of primary similarities among *i-nodes* belonging to cn_i and amongst *i-nodes* belonging to cn_j repectively. Let μ_k represent the mean of primary similarities connecting the *i-nodes* belonging to cn_i with the *i-nodes* belonging to cn_j. Then, we say that cn_i and cn_j are closely separated in terms of edge label similarity if $|\mu_i - \mu_j| <= MBOUND$, $|\mu_i - \mu_k| <= MBOUND$ and $|\mu_j - \mu_k| <= MBOUND$.

We set a value of 0.20 for MBOUND for all our runs on the synthetic and real datasets. This ensures that the standard deviation of the primary similarities in the final merged *c-node* doesn't increase a lot than the standard deviation of the individual *c-nodes*.

(2) Order of merging c-nodes. *c-nodes* which satisfy the two constraints mentioned above for structural similarity and edge label similarity are good for being merged together. Now, the problem is to decide the order in which to merge the candidate *c-nodes*. The order of merging of *c-nodes* can greatly affect the final structural and edge –label density of the clusters. In the merge phase, we use a quantity called cohesion to rank the *c-nodes* in the cluster initiator order. For a *c-node* cn_i, cohesion (η) represents how cohesively the *i-nodes* inside a cluster are connected to each other. Formally it is given by:

Procedure 1. Preliminary Phase of Clustering.

```
Input: List of i-nodes ∈ V (IP), MKNN relation matrix
(MKNN), Similarity matrix (SM)
Output: Cluster Label Array C labeling the preliminary
clusters for each i-node
begin
for i-node_i ∈ IP in decreasing order of Radius
     if cluster label C_i is not set for i-node i
       Set i as the cluster initiator;
       C_i ← θ;
   for i-node_j ∈ MKNN of i-node_i do
     if C_j has not been set, then
       C_j ← θ;
     else
      assign C_i as θ only if the standard deviation
      obtained of SM values in the resultant cluster is
      less than the standard deviation in i-node_j's
      current cluster;
     end
  end
  next θ;
end
end
```

$$\eta(cn_i)=\frac{Insim(cn_i)}{Insim(cn_i)+Cut(cn_i)} \quad (7)$$

We give *c-nodes* with a high cohesion value the first chance to choose amongst their MKNN neighbors for merging, before some other *c-nodes* with a relatively lower cohesion value chooses this *c-node* for merging. In figure 3, the *c-node* (R, S, T, U, V) has the best value of cohesion and thus, will be given the first chance to merge with it's only MKNN neighbor, (P, Q). Even though this neighbor *c-node* satisfies the edge label-similarity constraint, it is rejected for merging because of not satisfying the structural constraints. This merged cluster is again made a *c-node* to be used for merging in the next iteration of the merge phase.

(3) Gauranteeing Convergence. Using the cluster initiator order defined by cohesion, we continue the process of merging until we exhaust the whole list of *c-nodes*. This finishes one iteration of the merge phase. The clusters thus obtained as a result of merging are again made as *c-nodes* and MKNN neighbors are defined among these new set of *c-nodes* to iteratively continue with the merging process. However, *c-nodes* which failed to satisfy either one of the criterion concerning structural or edge label similarity as mentioned before, are not merged together. Further, we save this merge decision, so that for the *c-nodes* which did not merge together, their children may not be made MKNN neighbors again in future iterations. This in turn, helps in guaranteeing the convergence of the merging phase. When all possible merging of *c-nodes* get finished, we will reach a point where no *c-node* can find an MKNN neighbor to merge with. At this stage, we declare the set of *c-nodes* remaining as our final clusters. This finishes the merge phase. For the synthetic graph 1, the final set of clusters is displayed in figure 4. Further, the pseudo code for the merge phase is given in procedure 2.

5 Experimental Evaluation

We evaluate the performance of G-MKNN by comparing it with three representative graph clustering techniques, namely MCL, MCODE and Louvain algorithm. We use one synthetic dataset and three real world weighted protein-protein interaction (PPI) databases to evaluate our algorithm.

5.1 Description of Parameters Used

For running MCODE, we used the latest version (1.32) of its cytoscape [13] plugin. MCODE has three parameters, depth limit, node score cutoff, haircut and fluff. For our runs on the synthetic as well as real datasets, we used depth limit=4 and the rest of the parameters at their default values. For the implementation of MCL, we used the Network Analysis Tools (NeAT) web plugin [14]. MCL has a single parameter called inflation (Range:1.2 to 5) which can be used to tune the granularity of the clusters. For the synthetic and real datasets, we kept its value to be = 2. Louvain algorithm was implemented using its Matlab version freely available at [15]. For implementing GMKNN, we used MATLAB version 7.11. We set the value of K=4 for all the datasets.

5.2 Synthetic Dataset

We use synthetic dataset 1 as described in figure 2 before to compare the performance of G-MKNN with the other three algorithms. The clustering results for G-MKNN are presented in figure 4. G-MKNN captures the three dense core clusters represented by blue, brown and green colors, very efficiently. Further, all the low density clusters are also separated very well instead of merging them into the three core clusters.

Next, we compare our clustering results with MCL, Louvain method and MCODE. We use variance of primary similarity edges in the graph as a measure of edge-label based density. In order to assess that how well the clusters capture the structural similarity in the graph, we define a term call structural density (ρ_S).

$$\rho_S = \frac{Number\ of\ edges\ in\ the\ cluster}{All\ possible\ number\ of\ edges\ in\ the\ cluster}. \quad (8)$$

A high value of average ρ_S for all the clusters can be interpreted as high intra-cluster structural similarity.

Procedure 2. Merge Phase of Clustering.

```
Input: List of n c-nodes (CP) from preliminary phase, SM,
Cluster Label Array C for i-nodes (preliminary clusters)
Output: Final Cluster Label Array C for i-nodes
begin
Calculate CM|n|x|n| for c-nodes in CP;
Find MKNN neighbors for c-nodes using CM as metric;
Assign D←temporary cluster label array for c-nodes;
repeat
for c-node_i ∈ CP in decreasing order of cohesion
  if D_i is not set for c-node_i
       Set c-node_i as the cluster initiator;
       D_i ← λ;
       θ← cluster label of i-nodes ∈ c-node_i
   for c-node_j ∈ MKNN of c-node_i do
     if D_j has not been set, then
        if checks for structural similarity and edge
          label similarity are satisfied, then
          D_j ←λ;
          C(i-node_j) ← θ ∀ i-node_j ∈ c-node_j;
       end
     end
   end
   next λ;
  end
 end
Update CP with new set of c-nodes and recalculate CM;
until no merges take place for 3 successive iterations
end
```

In table 1, we present the values of average variance and average ρs for all the clusters obtained by G-MKNN and the algorithms MCL, Louvain and MCODE. It is to be noted that G-MKNN corresponds to the lowest value of average variance obtained for all the clusters. This helps to capture clique as well as non-clique structures in the graph. For example, the clique cluster (R, S, T, U, V) is found by all the four algorithms perfectly. However, it is only the G-MKNN algorithm, which is able to find the non-clique dense cluster (A, B, C, D, E, F, G) correctly. The average structural density ρs obtained by G-MKNN is less than that of MCL and MCODE. As pointed out before, the reason behind this is that our clustering criterion favors a balance between low variance as well as high structural density amongst the final clusters. Further, we can see that Louvain algorithm finds large sized clusters, which is a characteristic property of modularity based algorithms. Also, it assigns a cluster label to all the 39 nodes in the graph. On the other hand, the MCODE algorithm assigns a cluster label to only 21 nodes out of 39. G-MKNN is able to assign a cluster label to 38 out of 39 nodes without leading to very large sized clusters.

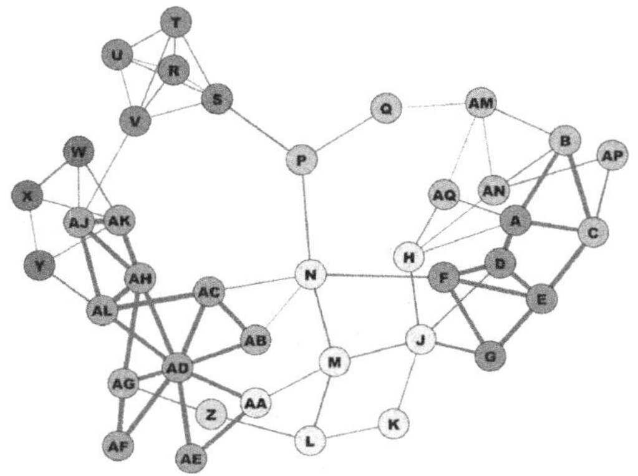

Fig. 4. Synthetic Graph 1 with clustering by G-MKNN

5.3 Real Datasets

Description of Real Datasets. We use three large scale weighted yeast protein-protein interaction datasets, Collins [16], Gavin [17], and Krogan database [18] to evaluate our algorithm. We mention the properties of these real PPI networks in table 2. We compare the set of predicted complexes obtained by G-MKNN and all the three comparison algorithms against a gold standard set has been derived from the SGD catalog of protein complexes [19]. This gold standard consists of 323 complexes comprising a total of 1279 proteins.

Evaluation Measures. In order to compare our predicted protein complexes with the gold standard set of complexes, we used 'Accuracy' as introduced by Brohee and van Helden [20]. Accuracy (Acc.) is calculated as a geometric mean of two other measures namely, clustering-wise sensitivity (Sn) and clustering-wise positive predicted value (PPV). The descriptions of these terms are the same as described in [20].

Table 1. Results for Synthetic Dataset 1

#nodes clustered	#clusters >=size 2	Max. size	Min. size	Avg. var.	Avg. ρ_s
G-MKNN					
38	6	10	1	0.003	0.53
MCL					
35	7	9	1	0.007	0.65
Louvain					
39	4	12	5	0.038	0.51
MCODE					
21	4	8	1	0.016	0.70

Table 2. Real PPI Datasets used

	Collins	Gavin	Krogan
#proteins	1622	1855	3672
#links	9074	7669	14,317

Results. We present the Sn, PPV and Acc. values obtained by different algorithms for SGD complexes in table 3. A high value of accuracy can be achieved in one of the following three ways (1) high Sn and a low PPV value (a small number of large sized clusters) or by (2) a high PPV and low Sn value (a large number of small sized clusters), or by (3) balanced Sn and PPV values (medium sized clusters). In the current scenario of predicted PPI complexes, case (3) means that the predicted clusters not only capture the proteins in the gold standard complexes well, they also have minimum extraneous noise proteins in the predicted complex. Thus, a desirable property of the predicted complexes is to achieve a balance between the Sn and PPV values.

As evident from table 3, we can see that the accuracy of G-MKNN is clearly better than that of Louvain algorithm and MCODE and similar to that of MCL for all the three datasets. However, when we look deeper at the Sn and PPV values, we see that our algorithm obtains a better balance between the Sn and PPV values. In order to further compare the performance of G-MKNN and MCL, we calculated the average variance and average structural density values of the top 20 clusters obtained by both the algorithms on real datasets in table 4. G-MKNN clearly obtains the lowest value of variance for all the datasets.

Table 3. Real Dataset Results for comparison with SGD complexes

Data	Method	#Clust.	Sn	PPV	Acc.
Collins	G-MKNN	130	0.46	0.54	0.50
	MCL	115	0.44	0.58	0.50
	Louvain	72	0.54	0.42	0.47
	MCODE	79	0.39	0.55	0.46
Gavin	G-MKNN	188	0.44	0.55	0.49
	MCL	163	0.40	0.61	0.49
	Louvain	47	0.55	0.29	0.40
	MCODE	110	0.38	0.57	0.47
Krogan	G-MKNN	267	0.47	0.55	0.51
	MCL	339	0.46	0.56	0.51
	Louvain	34	0.63	0.21	0.36
	MCODE	91	0.36	0.27	0.31

The difference is the most evident in Krogan dataset, where G-MKNN obtains both low avg. variance and high avg. structural density, thereby leading to more balanced clusters.

Table 4. Average variance and Structural Density comparison of G-MKNN vs. MCL

Data	**Method**	**Avg.Var. (top 20 clusters)**	**Avg.ρ_s (top 20 clusters)**
Collins	G-MKNN	0.016	0.72
	MCL	0.021	0.71
Gavin	G-MKNN	0.014	0.44
	MCL	0.016	0.76
Krogan	G-MKNN	0.064	0.48
	MCL	0.092	0.22

6 Conclusion

In this paper, we have presented a weighted graph clustering algorithm which aims to achieve a balance between structural and edge-label based density. A node-affinity measure called MKNN has been used to build a two phase iterative clustering algorithm. The first phase builds preliminary clusters which are iteratively merged together in the second phase. Experimental results on real world datasets show that we obtain comparable accuracy results with state-of-the art clustering algorithms like MCL and significantly better results as compared with Louvain algorithm and MCODE. Further, our top few predicted complexes obtain a low edge weight variance, thereby capturing clique as well as non-clique structures in the graph.

References

1. Girvan, M., Newman, M.E.J.: Community Structure in Social and Biological Networks. Proceedings of the National Academy of Sciences of the USA 99, 7821–7826 (2002)
2. Spirin, V., Mirny, L.A.: Protein Complexes and Functional Modules in Molecular Networks. Proceedings of the National Academy of Sciences of the USA, 12123–12128 (2003)
3. Hu, Z., Bhatnagar, R.: Clustering Algorithm based on Mutual K-Nearest Neighbor Relationships. Journal of Statistical Analysis and Data Mining 5, 100–113 (2011)
4. Guimer'a, R., Nunes Amaral, L.A.: Functional Cartography of Complex Metabolic Networks. Nature 433, 895–900 (2005)
5. Newman, M.E.J.: Fast Algorithm for detecting Community Structure in Networks. Phys. Rev. E. 69 (2004)
6. Clauset, A., Newman, M.E.J., Moore, C.: Finding Community Structure in very large Networks. Phys. Rev. E. 70 (2004)

7. Blondel, V.D., Guillaume, J.-L., Lambiotte, R., Lefebvre, E.: Fast Unfolding of Communities in Large Networks. Journal of Statistical Mechanics: Theory and Experiment 10 (2008)
8. Fortunato, S., Barth´elemy, M.: Resolution Limit in Community Detection. PNAS 104 (2007)
9. Bader, G.D., Hogue, C.W.V.: An Automated Method for finding Molecular Complexes in Large Protein Interaction Networks. BMC Bioinformatics 4 (2003)
10. Kalna, G., Higham, D.: A Clustering Coefficient for Weighted Networks, with Application to Gene Expression Data. J. AI Comm. -Network Anal. in Nat. Sci. and Eng. 20, 263–271 (2007)
11. van Dongen, S.: Graph Clustering by Flow Simulation. PhD thesis, University of Utrecht (2000)
12. Liu, G., Wong, L., Chua, H.N.: Complex Discovery from Weighted PPI Networks. J. Bioinformatics 25, 1891–1897 (2009)
13. Shannon, P., et al.: Cytoscape: A Software Environment for Integrated Models of Biomolecular Interaction Networks. Genome Res. 13, 2498–2504 (2003)
14. Brohée, S.: Using the NeAT Toolbox to compare Networks to Networks, Clusters to Clusters, and Network to Clusters. Methods Mol. Biol. 804, 327–342 (2011)
15. Scherrer, A.: (2008), `http://perso.uclouvain.be/vincent.blondel/research/louvain.html`
16. Collins, S.R., et al.: Toward a Comprehensive Atlas of the Physical Interactome of Saccharomyces Cerevisiae. Mol. Cell. Proteomics 6, 439–450 (2007)
17. Gavin, A.C., et al.: Proteome Survey reveals Modularity of the Yeast Cell Machinery. Nature 440, 631–636 (2006)
18. Krogan, N., et al.: Global Landscape of Protein Complexes in the Yeast Saccharomyces Cerevisiae. Nature 440, 637–643 (2006)
19. Cherry, J.M., et al.: SGD: Saccharomyces Genome Database. Nucleic Acids Res. 26, 73–79 (1998)
20. Brohée, S., van Helden, J.: Evaluation of Clustering Algorithms for Protein-Protein Interaction Networks. BMC Bioinformatics 7 (2006)

How Variability in Individual Patterns of Behavior Changes the Structural Properties of Networks

Somayeh Koohborfardhaghighi and Jörn Altmann

Technology Management, Economics, and Policy Program (TEMEP), College of Engineering
Seoul National University, Seoul, South Korea
skhaghighi@snu.ac.kr, jorn.altmann@acm.org

Abstract. Dynamic processes in complex networks have received much attention. This attention reflects the fact that dynamic processes are the main source of changes in the structural properties of complex networks (e.g., clustering coefficient and average shortest-path length). In this paper, we develop an agent-based model to capture, compare, and explain the structural changes within a growing social network with respect to individuals' social characteristics (e.g., their activities for expanding social relations beyond their social circles). According to our simulation results, the probability increases that the network's average shortest-path length is between 3 and 4, if most of the dynamic processes are based on random link formations. That means, in Facebook, the existing average shortest path length of 4.7 can even shrink to smaller values. Another result is that, if the node increase is larger than the link increase when the network is formed, the probability increases that the average shortest-path length is between 4 and 8.

Keywords: Network Properties, Network Growth Models, Small World Theory, Network Science, Simulation, Clustering Coefficient, Complex Networks.

1 Introduction

Small world theory comprises the idea of being connected to any other person by a chain of only five people in average [2, 3]. It has created a resurgence of interest in the field and motivated researchers to experimentally evaluate this theory. Despite controversies and empirical evidences on the existence of such pattern of connections [1], the variety in the obtained experimental results somehow should also be explainable. While the observed average shortest-path lengths in the range of 4 to 8 support the small world theory, it is still hard to draw solid conclusions on why the values differ.

In his famous experiments, Stanley Milgram was interested in computing the distance distribution of the acquaintance graph [2, 3]. The main conclusion outlined in Milgram's paper is that the average path length of individuals within the network is smaller than expected. The average path length was set to 6. Despite further empirical studies on this topic [4, 5, 6], the results obtained in various environments differ widely from an average path length of six.

D. Ślęzak et al. (Eds.): AMT 2014, LNCS 8610, pp. 49–60, 2014.

In social networks, the nodes and links represent the users and the social relations among the users, respectively. Assuming that a social network is a product of its constituents' interactions, the first question that comes to mind is what kinds of interactions take place within it and how these interactions can be categorized. The social distance between a source individual and a destination individual within the network can be utilized for categorization of their relations. This means that different degrees of friendships can be observed within the social circle belonging to each person. For instance, only degree-1 persons, who have either familial bonds or share a similar social character and personality with the respective person, are observed in the social distance equal to 1. These relationships can be established almost anytime (i.e., spare time, working time). In larger social distances such as friend-of-a-friend (FOAF) relation, the persons become more socially distant, but it is not necessarily concluded that individuals with similar characteristics cannot be found among them. The user is merely unaware of their presence. It is interesting to note that the social networking platforms including Facebook attempt to reduce these social distances as daily social activities take place online. Once a user specifies its social circle after registering on a social networking web site, the feature "Recommendation of a friend of a friend" enables the users to extend their social circles, and to change their social distance from 2 to 1. Although this is by itself very desirable, it is to be noted that such processes certainly have significant impact on the properties of a social network. According to this discussion, our hypothesis is: "*The interactions among constituents of a network can be regarded as dynamic processes which lead to changes in the structural properties of the network.*"

There are different social networks in the real world. Regardless of the relationships among the various constituents of a network, a network property such as degree distribution can be used to find similarities or differences among networks. Moreover, formation mechanism, based on these degree distributions, can result in different growth models. For example, the preferential attachment growth model, introduced by Barabási and Albert [9], is capable of generating power law degree distributions. In this model, new nodes are more willing to link to high-degree nodes during the growth of a network. The high-degree nodes can be imagined as nodes with better social skills compared to others [17]. The individuals having more activities in a social networking platform would find more friends or are even more likely to establish relationships with other people, whereas a huge number of users lack such capabilities. However, some other types of network formation mechanisms choose the attachment point for new nodes with the same probability for all nodes. Based on this random linking, a number of models have been offered in literature since the mid-19th century [10, 11, 12]. These models are able to create networks with bell shape or Poisson degree distributions.

As the process of network growth does not only depend on the method of new nodes entrances but also on the establishment of new links between existing nodes within the network, these interactions could definitely play a substantial role in the formation of networks as well. Based on this, our hypothesis can be extended to: "*Dynamic processes can be categorized into two groups: (1) The process that occurs during the growth of a network and represents the tendency of new users to establish links to other members upon entry into a network; (2) The process that occurs among users of a network in order to establish potential links.*"

Undoubtedly, adoption of a random network formation model for a growing social network is not an obvious practice. However, the significance of such random patterns can easily be justified. Although, as a member of a social network, we attempt to establish relationships with individuals of similar social character, we also realize that our social circle can easily be made more valuable through getting familiar with strangers. For example, an academic professional might meet a broad range of new people, with whom s/he shares similar characteristics, when attending conferences, presenting papers, and giving seminars. These random acquaintances have a positive influence on the performance of academics. If it is agreed here upon the existence of such random patterns, the question raises about the level of influence on the structural properties of a network. Our hypothesis can be extended to: "*The process which occur among existing users of a network in order to establish links is also divided into two categories: (a) The first category includes the establishment of potential links of FOAF type; (b) The second category includes the establishment of random potential links with other users, who are present in a network.*"

As this article engages with the idea that the human behavior (i.e., dynamic processes) is the key to formulate a network growth model, our major research question in this paper is how the level of influence of these dynamic processes impacts the structural properties of a network. With respect to our contribution, we propose a generative network growth model for analyzing the influence of the dynamic processes (i.e., the establishment of FOAF links and random links for existing nodes, as well as the establishment of random links for new nodes) on the structural properties of a network.

Based on our simulation analysis, these dynamic processes play a significant role in the formation of networks. The structural network properties of a network change. In particular, we analyzed the evolution of two structural properties of a network, namely the clustering coefficient and the average shortest-path length. The clustering coefficient is a fundamental measure in social network analysis, assessing the degree, to which nodes tend to cluster together, while the average shortest-path length provides a measure of how close individuals within the network are.

One of the main essential implications of the result derived from our simulations is the explanation for the different values of the average shortest path lengths that have been reported in empirical studies [4, 5, 6].

The remainder of this paper is organized as follows. In section 2, we discuss the theoretical background on the topic. In section 3, we detail our network formation model and its parameters. Simulation results and a discussion are presented in section 4. Finally, we present our conclusion and discuss the future work in section 5.

2 Theoretical Background

Our research work is based on literature on small world network topologies and network formation models.

2.1 Small World Network Topology

In the literature on network topology, we find different networks with their specific patterns of connections. Examples of these network topologies are random, regular, preferential attachment, and small-world networks. Despite the empirical studies of small world networks, the results obtained in various environments differ [4, 5, 6]. The following paragraphs are a summary of studies that tried to replicate Milgram's results with respect to small world networks.

Dodds et al. performed a global social search experiment to replicate the small-world experiment of Milgram and showed that social searches could reach their targets in a median of five to seven steps [5]. They classified different types of relationships and observed their frequencies and strength. The result of their analysis showed that senders preferred to take advantage of friendships rather than family or business ties. It was also indicated that the origin of relationships mainly appear to be family, work, and school affiliation. The strengths of the relationships were fairly high. Therefore, we can say that the most useful category of social ties were medium-strength friendships that originated in social environments.

Backstrom et al. repeated Milgram's experiment by using the entire Facebook network and reported the observed average shortest path length of 4.74, corresponding to 3.74 intermediaries or "degrees of separation" [4]. The study indicates the fact that various externalities such as geography have the potential to change the degree of locality among the individuals and, finally, increase or decrease the average shortest-path length.

Ugander et al. studied the anatomy of the Facebook social graph and computed several features [6]. Their main observations are that the degrees of separation between any two Facebook users are smaller than the commonly cited six degrees, and, even more, it has been shrinking over time. They also found that the graph of the neighborhood of users has a dense structure. Furthermore, they identified a globally modular community structure that is driven by nationality.

With our network formation model, we aim at providing an explanation for the differences found in the empirical studies. Our network formation model is capable to highlight the different processes that contribute to changes in the overall outcome.

2.2 Network Formation Models

There are two branches of literature that are related to network formation models. The first branch focuses on strategic network formation models, which are beyond the scope of this paper [13, 14, 15, 16]. The second one represents growth models of networks and is presented in this section.

From the view point of network growth, which is also our primary focus, several models have been proposed to produce predetermined structural properties. Proposals on network formation models have mainly been based on social networking service (SNS) networks with power law degree distributions. Examples of such SNS networks are Cyworld, Myspace, LiveJournal, and Facebook. In fact, based on observed features of the structures of these networks, different network formation models have been presented, accounting for those features.

Ishida et al. focused on modeling social network service networks, which can be characterized by three features, namely, preferential attachment, friends-of-a-friend (FOAF), and influence of special interest groups [17]. They borrowed the idea of preferential attachment from the fitness model [18]. That is to say, nodes have various capabilities to make links and communicate with others. The resulting network follows a power law degree distribution. The idea of friends of a friend is captured from the connecting-nearest-neighbor (CNN) model [20, 21]. In that model, a network grows based on the establishment of new links with friends of friends. In the same way, a random link formation strategy was adopted to model the establishment of links beyond the range of individuals' neighborhoods. It can be imagined to be equal to receiving benefits due to positive externalities from communication with individuals in other groups.

Consequently, it is reasonable to assume that the variability in individual patterns of behavior in social environments is the base for a model of network formation. However, the extent, to which such behaviors affect the network's structural properties, is still unclear and has not been discussed in literature. This is the significant difference to our research.

3 Simulation Model

3.1 Simulation Environment and Parameters

The experimental setting for our model is as follows: We conduct a multi-agent-based simulation in Netlogo [22], in order to verify our hypotheses. In our model, an agent represents an individual or a decision maker who is capable of joining the network and establishing links with others. The simulated network formation model is a generative model that is based on the ideas laid out in the introduction section. This model represents a growing network, in which new members follow the classical preferential attachment or the random growth model (uniform probability of attachment for all nodes) for connecting to other users. The probability of a new node joining the network is denoted as P_{GM}. Furthermore, existing users in this model have the ability to create new links. For this purpose, two parameters (i.e., P_{FOAF} and P_{RAN}) are used, representing the rate of establishment of links of FOAF type, and the rate of establishment of links of random type, respectively.

The value of P_{RAN} is assumed to be equal to 1- P_{FOAF}. Thus, if the value of the P_{FOAF} parameter equals 1, no random link formation process exists in the generative model. These parameters have values in the range from 0 to 1 and represent the probability for formation of potential links. If $P_{FOAF} = 0.5$, it signifies that the probability of random link formation or conversion of a link with social distance of 2 to a link with social distance of 1 is 50%. Such network modeling enables us to simulate and profoundly comprehend the dynamic transformations of a network and its effects on its structural properties.

The values of P_{GM}, which are set for the analysis, are 25%, 50%, or 75%. They are used either for representing the preferential attachment growth model or the uniform probability attachment growth model.

3.2 Structural Properties

In this paper, the structural properties used are the clustering coefficient (CC) and the average shortest path length (AVL) of a network.

The shortest path length is defined as the shortest distance between a node pair in a network [19]. Therefore, the average shortest-path length (AVL) is defined according to the following equation:

$$AVL = \frac{1}{\frac{1}{2}N(N-1)} \sum_{i \geq j} l_{ij} \tag{1}$$

where N is the number of nodes, and l_{ij} is the shortest-path length between node i and node j.

The clustering coefficient of a node i, C_i, is given by the ratio of existing links between its neighbors to the maximum number of such connections [19]. Thus, the clustering coefficient, C_i, is defined according to the following equation:

$$C_i = \frac{2E_i}{k_i(k_i - 1)} \tag{2}$$

where E_i is the number of links between node i's neighbors, and k_i is the degree of node i (i.e., the number of links connected to node i). Averaging C_i over all nodes of a network yields the clustering coefficient of the network (CC). It provides a measure of how well the neighbors of nodes are locally interconnected.

3.3 Simulation Environment

The graphical user interface of our simulation environment consists of a two-dimensional field that allows to set parameter values (e.g., simulation time, P_{GM}, P_{FOAF}, network growth model), visualize the formation of the network, and to calculate the structural properties of the network (i.e., AVL and CC) formed over the simulation time.

4 Simulation Results

4.1 Scenarios

The properties of the networks derived from our network growth model are computed over 10 suits of experiments and the average results are plotted on the diagrams shown in Figure 1 and Figure 2. The model was tested with thirty configurations and the results of the model were compared with each other. Figure 1 A-F show the results based on the preferential attachment growth model (P_{GM} = 25%, 50%, and 75%). Figure 2 A-F illustrate the results based on the uniform probability growth model (P_{GM} = 25%, 50%, and 75%).

The x-axes of Figure 1 and Figure 2 show the simulation periods, while the y-axes represent the CC values and the AVL values obtained with both network formation strategies, respectively.

By comparing the successive series of results shown in Figure 1 A-F, the CC values decrease and the AVL values increase with an increase in P_{GM} values. Furthermore, in the early stage of the simulation, the networks' clustering coefficients and average shortest path lengths have a significantly large variability in their values. However, as time goes by, the curves get much steadier.

Figure 2 A-F demonstrate a trend that is analogous to the one shown in Figure 1, though the CC values and AVL values are different from those in Figure 1. The difference in the results reflects the influence of the different network growth models (i.e., the preferential attachment strategy (Figure 1) and the uniform distribution attachment strategy (Figure 2)) on the structural properties. It can clearly be observed that the networks derived from the preferential attachment growth model always have smaller AVL and larger CC values than the networks derived from the uniform distribution attachment strategy. The ranges of the CC values and the AVL values of the preferential attachment growth model are [0.41, 0.58] and [3.1, 6], respectively. The ranges of the CC values and the AVL values derived from the uniform distribution attachment strategy are [0.28, 0.55] and [3.8, 7.9], respectively.

Furthermore, as our objective is also to investigate the influence of the social circle through the establishment of FOAF-type links and random-type links in addition to the influence of the preferential attachment and uniform distribution network growth models, we also assign the values 0.25, 0.5, 0.75 and 1 to the parameter P_{FOAF} (As mentioned, the parameter P_{RAN} is the complement of P_{FOAF} ($P_{RAN} = 1 - P_{FOAF}$)).

Figure 1 A-C and Figure 2 A-C show that the CC values consistently decline with a decrease of the value of FOAF-type links (i.e., with an increase of the value of random-type links) regardless of the P_{GM} value. Furthermore, the analysis of the simultaneous impact of the network growth model and the FOAF-type links suggest that the CC value reaches its minimum with a maximum value of the growth model and a minimum in the influence of FOAF-type links (Figure 1 C and Figure 2 C). Additionally, the CC values increase with the reduction in the influence of the growth model (Figure 1 A-B and Figure 2 A-B). It must also be noted that CC values reach their maximum in early simulation stages due to a small number of nodes in the network (i.e., small network size).

Although the trend of the influence of the network growth model and the parameter P_{FOAF} on AVL is similar, it seems that the influence is more intricate than in the case of CC. Although the AVL values decreases as a result of reducing the influence of FOAF-type links at constant influence rate of the network growth model in Figure 2 D and Figure 2 F, the remaining figures show some derivations from that behavior.

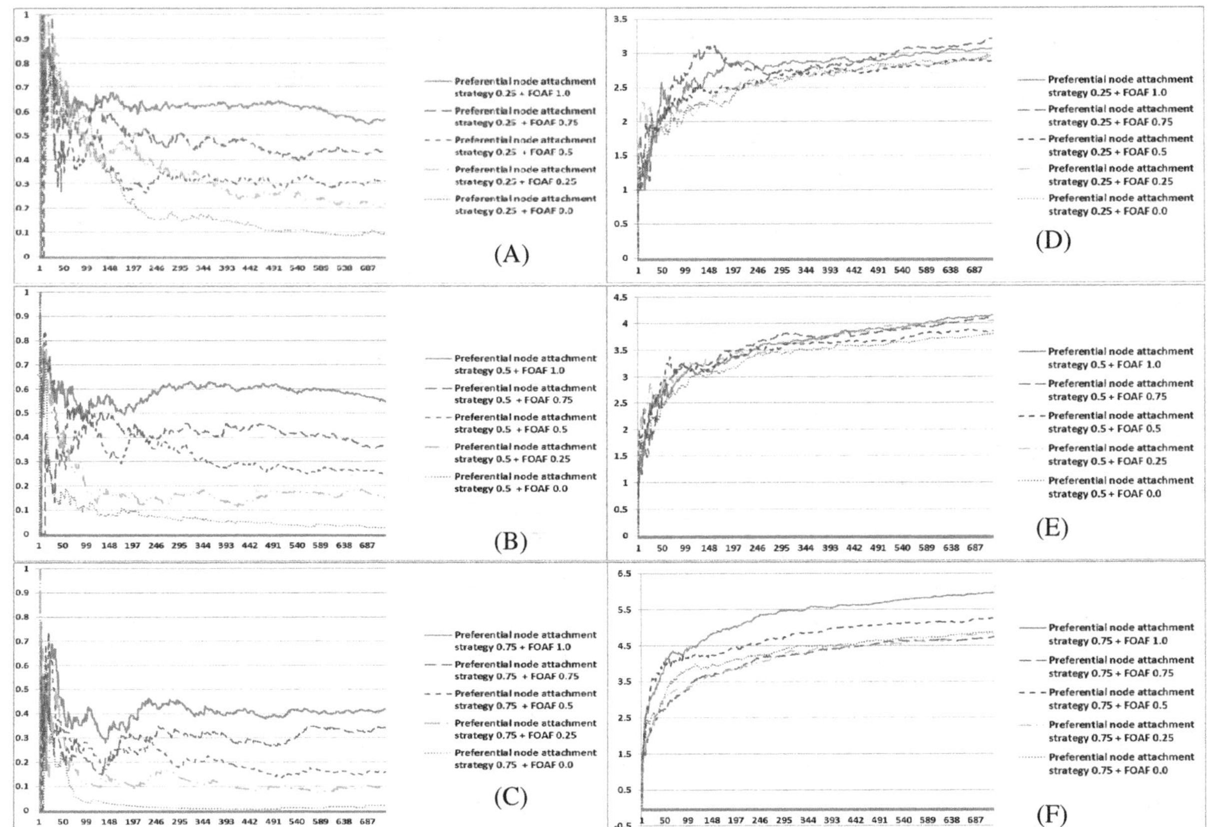

Fig. 1. Changes in the clustering coefficient, CC, with respect to the preferential node attachment strategy are shown in (A), (B), and (C). Changes in the average shortest path length, AVL, with respect to the preferential node attachment strategy are shown in (D), (E), and (F).

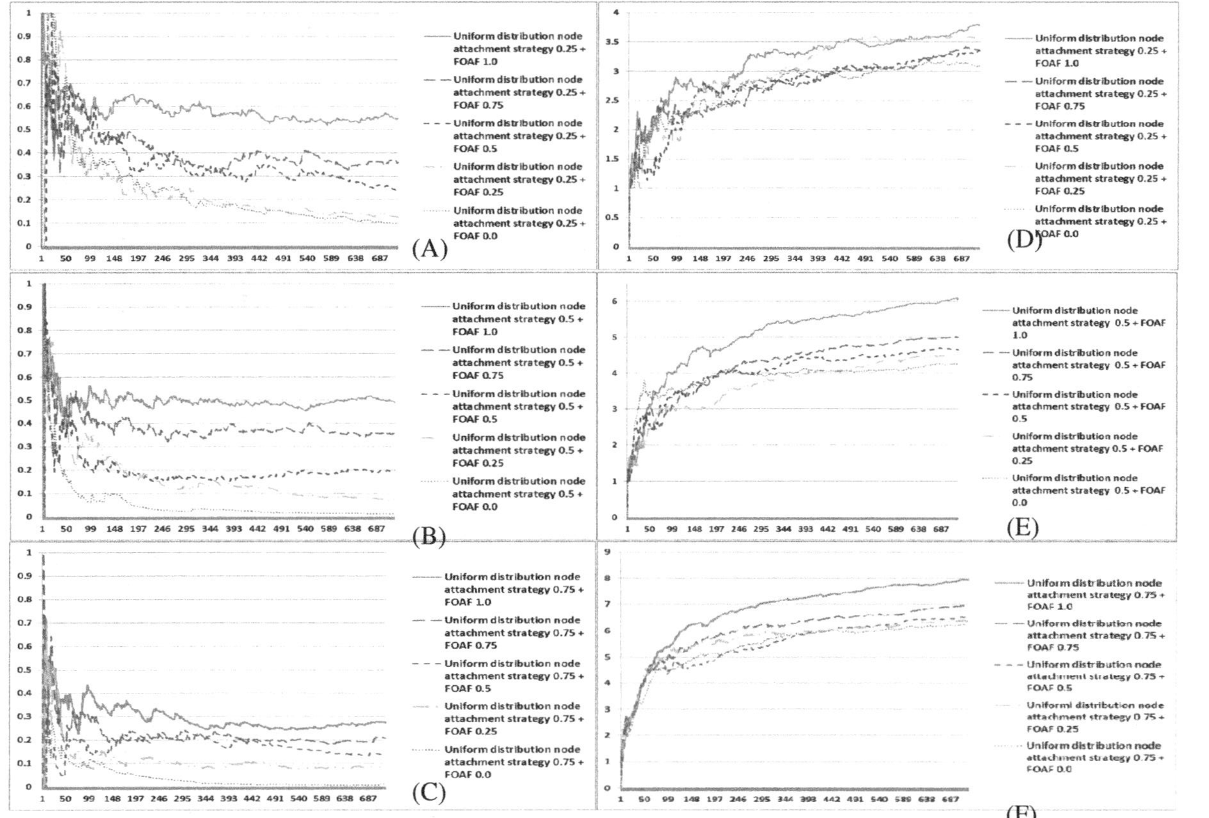

Fig. 2. Changes in the clustering coefficient, CC, with respect to the uniform distribution node attachment strategy are shown in (A), (B), and (C). Changes in the average shortest path length, AVL, with respect to the uniform distribution node attachment strategy are shown in (D), (E), and (F).

4.2 Discussion

Nowadays, no matter where we are in the world, social networking platforms are able to shrink our world. They make the world a smaller place by bringing people together. They form an important part of online activities and networking. The processes that are at work here are the bases for our research question in this paper. In particular, we ask how the rate of influence of these dynamic processes impacts the structural properties of a network (i.e., the clustering coefficient and the average shortest path length).

Despite the existence of some empirical studies on the topic, the variety of results of these studies led us to investigate how human behavior contributes to the formation of a social network. We utilized theories of sociology on human relations in a society to maneuver over the problem at hand. We base our work on the assumption that the extent of value change of the clustering coefficient and the average shortest-path length is the result of the variability in individual patterns of behavior in social environments.

Those patterns of behavior are related to situation-specific differences. For example, some people are particularly sociable. They can easily handle interpersonal relationships that come from random contacts (random-type links) and are able to extend their relationship boundaries. Other people need to be introduced via their friends (FOAF-type links). The extent, to which such behaviors affect the structural properties of networks, is still unclear and has not been discussed in literature until now.

Reported results of our simulation model are in line with the result of empirical studies and show the fact that the variety in the obtained results can be caused by different types and different levels of dynamic processes among the constituents of a network.

Our obtained results also indicate that, in social networking platforms such as Facebook, the average shortest path length (which is 4.7) can even be shrunk to smaller values if users show more patterns of random link formations. Therefore, we can expect that our world is getting smaller and smaller. Our results may also justify the result of the "communication project" mentioned in Stanley Milgram's papers, which was discussed in [1] and showed an average shortest path length of eight. If the individuals recruited for the project were not well-connected through random links to other people, the average number of individuals between the source user and the target user could increase.

Finally, our results confirm two points with respect to the effect of network growth models on AVL: First, networks derived from a preferential attachment growth model always have a smaller AVL than networks based on a uniform distribution growth model. Second, if the rate of network growth increases to larger values, the probability increases that the network's average shortest-path length lies between 4 and 8. That scenario could happen when the rate of network growth is more than the rate of establishment of potential links among the existing users of the network. These results also contribute to explaining the differences found in empirical studies. Experiments performed at different times, depending on the ongoing dynamic processes with different rates, will have different network properties.

5 Conclusions

This paper presents a comprehensive network formation model for analyzing the influence of the establishment of FOAF-type links and random-type links on structural properties (i.e., clustering coefficient and average shortest-path link) of networks. Our network formation model works based on the idea that human behavior is the key to formulate a network formation model.

Since social phenomena are complex, we followed an agent-based simulation approach, in order to depict a complex structure emerging from the interaction of many simple parts over time. Our model is able to capture and explain the structural network changes within a growing social network due to certain social characteristics of individuals.

We conducted numerical simulations to calculate and compare networks' clustering coefficient values and average shortest-path length values. Our network formation model was tested with thirty different parameter configurations and the results were compared with one another. The result of the comparison showed that different rates of variability in individual patterns of behavior in social environments lead to changes in the structural properties of networks.

As clearly observed, the comparisons confirm that networks derived from preferential attachment growth models always have smaller average shortest-path length values and larger clustering coefficient values compared to those of uniform distribution growth models. The comparisons also showed that there is an increase in the probability that a network's average shortest-path length lies between 3 and 4, if most of the dynamic processes within the network are the result of a random-type link formation. This result indicates that, in social networking platforms such as Facebook, the average shortest path length (4.7) can even shrunk to smaller values if users show behavior patterns of random-type link formation. However, bringing people closer together can only be achieved through the development of more innovative ideas and techniques in social networking platforms. Finally, another observation is that, if the rate of network growth increases to values of more than 50%, the probability increases that the network's average shortest-path length value is between 4 and 8. It means that, if most of the dynamic processes during the formation of a network are related to its growth, we can expect larger average shortest-path length among the population.

These results of our research provide an explanation for the differences in the average shortest-path lengths found in earlier empirical studies. These studies did not take into consideration the individual pattern of behavior, when they conducted their research. We also tested the effect of placing limitations on the size of personal networks on the structural properties of complex networks in one of our previous works [7]. Therefore, new empirical studies can now be conducted that consider these new results of our research.

As an extension to the current work, similar to [8], we are interested to investigate how network's characteristics positively or negatively affect individuals (e.g., how certain network features make contribution to user utility) that are located in the network. As another future extension of this study, we are also interested to investigate the effect of node's removal or deactivation on the obtained results.

Acknowledgements. This work was supported by Korea Institute for Advancement of Technology of the Ministry of Knowledge Economy within the ITEA 2 Project 10014, EASI-CLOUDS.

References

1. Kleinfeld, J.: Could it be a big world after all? The six degrees of separation myth. Society 36, 61–66 (2002)
2. Milgram, S.: The small world problem. Psychology Today 2, 60–67 (1967)
3. Travers, J., Milgram, S.: An experimental study of the small world problem. Sociometry 32(4), 425–443 (1969)
4. Backstrom, L., Boldi, P., Rosa, M., Ugander, J., Vigna, S.: Four degrees of separation. In: Proc. 4th ACM Int'l Conf. on Web Science, WebSci (2012)
5. Dodds, P.S., Muhamad, R., Watts, D.J.: An experimental study of search in global social networks. Science 301, 827–829 (2003)
6. Ugander, J., Karrer, B., Backstrom, L., Marlow, C.: The anatomy of the Facebook social graph (2011), accessed at `http://arxiv.org/abs/1111.4503`
7. Koohborfardhaghighi, S., Altmann, J.: How placing limitations on the size of personal networks changes the structural properties of complex networks. In: 6th Intl. Workshop on Web Intelligence & Communities (2014)
8. Koohborfardhaghighi, S., Altmann, J.: How structural changes in complex networks impact organizational learning performance. In: 6th Intl. W. on Emergent Intelligence on Networked Agents (2014)
9. Barabási, A.-L., Albert, R., Jeong, H.: Mean-field theory for scale-free random networks. Physica A 272, 173–187 (1999)
10. Erdős, P., Rényi, A.: On random graphs. Publicationes Mathematicae 6, 290–297 (1959)
11. Erdős, P., Rényi, A.: On the evolution of random graphs. Publications of the Mathematical Institute of the Hungarian Academy of Sciences 5, 17–61 (1960)
12. Erdős, P., Rényi, A.: On the strength of connectedness of a random graph. Acta Mathematica Scientia Hungary 12, 261–267 (1961)
13. Jackson, M.O., Wolinsky, A.: A strategic model of social and economic networks. Journal of Economic Theory 71, 44–74 (1996)
14. Galeotti, A., Goyal, S., Jackson, M.O., Vega-Redondo, F., Yariv, L.: Network games. The Review of Economic Studies 77(1), 218–244 (2010)
15. Rittenberg, L., Tregarthen, T.: Principles of Microeconomics (2009)
16. Jackson, M.O.: Social and economic networks. Princeton University Press, Princeton (2008)
17. Ishida, K., Toriumi, F., Ishii, K.: Proposal for a growth model of social network service. In: Web Intelligence and Intelligent Agent Technology (2008)
18. Bianconi, G., Barabási, A.-L.: Competition and multi-scaling in evolving networks. Europhysics Letters 54(4), 436–442 (2001)
19. Albert, R., Barabási, A.-L.: Statistical mechanics of complex networks. Reviews of Modern Physics 74(1), 47–97 (2002)
20. Davidsen, J., Ebel, H., Bornholdt, S.: Emergence of a small world from local interactions: Modeling acquaintance networks. Physical Review Letters 88(12) (2002)
21. Vazquez, A.: Growing network with local rules: Preferential attachment, clustering hierarchy, and degree correlations. Physical Review E 67(5) (2003)
22. Wilensky U.: NetLogo. Center for Connected Learning and Computer-Based Modeling, Northwestern University, Evanston, IL (1999) accessed at `http://ccl.northwestern.edu/netlogo`

A Scalable Boosting Learner for Multi-class Classification Using Adaptive Sampling

Jianhua Chen

Division of Computer Science and Engineering
School of Electrical Engineering and Computer Science
Louisiana State University
Baton Rouge, LA 70803-4020
jianhua@csc.lsu.edu

Abstract. Scalability has become an increasingly critical issue for successful data mining applications in the "big data" era in which extremely huge data sets render traditional learning algorithms infeasible. Among various approaches to scalable learning, sampling techniques can be exploited to address the issue of scalability. This paper presents our study on applying a newly developed sampling-based boosting learning method for multi-class (non-binary) classification. Preliminary experimental results using bench-mark data sets from the UC-Irvine ML data repository confirm the efficiency and competitive prediction accuracy of the proposed adaptive boosting method for the multi-class classification task. We also show a formulation of using a single ensemble of non-binary base classifiers with adaptive sampling for multi-class problems.

Keywords: Scalable Learning, Adaptive Sampling, Sample Size, Boosting, Multi-class Classification.

1 Introduction

In the "big data" era, data mining and machine learning algorithms have been applied to various real world applications such as electronic commerce and finance, healthcare, engineering and science, to entertainment and sports, and social media, in which extremely huge data sets are often the norm. The success of data mining applications depends critically on the development of scalable learning and knowledge discovery algorithms because the sheer size of data in many real world scenarios makes the traditional methods impractical. Sampling techniques can make important contributions to scalable learning and knowledge discovery. Therefore it is desirable to study smart sampling methods and their applications to learning. This paper aims at generalizing the scope of applications of sampling-based scalable learning method to handle multi-class classification tasks.

Random sampling is an important technique widely used in statistical analysis, computer science, machine learning and knowledge discovery. Efficient sampling has great potential applications to machine learning and data mining, especially when the underlying dataset is huge. For example, instead of using the entire huge data set for learning a target function, one can use sampling to get a subset of the data to construct a classifier.

D. Ślęzak et al. (Eds.): AMT 2014, LNCS 8610, pp. 61–72, 2014.

Designing a good sampling method for scalable learning is very much relevant to *Web Intelligence and Active Media*, because intelligent Web mining/Active Media systems often need to handle huge data sets collected from the Web, and such systems can benefit from smart sampling methods to gain scalability.

A key issue in designing a sampling scheme is to determine *sample size*, the number of sampled instances sufficient to assure the estimation accuracy and confidence. Well-known theoretical bounds such as the Chernoff bound and Hoeffding bound are commonly used for this purpose. Sample size could be determined a priori as in conventional *batch* sampling, or it could be *dynamically* determined as in *adaptive, sequential* sampling. Adaptive sampling decides whether it has seen sufficient samples based some criterion related to the samples seen so far. This adaptive nature of sequential sampling method is attractive from both computational and practical perspectives.

Earlier works in Computer Science on adaptive sampling include the methods in [18,19,20] for estimating the size of a database query. Adaptive sampling is also closely related to *active learning* [22]. In recent works on *Madaboost* [14,15,23,21], Watanabe et. al. proposed techniques for adaptive sampling and applied the methods to Boosting, an ensemble learning method. These works illustrated that adaptive sampling techniques can make important contributions to speed up and scale up ensemble learning, which typically requires much computation because of the need to construct the *ensemble* with many classifiers.

In recent works [12] [11] [10], we have proposed new adaptive sampling methods for estimating the mean of a Bernoulli variable, and applied such sampling methods in developing scalable algorithms for learning by boosting. We analyzed the theoretical properties of the proposed adaptive sampling methods [12] as well as that of the boosting algorithm. Experimental studies using bench-mark data sets from UC Irvine ML repository showed that our sampling-cased boosting learner is much more efficient and thus scalable, with competitive prediction accuracy, when compared with the relevant works in the literature [14,15].

The basic boosting algorithm only handles binary classification problems in its straight forward applications. To handle multi-class classification problems (in which the number of classes is more than two), most direct way to use boosting for such problem is to build multiple ensemble classifiers, one for each class label, and then use the collection of ensembles to make prediction. we show in this paper that the adaptive-sampling based boosting algorithm can be readily adapted in this fashion to handle multi-class classification problems with excellent efficiency and competitive prediction accuracy.

There are various research works in the literature [2,5,6,8,3,9,4] that proposed schemes for handling multi-class boosting problems. In particular, [8] proposed the *AdaBoost.HM* method which expanded the training dataset by a factor of K (the number of classes) and essentially reduced the multi-class problem to K binary one-vs-all classification problems. But the method indeed produced ONE ensemble of classifiers. More recent developments [2,6] proposed to avoid this reduction of multi-class problem to K binary classification problems and worked directly using multi-class base classifiers.

The works in these multi-class extensions of AdaBoost did not use sampling and thus build the boosting ensemble with all data points. We show in this paper that the sampling-based boosting method can be formulated in a fashion similar to that in [2] to tackle the task of multi-class classifications using a single ensemble of non-binary base classifiers.

The rest of the paper is organized as follows. In Section 2, we present a brief review of the adaptive sampling method for controlling absolute error, and the scalable boosting algorithm based on adaptive sampling. In Section 3, the multiple ensemble construction for multi-class classification problem using sampling-based boosting is demonstrated with experimental results with 3 bench-mark data sets from UC Irvine ML repository. We show the single ensemble sampling-based boosting method using the framework [2] for multi-class problems in Section 4. We conclude in Section 5 with remarks for future works.

2 Adaptive Sampling-Based Ensemble Learning

In this section we give a brief review of the newly developed sampling-based boosting learning method. Before that, we would first briefly describe the adaptive sampling method for estimating the mean of a Bernoulli variable which is the basis for the new boosting method.

2.1 Adaptive Sampling for Estimating the Mean of a Bernoulli Variable

In [12], we proposed a sampling scheme for estimating the mean $\mathbb{E}[X] = p$ of a Bernoulli random variable X in parametric estimation. In such setting, we draw i.i.d. samples $X_1, X_2, \cdots$ of the Bernoulli variable X such that $\Pr\{X = 1\} = 1 - \Pr\{X = 0\} = p$. An estimator for p can be taken as the *relative frequency* $\widehat{\boldsymbol{p}} = \frac{\sum_{i=1}^{\mathbf{n}} X_i}{\mathbf{n}}$, where $\mathbf{n}$ is the sample number at the termination of experiment. In the context of fixed-size sampling, the Chernoff-Hoeffding bound asserts that, for $\varepsilon, \delta \in (0, 1)$, the coverage probability $\Pr\{|\widehat{\boldsymbol{p}} - p| < \varepsilon\}$ is greater than $1 - \delta$ for any $p \in (0, 1)$ provided that $\mathbf{n} > \frac{\ln \frac{2}{\delta}}{2\varepsilon^2}$. Here ε is called the *margin of absolute error* and $1 - \delta$ is called the *confidence level*.

One problem dealt with in [12] is:

Problem: – Control of Absolute Error: Construct an adaptive sampling scheme such that, for *a priori* margin of absolute error $\varepsilon \in (0, 1)$ and confidence parameter $\delta \in (0, 1)$, the relative frequency $\widehat{\boldsymbol{p}}$ at the termination of the sampling process guarantees $\Pr\{|\widehat{\boldsymbol{p}} - p| < \varepsilon\} > 1 - \delta$ for any $p \in (0, 1)$.

The sampling scheme for tackling the above problem utilizes the function $\mathscr{U}(z, \theta)$:

$$\mathscr{U}(z, \theta) = \begin{cases} z \ln \frac{\theta}{z} + (1 - z) \ln \frac{1-\theta}{1-z} & z \in (0,\ 1),\ \ \theta \in (0,\ 1) \\ \ln(1 - \theta) & z = 0,\ \theta \in (0,\ 1) \\ \ln(\theta) & z = 1,\ \theta \in (0,\ 1) \\ -\infty & z \in [0,\ 1],\ \theta \notin (0,\ 1) \end{cases}$$

We use the notation $W(x)$ to denote the function $|\frac{1}{2} - x|$ to make the algorithm description concise. Let $0 < \varepsilon < 1$, $0 < \delta < 1$. The sampling scheme proposed in [12] proceeds as follows.

Algorithm 1.
Let $\mathbf{n} \leftarrow 0,\ X \leftarrow 0$ and $\widehat{\mathbf{p}} \leftarrow 0$.
While $\mathbf{n} < \frac{\ln \frac{\delta}{2}}{\mathscr{U}(w(\widehat{\mathbf{p}}),\ w(\widehat{\mathbf{p}})+\varepsilon)}$
Do
begin
Draw a random sample Y with parameter p.
Let $X \leftarrow X + Y,\ \mathbf{n} \leftarrow \mathbf{n} + 1$ and $\widehat{\mathbf{p}} \leftarrow \frac{X}{\mathbf{n}}$
end
Output $\widehat{\mathbf{p}}$ and $\mathbf{n}$.

We have conducted a preliminary theoretical analysis on the properties of our sampling method and the following theorem summarizes the results for the case of absolute error. The proof of the theorem is skipped here due to lack of space.

Theorem 1. *Let* $n_0 = \max\{\lceil \frac{\ln \frac{\delta}{2}}{\mathscr{U}(p+\varepsilon,\ p+2\varepsilon)} \rceil, \lceil \frac{\ln \frac{\delta}{2}}{\mathscr{U}(p-\varepsilon, p-2\varepsilon)} \rceil\}$. *Assume that the true probability* p *to be estimated satisfies* $p \le \frac{1}{2} - 2\varepsilon$. *Then with a probability of no less than* $1 - \frac{\delta}{2}$, *Algorithm 1 will stop with* $\mathbf{n} \le n_0$ *samples and produce* $\widehat{\mathbf{p}}$ *which satisfies* $\widehat{\mathbf{p}} \le p + \varepsilon$. *Similarly, if* $p \ge \frac{1}{2} + 2\varepsilon$, *with a probability no less than* $1 - \frac{\delta}{2}$, *the sampling algorithm will stop with* $\mathbf{n} \le n_0$ *samples and produce* $\widehat{\mathbf{p}}$ *which satisfies* $\widehat{\mathbf{p}} \ge p - \varepsilon$.

2.2 Scalable Boosting Learning with Adaptive Sampling

Boosting proceeds by constructing a sequence of hypotheses $h_1, h_2, ..., h_T$ in an iterative fashion such that the combination of these hypotheses will produce a strong classifier with high classification accuracy. The well-known Adaboost method [7] proceeds as follows. The algorithm is given a *fixed* data set D, and a probability distribution $\mathscr{D}(1)$ over the data set D, which is typically assumed to be uniform. Then at each boosting round $1 \le t \le T$, Adaboost will generate a hypothesis h_t which has minimal classification error over D according to the probability distribution $\mathscr{D}(t)$. A positive weight $\alpha^{(t)}$ is assigned to h_t which is proportional to its (weighted) classification accuracy on D. Moreover the distribution $\mathscr{D}(t)$ is updated to generate $\mathscr{D}(t+1)$ such that the data points misclassified by h_t would have their weights (probabilities) increased while the weights on other data points decreased.

One variation of Adaboost is the Madaboost approach[14]. The idea is to use sampling on D (according to distribution $\mathscr{D}(t)$) to construct h_t from a subset of training data S_t from D in evaluating the prediction accuracy of the base classifiers to make the boosting more efficient. A stopping condition that adaptively determines the sample size for S_t was proposed, and theoretical analysis and experimental results were presented in [14] showing that Madaboost can generate classifiers with comparable accuracies and better efficiency.

In [11] we showed that the adaptive sampling method for controlling absolute errors could be used for determining the sample size of S_t adaptively which would result in a much smaller sample size compared with the Madaboost method.

Remember that at each round t of Boosting, the adaboost algorithm selects the best hypothesis h^* with the *maximal* accuracy P_{h^*} with respect to the distribution $\mathscr{D}(t)$. Here $P_h = Pr\{V_h = 1\}$, it is the expected value of the (Bernoulli) random variable

V_h associated with hypothesis (classifier) h, such that $V_h = 1$ if $h(x) = c(x)$, and $V_h = 0$ otherwise, where each object x is drawn from D according to the distribution $\mathscr{D}(t)$. P_h is also called the *accuracy* of h. The *true error* of h (w.r.t. distribution $\mathscr{D}(t)$) denoted by $error_{\mathscr{D}}(t)(h)$, is defined as $1 - P_h$. Clearly when we try to *estimate* the accuracy of h, we will draw random samples $x_1, x_2, ..., x_n, ...$ from D (according to $\mathscr{D}(t)$) and thus observe values for the random variable V_h as ${V_h}^1, {V_h}^2, ..., {V_h}^n, ...$. The estimated value of P_h after seeing a set S of samples is $P_{h,S} = \frac{|\{x \in S:\ h(x) = c(x)\}|}{|S|}$.

When sampling is used to estimate the accuracy of each hypothesis h, the key issue is to determine a "reasonable" sample size for the estimation such that the size is sufficient to guarantee with high confidence that the selected hypothesis based on the samples is "close" enough to the best hypothesis h^*. There are various ways to define "closeness" between two hypotheses. But the most important issue in Boosting is that at least the "weak" hypothesis h_t selected at each round t should have accuracy above $1/2$. So one very modest requirement of "closeness" between the selected hypothesis h_t and the best one h^* is that if $P_{h^*} > 1/2$, then $P_{h_t} > 1/2$. Focusing on each individual hypothesis h, this requirement could be formulated as follows (per [14]). We introduce the utility function U_h for each hypothesis h and define $U_h = P_h - 1/2$. Here the utility function could take negative value in case $P_h < 1/2$. We want to estimate the utility function values close with high confidence. The sampling problem is then boiled down to the problem:

Problem: We want to design an adaptive sampling scheme such that when sampling is stopped with $n = |S|$, $\Pr\{|\boldsymbol{U_{h,S}} - U_h| \geq \varepsilon |U_h|\} \leq \delta$, with $U_{h,S} = P_{h,S} - 1/2$.

Note that $U_{h,S} - U_h = P_{h,S} - 1/2 - (P_h - 1/2) = P_{h,S} - P_h$ for the boosting problem. So we are trying to select a stopping rule on sample size $|S|$ such that $\Pr\{|\boldsymbol{P_{h,S}} - P_h| \geq \varepsilon |P_h - 1/2|\} \leq \delta$. The method for controlling absolute error with termination condition $\mathbf{n} \geq \frac{\ln \frac{\delta}{2}}{\mathscr{U}(w(\widehat{\boldsymbol{p_n}}),\ w(\widehat{\boldsymbol{p_n}})+\varepsilon)}$ can be adapted for the above problem. We will replace the $\widehat{\boldsymbol{p_n}}$ in the criterion by $P_{h,S}$ and the *fixed* ε above by $\frac{\varepsilon |P_{h,S} - 1/2|}{1+\varepsilon}$ which depends on the current estimation $P_{h,S}$.

This gives rise to the following hypothesis selection method for boosting by adaptive sampling: Keep sampling examples from the data set D (and add them to the set S of examples sampled) until there is a hypothesis h (base classifier) in hypothesis space H such that $U_{h,S} > 0$ and $n \geq \frac{ln \frac{\delta}{2}}{\mathscr{U}(W(P_{h,S}), W(P_{h,S}) + \frac{\varepsilon |U_{h,S}|}{1+\varepsilon})}$.

We have shown in [11] with experimental results using UC Irvine ML datasets that our adaptive boosting method described in this section achieves a significant reduction in sample size and thus much better efficiency while maintaining competitive prediction accuracy compared with the Watanabe method in [14].

3 Handling Multi-class Classification with Adaptive Boosting

In this section we show that a straightforward adaptation of our adaptive boosting method for binary classification problems can produce reasonably good results on multi-class classification problems.

The application of (adaptive) Boosting to handle multi-class problems is done by running the Boosting algorithm multiple times - once for each class label, with class labels for the data points modified in the obvious way for each run. So if there are $k > 2$ classes, we run the Boosting algorithm k times, and produce k ensembles of binary classifiers, each ensemble is used to predict whether an object belongs to the specific class or not.

When using the k ensembles to predict the class label for a new object, we return the class label j which gives the highest difference of the weighted vote for class j vs. weighted vote for not in class j. In case that no ensemble has classified the new instance as a positive example, the new instance is considered to be "unclassified".

We have implemented the above adaptation and applied it to handle multiple class classification problems. We report the empirical studies conducted to test the prediction accuracy and the computational efficiencies of our method. In particular we contrast the sample size and execution time between our method and the Watanabe method in [14].

Here in our experiments we used 3 bench-mark datasets from the UC Irvine ML database. The datasets are: Robot Wall-following dataset, the connect-4 game dataset and the theorem-proving heuristics dataset. The size, number of classes, and number of attributes, the hypothesis space size are shown in the following Table 1. We used the simple decision stump as base classifiers - each has a root node test with two outcome leaf nodes. H_{DS} denote the decision stump hypothesis space. The connect 4 dataset has categorical attributes, each with 3 possible values. The other two datasets have only continuous-valued attributes and we handled that by discretization into 6 intervals with equal-size binning.

Table 1. Data Sets Used in the Experimental Studies

Data Set	data size	numb of classes	number of attributes	$\|H_{DS}\|$
Wall	5456	4	24	276
Connect-4	67557	3	42	252
Theorem Proving	3059	6	51	516

The following Table 2 shows the prediction accuracy comparison of our sampling method for multi-class boosting in contrast to the Watanabe sampling method. The "B-Accu"("M-Accu") denotes the average prediction accuracy of each ensemble of binary classifiers, and that of the multi-class classifier, respectively. "New"("Wata") indicates our new method (and Watanabe method) respectively. In our experiments, T, the number of weak classifiers in each ensemble, is set to 10. The accuracy results are averages of 10 runs of using 50 percent of randomly selected data points as training data and the remaining data points as test data.

Table 2. Prediction accuracy (with 50 precent data as training)

Data Set	B-Accu(new)	B-Accu(Wata))	M-Accu(new)	M-Accu(Wata)
Wall	0.837	0.858	0.768	0.772
Connect-4	0.785	0.785	0.702	0.698
Theorem Proving	0.837	0.834	0.516	0.525

From the above Table 2, we see that our new method achieves competitive prediction accuracy compared with the Watanabe method. Thus we can *focus mainly in the sample size and execution efficiency* aspects of the methods.

Figures 1-3 show the comparison results between our method and the Watanabe method in terms of sample size and the execution time in seconds. Our implementation was written in C++ language and the program ran on an HP server computer with two quad core CPU running Linux. The execution time was recorded for learning one ensemble.

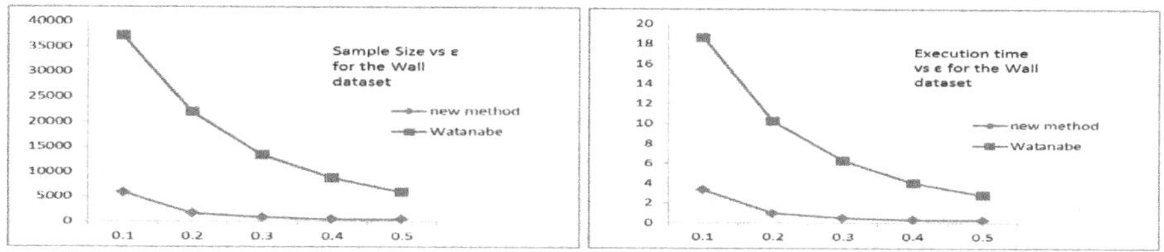

Fig. 1. Sample size (left) and execution time (in seconds) comparison for the wall dataset

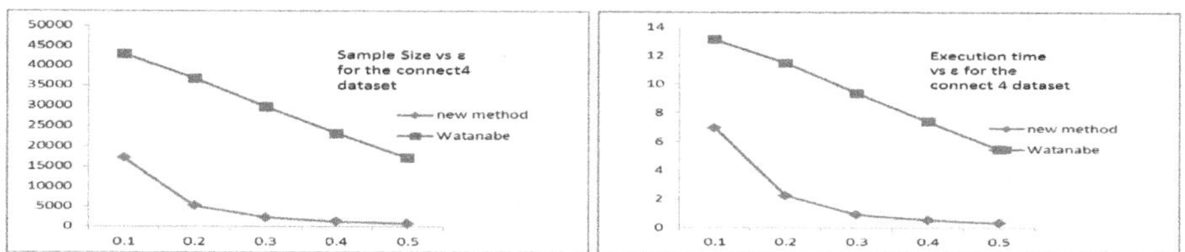

Fig. 2. Sample size (left) and execution time (in seconds) comparison for the connect4 dataset

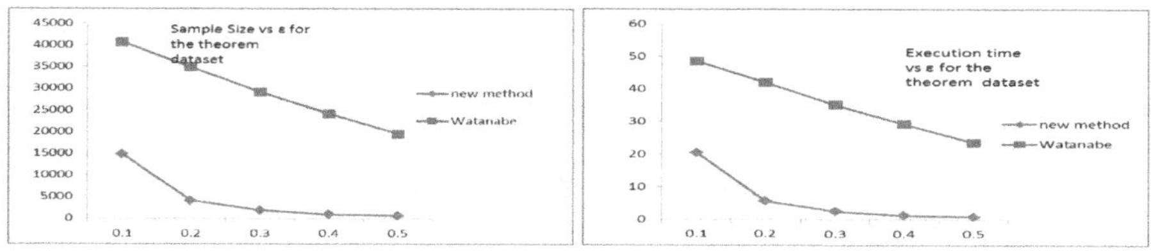

Fig. 3. Sample size (left) and execution time (in seconds) comparison for the theorem-proving dataset

4 Using a Single Ensemble for Multi-class Classification

In this section we describe the formulation of sampling-based boosting for multi-class problems using a single ensemble of non-binary base classifiers in the framework of [2].

Assume that the training dataset is $D = \{(x_1, y_1), (x_2, y_2), \cdots, (x_n, y_n)\}$ where each y_i is from the set of class labels $\{1, 2, ..., K\}$.

The SAMME algorithm proposed in [2] is shown below ("Ind" denotes the indicator function that returns 1 or 0 depending on whether the argument is true or false):

Algorithm SAMME:

Step 1. Initialize the observation weights w_i = 1/n, i = 1, 2, . . . , n.

Step 2. For t = 1 to T:

(a) Fit a classifier $h_t(x)$ to the training data using weights w_i.

(b) Compute $err^{(t)} = \sum_{i=1}^{n} w_i \, Ind(h_t(x_i) \neq y_i)/\sum_{i=1}^{n} w_i$

(c) Compute $\alpha^{(t)} = \log \frac{1-err^{(t)}}{err^{(t)}} + \log(K-1)$

(d) Set $w_i \leftarrow w_i EXP(\alpha^{(t)} Ind(h_t(x_i) \neq y_i))$ for $i = 1, 2, \cdots, n$.

(e) Re-normalize w_i.

Step 3. Output $C(x) = argmax_j \sum_{t=1}^{T} \alpha^{(t)} * Ind(h_t(x) = j)$ for a new instance x.

We note that SAMME is very similar to AdaBoost, with the major difference in the formula for computing the $\alpha^{(t)}$ (score for classifier h_t): the extra term $\log(K-1)$. This will make the $\alpha^{(t)}$ positive as long as the prediction accuracy of h_t, $1-err^{(t)}$ is strictly bigger than $1/K$, namely, h_t does better than random guessing. Requiring that each weak classifier performs better than random guessing is a reasonable constraint that is easier to satisfy than $1 - err^{(t)} > 1/2$ for each h_t in the case of multi-class problems. The authors in [2] also introduced a variant of SAMME to allow the output to be real valued confidence-rated predictions. The theoretical foundation of the 2-class and K-class versions of boosting as forward stage-wise additive modeling with exponential loss functions have been discussed in [3,2].

Here we want to design a sampling-based multi-class boosting learner with non-binary base classifiers. We can easily see the sampling method described in Section 2.2 can be adapted to tackle the task. The previous utility function $U_{h,S} = P_{h,S} - 1/2$ should be modified: Indeed, if define the new utility function for a hypothesis h with respect to sample S to be $U_{h,S} = P_{h,S} - 1/K$, then the sampling problem is simply:

We want to determine a stopping criterion such that when sampling is stopped with $n = |S|$,

$$\Pr\{|\boldsymbol{P_{h,S}} - P_h| \geq \varepsilon|P_h - 1/K|\} \leq \delta.$$

Clearly using the same technique shown in Section 2.2, we obtain the following hypothesis selection method for boosting by adaptive sampling for multi-class problems: Keep sampling examples from the data set D (and add them to the set S of examples sampled) until there is a hypothesis h (base classifier) in hypothesis space H such that $U_{h,S} > 0$ and

$$n \geq \frac{ln\frac{\delta}{2}}{\mathscr{U}(W(P_{h,S}), W(P_{h,S}) + \frac{\varepsilon|U_{h,S}|}{1+\varepsilon})}$$

Note that the above termination condition appears to be the same as in the case of binary classification; the only difference is the definition of the utility function here is $U_{h,S} = P_{h,S} - 1/K$, with $K > 2$. Combining the ideas of sampling-based Boosting and the framework of SAMME, we obtain the following sampling-based multi-class boosting Algorithm 2 (next page). Here "consistent(x,y)" returns $+1$ if $x = y$, returns -1 if $x \neq y$.

Algorithm 2:
Step 1. Initialize the distribution $\mathscr{D}(1)$ over the training data, typically setting weights w_i = 1/n, i = 1, 2, . . . , n.
Step 2. For t = 1 to T:
(a) Use adaptive sampling to draw a sample S_t from D according to distribution $\mathscr{D}(t)$ to select a weak classifier $h_t(x)$ such that the sample error of h_t on S_t is strictly below $\frac{K-1}{K}$.
(b) Set $err^{(t)}$ to be the sample error of h_t on S_t.
(c) Compute $\alpha^{(t)} = \log \frac{1-err^{(t)}}{err^{(t)}} + \log(K-1)$
(d) Update the distribution $\mathscr{D}(t)$ to obtain $\mathscr{D}(t+1)$:
Set $w_i \leftarrow Min\{1/n, w_i \cdot EXP(-\alpha^{(t)} \cdot consistent(h_t(x_i), y_i))\}$
for $i = 1, 2, \cdots, n$. //Note: this step is done only conceptually
(e) Re-normalize w_i //Note: again, this step is only done conceptually.
Step 3. Output $C(x) = argmax_j \sum_{t=1}^{T} \alpha^{(t)} * Ind(h_t(x) = j)$ for a new instance x.

Note that the above Algorithm 2 would not actually calculate the weights w_i for *each* data point because if so, we would have to go through all data points in each boosting round, and that would defeat the purpose to use sampling to gain efficiency. Instead, as done in [14] and [11] the weights w_i are actually updated and used in Step 2(a) only on the sampled data points.

We have implemented Algorithm 2 and Samme. We have conducted a preliminary study comparing the prediction accuracy and computational efficiency of Algorithm 2 and its non-sampling counterpart Samme. The results of the preliminary experiments suggest that Algorithm 2 produces competitive prediction accuracy while using much lower number of samples for the learning task, and is much more efficient.

In the study, we still used the 3 UC Irvine benchmark datasets *Wall, Connect4, Theorem-Proving*. The hypothesis space still consists of the *Decision Stump* classifiers, except that the leaf nodes take class labels from a set with more than 2 elements - so actually the hypothesis space is larger. For example, the hypothesis space size $|H_{DS}|$ for the "Theorem Proving" dataset is now 9180 in contrast to 516 in the binary classifier case. The hypothesis space size $|H_{DS}|$ for the "Connect4" dataset is now 756 compared with 252 for the binary classifier case. For the "Wall" dataset, $|H_{DS}| = 1728$ in contrast to $|H_{DS}| = 276$ for the binary classifiers. In Algorithm 2, we chose $\varepsilon = \delta = 0.5$. For both Samme and Algorithm 2, we vary the parameter T, the number of classifiers in the ensemble, taking prediction accuracy and execution time measures for $T = 20$, $T = 50$, $T = 100$, $T = 150$ and $T = 200$. The accuracy results are averages of 10 runs of 4-fold cross-validation.

As shown in Table 3, prediction accuracies of Samme and Algorithm 2 are pretty much comparable. In fact on the "wall" and "Theorem Proving" data sets, our algorithm 2 achieved higher prediction accuracy than Samme. Samme performed slightly better for $T = 150$ and $T = 200$ on the "connect4" data set.

It is also instructive to compare the prediction accuracy of the single ensemble multi-class learning method (i.e., Algorithm 2 and Samme) with the multiple-ensemble method presented in the previous section. Apparently the multiple ensemble method achieved higher prediction accuracy using the same type of classifiers (decision Stump) for the "Wall" and "Theorem Proving" datasets. However one needs to keep in mind that

Table 3. Prediction accuracy (4-fold cross validation)

Data Set/Method	T= 20	T = 50	T=100	T = 150	T = 200
Wall:Samme	0.558	0.594	0.588	0.614	0.621
Wall: New	0.649	0.651	0.666	0.662	0.664
Connect4: Samme	0.661	0.683	0.703	0.716	0.728
Connect4: New	0.670	0.681	0.711	0.706	0.719
Theorem Proving: Samme	0.339	0.369	0.358	0.350	0.377
Theorem Proving: New	0.389	0.384	0.411	0.408	0.417

the multiple binary classifiers approach can lead to quite a number of "unclassified" test data points which are labeled by each of the binary classifiers as "negative example" of each class. The single ensemble multi-class method, on the other hand, labels every test data point.

A comparison about the computational efficiency of Samme and Algorithm 2 shows that Algorithm 2 is much more efficient, because it uses much smaller number of samples in selecting a good classifier. The computational time comparison results are shown in the following Figures 4-6. Again here the time (in seconds) are recorded for learning one ensemble.

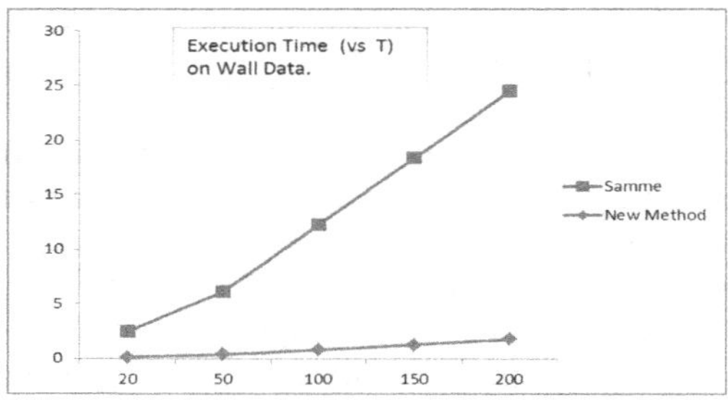

Fig. 4. Execution time (in seconds) comparison for the wall dataset

Looking at the Figures 4 - 6, we notice some interesting observations. For the "Connect4" dataset, the non-sampling based Samme algorithm is much more time-consuming. This is because the huge size of the data - the dataset has 67757 records and 4-fold cross validation uses a training dataset of more than 50,000 records. So it takes a lot of time to update the weights on data points in the boosting algorithm, which makes the non-sampling based boosting rather slow. Our Algorithm 2 used about 250 data sample points in selecting one classifier at each boosting round, which clearly made the algorithm much more efficient. So we observe that Samme is about 150 times slower than Algorithm 2 for this dataset. For the "Wall" dataset, Samme is about 15 times slower. For the "Theorem Proving" dataset, the situation is somewhat different: Samme takes about 3.5 times of the execution time for Algorithm 2. Here the average sample size for Algorithm 2 to select a good base classifier in each boosting round is around 330, which is about 15% of the training dataset size in 4-fold cross validation (the "Theorem Proving" dataset has

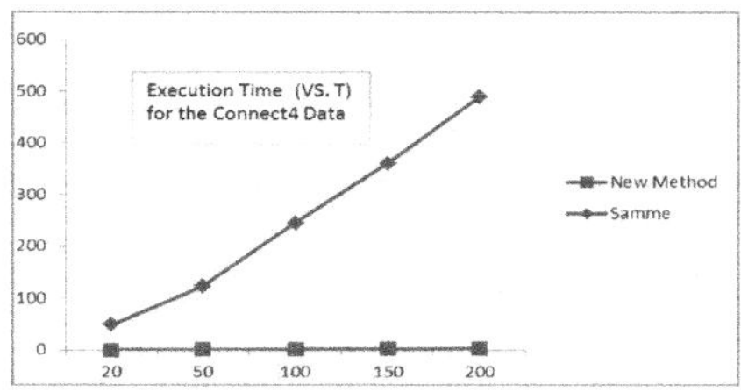

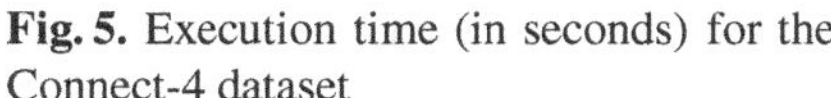
Fig. 5. Execution time (in seconds) for the Connect-4 dataset

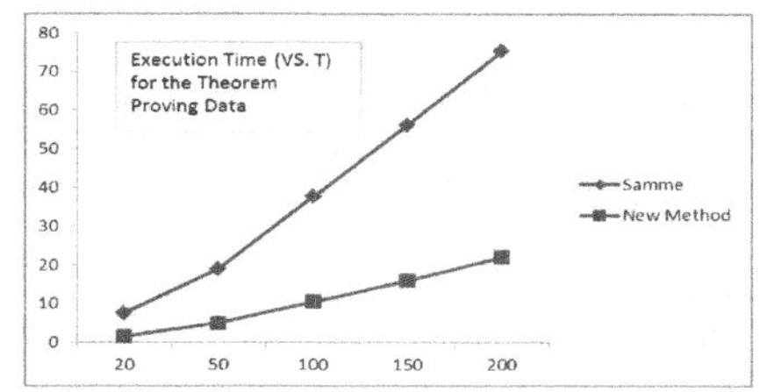

Fig. 6. Execution time (in seconds) for the Theorem-Proving dataset

size 3059 and 3/4 of that would be used as training in each of the 4-folds). Because the other computing costs in Algorithm 2 (such as the time to generate sample data points according to the underlying distribution), Algorithm 2 cannot run in time which is 15% of the time for Samme. Nevertheless, Algorithm 2 still achieved significant reduction in computational time for this dataset compared with Samme.

This observations indicate that the computational advantage of Algorithm 2 is more significant when the dataset is large in size and the boosting round parameter T is also large.

5 Conclusions and Future Work

In this paper we present an empirical study on applying a newly developed sampling-based boosting learning method for multi-class (non-binary) classification. Preliminary experimental results using bench-mark data sets from the UC-Irvine ML data repository confirm the much improved efficiency and competitive prediction accuracy of the proposed adaptive boosting method for the multi-class classification task, in comparison with other sampling-based boosting method.

We acknowledge the preliminary nature of our experimental studies because the limited number of datasets used in the experiments. In particular we did not carry out experiments on the datasets that were used in several relevant works cited here, mainly due to time limit. More extensive empirical studies are definitely desirable and we hope to address this in future work.

We also show in this paper a formulation and algorithm of using a single ensemble of non-binary base classifiers with adaptive sampling for multi-class problems. Preliminary experimental study shows its competitive classification accuracy performance and much better computational efficiency, as compared with Samme, which is not based on sampling. Again, further theoretical and experimental analysis of the proposed method would be conducted in the future to gain a better understanding of its strength and limitations.

Acknowledgments. This work is supported in part by the Louisiana Board of Regent Grant LEQSF-EPS(2013)-PFUND-307. The author would like to thank Seth Burleigh for his efforts in implementing early versions of the computer program related to this study.

References

1. Chernoff, H.: A measure of asymptotic efficiency for tests of a hypothesis based on the sum of observations. Ann. Math. Statist. 23, 493–507 (1952)
2. Zhu, J., Rosset, S., Zou, H., Hastie, T.: Multi-class AdaBoost. Statistics and its Interface 2, 349–360 (2009)
3. Friedman, J., Hastie, T., Tibshirani, R.: Additivel Logistic Regression: A Statistical View of Boosting. Annals of Statistics 28, 337–407 (2000)
4. Mukherjee, I., Shapire, R.: A Theory of Multiclass Boosting. Journal of Machine Learning Research 14, 437–497 (2013)
5. Sun, P., Reid, M.D., Zhou, J.: AOSO-LogitBoost: Adaptive one-vs-one LogitBoost for Multiclass Problem. In: International Conference on Machine Learning (ICML) (2012)
6. Kegl, B.: The Return of AdaBoost.MH: Multi-class Hamming Trees. arXiv:1312.6086 [cs.LG] (preprint)
7. Freund, Y., Schapire, R.: Decision-Theoretic Generalization of on-Line Learning and an Application to Boosting. J. of Computer and System Sciences 55(1), 119–139 (1997)
8. Schapire, R., Singer, Y.: Improved Boosting Algorithms using Confidence-rated Prediction. Machine Learning 37(3), 297–336 (1999)
9. Allwein, E., Schapire, R., Singer, Y.: Reducing Multiclass to Binary: A Unifying Approach for Margin Classifier. Journal of Machine Learning Research 1, 113–141 (2000)
10. Chen, J., Xu, J.: Sampling Adaptively using the Massart Inequality for Scalable Learning by Boosting. In: Proceedings of ICMLA Workshop on Machine Learning Algorithms, Systems and Applications, Miami, Florida (December 2013)
11. Chen, J.: Scalable Ensemble Learning by Adaptive Sampling. In: Proceedings of International Conference on Machine Learning and Applications (ICMLA), pp. 622–625 (December 2012)
12. Chen, J., Chen, X.: A New Method for Adaptive Sequential Sampling for Learning and Parameter Estimation. In: Kryszkiewicz, M., Rybinski, H., Skowron, A., Raś, Z.W. (eds.) ISMIS 2011. LNCS, vol. 6804, pp. 220–229. Springer, Heidelberg (2011)
13. Chen, X.: A new framework of multistage parametric inference. In: Proceeding of SPIE Conference, Orlando, Florida, vol. 7666, pp. 76660R1–76660R12 (April 2010)
14. Domingo, C., Watanabe, O.: Scaling up a boosting-based learner via adaptive sampling. In: Terano, T., Liu, H., Chen, A.L.P. (eds.) PAKDD 2000. LNCS, vol. 1805, pp. 317–328. Springer, Heidelberg (2000)
15. Domingo, C., Watanabe, O.: Adaptive sampling methods for scaling up knowledge discovery algorithms. In: Proceedings of 2nd Int. Conference on discovery Science, Japan (December 1999)
16. Frey, J.: Fixed-width sequential confidence intervals for a proportion. The American Statistician 64, 242–249 (2010)
17. Hoeffding, W.: Probability inequalities for sums of bounded variables. J. Amer. Statist. Assoc. 58, 13–29 (1963)
18. Lipton, R., Naughton, J., Schneider, D.A., Seshadri, S.: Efficient sampling strategies for relational database operations. Theoretical Computer Science 116, 195–226 (1993)
19. Lipton, R., Naughton, J.: Query size estimation by adaptive sampling. Journal of Computer and System Sciences 51, 18–25 (1995)
20. Lynch, J.F.: Analysis and application of adaptive sampling. Journal of Computer and System Sciences 66, 2–19 (2003)
21. Watanabe, O.: Sequential sampling techniques for algorithmic learning theory. Theoretical Computer Science 348, 3–14 (2005)
22. Hanneke, S.: A bound on the label complexity of agnostic active learning. In: Corvallis, O.R. (ed.) Proceedings of the 24th Int. Conf. on Machine Learning (2007)
23. Watanabe, O.: Simple sampling techniques for discovery sciences. IEICE Trans. Inf. & Sys. ED83-D, 19–26 (2000)

Scaling of Complex Calculations over Big Data-Sets*

Marek Grzegorowski

Faculty of Mathematics, Informatics and Mechanics, University of Warsaw, ul. Banacha 2, 02-097 Warsaw, Poland
marekgrzegorowski@wp.pl

Abstract. This article introduces a novel approach to scale complex calculations in extensive IT infrastructures and presents significant case studies in SONCA and DISESOR projects. Described system is enabling parallelism of calculations by providing dynamic data sharding without necessity of direct integration with storage repositories. Presented solution doesn't require to complete a single phase of processing before starting the next one, hence it is suitable for supporting many dependent calculations and can be used to provide scalability and robustness of whole data processing pipelines. Introduced mechanism is designed to support case of still emerging data, thereby it is suitable for data streams e.g. transformation and analysis of data collected from multiple sensors. As will be shown in this article, this approach scales well and is very attractive because can be easily applied to data processing between heterogeneous systems.

1 Introduction

Ensuring scalability of calculations performed over large datasets is no longer a purely academic debate but the real problem faced by the majority of firms. Problem of the collection, storage, transformation and analysis of big data has become a top topic not only in computer science but also for companies from media, telecommunication or financial markets. Relational approach to store and process data has become unable to provide adequate performance since big data has emerged. Example of such situation is process of feeding data-warehouses (DWH). Extract Transform Load (ETL) processes responsible for transferring data to data-warehouses and construction of data-marts performs longer and longer and hardware scaling become insufficient.

The problem of data storage is not only associated with the ever-increasing volumes of data but also with their various formats. The need for storage of relational records have been expanded to documents, ontologies, semi-structural data etc. Together with data representation also character of performed operations has changed and thus characteristics of most queries. Many times we need to store redundant information in multiple formats and create dedicated indexes to enable efficient query execution so we increase variety of used technology e.g. inverted indexes and n-grams for text search[12].

* This research was partly supported by Polish National Science Centre (NCN) grant DEC-2011/01/B/ST6/03867, as well as Polish National Centre for Research and Development (NCBiR) grant PBS2/B9/20/2013 in frame of Applied Research Programmes. This publication has been co-financed with the European Union funds by the European Social Fund.

D. Ślęzak et al. (Eds.): AMT 2014, LNCS 8610, pp. 73–84, 2014.

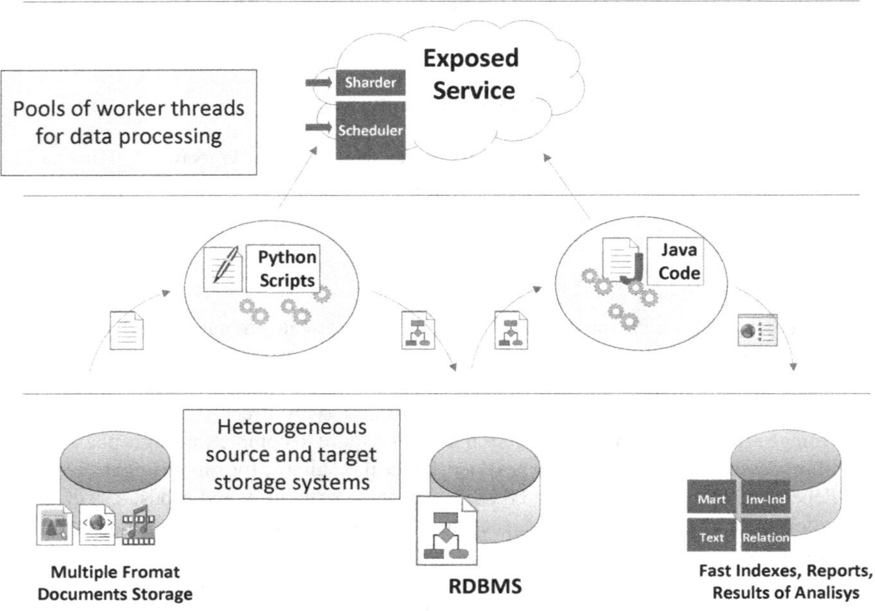

Fig. 1. A high-level perspective of communication between the presented service and pooled worker threads, including the data flow. It is worth to notice, that in order to shard data and to schedule tasks service doesn't require direct access to the data. Figure presents simplified fragment of document processing performed in SONCA system(see. section 4). Multiple python threads transform collected documents to common format. Multiple java threads analyse documents common format and create specific indexes. Both phases are performed simultaneously, each phase is run over multiple threads. Construction of python and java programs fits to algorithm 1. Process is supported by the "Exposed Service" that is the contribution of this paper.

In the wake of the ever-changing customer needs, a great diversity of available software solutions has appeared. Over many years, big corporations have deployed outstanding number of heterogeneous programs which were developed in different technologies and programming languages like: Cobol, PL/SQL, C#, Jee etc. but must cooperate within same software infrastructure. This situation has led to the creation of new paradigms of system construction such as Enterprise Application Integration (EAI) and Service Oriented Architecture (SOA) [2], which introduce a transparent way of integration of heterogeneous computer systems through exposed services.

Article discusses the problem of supporting scalability of complex calculations over big datasets with very strong emphasis on data sharding, parallelization of performed calculations and ability to work in distributed, heterogeneous software infrastructures. Contribution of this work is general architecture and design of novel approach to support scalability of entire pipelines of dependent calculations. Presented mechanism allows scaling well-defined class of operations that includes the majority of calculations used in computer systems and addresses all problems mentioned in above paragraphs.

Described solution was deployed in project SONCA (see fig. 1) and currently is part of DISESOR system architecture (see section 4). The presented approach is perfectly suited to data streams since emergence of new data automatically triggers further calculations. Presented idea allows to scale many dependent functions and allows to publish partial results before completion of whole processing.

This paper is organised as follows. In section 2 is defined a class of functions that can be scaled by described approach. Subsequently, in section 3 is presented API and high-level architecture of exposed service. Section 4 contains an extensive case studies with performance measurements. Finally, in section 5 this idea is compared to other solutions, and in section 6 is summary and plan for future research.

2 Problem Description

Calculations for data sets should take entire collection as input but the output data may be a set (in case of transformation and enrichment of the input data) or a single value (e.g. aggregation of input data - sum, average or min-max values). To present the problem more formally let's define calculation as a function $\mathscr{F}$ which domain is the power set $P(\mathscr{S})$ and co-domain is $\mathscr{T}$. An example of program corresponding to the function $\mathscr{F} : P(\mathscr{S}) \to \mathscr{T}$ could be OCR parser which recognizes text on pictures. Then: $\mathscr{S}$ corresponds to "Pictures", $\mathscr{T}$ corresponds to "Text" and $\mathscr{F}$ to "C++ OCR code". Let's define what properties must meet the function $\mathscr{F}$, to scale it via this solution:

- Notation: Let $S = \{s_1, s_2, s_3, s_4, \dots\}$, $S \in P(\mathscr{S})$ and $T \in \mathscr{T}$
- S can be represented as: $S = S_1 \bigcup S_2 \bigcup S_3 \bigcup \dots, \forall S_i, S_j \subseteq \mathscr{S} : S_i \bigcap S_j = \emptyset$
- There must be: a function $\mathscr{F}' : P(\mathscr{S}) \to \mathscr{T}$ and commutative and associative operator $\bigoplus$ to aggregate partial results (if $\mathscr{F}$ correspond to e.g. ETL process we could use $\bigcup$ as operator), then: $\mathscr{F}(S) = \mathscr{F}'(S_1) \bigoplus \mathscr{F}'(S_2) \bigoplus \mathscr{F}'(S_3) \bigoplus \dots$
- Instead of $\bigoplus$ we can define function $\mathscr{F}'' : P(\mathscr{T}) \to \mathscr{T}$, which hold operator properties and $\mathscr{F}''(T_1 \bigcup T_2) = \mathscr{F}''(\mathscr{F}''(T_1 \bigcup T_2)) = \mathscr{F}''(\mathscr{F}''(T_1) \bigcup T_2)$, then:
- $\mathscr{F}(S) = \mathscr{F}''(\mathscr{F}'(S_1) \bigcup \mathscr{F}'(S_2) \bigcup \mathscr{F}'(S_3) \bigcup \dots)$
- The way in which set $S \subseteq \mathscr{S}$ was split into $S_1, S_2, S_3 \dots$ shouldn't affect the result.

Although, function argument $S \in P(\mathscr{S})$ should be defined before function is applied to it, in this case input set S is split into sub-sets: $S = S_1 \bigcup S_2 \bigcup S_3 \bigcup \dots$, where $S_i \subseteq S$, thus we don't need to have the entire input set S predefined to start calculation. In other words, we can start calculation with knowledge about few firs subsets S_1,$S_2 \dots$, rest of them can be defined during processing. This allows us to handle situation when new data occur during long-term calculations as in case of stream data.

Let's take a look at real-life data mining problem of collection and analysis readings from multiple sensors. Firstly, collected readings are persisted in source repository and marked with time-stamp, then are cleared and lastly rule-based decision-model that predicts potential risks is applied to them. Let $S = \{s_1, s_2 \dots\} = S_1 \bigcup S_2 \bigcup S_3 \bigcup \dots$ corresponds to still incoming data from sensors. Let $\mathscr{F} : P(\mathscr{S}) \to P(\mathscr{T})$ corresponds to data clearing transformation and $\mathscr{G} : P(\mathscr{T}) \to \mathscr{D}$ be the function that calculates risk for given set of sensor reading. Using proposed solution we can parallelize both functions $\mathscr{F}$ and

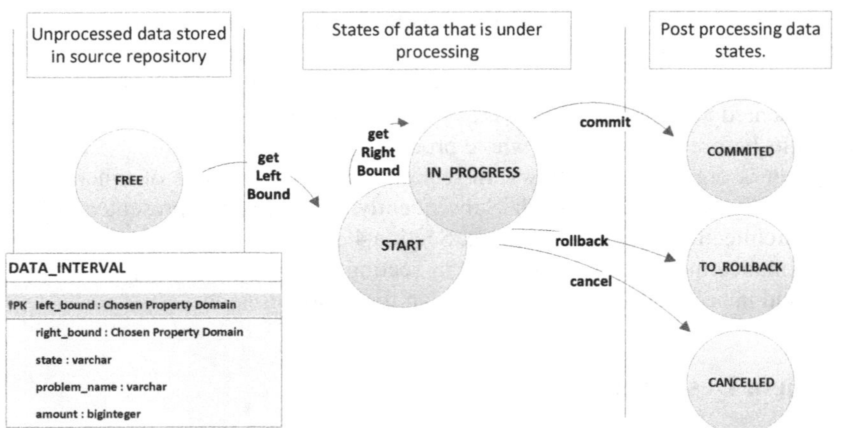

Fig. 2. Graph presents changes of DATA_INTERVAL states after service API calls. Data-shards in START or IN_PROGRESS state are stored in cur_proc_shards table since they are currently under processing by some worker. Data in START and IN_PROGRESS states can be cancelled and rolled-back but COMMITED state can be reached only through IN_PROGRESS. Occurrence of any of post processing states triggers algorithm for reassigning checkpoint (see. table 1). Since DATA_INTERVAL shards are created dynamically, pre_proc_shards table can be initially empty. However, FREE shards may be create during prediction of leftBound (JUMP) in getLeftBound method or during getRightBound method execution when there are some rows in pre_proc_shards tabele with leftBound greater than rightBound for processed shard.

$\mathscr{G}$ at the same time, hence we can scale whole process of data collection, cleaning and analysis and allow publishing of partial results and handle still incoming data.

While above application of presented idea to data transformation is straight-forward, aggregation functions would require additional explanation. In case of sum, min and max functions it is enough that source and target streams are the same. Let $S = S_1 \bigcup S_2 \bigcup S_3 \bigcup \ldots S_n$ be multi-set of integers. Let $\mathscr{F}$ and $\bigoplus$ correspond to algebraic sum of elements. In that case, it is enough that domain and co-domain of $\mathscr{F}$ are represented by the same multi-set S and partial results of $\mathscr{F}(S)$ are added to the end of stream S.

To support processing of function average $\mathscr{AVG} : P(\mathscr{N}) \rightarrow \mathscr{N}$, partial results of $\mathscr{F} : P(\mathscr{N}) \rightarrow \mathscr{N}$ x $\mathscr{R}$ x $\mathscr{N}$ should be tuples as follow: <number of elements, partial average value, partial sum of elements> [1] and should be stored in intermediary stream of tuples $M = M_1 \bigcup M_2 \ldots$. We must define additional average function: $\mathscr{F}' : P(\mathscr{N} x \mathscr{R}\ x\ \mathscr{N}) \rightarrow \mathscr{N}$ x $\mathscr{R}$ x $\mathscr{N}$, where $\mathscr{F}'$ corresponds to $\bigoplus$ operator and both source and target of this function is stream represented by M. Thanks to presented solution we can implement functions: $\mathscr{F}$ and $\mathscr{F}'$ in different technologies and run them simultaneously over multiple parallel threads. In both cases, last emitted number/tuple would contain final value of performed calculation.

[1] Since we emit information about partial average value, hence enforcing constant number of elements of S_i and M_i would simplify calculation.

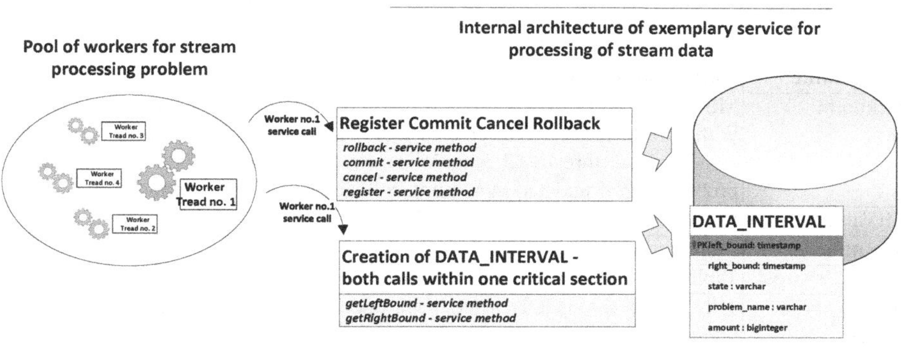

Fig. 3. Architecture of presented services for data on the fly sharding and job scheduling. Solution is designed to support pipelines of data-processing. The behaviour of commit, cancel and rollback methods is straight forward - those functions change state of appropriate record and update checkpoint (see. table 1). Pseudo code of getLeftBound and getRightBound methods has been described as algorithms: 2 and 3. Algorithm 1 shows pseudo code of exemplary worker program.

Functions that performs calculation on data can be implemented in any programming language. Prepared program would consist of three main phases: retrieving data from the source repository; performing calculation; uploading result to the target repository. In order to split initial data set into sub-sets there must be natural or artificially created linear order on source data-set and each newly stored object must be greater than any of existing object. Subsets to be processed as shards are created dynamically during processing by splitting ordered elements into left-closed, right-open non-overlapping intervals. It is also necessary that the source repository exposes specific API to query data, which could be expressed by the following SQL statement:

```
SELECT * FROM source_data WHERE timestamp >= paramTimestamp
ORDER BY timestamp LIMIT paramAmount;
```

3 Solution Architecture

In order to present general idea of this solution let's define artificial problem A of processing stream data of still incoming objects which are continuously stored in source repository. Every stored object is marked with time-stamp of creation which in turn is the property that defines stream order that satisfies requirements defined in section 2. We can imagine that there are three sets of elements: A, B and C defined over the same domain. Set A contains raw, not processed data, set B contains data which is currently under processing and set C consists of already processed data. We can imagine processing as moving elements form set A through set B up to set C. Elements of set

Table 1. Internal parameters of the system, all are stored separately for every registered problemName. Values of parameters may vary in time to adjust to problem specific characteristics.

setting name	**configuration setting description**
checkpoint	Minimal, possible value returned by getLeftBound algorithm. Every object with chosen property lower than this value has been already processed by some worker thread. Checkpoint is stored separately for every registered problem and is updated in commit, rollback and cancel methods.
shiftValue	Used as shift during prediction of leftBound in algorithm 2.
table name	**table description**
pre_proc_shards	Table contains data intervals that hasn't been processed yet.
currently_proc _shards	This table contains left-closed, right-open ranges of data which are currently under processing (see. Fig. 2). This table is going to be intensively queried so it should be as small as possible - that is the main reason for de-normalization of DATA_INTERVAL entity into three tables.
post_proc_shards	This table stores DATA_INTERVAL shards of type: <leftBound, rightBound) that has been already processed. In practice, it is sufficient to enrich interval representation with additional information and eventually DATA_INTERVAL entity may look as proposed on Fig. 2.

A are created dynamically during processing as non-overlapping intervals according to defined linear order. At any time of calculation there exists checkpoint value and every element smaller than it is surely in set C - this idea is very similar to loop invariant. In described system sets: A, B and C are implemented as database tables (see. table 1).

This solution ensure that every part of data would be processed only once and allow to process data in parallel. Presented approach gives the possibility to resume calculation of fault packages of data. What more, the number of concurrently working processes may vary during processing to adjust expected performance. Changing number of worker threads does not affect the final result. Prepared solution is designed and exposed as a service, so that it can be easily applied in majority of software infrastructures. The exposed service API is as follows:

1. register(problemName) – Registers worker thread under problemName which will be used during every call as a semaphore and scope of calculation.
2. getLeftBound(problemName, dataAmount) – Returns lower (left) bound of left-closed, right-open data interval. Returned value is unique for given problem scope and is treated as identifier of interval. This method takes as arguments registered problemName and maximum data amount that can be processed by this worker.
3. getRightBound(problemName, leftBound, proposedRightBound) – This method takes as arguments registered problemName and leftBound which compose interval identifier and proposition of right bound established by worker from retrieved data and returns upper (right) bound of data interval: <leftBound, rightBound).
4. commit(problemName, leftBound) – Commits after successful processing.
5. rollback(problemName, leftBound) – Marks interval as erroneous and unprocessed.

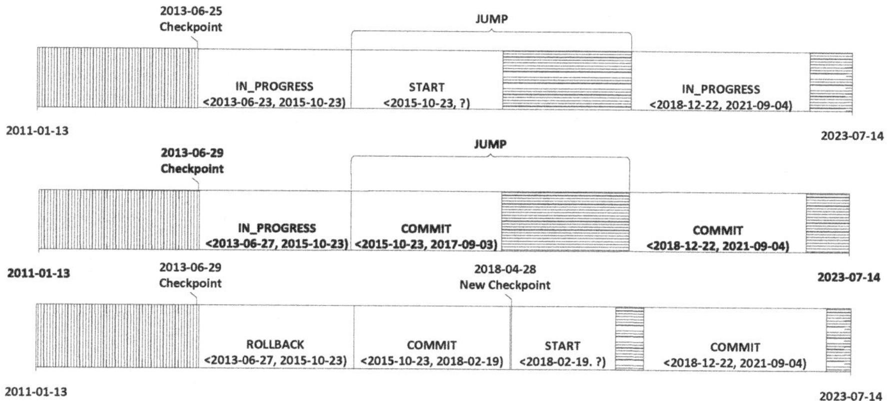

Fig. 4. Internal system states during serving three worker threads. The area covered by the vertical lines represents data that have already been processed. Horizontal lines represents unprocessed data. White boxes represent data intervals that are currently under processing. JUMP on the first and second time-line corresponds to prediction of left bound in algorithm 2. It can be noticed that both commit and rollback methods triggers reassignment of checkpoint (see table 1).

4 Case Study

This section presents two most significant case studies within SONCA and DISESOR projects and extended use case where three worker threads perform calculation in parallel. Fig.4 presents consecutive states of the system for the following sequence of calls:

1. First time-line shows situation after following sequence of service calls:
   (a) Thread 1 has called getLeftBound(...) and getRightBound(...).
   (b) Thread 2 has called getLeftBound(...).
   (c) Thread 3 has called getLeftBound(...) and getRightBound(...).
2. Second time-line is continuation of the first one and presents system state after:
   (a) Thread 2 has called getRightBound(...).
   (b) Thread 3 has called commit(...).
   (c) Thread 2 has called commit(...).
3. Third time-line starts from the point where the first one ends:
   (a) Thread 2 has called getLeftBound(...).
   (b) Thread 1 has called rollback(...).

First important appliance of presented solution is SONCA system [9], which is part of the SYNAT[1] project. The main aim of SONCA is to enrich scientific articles with semantic information about relations between various types of objects exposed from stored documents [16,10], clustering of same or similar documents or merging fragmentary information from many instances of same: articles, references. citations etc. Currently there are stored almost 4 000 000 medical and scientific articles, journals

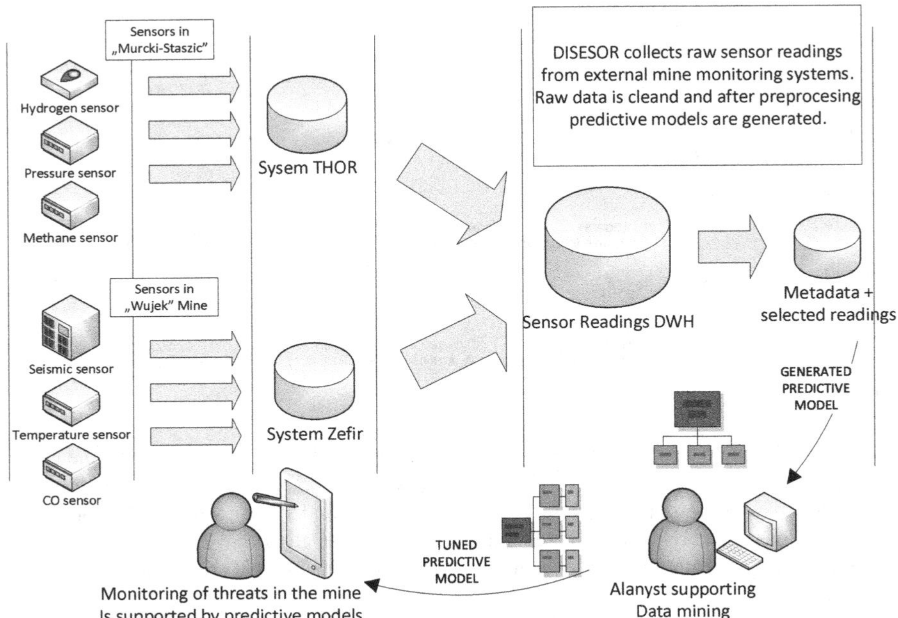

Fig. 5. General view of DISESOR system architecture. External mine monitoring systems like THOR and Zefir collect readings directly from multiple sensors placed in coal mines. DISESOR fetches data from those systems and allows building of predictive data mining models in purpose to support risk prediction. Presented solution is supposed to support all phases of data transformation, processing and appliance of predictions.

and books. Processing of such an amount of data takes a lot of time and is vulnerable to various types of faults and errors. To allow robustness and sufficient throughput of developed system it was necessary to use hybrid storage approaches. By usage of presented approach it was possible to scale transformation of XML-documents form NoSQL storage into SQL-based data model without necessity to change old code. Table 2 presents overall time spent on execution of exposed API methods during five hour long data transformation and was performed respectively by 1, 2, 3 or 4 parser threads.

Another important appliance of presented solution is DISESOR project. DISESOR is a decision support system designed for monitoring potential threats in coal mines. System collects data from multiple types of sensors like methane gas sensor, CO_2 sensor, seismic sensor, machines and engines monitoring devices and many others. The application design takes into account the data cleaning process, the process of building data mining models and on-line predictive reasoning for the latest data readings. As the most important use cases of the DISESOR system can be indicated:

Table 2. Table contains comparison of times spent on task scheduling and data sharding for up to 4 parallel threads during five hour long transformation of files from MongoDB to Infobright. Time spent on execution of getRightBound and getLeftBound methods was shown separately because those methods must be executed in critical section. Implementation of presented service was prepared using Java and was deployed no Jboss application server. Data intervals were implemented as records in PostgreSQL table. Experiment was performed 10 times for each number of concurrent threads on hardware infrastructure that is composed of two physical servers ($2xIntel\ Xeon^{®}\ CPU$ 2.67GHz *and* 64$GB\ RAM$), dedicated for storage and calculations.

	Number of worker threads			
	1	2	3	4
work period duration	5 h	5 h	5 h	5 h
overall time spent on scheduling	24,9 s	27,9 s	38,1 s	36,9 s
overall time spent on getLeftBound method	5,5 s	6,2 s	8,9 s	8,9 s
overall time spent on getRightBound method	6,6 s	7,6 s	10,5 s	9,5 s

- Assessment of seismic hazard probabilities in the vicinity of the mine
- Forecasting dangerous increase in the methane concentration in the mine shafts
- Forecasting of the possible ranges of the sensor readings in advance
- Detection of endogenous fires and conveyor belts fires
- Detecting anomalies in the consumption of media such as electricity
- Diagnostics of machines: roadheaders and longwall shearers.

The nature of the collected and analysed data, fact that sensor readings are still incoming (as in case of stream data), necessity of frequent updates of data mining models and providing that on-line predictions must take place continuously makes that presented in this article idea perfectly fits the requirements of the DISESOR project. DISESOR is currently in the phase of analysis and high-level design.

5 Related Works

This section provides a comparison of the proposed mechanism to one of most popular parallel computing frameworks: Map-Reduce [6] which relies on few other projects: GFS [8], simplified database Bigtable and lock service Chubby [5]. There appeared many implementations of this framework like Hadoop [2] which provides full software stack for Map-Reduce: Hadoop Distributed File System, storage HBase [3] and high level programming languages translating relational algebra to Map-Reduce jobs: Hive and Pig. The strength of the Map-Reduce framework lies primarily in its simplicity, it allows coders to write programs that can run in parallel without necessity of putting much attention to the management of shared resources. However, usage of the Map-reduce imposes the need to work in a homogeneous software environment composed of systems highly integrated with a given Map-Reduce implementation e.g. Cloudera [4], MapR

[2] See `hadoop.apache.org`

[3] See `hbase.apache.org`

[4] See `www.cloudera.com/content/cloudera/en/home.html`

[5] or Hortonworks [6] and in order to run Map-Reduce job data have to be stored in appropriate form in specific storage system like HDFS and thus Map-Reduce is hardly applicable as a calculation engine for data stored in other systems.

Another important limitation of Map-Reduce framework in comparison to presented approach is necessity to complete the individual phase of the calculation before proceeding to the next phase what is a problem in case of e.g. recursive queries [4]. Process oriented calculations are investigated in the associated system: Pregel [11], which is dedicated for graph processing and Prelocator[14] which renovates parallel DBMS concept [13,7] to enable incremental updates of inverted-index. Prelocator implements snapshot transaction isolation semantics [3] by ensuring data versioning with usage of time-stamp dimension and relying on strictly monotonic order - that assumption is common with this solution. What distinguishes discussed idea in comparison with other job schedulers is a novel approach to assign tasks to workers. While most of centralized scheduling systems [6],[17] involves a strong supervision over workers, the proposed system prefers a loosely coupled approach.

The general model that defines bridge between hardware and software for parallel processing was introduced by L. G. Valiant in 1990 in [17] and named bulk-synchronous parallel (BSP) model. BSP was proposed for shared-nothing distributed systems architecture that have superseded shared-disks and shared-memory solutions which in turn didn't scale well because of slowdown that each new process imposes on all existing during access to shared resources (interference) [7]. Valiant pointed that scaling parallel algorithms should concern issues and problems connected with communication and network bandwidth rather then with missing resources on single machine. Similar conclusions about need for analysis of Map-Reduce algorithms in terms of generated communication overhead were reached by authors of [15]. Authors noticed also that Map-Reduce is vulnerable to skew data e.g. heavy-hitters problem.

6 Conclusions and Future Reaserch

A novel approach to shard data and hence to support scalability of complex calculations was proposed. As has been shown in this article, presented approach scales well and is very attractive because of its ease of implementation and ability to collaborate with multiple systems regardless differences in technologies. Presented idea allows to scale processing of entire pipelines of dependent calculations among a given number of concurrent or parallel processes, since it doesn't impose to complete a single phase of processing before starting the next one. System is able to dynamically shard data during processing, thus can adjust appropriate size of shard both to the problem and to computing capabilities of individual worker node. The full potential of this approach still haven't been reached and thereby there is a lot of room for elaboration of extensions and modifications. Further research and development could investigate recognizing failures of worker threads, separating data that is the cause of error and re-arranging unfinished tasks. There is also lot's of room to arrange AI methods in improving predictions of leftBound(see. jump in algorithm 2) basing on collected statistics of data density.

[5] See `http://www.mapr.com/`

[6] See `http://hortonworks.com`

Algorithm 1. Pseudo-code of worker process including business logic (in comments) and calls to exposed SERVICE API.

```
Data: problemName: String
SERVICE.register(problemName);
while process should work further do
    leftBound ← SERVICE.getLeftBound(problemName, AmountOfData);
        /* Code to fetch data from source repository */
    rightBound ← SERVICE.getRightBound(problemName, leftBound, rightBound);
        /* Code performing calculation and uploading data */
    if there was no error then
        SERVICE.commit(problemName, leftBound);
    else
        SERVICE.rollback(problemName, leftBound);
```

Algorithm 2. Pseudo code of getLeftBound method.

```
Data: problemName: Semaphore
Result: leftBound: Timestamp
enterCriticalSection(problemName)
if pre_proc_shards is not ∅ then
    leftBound ← pre_proc_shards[0].letfBound
else
    if cur_proc_shards = ∅ then
        leftBound ← config[checkpoint]
    elseMust perform JUMP
        leftBound ← MAX(cur_proc_shards.leftBound) + shiftValue
cur_proc_shards.insert(leftBound, rightBound ← null, problemName)
leaveCriticalSection(problemName)
return leftBound;
```

Algorithm 3. Pseudo code of getRightBound method.

```
Data: inLeftBound, inRightBound: Timestamp; inName: Problem name semaphore;
Result: rightBound: Timestamp
enterCriticalSection(problemName)
thisShard ← shard ∈ pre_proc_shards that leftBound = inLeftBound
nextShard ← MIN shard ∈ cur_proc_shards that shard.leftBound > inLeftBound
if nextShard.leftBound = inRightBound then
    rightBound ← inRightBound
else if nextShard.leftBound < inRightBound then
    rightBound ← nextShard.leftBound
else
    rightBound ← inRightBound
    cur_proc_shards.insert(inRightBound, nextShard.leftBound, inName)
thisShard.rightBound ← rightBound; thisShard.state ← IN_PROGRESS
leaveCriticalSection(problemName)
return rightBound;
```

References

1. Bembenik, R., Skonieczny, L., Rybiński, H., Niezgodka, M. (eds.): Intelligent Tools for Building a Scient. Info. Plat. SCI, vol. 390. Springer, Heidelberg (2012)
2. Bennett, K., Layzell, P., Budgen, D., Brereton, P., Macaulay, L., Munro, M.: Service-based software: The future for flexible software. In: Proceedings of the Seventh Asia-Pacific Software Engineering Conference. IEEE Computer Society, Washington, DC (2000), http://dl.acm.org/citation.cfm?id=580763.785797
3. Berenson, H., Bernstein, P., Gray, J., Melton, J., O'Neil, E., O'Neil, P.: A critique of ansi sql isolation levels. SIGMOD Rec. 24(2), 1–10 (1995), http://doi.acm.org/10.1145/568271.223785
4. Boniewicz, A., Wiśniewski, P., Stencel, K.: On redundant data for faster recursive querying via orm systems. In: FedCSIS, pp. 1439–1446 (2013)
5. Burrows, M.: The chubby lock service for loosely-coupled distributed systems. In: Proceedings of the 7th Symposium on Operating Systems Design and Implementation, OSDI 2006, pp. 335–350. USENIX Association, Berkeley (2006), http://dl.acm.org/citation.cfm?id=1298455.1298487
6. Dean, J., Ghemawat, S.: Mapreduce: Simplified data processing on large clusters. In: Proceedings of the 6th conference on Symposium on Opearting Systems Design & Implementation, OSDI 2004, vol. 6, p. 10. USENIX Association, Berkeley (2004), http://dl.acm.org/citation.cfm?id=1251254.1251264
7. DeWitt, D., Gray, J.: Parallel database systems: The future of high performance database systems. Commun. ACM 35(6), 85–98 (1992), http://doi.acm.org/10.1145/129888.129894
8. Ghemawat, S., Gobioff, H., Leung, S.T.: The google file system. SIGOPS Oper. Syst. Rev. 37(5), 29–43 (2003), http://doi.acm.org/10.1145/1165389.945450
9. Grzegorowski, M., Pardel, P.W., Stawicki, S., Stencel, K.: Sonca: Scalable semantic processing of rapidly growing document stores. In: ADBIS Workshops, pp. 89–98 (2012)
10. Janusz, A., Slezak, D., Nguyen, H.S.: Unsupervised similarity learning from textual data. Fundam. Inform. 119(3-4), 319–336 (2012)
11. Malewicz, G., Austern, M.H., Bik, A.J., Dehnert, J.C., Horn, I., Leiser, N., Czajkowski, G.: Pregel: A system for large-scale graph processing. In: Proceedings of the 2010 ACM SIGMOD International Conference on Management of Data. ACM, New York (2010), http://doi.acm.org/10.1145/1807167.1807184
12. Manning, C.D., Raghavan, P., Schütze, H.: Introduction to Information Retrieval. Cambridge University Press, New York (2008)
13. Ozsu, M.T., Valduriez, P.: Principles of Distributed Database Systems, 3rd edn. (2011)
14. Peng, D., Dabek, F.: Large-scale incremental processing using distributed transactions and notifications. In: Proceedings of the 9th USENIX Conference on Operating Systems Design and Implementation, OSDI 2010, pp. 1–15. USENIX Association, Berkeley (2010), http://dl.acm.org/citation.cfm?id=1924943.1924961
15. Rajaraman, A., Ullman, J.D.: Mining of massive datasets. Cambridge University Press, Cambridge (2012), http://www.amazon.de/Mining-Massive-Datasets-Anand-Rajaraman/dp/1107015359/ref=sr_1_1?ie=UTF8&qid=1350890245&sr=8-1
16. Ślęzak, D., Janusz, A., Świeboda, W., Nguyen, H.S., Bazan, J.G., Skowron, A.: Semantic analytics of PubMed content. In: Holzinger, A., Simonic, K.-M. (eds.) USAB 2011. LNCS, vol. 7058, pp. 63–74. Springer, Heidelberg (2011)
17. Valiant, L.G.: A bridging model for parallel computation. Commun. ACM 33(8) (August 1990), http://doi.acm.org/10.1145/79173.79181

Music Data Processing and Mining in Large Databases for Active Media

Piotr Hoffmann and Bożena Kostek

Gdańsk University of Technology, Faculty o Electronics, Telecommunications and Informatics, Audio Acoustics Laboratory, Gdańsk, 80-233, Poland
phoff@sound.eti.pg.gda.pl, bokostek@audioacoustics.org

Abstract. The aim of this paper was to investigate the problem of music data processing and mining in large databases. Tests were performed on a large database that included approximately 30000 audio files divided into 11 classes corresponding to music genres with different cardinalities. Every audio file was described by a 173-element feature vector. To reduce the dimensionality of data the Principal Component Analysis (PCA) with variable value of factors was employed. The tests were conducted in the WEKA application with the use of *k*-Nearest Neighbors (*k*NN), Bayesian Network (Net) and Sequential Minimal Optimization (SMO) algorithms. All results were analyzed in terms of the recognition rate and computation time efficiency.

Keywords: Index Terms — Music processing, active media, PCA, WEKA, decision algorithms, *k*-Nearest Neighbors (*k*NN), Bayesian Network, Sequential Minimal Optimization (SMO) SMO, classification effectiveness, computation time.

1 Introduction

The most cited definition of active media technology refers to the development of autonomous computational or physical entities capable of perceiving, reasoning, adapting, learning, cooperating, and delegating in a dynamic environment. In this context one may point out processing and mining of music in large repositories as a growing area of interest for active media. In this sense the stress should be on 'large' because smaller databases could easily be managed by human resources. Most prior research done into the audio genre recognition within the field of Music Information Retrieval were based on rather small music databases with a few classes of music genres [12, 20, 25]. Even though Tzanetakis' and Cook's [25] publication presented a significant contribution to automatic music genre classification, the database they used consisted of 1000 audio excerpts. In many cases such databases are created manually which means that audio files are correctly assigned to the corresponding music genre. The latter aspect may have a very positive impact on the classification experiment effectiveness, because the user has a control over the training and test sets. In this study, the authors use a large database of audio files (over 30000), which have been collected by the music robot automatically. This may imply frequent erroneous assignment of an audio file to the given genre.

D. Ślęzak et al. (Eds.): AMT 2014, LNCS 8610, pp. 85–95, 2014.

On the assumption that a live music database is automatically expanded, it is of great importance that new records should be integrated with appropriate genres. For this purpose one may envision active rule-based decision system that maintain the consistency of the database by dealing with correct assignment or triggering repairing actions in the opposite case. One of the actions could be associated with recalling additional descriptors from a set of spare features to improve music system maintenance and performance. This may especially be valuable in music social services, when the system tracks also music listening habits. This may even result in recreating or renaming music genres according to listeners' preferences. It is essential, however, that testing a large database should first take place in a controlled environment.

The purpose of this paper is to present the music database and music recommendation service created within the carried out Synat project [6], and then to describe the preprocessing stage consisting in purposeful music parametrization and the usage of PCA in the process of reducing data dimensionality. The analysis of the effectiveness of music genre classification was performed for three classifiers: *k*NN (*k*-Nearest Neighbor), Bayesian Networks, SMO (Sequential minimal optimization) algorithms. Tests were performed in the WEKA environment [27]. The experimental setup was tested from the practical perspective, i.e. an effective approach to computation time and to less false classification rate.

It should be noted that these findings may be interesting for automatic recognition of other audio signals, content-based indexing, or signals from surveillance systems that may form active media [3, 4, 5, 7, 8, 9, 13, 14, 18, 23, 26].

2 Preprocessing

The music database, which was collected during our research, stores approximately 52,000 30-second long excerpts of songs. They were classified as one of the 22 following genres: Alternative Rock, Blues, Broadway&Vocalists, Children's Music, Christian&Gospel, Classic Rock, Classical, Country, Dance&DJ, Folk, Hard Rock&Metal, International, Jazz, Latin Music, Miscellaneous, New Age, Opera&Vocal, Pop, Rap&Hip-Hop, Rock, R&B, and Soundtracks. The majority of the files were collected by the music robot from the Internet. For the experiments carried out within this study the Synat database containing 32 110 music excerpts of 11 music genres was used. Such genres as for example Miscellaneous or Children's Music, also gathered by the music robot, were not used in the experiments. ID3 tags of excerpts were automatically assigned to songs by the music robot. The tags were saved in a fully automatic way without human control. Table 1 contains cardinalities of the sets used in the experiments [10].

Music parameterization is the main step taken to assure high effectiveness of music genres classification [6]. Currently, the most frequently used parameterization standard is MPEG 7 [2, 17, 29, 30]. This standard enables to describe audio and video files in a parametric manner. Basic features of the data reported in the MPEG 7 standard make a very comprehensive parameter vector which in the context of musical genres recognition may not always mean good results [30] as such a large amount of information certainly increases computation time. Systems used for automatic musical genres classification should ensure the highest efficiency with the fastest possible computation time. The latter can be cut down at the preprocessing stage by applying data reduction. Even though the key issue in music genre recognition is parameterization which should result in appropriate feature vector, the size of the vector should then be minimized, i.e. strongly correlated parameters should be represented by only one parameter, so that the feature vector obtained contains only orthogonal features.

For the purpose of experiments, the parameterization process was performed using the MPEG 7 standard [17]. The created music database contained feature vectors composed of 191 descriptors. The list of the calculated parameters included (among others): Temporal Centroid, Spectral Centroid, Audio Spectrum Envelope, Audio Spectrum Spread, Spectral Flatness Measure. A complete list of these parameters was presented in an earlier paper of the authors and its quality was checked in the ISMIS conference contest [11]. However, the carried out tests were based on 173-parameter feature vectors. The 191 feature vector was reduced by removing higher frequency Audio Spectrum Envelope and the Spectral Flatness Measure descriptors, which was a consequence of the limited analysis band to 8 000 Hz.

The result of this parametrization process is a big parameter vector file. For the whole Synat database it resulted in 32 110 audio files x 173-element MPEG7 feature vector, i.e. 5 555 030 parameters number altogether. To assure shorter computation time, some known methods to reduce the dimensionality of data such as for example the Principal Component Analysis (PCA), Sammon Mapping, Multidimensional Scaling might be employed. The experiments described in this paper were carried out using the PCA method [24, 28]. The role of PCA is to determine new variables forming a smaller subset of descriptors representing a similarity pattern for the observations [1]. It is well known that employing PCA results in a new set of orthogonal variables in the space of features that contain the same information as the original set. The first variable has the largest possible variance. The second component is calculated under the constraint of being orthogonal to the first component and to have the largest possible inertia. Next variables are computed in the same way. The resulting set of vectors forms an orthogonal base in the feature space which better reflects the variability among observations. The values of these new variables are called factor scores [1].

In the first stage mean values were calculated from the parameterized data. These values were later used to calculate the covariance matrix array. In experiments Pearson correlation matrix was used to determine connection between each of 173 variables. In the next step eigenvalues were calculated from the covariance matrix of 173 parameters which were used in the final stage of the PCA reduction to create a new parameter vector, i.e. factors.

Further, descriptors were normalized to <-1;1> and they subsequently formed the input for the decision algorithms. In this paper, we describe tests performed with the following three algorithms: *k*NN (*k*-Nearest Neighbor), Bayesian Networks, SMO (Sequential minimal optimization algorithm).

Table 1. Number of excerpts in the Synat database

Genre:	**Synat**
Pop	5976
Rock	4957
Country	3007
R&B	2907
Rap & Hip-Hop	2810
Classical	2638
Jazz	2543
Dance & Dj	2289
NewAge	2122
Blues	1686
Hard Rock & Metal	1175
$\sum$	32110

The *k*-Nearest Neighbor algorithm is the simplest one, and as such is very commonly used for classification. We utilized the minimal distance that uses the Euclidean distance function. Its aim is to predict class membership of objects. The decision is based on the *k*-closest objects. An object is classified by the majority vote.

Sequential Minimal Optimization is a training algorithm for Support Vector Machine (SVM). It was proposed to solve the quadratic programming (QP) problem. This method is proposed for binary classification problems. The decision in a multi-class problem (like in music genre classification) can be done in parallel, each classifier is responsible for predicting the 0/1 association for each corresponding class. Usually the simple majority voting is used to get a final decision. Basically, SMO searches for a hyperplane through the feasible region that separates data in a high-dimensional feature space. The primary profits of using SMO for training the Support Vector Machine are much faster computation time and much easier implementation in various environments [21].

The Bayesian belief networks are based on the foundations of decision theory and probability. Bayes' theorem expresses the calculation of performance necessary to satisfy the veracity of the earlier indications of the use of new knowledge. The standard definition interprets a Bayesian network as a tool for modeling and reasoning with uncertain beliefs. Bayesian network is an acyclic directed graph in which there are nodes and arcs. The nodes represent random variables and arcs form the relationships between variables. Bayesian networks are mathematically defined strictly in terms of probabilities and conditional independence statements [22]:

$$P\left(a|C^{(j)}\right) = \frac{p\left(C^{(j)}|a\right) * p(a)}{p(C^{(j)})} \tag{1}$$

$p(a|C^{(j)})$ – a priori conditional probability distribution
$p(a)$ – the probability density function
$p(C^{(j)})$ – probability that object belongs to class $C^{(j)}$

This algorithm was chosen for its ability to minimize the probability of misclassification.

3 Experiments

This Section describes the results of the experiments in which the PCA method was used along with three decision algorithms. All classification tests were performed in the WEKA environment. Both the effectiveness of classification and the computation time of decision algorithms were examined. The main point of a test was to check, in which way a reduced number of parameters influenced the effectiveness of a decision. The number of the PCA factors was chosen by percentage of information variability of data of the 173-parameter set. The selected PCA factors are listed here: 6 PCA factors – 50 % of variability, 19 PCA factors – 70 % of variability, 33 PCA factors – 80 % of variability, 57 PCA factors – 90 % of variability, 82 PCA factors – 95 % of variability. It should be noted that even 82 factors described 95% of the information from 173-parameter MPEG 7 descriptors. The calculation time was measured as the time WEKA [27] needed to return a decision when cross-validation of 10 folds is used. Computation time did not include the PCA processing. Classification effectiveness was calculated using the cross-validation method. The data set was divided into 10 sets. Each set became the test set in each of round of classification. In total there were 10 rounds conducted for each test. Sum of all results in the test was named as classification effectiveness. The input of the PCA algorithm was a set of 173 previously mentioned audio parameters. The PCA analyses were always performed on the data set without the test data. In consequence the test set had cardinality of 3210 excerpts and the learning set contained 28890 files. The test set and the learning set were changed every round.

The selected decision algorithms belong to three different classes. The *k*NN is a minimal distance algorithm, the Bayesian Network is the probability parametric and SMO is a binary algorithm. The study performed answered the question which of the examined algorithms is the best and the fastest in the classification of a 30000 audio, as the whole setup should be evaluated from a practical perspective. files. As mentioned before, the *k*NN algorithm was used with Euclidean distance function, all of data were normalized to <-1;1> before classification. The kernel used for SMO algorithm was PolyKernel, and K2 for Bayesian Networks was employed.

It is worth noticing that the effectiveness of the classification even without the PCA method was above 80% for all of the chosen classifiers (Figure 2). After applying PCA, the results were even better. Even the use of 19 PCA factors ensured the effectiveness level above 90%, and as we can see in Figure 1 with a much shorter computation time. Computation time does not include the PCA algorithm operation time. The best result even for 6 factors was reached by the *k*NN algorithm. It classified

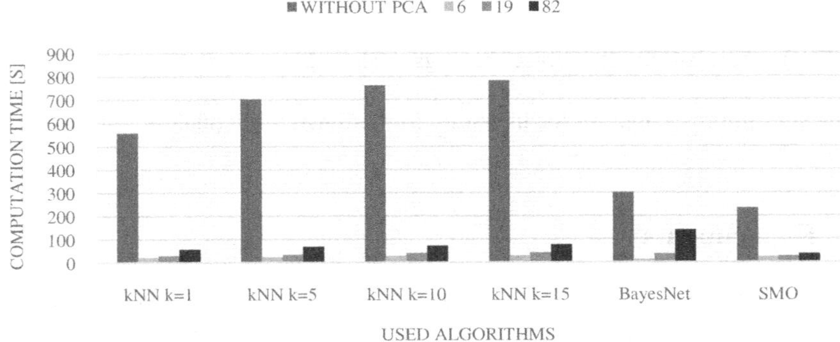

Fig. 1. Computation time in terms of algorithms employed

music genres with 90% efficiency. The very good classification accuracy was achieved as a result of complex parameterization of musical signal.

The computation time of the *k*NN algorithm without the PCA method applied was very long and took more than 9 minutes. The reduction of the PCA factors decreased the computation time to below one minute. As may be expected, a high number of parameters resulted in a long computation time. As commonly known, SMO is the optimized algorithm for training Support Vector Machine. This was also proved in experiments. SMO algorithm was the faster one in comparison.

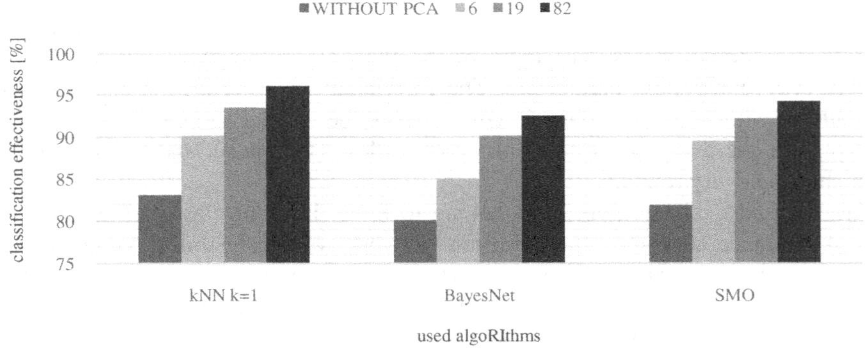

Fig. 2. Classification effectiveness of the algorithms employed

The worst results were achieved for the Bayesian Network classifier, although achieving the classification effectiveness above 80 % is a good result. The *k*NN and SMO classifiers recognized music genres faster and with better effectiveness. In Figure 3, the classification effectiveness for all tested PCA factors is shown.

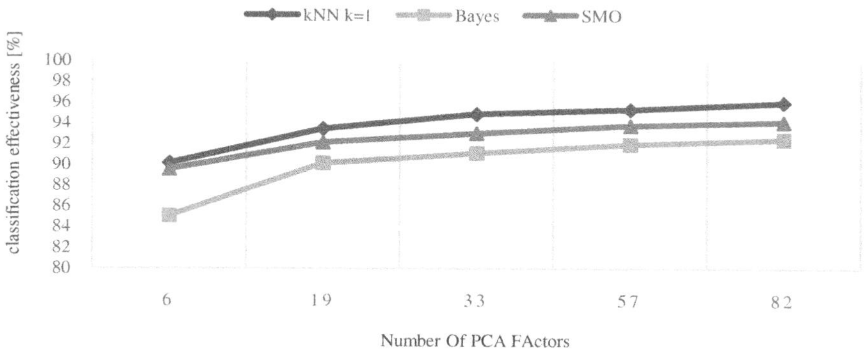

Fig. 3. Classification effectiveness for the *k*NN, Bayesian Network and SMO algorithms in a function of the PCA factor number

Figure 4 shows the computation time of the algorithms as the number of the PCA factor function. The computation time grows with the number of factors for all algorithms. For the Bayesian Network algorithm the increase of computation time was the largest.

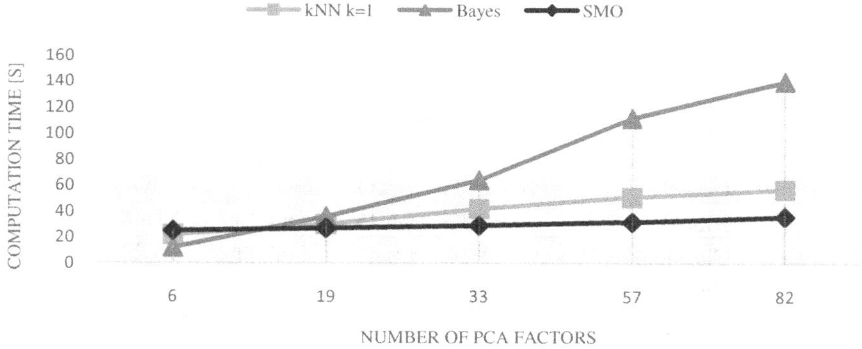

Fig. 4. Computation time shown in terms of the number of the PCA factors

For the most efficient classifier – *k*NN – an additional test was performed for a variable number of *k* parameters. The detailed results are shown in Figure 5. Four *k* numbers were tested: 1, 5, 10, 15. The best result was achieved for the *k*NN algorithm (*k*=10). The change of the *k* parameter did not raise significantly the accuracy of classification. But it is worth saying that for all of the tested PCA factor numbers, the *k*NN algorithm achieved very good efficiency. Even using *k*=1 parameter in the *k*NN algorithm resulted in better efficiency then SMO or Bayesian Networks algorithms but with comparable time or shorter.

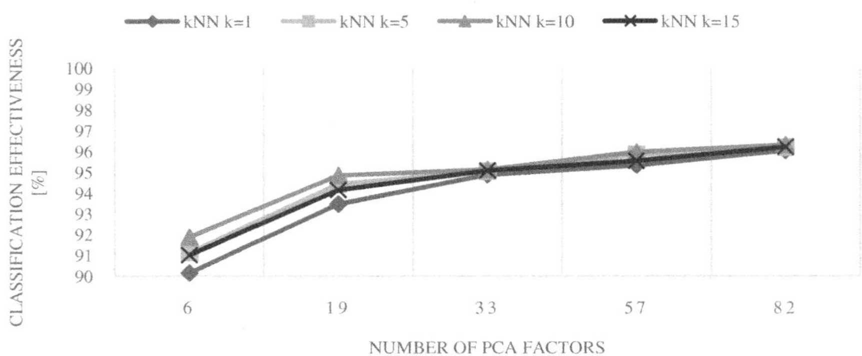

Fig. 5 Classification effectiveness of the *k*NN algorithm (variable number of *k* parameter)

In Table 2, the recognition of each of the eleven genres with and without PCA is compared. Most of music genres were recognized with 10% higher efficiency after using the PCA algorithm. Two music genres which were recognized with the highest efficiency were Classical and Hard Rock. These results are consistent with previous experiments [10].

Table 2. Classification effectiveness for 11 tested genres before and after the PCA-based data dimensionality reduction

[%]	19 PCA FACTORS		WITHOUT PCA	
Genre	*k*NN *k*=1	SMO	*k*NN *k*=1	SMO
Blues	92.32	91.45	83.23	81.48
Classical	97.14	98.43	86.44	88.23
Country	90.47	89.40	81.54	79.44
DanceDj	91.58	90.14	80.25	80.18
HardRock	98.75	98.10	86.45	88.05
Jazz	94.32	90.17	83.13	80.14
NewAge	92.14	88.14	82.18	78.13
Pop	94.80	93.10	84.60	83.54
RB	90.12	90.40	80.26	79.54
Rap	90.87	90.79	80.32	78.10
Rock	95.40	93.44	85.41	83.47

To summarize, it may be concluded that reducing the redundancy of information can result in higher efficiency and lower computation time. The PCA method is only a linear combination of basic parameters and eigenvalues, so it is not as resource-consuming as decision algorithms working on large arrays. The most efficient was

PCA with 19 factors used along with the *k*NN or SMO algorithms. To remind, the size of the original vectors resulted in 5 555 030 parameter numbers. Applying PCA reduced this number to 610 090, which not only did not lower the efficiency, but resulted in a much faster computation time.

4 Conclusions

A large database collected randomly requires a very good parametrization module. The MPEG 7-based standard describes audio files very utterly, but it also generates a very long feature vector which increases computation. In this paper, the PCA method was used in order to reduce the dimensionality of data, and it was tested in the context of the music genre recognition efficiency and time of computation.

The classification effectiveness achieved in the experiments is very good (above 90%) for the whole data set. In other publications related to music genre classification [15, 16, 19, 31], tests that used *k*NN and SMO algorithms performed worse. The main reason for such good results may be a unique parameterization module which was applied along with the PCA dimensionality reduction. Very good results have been achieved through effective parameterization, accurately describing the test set. Data after analyzing the PCA were very well separable and were described by a much smaller number of data without losing much information.

Applying PCA for music genre recognition assured better recognition within a shorter computation time. Reducing 173 parameters to 19 factors raised the effectiveness of recognition by about 10% and decreased the computation time from 550 to 60 seconds (about 9 times shorter).

The future work should focus on testing other decision algorithms common for ISMIR [12] or MIREX Music Information Retrieval communities [20] with regard to recognition effectiveness [15, 16, 19, 31]. Also, the same routine may be applied to an expanding live music database, checking whether this type of feature vector, as well the whole approach is universal or it may require active rule-based system that would enable to choose different feature vectors or rename music genre.

Acknowledgements. This work was partially supported by the grant no. PBS1/B3/16/2012 entitled „Multimodal system supporting acoustic communication with computers" financed by the Polish National Centre for Research and Development and the company Intel Technology Poland.

References

1. Abdi, H., Williams, L.J.: Principal Component Analysis. Wiley Interdisciplinary Reviews: Computational Statistics 2 (2010)
2. Chang-Hsing, L., Hwai-San, L., Jau-Ling, S., Kun-Ming, Y.: Automatic Music Genre Classification Based on Modulation Spectral Analysis of Spectral and Cepstral Features. IEEE Transaction on Multimedia 11(4) (2009)

3. Cichowski, J., Czyżewski, A., Kostek, B.: Analysis of impact of audio modifications on the robustness of watermark for non-blind architecture. Multimedia Tools and Applications (2012), doi:10.1007/s11042-013-1636-0
4. Haque, M.A., Kim, J.-M.: An analysis of content-based classification of audio signals using a fuzzy c-means algorithm. Multimedia Tools and Applications 63(1), 77–92 (2013)
5. Kiktova-Vozarikova, E., Juhar, J., Cizmar, A.: Feature selection for acoustic events detection. Multimedia Tools and Applications (2013)
6. Kostek, B.: Music Information Retrieval in Music Repositories. In: Skowron, A., Suraj, Z. (eds.) Rough Sets and Intelligent Systems. ISRL, vol. 42, pp. 463–489. Springer, Heidelberg (2013)
7. Kostek, B.: Perception-Based Data Processing in Acoustics, Applications to Music Information Retrieval and Psychophysiology of Hearing. SCI, vol. 3. Springer, Heidelberg (2005)
8. Kostek, B.: Soft Computing in Acoustics, Applications of Neural Networks, Fuzzy Logic and Rough Sets to Musical Acoustics. STUDFUZZ. Physica Verlag, Heidelberg (1999)
9. Kostek, B., Czyzewski, A.: Representing Musical Instrument Sounds for their Automatic Classification. J. Audio Eng. Soc. 49, 768–785 (2001)
10. Kostek, B., Hofmann, P., Kaczmarek, A., Spaleniak, P.: Creating a Reliable Music Discovery and Recommendation System. In: Intelligent Tools for Building a Scientific Information Platform: From Research to Implementation. Springer Verlag (2013)
11. Kostek, B., Kupryjanow, A., Zwan, P., Jiang, W., Raś, Z.W., Wojnarski, M., Swietlicka, J.: Report of the ISMIS 2011 Contest: Music Information Retrieval. In: Kryszkiewicz, M., Rybinski, H., Skowron, A., Raś, Z.W. (eds.) ISMIS 2011. LNCS, vol. 6804, pp. 715–724. Springer, Heidelberg (2011)
12. Kotropoulos, C., Benetos, E., Panagakis, E.: Music genre classification: A multilinear approach. In: ISMIR (2008)
13. Kupryjanow, A., Czyzewski, A.: Real-Time Speech Signal Segmentation Methods. J. Audio Eng. Soc. 61(7/8), 521–534 (2013)
14. Łopatka, K., Czyżewski, A.: Automatic Regular Voice, Raised Voice, and Scream Recognition Employing Fuzzy Logic. In: 132nd Audio Eng. Soc. Convention, preprint 8636, Budapest, Hungary, April 26-29 (2012)
15. Mei-Lan, C.: Automatic Music Genre Classification, Graduate Institute of Biomedical Electronics and Bioinformatics, National Taiwan University, Taipei, Taiwan
16. Mlynek, D., Zoia, G., Scaringella, N.: Automatic Genre Classification of Music Content. IEEE Signal Processing Magazine (2006)
17. MPEG 7 standard, `http://mpeg.chiariglione.org/standards/mpeg-7`
18. Ntalampiras, S., Potamitis, I., Fakotakis, N.: Acoustic Detection of Human Activities in Natural Environments. J. Audio Eng. Soc. 60(9), 686–695 (2012)
19. Qi, L., Tao, L., Mitsunori, O.: A Comparative Study on Content-Based Music Genre Classification. In: Proc. SIGIR (2003)
20. Pertusa, A., Rauber, A., Inesta, J., Lidy, T.: Combining audio and symbolic descriptors for music classification from audio. Music Information Retrieval Information Exchange, MIREX (2007)
21. Ragni, A., Gautier, C., Gales, M.J.F.: Support Vector Machines for Noise Robust ASR, Cambridge University Engineering Department Trumpington St., Cambridge, U.K IEEE ASRU (2009)
22. Remco, R.: BayesianNet algorithm description, University of Waikato (2007), `http://www.cs.waikato.ac.nz/~remco/weka_bn/`

23. Schnitzer, D., Flexer, A., Widmer, G.: A fast audio similarity retrieval method for millions of music tracks. Multimedia Tools and Applications (2012)
24. Shlens, J.: A Tutorial on Principal Component Analysis, Version 2 (December 10, 2005)
25. Tzanetakis, G., Cook, P.: Musical Genre Classification of Audio Signals. IEEE Transactions on Speech and Audio Processing 10(5) (July 2002)
26. Tzacheva, A.A., Bell, K.J.: Music Information Retrieval with Temporal Features and Timbre. In: An, A., Lingras, P., Petty, S., Huang, R. (eds.) AMT 2010. LNCS, vol. 6335, pp. 212–219. Springer, Heidelberg (2010)
27. Weka Mining Software, `http://www.cs.waikato.ac.nz/ml/weka/`
28. Wold, S., Esbensen, K., Geladi, P.: Principal component analysis. Chemometrics and Intelligent Laboratory Systems
29. Xu, B., Liang, J., Wang, L., Huang, S., Wang, S.: Music Genre Classification Based on Multiple Classifier Fusion, Chinese Academy of Sciences, Beijing, China (2008)
30. Yao-Chang, H., Shyh-Kang, J.: An Audio Recommendation System Based on Audio Signature Description Scheme in MPEG-7 Audio. In: 2004 IEEE International Conference on Multimedia and Expo, ICME (2004)
31. Zheng, J., Oussalah, M.: Automatic System for Music Genre Classification University of Birmingham, Electronics, Electrical and Computer Engineering (2006)

Mining False Information on Twitter for a Major Disaster Situation*

Keita Nabeshima[1], Junta Mizuno[2], Naoaki Okazaki[1,3], and Kentaro Inui[1]

[1] Graduate School of Information Sciences, Tohoku University / Miyagi, Japan
[2] Resilient ICT Research Center, NICT / Miyagi, Japan
[3] Japan Science and Technology Agency (JST) / Tokyo, Japan
nabeshima05@gmail.com, junta-m@nict.go.jp,
{okazaki,inui}@ecei.tohoku.ac.jp

Abstract. Social networking services (SNS), such as Twitter, disseminate not only useful information, but also false information. Identifying this false information is crucial in order to keep the information on a SNS reliable. The aim of this paper is to develop a method of extracting false information from among a large collection of tweets. We do so by using a set of linguistic patterns formulated to correct false information. More specifically, the proposed method extracts text passages that match specified correction patterns, clusters the passages into topics of false information, and selects a passage that represents each topic of false information. In the experiment we conduct, we build an evaluation set manually, and demonstrate the effectiveness of the proposed method.

1 Introduction

Social networking services (SNS), such as Twitter and Facebook, are useful not only for communicating with friends but also for disseminating information. Disseminating information is enhanced by functions such as *retweet* (reposting messages from other users on Twitter) and *share* or *like* (on Facebook). Generally speaking, these mechanisms are useful for spreading worthwhile information rapidly. However, spreading false information becomes a serious problem in emergency situations. For instance, during Hurricane Sandy, a tweet "*BREAKING: Con Edison has begun shutting down ALL power in Manhattan*" caused panic[1].

During the 2011 East Japan Earthquake and Tsunami Disaster, more than 80 pieces of false information were disseminated. Typical examples included:

- *The Cosmo Oil explosion is causing toxic rain.*

* This paper was partly supported by the Japan Society for the Promotion of Science (JSPS) KAKENHI Grants No. 23240018 and 23700159 and by the Precursory Research for Embryonic Science and Technology (PRESTO), Japan Science and Technology Agency (JST). We are grateful to Twitter Japan for its provision of invaluable data.

[1] http://techcrunch.com/2012/10/31/panic-inducing-rumors-over-twitter-during-a-hurricane-should-be-illegal/

D. Ślęzak et al. (Eds.): AMT 2014, LNCS 8610, pp. 96–109, 2014.

– *Drinking Isodine*[2] *protects one against radiation.*

Seeing this kind of information in an emergency situation, people are apt to share it without verifying its credibility in the hope of saving human lives. One tweet consolidation site dedicated to collecting/correcting information on the Tohoku Earthquake[3] found that during January 2012, 10 months after the event, more than 10 misinforming posts related to the earthquake. This indicates a strong need for misinformation alerts in normal times as well as in times of disaster.

In this study, we propose an approach for generating such alerts by extracting and aggregating false information on Twitter. In order to be able to recognize false information rapidly, we focus on a tweet that corrects a piece of information. For example, the following tweet corrects the information on Isodine.

– *It is counterfactual that drinking Isodine protects one against radiation.*

This tweet can be split into two parts: false information and the phrase rebutting the information (the underlined part).

Examining correction patterns in a corpus, we develop rules by which pieces of information are corrected or refuted. After extracting false-information candidates, we score keywords in the candidates in terms of the strengths of their connections with correction patterns. In order to avoid redundant false information, the keywords are clustered based on semantic similarity. Finally, we select a small set of sentences that explain the false information. In our experiments, we evaluate the coverage and precision when the proposed method was used to extract false information from roughly 180 million tweets posted during the 2011 East Japan Earthquake and Tsunami Disaster.

2 Related Work

Information Credibility Analysis. There have been several studies conducted on analyzing information credibility. Fact-Finders is the most popular algorithm by which information credibility is evaluated by considering information content and sources based on link analysis [6]. In [11] Fact-Finders is extended to consider related knowledge and contextual details. In [5] comments are ranked on a SNS automatically based on the community's expressed preferences. Then the quality of the page is evaluated by high-ranked comments. To capture the aspects of credibility and quality, Open Information Extraction is used to count a number of facts in a sentence [7].

Dispute Finder [2] provides users browsing the Web with a list of known conflicts whenever a disputed statement is encountered. Dispute Finder builds a database of disputed claims by enabling its users to identify disputed claims on the Web and links them to a trusted source of rebuttal. The goals of our system and Dispute Finder are similar but the latter relies on crowdsourcing to build its dispute database, which limits its automation potential.

[2] In Japan, 'Isodine' commonly refers to antiseptics containing povidone iodine.

[3] `https://twitter.com/#!/jishin_dema`

False Information Extraction on Twitter. In this subsection, we review previous research on detecting false information on Twitter. A method for classifying a group of tweets related to misinformation (e.g, tweets containing the terms "Barack Obama" and "Muslim") into those including explicit expressions of misinformation (e.g., "Barack Obama is a Muslim"), and those that do not (e.g., "Barack Obama met with Muslim leaders"), is proposed in [12]. This method assumed that mining misinformation from a large volume of tweets and obtaining a group of misinformation-related tweets was outside its scope.

In [1] the credibility of news propagated on Twitter is analyzed. Tweets related to news are classified as credible or not by using machine learning. The paper assumes that there are several factors in social media by which credibility can be evaluated. Based on this hypothesis, they designed four groups of features: tweet messages, tweet users, topic of the messages, and propagation tendency. They found that credible news include URLs and have a deep propagation tree.

In [13] a method for detecting astroturfing, smear campaigns and misinformation in tweets related to U.S. political elections is proposed. Information diffusion networks are generated based on memes such as hashtags, mentions and URLs. It argues several types of diffusing true information.

In all of these studies, each tweet text is taken as a unit with the focus on determining whether it contains misinformation or correct information by providing specific information[4] without identifying the misinforming phrase in the text. To the best of our knowledge, our study is the first investigation of a comprehensive collection of misinformation from a large volume of tweet data.

Contradiction Recognition. When a tweet contradicts another tweet, it is possible that either one of them is false information. Therefore, we could formalize false information extraction as the task of recognizing contradictions. This task is known to be difficult [4]. In [9] authors reported rates of precision and recall of 23% and 19%, respectively, when detecting contradictions in the RTE-3 dataset [3]. However, as it was constructed artificially, the RTE-3 dataset does not reflect real contradictions and the authors collected contradictions in a real-world setting. Their system was evaluated using this new dataset, and they reported that the performance was limited.

In [10] there is also focus on recognizing contradictions. The authors classified contradiction types into two groups, with seven categories: Antonym, Negation, Numeric, Factive, Structure, Lexical, and World Knowledge. They defined feature sets based on this classification. Although their results were promising for the RTE-3 dataset, the performance dropped when their system was applied to other datasets. They concluded that adapting to other datasets is a difficult challenge for contradiction recognition.

The research described in [14] addresses contradiction recognition between tuples that represent entities in sentences and the relation between them (e.g., *was_born_in*(*Mozart*, *Salzburg*)). For example, *was_born_in* (*Mozart*,

[4] For example, "Be careful! All kinds of misinformation are circulating on Twitter" contains "misinformation" but does not correct any specific information.

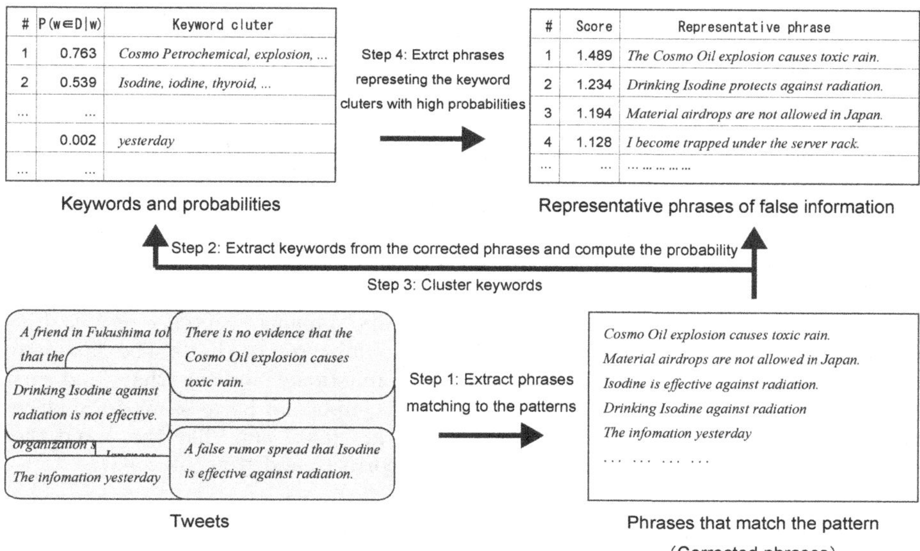

Fig. 1. An overview of the proposed method. The upper left table shows a ranking of topic keywords. The upper right table is a list of the extracted false information.

Salzburg) and *was_born_in* (*Mozart*, *Vienna*) are contradictory, and *visited* (*Mozart*, *Salzburg*) and *visited* (*Mozart*, *Vienna*) are not. They detected the contradiction by using the functionality of $was\_born\_in(x, y)$.

In [16] the authors propose to align units in two sentences and to label local semantic relations defined in Natural Logic [8]. This allows to infer a relation between sentences. The best performance of contradiction recognition was achieved in NTCIR-10 RITE-2 [15], but it (28.57 of F1) is still insufficient for our purpose.

In addition, it is not feasible to recognize contradictions between tweets on Twitter. When N tweets are given, we need to recognize contradictions N^2 times. Moreover, when a new tweet is posted, it is necessary to recognize contradictions between that tweet and existing tweets N times. In contrast, our approach is able to recognize contradiction by using one tweet.

3 Extracting False Information

In order to extract false information, we first assume that some users on Twitter provide the correct information for a false tweet. For example, with regards to the Cosmo Oil fire, many tweets rumored that harmful substances would be coupled with any rainfall in the area in question. In response to this information, the following tweets provided some information about this scenario being false.

- *It is untrue that toxic rain will fall due to the Cosmo Oil explosion.*
- *Be wary of the rumor that rain containing chemicals will fall.*

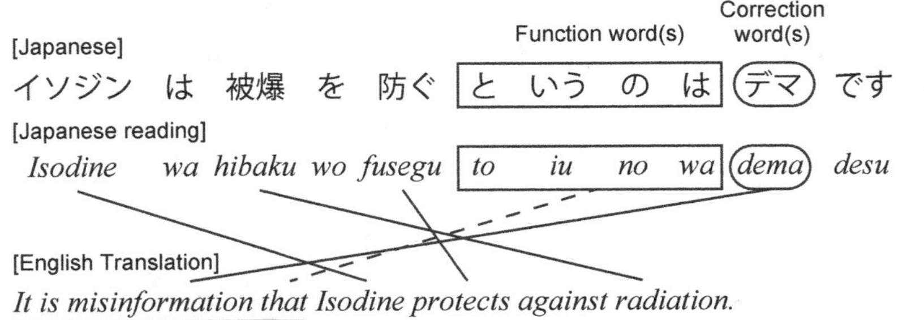

Fig. 2. Example of a correction pattern. The original Japanese sentence and its English translation; the correction pattern is "(it is) a false rumor that". The Japanese correction pattern is decomposable into two parts: functional word(s) "that" and correction word(s) "false rumor". The decomposition is important because we can replace the functional words with other expressions such as "that" and "like that" and the correction words with other synonyms such as "incorrect" and "unconfirmed".

The correcting tweet consists of both a correction pattern (the underlined part) and the false information. To extract false information, it is a reasonable approach to discover a correction pattern in the tweets. The goal of the proposed method in this section is to construct an extractor for estimating the location where false information is written by using the correction patterns. Additionally, we comprehensively locate correction patterns in order to gather false information via a collection of tweets. Figure 1 shows the flow of the proposed method, which essentially comprises of four steps which are described below.

Step 1: Extraction of Corrected Phrases. In this step, we search for tweets containing corrected phrases. Co-occurring with a correction pattern, they result in a *misinformation* or *false information* candidate, as in the sentence:

– *It is misinformation that Isodine protects against radiation.*

The underlined portion is the correction pattern and the non-underlined is the corrected phrase. Corrected phrases are noun phrases (e.g., "That Cosmo Oil explosion") and simple or complex sentences (e.g., "Isodine protects against radiation" and "Toxic rain will fall due to the Cosmo Oil explosion", respectively).

We represent correction patterns as concatenations of functional words and correction words. To gather correction patterns of excellent quality, we investigated tweets related to 15 kinds of well-known false information, and formulated sets of functional words and correction words appearing in the tweets. We selected correction words such as "rumor" and "false information", which clearly correct information. We designed 550 correction patterns as combinations of 11 functional words and 55 correction words. Table 1 presents their subset.

When using the correction patterns for extracting corrected phrases, we apply Japanese morphological analysis to split a sentence into words. If a tweet contains a correction pattern, we extract the sequence of words appearing prior to the correction as a corrected phrase. We remove conjunctions and adverbs at the

Table 1. Correction patterns as concatenations of functional and correction words

Functional words	などの (etc.), なんて (such as), とか (such as), のような (like that), というのは (that)
Correction words	デマ (false rumor; misinformation), ガセ (bogus information), ソースがない (unknown source), 嘘 (lie), 未確認 (unconfirmed), 誤情報 (false information), 事実はありません (counterfactual), 誤報 (false report), 根拠のない (unfounded; groundless),

beginning of a sentence because they are unnecessary to an explanation of the false information. By applying this process to all tweets in a corpus, we extract a set of corrected phrases (denoted D hereafter).

Step 2: Keyword Extraction. Some of the corrected phrases extracted in Step 1 simply refer to the misinformation rather than stating it, as in "The information yesterday" in a sentence, "The information yesterday was incorrect." Such phrases are not informative to users. To avoid them, the conditional probability that a noun phrase w appears in the set of corrected phrases D is

$$P(w \in D|w) = \frac{\#\text{ tweets where } w \text{ co-occurs with CPs}}{\#\text{ tweets containing } w} \tag{1}$$

A noun phrase occurring frequently with correction patterns and infrequently in other tweets gains a high conditional probability, and has the potential of being a misinformation keyword. On the contrary, a noun phrase occurring frequently not only with correction patterns but also in other tweets, has a low conditional probability. Therefore, Equation 1 yields a low probability for the phrase "the information yesterday" because the words "yesterday" and "information" are common words. Due to this fact, we extract the top 500 words yielding the highest probability as misinformation keywords. In addition, we remove noun phrases occurring less than 10 times in the corpus.

Step 3: Keyword Clustering. Misinformative phrases pertaining to the same information may differ considerably in wording and information quantity, as in "Rain containing hazardous substances from the Cosmo Oil fire will occur" and "The Cosmo Oil explosion is toxic." These phrases need to be consolidated to avoid redundancy when extracting misinformation. For this reason, we cluster the keywords extracted in Step 2.

We use the cosine between context vectors (i.e., the pointwise mutual information (PMI) of co-occurrence in sentences between the keywords and the content words – nouns, verbs, and adjectives) to measure the inter-keyword similarity. This is based on the distributional hypothesis that words surrounded by similar words are also similar. In our experiment, we obtained 189 clusters from 500 keywords by performing complete-link (furthest-neighbor) clustering. We choose a representative keyword of a cluster as that yielding the highest conditional probability among the cluster members.

Table 2. A subset of the 60 gold instances

There is need for a major brownout in the Kansai region.
The author of the manga series "One Piece" donated 1.5 billion yen.
The emperor fled to Kyoto.
Boric acid protects one against radiation.
Staff at Futaba Hospital fled, leaving patients unattended.
The village of Hanayama in Miyagi is marooned.
A commemorative T-shirt has been made in Korea.
The Democratic Party of Japan bought up all supplies of pot noodles.

Step 4: Representative Phrase Selection. For each cluster and its representative keyword obtained in Step 3, we produce the best explanation of the false information corresponding to the cluster. Here, the 'best' explanation is the phrase that describes the false information precisely and concisely. For example, the second of the following phrases is too short to explain the necessary information, and the third phrase has a portion that is not important ("raincoat and umbrella"). The goal of Step 4 is to select representative phrases:

– *After the Cosmo Oil fire, harmful substances will fall with the rain.*
– *Matters related to Cosmo Oil.*
– *It is true that there has been an explosion at Cosmo Oil; raincoats and umbrellas seem to be required to protect one from the harmful rain that is falling.*

To select corrected phrases of suitable length that can provide a sufficient description of the false information, we compute the score:

$$\text{Score}_p(s,t) = \text{hist}_t(\text{len}_s) \textstyle\sum_{w \in C_s} \text{PMI}(t,w), \tag{2}$$

where s denotes the corrected phrase, t is a representative keyword of the misinformation cluster, C_s indicates the set of content words in s and len_s is the number of words in s. The term $\text{hist}_t(\text{len}_\text{s})$ represents the relative frequency of occurrences of sentences consisting of len_s words with the keyword t. $\text{PMI}(t,w)$ is the PMI of t and w co-occurrence.

Equation 2 is a product of $\text{hist}_t(\text{len}_\text{s})$ and $\sum_{w \in C_s} \text{PMI}\ (t,w)$. The first component yields a higher value if the corrected phrase s has the standard number of words in the collection of corrected phrases containing the keyword t. The assumption here is that each topic of false information has an adequate length to explain the information. Equation 2 does not favor phrases that are too short or too long because in such cases the second component $\text{hist}_t(\text{len}_\text{s})$ is small.

The more relevant words appear in the phase s, the higher value the latter component gives. The relevance of a word w to the topic t is measured by $\text{PMI}(t,w)$. To sum up, Equation 2 yields a high score for phrases containing an adequate number of relevant words. For each cluster obtained in Step 3, we choose the phrase with the highest score as the description for the keyword t.

4 Evaluation of Correction Patterns

In this section, we evaluate the coverage and accuracy of the correction patterns (CPs). If the false information is not extracted in the first step, there is no way to extract it later, so this evaluation shows the upper-bound of our method.

Table 3. Types of corrected phrases and causes of failure to extract misinformation

Phrase type:	#
(a) Sufficient content for recognition as phrases with corrected information	76
(b) Insufficient content for recognition as phrases with corrected information	42
(c) Phrases erroneously extracted that are instances of ambiguous patterns	24
(d) Phrases erroneously extracted that are instances of unclear author intent	8
Total:	150
Cause of failure to extract:	**#**
(e) Unseen correction pattern	7
(f) No corrective tweet	3
Total:	10

Experimental Setting. The corpus used for the false information extraction evaluation consisted of 179,286,297 tweets posted between March 11 and March 18, 2011, which were provided by Twitter Japan at the Great East Japan Earthquake Big Data Workshop[5]. To the date, no corpus composed of misinformation tweets circulated during the Great East Japan earthquake has been compiled. To create a reference dataset, we collected all the instances of misinformation from four misinformation consolidation websites[6] and chose from them 60 instances of misinformation that were determined to have been posted during the week following the Tohoku Earthquake. During the CP performance evaluation, these 60 instances were compared with approximately 20,000 phrases that were extracted by our CPs. In this evaluation, these 60 instances were treated as the "ground truth". Table 2 shows a subset of the selected instances.

The CPs were evaluated for precision and recall. For an evaluation of precision, we took 150 samples randomly selected from approximately 20,000 corrected phrases. Precision was defined as a proportion of those samples that were recognized by the CPs as instances of information correction or refutation made by their posters. Recall was defined as a proportion of the 60 valid instances that were recognized from the set of approximately 20,000 corrected phrases.

Results and Analysis. The precision of the misinformation extraction was 79% (118/150) and the recall was 83% (50/60). The corrected phrases that were extracted consisted of four types, as shown in Table 3. Types (a) and (b) were identified as valid in the evaluation. Type (b) is of special interest because it comprises phrasing instances in which the misinformation is either not explicitly stated (e.g., "*The information yesterday*" in "*The information yesterday* <u>*was not true*</u>", where the CP is underlined) or insufficiently expressed (e.g., "*that Isodine affair*" in "*I heard that Isodine affair* <u>*was a case of disinformation*</u>"). Presumably, the phrases of type (b) can be eliminated by the conditional probability ranking and representative phrase selection performed in Steps 2 and 4.

[5] `https://sites.google.com/site/prj311/`

[6] The following four websites: `http://www.kotono8.com/2011/04/08dema.html`, `http://d.hatena.ne.jp/seijotcp/20110312/p1`, `http://hara19.jp/archives/4905` and `http://matome.naver.jp/odai/2130024145949727601`

Those of types (c) and (d) were extracted incorrectly. The type (c) comprises instances in which the corrected phrase was extracted by erroneous CP application (e.g., "*In times of disaster such as this*" in "*In times of disaster such as this, disinformation flows freely*"). The type (d) comprises phrases in which the attitude of the writer with regard to the correction is ambiguous (e.g., "*I have a good idea to spread a rumor that fundraising makes you popular*").

We also examined the 10 instances of failure to extract misinformation and, as shown in Table 3. The type (e) involved corrective phrasing that was not covered by the existing CPs, such as the underlined portion of the statement, "*No information source is given to show that the Emperor actually prayed for 24 hours*". Extraction of this type will be possible with the addition of new CPs. The type (f) comprises several instances in which tweets purveying misinformation were present among the tweet collection used in this study, but related corrective tweets were not. Extraction of such misinformative tweets by the proposed method would be difficult, as our method assumes the occurrence of correction tweets, but such instances were small in number.

5 Evaluation of Misinformation Consolidation

Next, we evaluate Steps 2 to 4 from Section 3. This evaluation consists of determining whether these steps, when applied to the corrected phrases extracted in Section 4, effectively exclude the type (b) corrected phrases from the extracted phrase set. We also check whether the selected representative phrases contain appropriate descriptions of misinformation. For our error analysis, we classify instances which could not be extracted from tweets or were incorrectly extracted.

Experimental Settings. We assessed the misinformation extracted by the proposed method by manually examining each instance to determine whether it was equivalent in content to any of the reference instances. For some of the extracted misinformation, no similar instances were found. In those cases, we manually investigated the information using Web search engines to determine whether it actually was a case of misinformation. Additionally, in cases when two or more instances of the extracted information were essentially the same, we counted them as one correct instance of extraction.

The evaluation metrics are twofold: 1) we investigated whether the phrases are in the 60 gold instances (Acc); and, 2) in addition to Acc, false positives in Acc are manually checked to ascertain whether the phrases are false information (Acc(man)). In the evaluation metric Acc(man), there are some phrases which are not contained in the 60 gold instances, but are clearly false information. As there is currently no complete list of false information, we manually checked the evaluation metric Acc(man). Recall is calculated by the 60 gold instances. The above are calculated by the following equations.

$$\begin{aligned}
\text{Acc} &= \#\ \text{phrases in gold instances (w/o overlap)} / N \\
\text{Acc(man)} &= \#\ \text{manually determined phrases (w/o overlap)} / N \\
\text{Recall} &= \#\ \text{phrases in gold instances (w/o overlap)} / 60\ \text{gold instances}
\end{aligned}$$

Table 4. Evaluation results

N	Acc	Acc (man)	Recall
25	0.44 (11/25)	0.68 (17/25)	0.18 (11/60)
50	0.34 (17/50)	0.60 (30/50)	0.28 (17/60)
75	0.36 (27/75)	0.59 (44/75)	0.45 (27/60)
100	0.31 (31/100)	0.54 (54/10)	0.52 (31/60)
189 (Upper bound)	—	—	0.63 (38/60)

Experimental Results. Table 4 shows the results of the evaluation. With an N value of 100, approximately 30% of the information instances extracted by the proposed method were found to be present in the gold set. In addition, approximately 20% of the extracted instances were found to be actual instances of information, and thus correct, even though they were not present in the gold set. Therefore, it can be said that the proposed method extracted misinformation with a precision of approximately 50%. Among the incorrect answers, approximately half involved redundant expressions of essentially the same misinformation phrased differently. In summary, approximately 70% of the misinformation extracted by the proposed method represented a correct answer.

We next discuss recall that is not limited to the top N. "N=189 (Upper bound)" is the recall when a representative phrase is extracted from all 189 clusters generated in Step 3 obtained by clustering 500 keywords. Fifty instances covered by the set of corrected phrases are expected to be improved by the process in the latter part such as clustering or selecting keywords, but the other 10 instances are difficult cases needed for changing the extracting method based on CPs.

Error Analysis. In this subsection, we analyze the errors in the experimental result. Investigation of the causes of the inaccuracy in output represented by the 69 incorrect answers present among the top 100 extracted misinformation instances showed that they could be classified into six types. These are listed in Table 5, together with the number of incorrect answers attributable to each type. Types (a) to (d) involve instances that were easily recognized as errors, but types (e) and (f) involve instances that would be difficult for humans to characterize as either true information or misinformation.

Types (g) and (h) involve instances that are not listed in the instances of misinformation from four misinformation consolidation websites, but were determined as misinformation by manually investigating them on the Web. The eight types of cause, and potential means of avoiding them, are as follows:

(a) Some errors are caused by keyword extraction. In the following example, "malicious actions" is a keyword because it often occurs with a correcting word "rumor"; therefore, the conditional probability of Eq. 1 is increased. However, some errors are caused by such keywords without details of the false information.

– *I cannot believe that some people believe the false rumor "there were some malicious actions".*

(b) Among the top 100 instances of extracted phrases, there are some redundancies in the form of different phrases that have essentially the same content. In the following examples, keywords in parentheses are used in the selection process.

– *Due to the explosion of the (Cosmo Oil Chiba Refinery LPG tank in Ichihara City), residents of Chiba Prefecture and its neighboring regions will be subjected to toxic rain.*
– *Due to (the Chiba Prefecture petrochemical complex explosion), substances that have an adverse affect on human health will be released into the air and fall as acid rain.*

These instances were not assigned to the same cluster from Step 3. While the current method takes words that co-occur in corrected phrases as their features, this redundancy may be eliminated by adding basic information of the keywords.

(c) Information of uncertain content involves instances in which the selected phrase states the misinformation inadequately, as in the following example:

– *Deaths due to starvation and hypothermia have occurred.*

The gold set included the sentence "In Iwaki City, death by starvation and hypothermia have occurred," but the above representative statement is less specific and was therefore considered to be uncertain in content. Tweets containing such phrases were small in number, and may therefore be excluded by setting a threshold number for this purpose.

(d) The following was falsely extracted as a misinformation:

– *The top of Tokyo Tower has bent.*

Through investigation of tweets relevant to this example, many people considered this to be misinformation as its content seems widely implausible even if a photograph were provided as evidence. Therefore, many correcting tweets had been posted. Because the phrase has a high frequency of being ranked in our approach if many tweets correct its information, this error has been extracted. Considering our purpose is extracting false information, it is a more serious error than types (a) – (c). However, some people who were originally suspicious of the information after first posting it ended up posting, as a follow-up, that the information was indeed not false.

– *I found it is not rumor that Tokyo Tower has bent.*
– *I thought it was a rumor that Tokyo Tower had bent, but it is true.*

In this way, there are some tweets correcting the correction. It is considered that determining double negation will improve the situation.

(e) We found some instances in which a search of several websites yielded no indication of whether they involved misinformation, as in the following example:

– *Suntory unlocks vending machines to dispense products free of charge.*
– *Fish is left over at Tsukiji market.*

(f) In some instances of (e), expressions comprising the prediction of a future event were extracted, such as the following:

Table 5. Types of errors that lowered the accuracy and recall

Error type (accuracy):	#	%
(a) Topic extraction error	6	8.70
(b) Clustering error I	22	31.9
(c) Information of uncertain content	5	7.25
(d) Extraction of correct information	1	1.45
(e) Validity unclear	6	8.70
(f) Prediction of future events	6	8.70
(g) False information not listed in 4 sites (past)	9	13.0
(h) False information not listed in 4 sites (now)	14	20.3
Total:	69	100.0
Error type (recall):	#	%
(i) Clustering error II	2	10.5
(j) Low ranking	17	89.5
Total:	19	100.0

– *A nuclear explosion will occur in Fukushima.* – *Mt. Fuji will erupt.*

The truth is not clear at this time. Many extracted phrases consist of information which causes anxiety among people as in the example just described. We considered that this example is extracted because there were many people seeking to prevent panic, who posted correcting tweets.

(g) This involves instances that are not listed on the four misinformation consolidation websites, but which are determined as false information through manual investigation. Furthermore, this information arose before the span of the tweet corpus we used in this paper, as in the following examples:

- *When the Great Kanto earthquake occurred, Koreans poisoned the well water.*
- *The most destructive earthquake tremor occurred during the Great Hanshin-Awaji Earthquake several hours after the first tremor.*

These examples of past disasters have been regarded as false information.

(h) This involves instances that are not listed on the four misinformation consolidation websites, but determined as false information through manual investigation. This false information arose during the span of the tweet corpus we used in this paper, as in the following examples:

– *Rescue dog from Korea ran away.* – *There is some risk of infection.*

Among the 60 gold instances, 19 cannot be extracted by our method. Our investigation into the causes showed that they were of the following two types, which are listed in Table 5, together with the number that occurred in each type.

(i) In some instances, the candidates were extracted by the CPs, but were mistakenly merged with other misinformation instances during the clustering process. However, they do not pose a substantial problem since their number was small in comparison with the total quantity of extracted misinformation.

(j) In some instances, candidates were extracted by the CPs but were not extracted as keywords because of their low conditional probability. One example of this is in the misinformation, " A man pretending to be from Tokyo Electric

Power appeared on the scene". The keyword "Tokyo Electric Power" frequently occurs in statements that do not involve misinformation, and its conditional probability of exhibiting misinformation was therefore estimated to be low. Accordingly, a means of scoring for corrected phrases themselves, rather than for independent keywords, is necessary to eliminate this problem.

We refer to the method to improve performance. (a) to (d), (i), and (j) are extraction errors as mentioned above. (b) and (i) are caused by clustering error. These errors can be reduced by adding more features, and introducing other methods to calculate contextual similarity such as recognizing textual entailment. (a), (c), and (j) are caused by scoring error. There are two ways to reduce these errors. One way is to expand the correction patterns. When there is more variation of the patterns, the conditional probability by Equation 1 can be correctly calculated. The second way is to introduce other measures into Equation 2 such as tf-idf and term frequency in a short period. (d) is a very difficult problem for our method when using correcting information from users.

We refer to the method to improve performance. (a) to (d), (i), and (j) are extraction errors as mentioned above. (b) and (i) are caused by clustering error. These errors can be reduced by adding more features, and introducing other methods to calculate contextual similarity such as recognizing textual entailment. (a), (c), and (j) are caused by scoring error. There are two ways to reduce these errors. One way is to expand the correction patterns. When there is more variation of the patterns, the conditional probability by Equation 1 can be correctly calculated. The second way is to introduce other measures into Equation 2 such as tf-idf and term frequency in a short period. (d) is a very difficult problem for our method when using correcting information from users.

6 Conclusion

In this paper, we proposed a method to extract and aggregate false information on Twitter during the 2011 East Japan Earthquake and Tsunami Disaster. We focused on expressions that correct or refute false information. In our experiments, we used roughly 180 million tweets for our evaluation data set and 60 instances from four misinformation consolidation websites as gold data. In our first experiment, the precision and recall of the manually generated correction patterns were both approximately 80%. If the false information is not extracted in the first step, then there is no way to extract this false information in the later steps. This encourages us to acquire correction patterns automatically from larger corpora in our future work. In our second experiment, more than half of the gold instances were successfully extracted and 23 phrases that were not contained in the gold data were found.

In future work, we plan to use authoritative information on Twitter. In this paper, we focused on correction patterns, so we did not use any authoritative information such as the number of follows, followers, and number of posts. We also plan to apply our method to other tweets under general situations in order to evaluate the coverage of correction patterns and phrase extraction strategy.

References

1. Castillo, C., Mendoza, M., Poblete, B.: Information credibility on Twitter. In: Proceedings of the 20th International Conference on World Wide Web, pp. 675–684 (2011)
2. Ennals, R., Trushkowsky, B., Agosta, J.M.: Highlighting disputed claims on the web. In: Proceedings of the 19th International Conference on World Wide Web, pp. 341–350 (2010)
3. Giampiccolo, D., Magnini, B., Dagan, I., Dolan, B.: The third pascal recognizing textual entailment challenge. In: Proceedings of the ACL-PASCAL Workshop on Textual Entailment and Paraphrasing, pp. 1–9 (2007)
4. Harabagiu, S., Hickl, A., Lacatusu, F.: Negation, contrast and contradiction in text processing. In: Proceedings of the 21st National Conference on Artificial Intelligence, pp. 755–762 (2006)
5. Hsu, C.F., Khabiri, E., Caverlee, J.: Ranking comments on the social web. In: International Conference on Computational Science and Engineering, CSE 2009, vol. 4, pp. 90–97 (2009)
6. Kleinberg, J.M.: Authoritative sources in a hyperlinked environment. Journal of the ACM (JACM) 46(5), 604–632 (1999)
7. Lex, E., Voelske, M., Errecalde, M., Ferretti, E., Cagnina, L., Horn, C., Stein, B., Granitzer, M.: Measuring the quality of web content using factual information. In: Proceedings of the 2nd Joint WICOW/AIRWeb Workshop on Web Quality, pp. 7–10 (2012)
8. MacCartney, B., Galley, M., Manning, C.D.: A phrase-based alignment model for natural language inference. In: Proceedings of 2008 Conference on Empirical Methods in Natural Language Processing, pp. 802–811 (2008)
9. de Marneffe, M.C., Rafferty, A.N., Manning, C.D.: Finding contradictions in text. In: Proceedings of ACL 2008: HLT, pp. 1039–1047 (2008)
10. de Marneffe, M.C., Rafferty, A.R., Manning, C.D.: Identifying Conflicting Information in Texts. In: Handbook of Natural Language Processing and Machine Translation: DARPA Global Autonomous Language Exploitation (2011)
11. Pasternack, J., Roth, D.: Making better informed trust decisions with generalized fact-finding. In: Proceedings of the Twenty-Second International Joint Conference on Artificial Intelligence, pp. 2324–2329 (2011)
12. Qazvinian, V., Rosengren, E., Radev, D.R., Mei, Q.: Rumor has it: Identifying misinformation in microblogs. In: Proceedings of the Conference on Empirical Methods in Natural Language Processing, pp. 1589–1599 (2011)
13. Ratkiewicz, J., Conover, M., Meiss, M., Gonçalves, B., Patil, S., Flammini, A., Menczer, F.: Truthy: Mapping the spread of astroturf in microblog streams. In: Proceedings of the 20th International Conference Companion on World Wide Web, pp. 249–252 (2011)
14. Ritter, A., Soderland, S., Downey, D., Etzioni, O.: It's a contradiction – no, it's not: A case study using functional relations. In: Proceedings of 2008 Conference on Empirical Methods in Natural Language Processing, pp. 11–20 (2008)
15. Watanabe, Y., Miyao, Y., Mizuno, J., Shibata, T., Kanayama, H., Lee, C.W., Lin, C.J., Shi, S., Mitamura, T., Kando, N., Shima, H., Takeda, K.: Overview of the recognizing inference in text (RITE-2) at NTCIR-10. In: Proceedings of the NTCIR-10 Conference, pp. 385–404 (2013)
16. Watanabe, Y., Mizuno, J., Inui, K.: THK's natural logic-based compositional textual entailment model at NTCIR-10 RITE-2. In: Proceedings of the NTCIR-10 Conference, pp. 531–536 (2013)

Finding Cyclic Patterns on Sequential Data

António Barreto and Cláudia Antunes

Department of Computer Science and Engineering,
Instituto Superior Técnico, Universidade de Lisboa
Av. Rovisco Pais, 1049-001 Lisboa, Portugal
{antonio.barreto,claudia.antunes}@tecnico.ulisboa.pt

Abstract. The need for the study of dynamic and evolutionary settings made time a major dimension when it comes to data analytics. From business to health applications, being able to understand temporal patterns of customers or patients can determine the ability to adapt to future changes, optimizing processes and support other decisions. In this context, different approaches to Temporal Pattern Mining have been proposed in order to capture different types of patterns able to represent evolutionary behaviors, such as regular or emerging patterns. However, these solutions still lack on quality patterns with relevant information and on efficient mining methods. In this paper we propose a new efficient sequential mining algorithm, named PrefixSpan4Cycles, for mining cyclic sequential patterns. Our experiments show that our approach is able to efficiently mine these patterns when compared to other sequential pattern mining methods. Also for datasets with a significant number of regularities, our algorithm performs efficiently, even dealing with significant constraints regarding the nature of cyclic patterns.

Keywords: temporal pattern mining, cyclic patterns, sequential pattern mining.

1 Introduction

The increasing collection of data in the most varied domains has created the opportunity and the need for analyzing this data in order to obtain an information gain, when compared to the commonly discovered information. Temporal information usually appears in several forms, either associated with the occurrence of the events or their insertion into the database. Together, these aspects have created the possibility of understanding the evolution of the data and, consequently, the domain.

Sequential Pattern Mining and Temporal Data Mining are the two most relevant streams of research that seek the use of temporal related data in order to disclose trends and behaviors from the data. Although recent work in these fields already provides some temporal related solutions [12,13,14,15], there are still two main issues that remain a challenge. First, the efficiency of the mining methods in producing temporal relevant patterns is impaired, given the nature of the patterns that usually don't follow the downward-closure property from the *Apriori* algorithm [5]. Therefore, temporal-related information requires more complex structures to represent the same pattern information. Second, most of the existing methods consider the database as a

D. Ślęzak et al. (Eds.): AMT 2014, LNCS 8610, pp. 110–121, 2014.

population instead of a set of different instances, which in terms of the representativeness of the discovered patterns does not guarantee the support of the patterns based on the individual instances, possibly creating some false expectations.

Therefore, we propose a new mining approach for mining cyclic patterns with regular periodicities, in order to capture regular behaviors from sequential data. Our approach presents improvements both in terms of patterns quality and in their representativeness across the dataset, being mined based on individual instances. We also discuss the extensibility of our approach in order to consider other types of temporal patterns.

The paper is organized as follows. In Section 2 we introduce the problem of sequential and temporal pattern mining, together with some recent research in those fields regarding the mining of regularities. In Section 3, we formulate our problem introducing the main concepts regarding the data model and the notion of cyclic sequential patterns, and in Section 4 we propose a new algorithm – the PrefixSpan4Cycles. Later, we present the conducted experiments with results for two datasets and analyze different metrics (Section 5) and finally a resume of the most important contributions of this paper is presented in Section 6.

2 Background

The potential of temporal related information on understanding evolutionary behaviors, together with the evidence from earlier pattern mining technics, lead to the development of new methods that are able to deal with this temporal information during the mining process. However, solutions have mostly been explored based on transactional data models and on Apriori-based algorithms, which introduce the limitations previously mentioned.

In fact, one of the first fields of research that somehow approached the temporal dimension was the field of Sequential Pattern Mining (SPM). SPM was initially introduced with the goal of mining frequent subsequence patterns from data sequences, which are sets of ordered events that may or not have some temporal association [10]. In the field of SPM several approaches have been proposed either regarding the type of patterns found or the mining method itself. PrefixSpan [7] is one of the SPM algorithms that makes use of the concept of projected databases introduced by [11], adopting a divide and conquer strategy in order to reduce the search space at each stage of the algorithm. This approach allows for a better memory and computational complexity management, when compared to some candidate generation algorithms such as SPAM [1] or GSP [2].

Also, regarding the mined patterns, some variations have also been proposed. GSP [2] and GenPrefixSpan introduced the notion of gap into the mining process, allowing for a minimum and maximal gap in the occurrence of the items in the pattern, which allowed for more constrained pattern representation.

However, these SPM approaches are limited in what to the representation of temporal relations respects. Most of these techniques are only able to represent the frequent order among the items of a pattern, which is not enough to improve the predictive power or understand the evolutionary behavior of the domain. Recent work in [4] introduced the concept of Multiple Minimum Repetition Support (MMRS) in addiction to the work of [3] that introduced the Minimum Repetition Support (MRS) in order to detect regularities in the data. These approaches define cyclic patterns as patterns that verify the specified MRS, ensuring that for a pattern to be cyclical it must occur at least MRS times in the same sequence. This is a complementary support for the ordinary support concept and guarantees that the patterns found present some regular behavior. However, in our opinion, these regularities are not cycles since the distance among repetitions is not constant, neither follows any known periodicity. Regarding the MMRS, it brings an additional feature related with the repetitions of the pattern items, allowing for the specification of different repetition supports for the different elements in the pattern. This allowed for a better user specification of the type of patterns of its interest, however maintaining the same restriction of the unknown type of repetition of the pattern.

Also, the last work presented a new mining approach based on PrefixSpan against the approach of [3] that opted for a candidate generation approach based on the AprioriAll algorithm [9]. Even though no comparisons are performed in between these two approaches, we believe, based on the comparisons between the original algorithms (PrefixSpan and AprioriAll) that the approach presented in [4] benefits from the pattern growth algorithms properties and the approach in [3] suffers from the drawbacks of apriori-based algorithms, such as the heavy candidate generation phase and the multiple scans to the database.

On the other hand, [6] presents interesting results in terms of patterns' representation. In this approach cyclic patterns are represented based on their duration, period and the first instant where they occur, introducing additional information to the comprehension of its behavior once we can estimate how likely the pattern is to occur in future time instants. However, the use of a transactional data model and an apriori-based mining method creates significant constraints to both efficiency and support of the patterns in terms of instances.

The limitations of the previously presented works, and also the lack of solution in sequential pattern mining for mining interesting temporal related patterns created the opportunity to develop a new approach for mining quality cyclic patterns in an efficient way. Though, in next section, we present our approach for mining sequential cyclic patterns, bringing together the best properties of the work in [6] and [4].

3 Problem Statement

For the definition of our problem, let an *item* be an element from an alphabet Σ. An *itemset I* is an ordered set of items, that occur simultaneously, for example in the same transaction. An *event*, *e*, is a tuple (*I*, *t*), where *e.I* represents its itemset and *e.t* its time point.

Consider the data on *Table I*: Σ is composed with the literals from '*a*' to '*k*' and an example of an event is the first occurrence for the customer with the *S_id* = 3 represented by (*ac*:1) where in this case the event is formed by an itemset {*a*,*c*} and a time instant = 1.

Table 1. Itemset Sequence Database

S_id	Sequence
1	*<(a:1)(b:2)(d:3)(e:4)(ae:5)(f:6)(g:7)(h:8)(a:9)(b:10)(h:11)(i:12)(a:13)(j:14)(k:15)(l:16)(a:17)(b:18)>*
2	*<(f:1)(ce:2)(g:3)(c:4)(:5a)(ce:6)(b:7)(d:8)(f:9)(ce:10)>*
3	*<(ac:1)(b:2)(d:3)(e:4)(a:5)(f:6)(g:7)(h:8)(ae:9)(b:10)(i:11)(j:12)(a:13)(k:14)(l:15)(i:16)(ac:17)(b:18)>*
4	*<(h:1)(i:2)(ce:3)(a:4)(c:5)(j:6)(ce:7)(b:8)(d:9)(k:10)(ce:11)>*

In this context, we can define an itemset sequence, *s*, as an ordered set of itemsets, say $s = <I_1, ..., I_n>$. If those itemsets have occurred on a sequence of time ordered events, than we can create a temporal itemset sequence.

Definition 1. A *temporal itemset sequence* is a sequence of events, $s = <e_1,...,e_n>$, where $\forall i \in \{1..n-1\}: e_i.t < e_{i+1}.t$.

In the same example (Table 1), the represented sequences are ordered according to the itemsets time points, exemplifying temporal itemset sequences.

Whenever a temporal itemset sequence presents repetitions equally distributed along the time, we call it a cyclic temporal sequence. In order to define it, we need to define a cycle.

Definition 2. A *sequential cyclic pattern* is a triple (s, ρ, δ), where *s* is an itemset sequence and $\phi=(\rho, \delta)$ is a *cycle*, where ρ corresponds to the period of the cycle – the gap between two consecutive repetitions of *s*; and δ to its duration – the number of occurrences of *s*.

When considering temporal information, the duration of a cycle corresponds to the temporal distance between the repetitions of the pattern.

Considering the previous definitions and the usual definition of itemset subsequences [2], we can now introduce the notion of subsequence when applied to itemset sequence cyclic patterns.

Definition 3. An itemset sequence *S contains* a sequential cyclic pattern (s, ρ, δ), if *s* is a contiguous subsequence of *S*, δ times, and each occurrence of *s* in *S* is equidistant, with ρ the distance between occurrences.

Remember that the definition of a contiguous subsequence implies that the consecutive elements in the pattern occur exactly in the same manner in the sequence. For example, the sequence in *Table 1* with *S_id* = 1 contains the cyclic pattern $p = (\{(a)(b)\}, 6, 3)$ once the subsequence $(a)(b)$ is contiguous in the database sequence (has no more elements between the ones in the pattern) and it occurs three times in the sequence ($\delta=3$) with an equal distance ($\rho=6$).

In order to prevent every itemset sequence to be considered as a trivial cyclic pattern, with just two occurrences, we require that $\delta>2$. Naturally, we may be interested in cycles with more repetitions (longer cycles). We call *minimum duration* to the minimum number of occurrences that an itemset must present to be considered a cyclic pattern.

According to the previous definitions, we state our problem as that of given a database of itemset sequences, a minimum support threshold σ (as usual in pattern mining), and a minimum duration δ, discover the complete set of frequent sequential cyclic patterns that verify both σ and δ.

Definition 4. Given a minimum support threshold σ and a dataset *D*, a sequential cyclic pattern is said to be *frequent* if it is contained in at least σ sequences in the dataset *D*.

4 Algorithm

In this section we will introduce the *PrefixSpan4Cycles* algorithm, for mining sequential cyclic patterns from a database of sequences. The *PrefixSpan4Cycles* is an extension of the *PrefixSpan* algorithm [7], and adopts a similar divide and conquer strategy in order to reduce the search space at each step of the algorithm. This is done by creating multiple projected databases for the database sequences in order to progressively extend the sequential patterns

In general, our algorithm follows the same phases as PrefixSpan, though it presents some additional procedures in order to build and verify the periodicities of the discovered patterns. At each recursion step, the algorithm extends already discovered sequential cyclic patterns with an additional frequent item. Later, it verifies if the extended pattern continues to be a sequential cyclic pattern. In the case this is true, then the algorithm calls a new recursion step.

In particular, in the initial phase of the algorithm (Fig. 1), given a sequence database *DB*, a minimum support value σ, and minimum duration value δ, the algorithm first constructs the list of frequent items of size 1 (line 1), that verify the minimum support value and then, for each of those frequent items, it runs the recursive method with the item's projected database (line 3). For the example of *Table I* and considering a minimum support value of 50% (corresponding to the verification of the pattern in at least two sequences of the dataset) and also a minimum duration value of 3, meaning the repetition of the pattern for at least twice in the same sequence (corresponding to three occurrences of the pattern), in this initial phase we would get {*a*,*b*,*c*,*e*} as frequent items.

This result already presents some differences from the original algorithm, once it considers the new constraint variable of the minimum duration in order to calculate the frequent elements. Therefore, we can state that only the items {*a*,*b*,*c*,*e*} verify both the support and minimum duration constraints, since they occur at least three times (minimum duration) in two of the database sequences (support).

```
PrefixSpan4Cycles(Database DB, Support σ, Duration δ)
1   f_list←findFrequentItems(σ, DB, δ)
2   for each b in f_list do
3      L←L∪run(b,σ, δ, genProjDB(b,DB))
4   return L
```

Fig. 1. PrefixSpan4Cycles main function

In the iteration phase, the first step of constructing the projected databases is quite similar to the one in the original algorithm [7]. For the 1-length frequent items, the projected databases are constructed by mapping all the occurrences of those frequent elements in the database sequences, projecting the postfix-sequences having as prefix those same elements. This is done considering only the sequences of the database that contributed to the support of that element and by mapping the offsets of the occurrences of the items in those corresponding sequences. In this initial phase the entire sequence is swept in order to capture all the occurrences of the elements in the sequence.

Taking the frequent element *a* as example, its projected database will be composed by the sequences: *<1:{0,4,8,12,16}>*, *<3:{0,4,8,12,16}>* where the first identifiers *{1,3}* correspond to the unique identifiers of the sequences in the original database, and the set of integers *({0,4,8,12,16},{0,4,8,12,16})* are the corresponding offsets for each occurrence of *a* in the sequences. As we can see, the sequences {2,4} are not mapped into *a*-projected database, since the occurrences of that item in those sequences do not verify the minimum duration constraint (only one occurrence in each sequence), thus not contributing to the support of the item. Regarding the structure of the projected databases, we use a simplistic representation by keeping only the reference to the original sequence and the set of offsets, instead of duplicating the postfix sequences. This allows for a better memory performance management, one of the main drawbacks pointed to this type of methods [8]. This main function returns the complete list of patterns *L*, generated by the recursive method *'run'* (lines 3 and 4), which verifies a cyclic behavior according to the constraint δ.

The recursive method (Fig. 2) presents the depth-first search strategy of the algorithm. Here is where, given a pattern α, a minimum support and duration values and the projected database for α, the new frequent patterns 'αb' are generated, being 'b' a frequent item that occurs (concurrently or contiguously) after α. Also, the cycles for the frequent pattern α are disclosed. Therefore, and similarly to the main method, in a first step the new frequent items are calculated (line 1), now based on the projected database of the previous frequent element α, which reduces the search space for the possible new patterns. In this step, frequent items after α are found, again making use of the δ constraint to reduce the number of items considered.

```
Run(Pattern α,Support σ,Duration δ,ProjDB db)
1   f_list←findFrequentItems(σ, db, δ)
2   for each b in f_list do
3      β ← αb or (αb)
4      L ← L ∪ run(β, σ, genProjDB(β, db))
5   L' ← generateCycles(α,db, δ, σ)
6   L ← L' ∪ L
7 return L
```

Fig. 2. Recursive method

Maintaining the example of the frequent item *a*, and considering its projected database, the result of the `findFrequentItems` for that item will be only the item *b*, this is because it is the only frequent item that occurs at least δ times in a position that is contiguous to the offsets of the *a*-projected database. In this particular case (*Table I*), we can verify that *b* occurs in the offsets {1,9,17}, for both sequences {1,3}, being contiguous to the offsets {0,8,16} of *a*-projected database.

Moreover, when considering itemsets, this method guarantees that all the frequent items that occur in the same position of the element in analysis are also considered. This happens in the case of the frequent item *c* whose projected database generates the frequent itemset *(ce)* being the item *e* frequent in the same offsets of the item *c*. The second phase of the recursive method is then responsible for creating the new frequent patterns (line 3), either as an itemset or an itemset sequence, and run the method recursively with the new projected databases for the new patterns (line 4). Finally, in the last step, and when there are no more frequent items to explore, the patterns' cycles are generated according to the projected database for each pattern. Again, we should note that the projected databases are composed by the reference to the sequence and the offsets of the occurrence of the pattern in the database sequences, allowing us to determine the periods between every occurrence of those patterns.

According to the projected databases for the pattern α, the method GenerateCycles (Fig. 3) is responsible for generating the possible cycles for that pattern. The method goes as follows: for every projected database sequence, the periods for the pattern are computed (line 3), which corresponds to the difference between the set of offsets of the pattern's occurrences. Recalling the example of pattern a, the result of getting the set of periods would be (<1:{3,3,3,3}>, <3:{3,3,3,3}>) for both sequences {1,3}.

With this set of periods, we are able to determine the complete set of cycles of the pattern, since it contains the complete set of occurrences of the pattern. This information is important in order to guarantee that we are able to capture the complete set set of cycles for a specific pattern and also to consider the pattern frequent during the execution of the algorithm. As mentioned in Section 3, this type of patterns does not verify the prefix-anti-monotonic property: if there is a frequent k-length cyclic pattern there is no guarantee that all its sub-patterns are also frequent with the same cycle.

This, translated into the way the algorithm computes, means that even if a frequent k-length pattern does not verify any cyclic pattern we cannot neglect its (k+1)-length frequent patterns, so we do not miss possible patterns. Therefore, we should guarantee that we explore the super patterns in an efficient way.

A first approach could be based on the generation of all possible subsequences of the k-length pattern offsets sequence (for all projected database sequences), guaranteeing the analysis of all possible (k+1)-length pattern sequences. However, this approach would require the creation of new projected databases with all those subsequences, increasing both memory and computational requirements together with the fact that most of the generated subsequences would not produce any results, even though they would have to be analyzed. To overcome this problem we do not make (k+1)-length patterns dependent on k-length discovered patterns. In our approach we simply project one subsequence, containing all the occurrences of the new (k+1)-length pattern, for each of the projected database sequences of the k-length pattern. This guarantees that we are able to identify all the possible cycles for the super pattern while reducing the number of possibilities that we have to test to generate those cycles.

Consider the sequence with *S_id = 1* from *Table I*, given the set of offsets for the item *a : {0,4,8,12,16}*, in order to guarantee all possible super-patterns we should verify, according to the first approach, the set of combinations formed by the offsets {0,4,8,12,16}, {0,4,8,12}, {0,4,8}, {4,8,12,16}, and so on until we got all possible combinations. However, many of these combinations are unlikely to produce new patterns being, anyway, verified for that. By projecting only the global offset sequence for the super pattern, according to the implemented approach, we only verify one candidate sequence, for example for the pattern *ab*, the sequence {1,9,17}. After determining its set of periods we are able to compute the existing cycles.

```
GenerateCycles(Pattern α, Support σ, Duration δ, ProjDB db)
1   cycle_list ← empty
2   for each s in db do
    //Pattern periods
3      p_list ← s.getPeriods()
    //Generates set of cycles
4      cycle_list←cycle_list∪getCycles(p_list)
 //The frequent cycles
5   f_cycle ← cycle_list.getFrequent(σ)
 //The list of cycles for α
```

Fig. 3. Cycle generation method

In this step, after determining the periods for the set of offsets, the possible cycles are determined and added to the set of cycles of the pattern (line 4). For those cycles that verify the minimum support required (line 5) they are added to the set of cyclic patterns represented by the items and the set of cycles that the pattern verifies (line 6). For example, given a set of periods = {3,3,3,4,5,5,5} we derive the cycles {(3,3),(5,3)} respectively for a cycle with period 3 and duration 3 and another with period 5 and duration 3. The computed patterns are recursively returned to the previous iteration until the complete set of patterns is achieved in the main iteration of the algorithm (Fig. 1, line 3).

5 Experimental Results

In order to evaluate the performance of the proposed solution, we adopted an experimental approach based on the analysis of two different datasets, the *LastFM* and the *HouseholdPower* datasets. First, we will provide a brief description of those datasets and then introduce our experimental results. In the experimental results, we both evaluate the performance of the algorithm against the GenPrefixSpan and PrefixSpan algorithms and the quality of the discovered patterns.

The LastFM dataset reports the listening habits of 1000 LastFM users, spanning a temporal interval of around four years. We have pre-processed this dataset in order to obtain the listening habits for each user at month granularity, meaning that we have the set of sequences where each one represents a pair (user, month) with the listening habits for that user in that specific month. With this, we have a dataset with 22478 instances each one with a sequence with 30 or 31 itemsets, corresponding to the days of the month. The itemsets contain the artists listened by each user in each day, from a set of 95090 different artists. The itemset size (number of items) is very irregular throughout the sequences, varying from 16 to 377 items per basket. This information shows that this is a very sparse dataset, considering not only the size of the itemsets but also the alphabet size. The HouseholdPower dataset reports the measures of the expended power (in kilowatt) for every minute of a day. The dataset was also pre-processed in order to comprise the records for approximately one year (379 days), with each record having 1440 measures, corresponding to the total number of minutes of a day (24 hours). The dataset has a total of 99 different measures.

The experiments were performed according to different support measures and analyzing the following metrics: computational time spent, number of computed patterns, variation of the minimum and maximum cycle duration and variation of the patterns size. Regarding the support, we measured the results for the values of 30%, 20%, 15%, 10%, 7,5%, 5%, 2,5% and 1% which allowed us to understand the behavior of the algorithm and the variation in terms of generated patterns. We also provide a comparison between our algorithm and the GenPrefixSpan. The results for the LastFM dataset were expected to be quite poor, given the previous presented characteristics (Section 5), which are somehow shown by the results in Fig. 4 and 5.

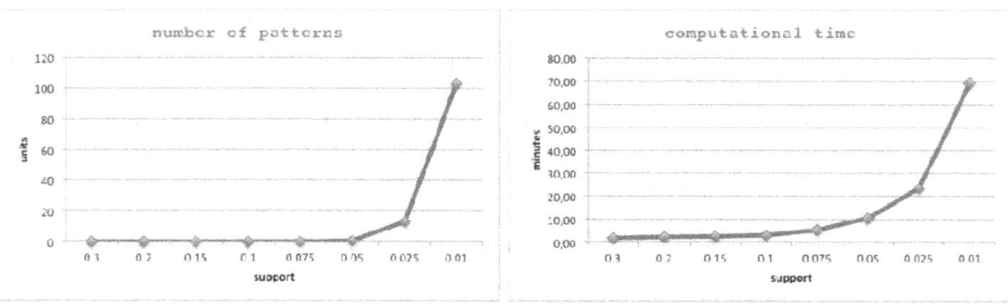

Fig. 4. Number of computed patterns (LastFM)

Fig. 5. Time spent on computation (LastFM)

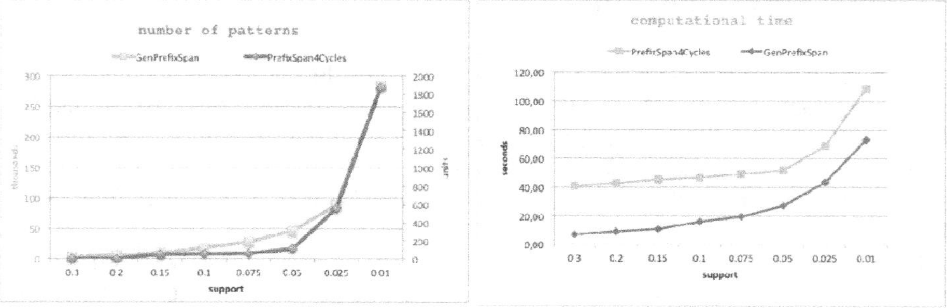

Fig. 6. Number of computed patterns (HouseholdPower)

Fig. 7. Time spent on computation (HouseholdPower)

With respect to the number of generated patterns we only got significant results for low support values (< 2,5%, corresponding to 562 sequences). For a support value of 1% the number of patterns was 103 and the maximum discovered cycle was (0, 5) meaning a periodic pattern with period 0 and duration 5 (five consecutive occurrences of the itemset).

In this specific case, the pattern translates a contiguous behavior, where the users have listened the same artists for 5 consecutive days. The majority of the discovered patterns reveal contiguous behaviors (period zero), despite some patterns with period = 1 were also discovered with less expressivity (9 out of 103). Regarding computational results, the sparseness of the data in terms of attributes and its significant size resulted in a computational time in the order of minutes. However we can understand that the relation between the computational time spent and the patterns generation follows the same trend.

The results for the HouseholdPower dataset show how the algorithm performs for a smaller dataset with more regularities and provide a comparison between the PrefixSpan4Cycles and the GenPrefixSpan algorithm [8]. Regarding the number of generated patterns (Fig. 6), we can understand that there is a significant difference between the algorithms. While the GenPrefixSpan, produces patterns in the order of thousand units, with almost 300 thousand patterns for a support of 1%, our algorithm produces a maximum of 2 thousand patterns for the same support. These results are achieved when running the GenPrefixSpan with gap constraint of zero, meaning that it only produces contiguous patterns, the same as in our approach. Therefore, the cyclic patterns discovered represent less than 1% of the patterns discovered in the GenPrefixSpan, which reveals the lack of specificity of the method and suggests a complex post-processing task in order to disclose regularities.

Also, with this dataset we are able to understand the behavior of the algorithm when there are a significant number of patterns to be mined (contrary to the LastFM dataset). The execution time for the same support show an evolution close to the evolution of the number of patterns found, which is consistent with the previous dataset. However, in this experiment the results were in the order of seconds, which shows the efficiency of the algorithm in a different type of dataset. When compared to the execution time of the GenPrefixSpan, our algorithm performs in slightly more time, which was expected due to the process of the cycle generation phase that requires extra processing. However, when comparing the results and the time required, the increase in the execution time is not significant given the quality of the expected patterns (it would take longer to analyze the results of the GenPrefixSpan in order to disclose the cyclic patterns).

In respect to the patterns' quality, in this dataset we achieved a maximum duration cycle of 43 occurrences and multiple periods, from 0 to 40. In the context of the energy domain this could be interesting to evaluate since these patterns represent periodic energy expenditures that can be associated with several behaviors. For example, the discovered pattern $p = (\{(.2)^{10}\}, 30, 3)$ (with the itemset sequence having size ten, representing ten repetitions of the item (.2) allow us to understand an expenditure of 0.2KW during ten minutes (the granularity of the data) every 30 minutes (the period of the pattern) for three times (the duration of the pattern).

When considering a gap constraint different from zero, meeting the behavior of the PrefixSpan, the experiments were not possible to perform, for both datasets, due to the memory requirements of the algorithms. This also supports the ability of our approach to efficiently mine this specific type of patterns. In a global analysis the results presented consist performance execution times and when compared to other sequential pattern mining methods showed their efficiency in mining this type of patterns with cyclic regularities, overcoming the pre-processing phase that would be required in other methods.

6 Conclusions

The comprehension of behaviors and trends is one of the major challenges in various domains and still lack on efficient and quality solutions regarding temporal pattern mining techniques. This increasing need for up to date information, able to predict future events, brought the complexity of the temporal dimension into the mining process, which has shown to be quite hard to define and adapt to the existing techniques.

In this paper we proposed a new solution for mining cyclic patterns, aiming to address those challenges. We proposed a new sequential algorithm for mining sequential cyclic patterns, patterns that occur multiple times in a sequence and have a regular period in between those occurrences. Sequential cyclic patterns translate regular behaviors that are likely to happen with a specific frequency, providing new information over the events that might occur in the future. Our approach introduces new information to this type of patterns, when compared to other related works, and also a new mining approach based on a pattern growth algorithm, the PrefixSpan.

The conducted experiments, using two different datasets, have also shown the potentialities of the algorithm in mining the sequential cyclic patterns, showing also some drawbacks when dealing with sparse datasets. Still, the algorithm behaves consistently in terms of computational time and patterns generated, showing a similar trend in these metrics. When compared to the GenPrefixSpan algorithm [8], our approach showed good results considering the extra processing required by this algorithm in order to achieve the same results as ours.

The future work goes through the definition of the previously introduced new cyclic patterns (converging and diverging) using the same algorithm structure and through the more specific analysis of this approach comparing it to similar algorithms (that mine regularities) in order to prove its efficiency.

Acknowledgments. This work is partially supported by FCT Fundação para a Ciência e a Tecnologia, under research project D2PM (PTDC/EIA-EIA/110074/2009).

References

1. Ayres, J., Flannick, J., Gehrke, J., Yiu, T.: Sequential pattern mining using a bitmap representation. In: Proceedings of the Eighth ACM SIGKDD International Conference on Knowledge Discovery and Data Mining, pp. 429–435. ACM (2002)

2. Srikant, R., Agrawal, R.: Mining sequential patterns: Generalizations and performance improvements. In: Apers, P., Bouzeghoub, M., Gardarin, G. (eds.) EDBT 1996. LNCS, vol. 1057, pp. 1–17. Springer, Heidelberg (1996)
3. Toroslu, I.H.: Repetition support and mining cyclic patterns. Expert Systems with Applications 25(3), 303–311 (2003)
4. Hu, Y.H., Chiang, I.C.: Mining cyclic patterns with multiple minimum repetition supports. In: 2011 Eighth International Conference on Fuzzy Systems and Knowledge Discovery (FSKD), vol. 3, pp. 1545–1549. IEEE (2011)
5. Agrawal, R., Srikant, R.: Fast algorithms for mining association rules. In: Proc. 20th Int. Conf. Very Large Data Bases, VLDB, vol. 1215, pp. 487–499 (September 1994)
6. Pina, S.M., Antunes, C.: (TD)2PaM: A Constraint-Based Algorithm for Mining Temporal Patterns in Transactional Databases. In: Correia, L., Reis, L.P., Cascalho, J. (eds.) EPIA 2013. LNCS (LNAI), vol. 8154, pp. 390–407. Springer, Heidelberg (2013)
7. Pei, J., Pinto, H., Chen, Q., Han, J., Mortazavi-Asl, B., Dayal, U., Hsu, M.C.: Prefixspan: Mining sequential patterns efficiently by prefix-projected pattern growth. In: 2013 IEEE 29th International Conference on Data Engineering (ICDE), p. 0215. IEEE Computer Society (April 2001)
8. Antunes, C., Oliveira, A.L.: Generalization of pattern-growth methods for sequential pattern mining with gap constraints. In: Perner, P., Rosenfeld, A. (eds.) MLDM 2003. LNCS (LNAI), vol. 2734, pp. 239–251. Springer, Heidelberg (2003)
9. Agrawal, R., Srikant, R.: Mining sequential patterns. In: Proceedings of the Eleventh International Conference on Data Engineering, pp. 3–14. IEEE (March 1995)
10. Mabroukeh, N.R., Ezeife, C.I.: A taxonomy of sequential pattern mining algorithms. ACM Computing Surveys (CSUR) 43(1), 3 (2010)
11. Han, J., Pei, J., Mortazavi-Asl, B., Chen, Q., Dayal, U., Hsu, M.C.: FreeSpan: Frequent pattern-projected sequential pattern mining. In: Proceedings of the Sixth ACM SIGKDD International Conference on Knowledge Discovery and Data Mining, pp. 355–359. ACM (August 2000)
12. Wang, W., Yang, J., Muntz, R.: TAR: Temporal association rules on evolving numerical attributes. IEEE (2001)
13. Dong, G., Li, J.: Efficient mining of emerging patterns: Discovering trends and differences. ACM (1999)
14. Ozden, B., Ramaswamy, S., Silberschatz, A.: Cyclic association rules. IEEE (1998)
15. Böttcher, M., Höppner, F., Spiliopoulou, M.: On exploiting the power of time in data mining. ACM SIGKDD Explorations Newsletter 10(2), 3–11 (2008)

Exploring Temporal Dependencies to Perform Automatic Prognosis

Daniel Cardoso and Cláudia Antunes

Department of Computer Science and Engineering,
Instituto Superior Técnico, Universidade de Lisboa
Av. Rovisco Pais, 1049-001 Lisboa, Portugal
{daniel.cardoso,claudia.antunes}@tecnico.ulisboa.pt

Abstract. The use of data mining techniques in healthcare has been noticing an increased relevance over the last few years, being applied with a variety of objectives, with the most common one being the automatic diagnostic process. In this process, data mining techniques have achieved interesting and successful results. However, when it comes to prognosis the same quality of results is not being achieved. We argue that this happens thanks to the inability of the used techniques to capture the inherent temporal dependencies present on the data. Specifically, the temporal evolution of a patient is not being taken into account when performing prognosis. In this paper, we propose a different approach, independent of the domain, to address this issue. We present our preliminary results on two different datasets that show an improvement in the overall precision of the prognosis.

Keywords: Prognosis, Classification, Temporal dependencies.

1 Introduction

The advances on computer-based medical systems have benefited substantially from the high performance of data mining techniques, in particular from classification. Indeed, these techniques, on some extent, have been supporting *evidence-based medicine* [1], by providing robust and accurate methods to generalize the results obtained from clinical trials.

Classification methods have been shown to perform with high levels of accuracy in healthcare, in particular in cancer research [2], but also in more modern diseases, like Alzheimer's [3], and AIDS [4]. Despite the success of those techniques, they are mostly appropriate to analyze *tabular data*, described by a set of independent variables, in same manner describing a static snapshot of some entity's status. Moreover, they are not adequate to address other structured data, such as graphs and temporal data, where variables depend on each other. Indeed, while presenting high levels of accuracy in diagnosing systems, the same techniques present considerable lower results in prognosis tasks.

Actually, prognosis may be defined as the prediction of an outcome in a future instant, based on the data collected in the past and along time. In healthcare, prognosis is usually the task of predicting patient evolution, given a set of consultation records,

D. Ślęzak et al. (Eds.): AMT 2014, LNCS 8610, pp. 122–133, 2014.

composed by a single set of variables assessed repeatedly along time, in different snapshots. While in a single snapshot, methods may assume some level of independency among variables; this assumption is clearly unlikely in a set of snapshots, where the same variable is measured along different instants of time. Despite this dependency among snapshots, a large number of classification-based approaches have been proposed for prognosis (see [2], [5], [6], for example). However, the results on prognosis and diagnosis were considerably different, which in our opinion is due to the dependency among the different values for the same variable along time.

In this paper, we propose to explore the evolution verified in each variable that compose the snapshots, to estimate their future values, and from them to predict the outcome by using tabular classifiers. In order to validate our claim, we formalize the problem addressed, and present some experiments on two real datasets.

After the formalization of the prognosis problem, we review a set of case studies on several different diseases, with the most well known classification techniques (section 2). In section 3, we describe our approach, and propose two distinct implementations of it, followed by a description of some preliminary experiments that compares the accuracy of both traditional classifiers and our approach on two datasets for two different diseases (section 4). The paper concludes with a discussion of the improvements achieved, the issues constraining those improvements and proposing some guidelines for the next steps (section 5).

2 Computer-Based Prognosis

In the medical context, diagnosis is the use of patient's data, demographic and clinical, in order to understand and classify the current health condition of a patient [7]. From a formal point of view, and in the computer-based context, let $\mathcal{A}$ be a set of variables (either known as attributes) and C a set of possible classes. Given an instance $\bar{x}_i$ described by a set of m variables from $\mathcal{A}$, say $\bar{x}_i = (x_{i,1}, \dots, x_{i,m})$), the goal is to discover the most probable value y_i, which corresponds to its class or status, with $y_i \in C$, as in (1).

$$\bar{x}_i = (x_{i,1}, \dots, x_{i,m}) \rightarrow y_i \tag{1}$$

In a classification context, this is done in two steps: first by producing a classification model M_D, based on a set of known pairs (x_i, y_i) – the training dataset, and second, by applying the discovered model to each instance to classify.

On the other hand, *prognosis* is the foreseeing or prediction of the risk or probability of a certain health event happening, in the future, using the clinical and non-clinical data. It is the medical prediction of how the pair patient disease is going to evolve in a specified period of time.

To do this prognosis, a physician will use data that relates the patient to a certain part of the population, i.e. demographic data, as well as the patient's and patient's family clinical history. This means that the evolution of the patient is important in the prediction of his next state. Simply putting, if a patient is showing improvement in a certain factor that is responsible for some disease, it is more probable that his prognosis related to that disease is better than if it the patient had the same value but that factor was deteriorating.

As previously stated, in the process of making a prognosis a physician uses the medical history of a patient. This includes the different states a patient has been in the form of various clinical analyses he had done in different points over time. The need to use this sequential information shows the utmost importance that time has when predicting someone's survivability, risk of recurrence.

Considering all of this, then the prognosis task can be formalized as follows:

Let a patient be represented by a sequence of pairs, $(\overline{\mathrm{x}}_i^1, y_i^1)$ ... $(\overline{\mathrm{x}}_i^n, y_i^n)$, then the goal is to predict his y_i^{n+1} value – equation (2). Note that the different values for y_i^t may be observable (available) or non-observable at time instant t for instance i.

$$(\overline{\mathrm{x}}_i^1, y_i^1) \dots (\overline{\mathrm{x}}_i^n, y_i^n) \rightarrow y_i^{n+1} \tag{2}$$

2.1 State-of-the-Art

A survey of the works on computer-aided multivariate prognosis shows that it has been approached by tabular classifiers, from decision trees to artificial neural networks, passing through Bayes classifiers, Support Vector Machines and regression [8,9]. This is in accordance to some authors point of view [10], that argue that the development, validation and impact assessment of both diagnosis and prognosis can be mutatis mutandis applied. Indeed, they only differ on the amount of time until the outcome assessment: the present for diagnosis, and a future point in time for prognosis.

Work on computer-based prognosis dates back to 1980 [11] where a regression analysis was used to find the predictive power of 17 features when predicting the survival of breast cancer patients. From then on, several different methods have been applied with particular emphasis on the use of decision trees, neural networks and regression techniques. **Table I** summarizes some of those approaches, distributed by the most addressed diseases. A common characteristic of those approaches was the use of a hard pre-processing phase, where different methods were applied trying to mitigate the inefficiency of tabular procedures to capture the relations among variables.

From the different studies, it is clear that there is no best technique to perform overall prognosis and that the result of a technique depends highly on the data being used [2]. In other words, there is no general solution that can be used in more than one dataset maintaining their performance.

Additionally, none of the identified works contemplate temporal information, neither use the evolution of the patients. Moreover, it seems that the improvements are not made incrementally, with just a few being based on previous works.

An exception to this general picture is the use of Cox models ([7], [5]), which try to estimate the impact of variable changes in the time of survival [12]. Along with these methods, the advances on the analysis of time series are undeniable. *Time series*

Table 1. Classification-based Prognosis for Different Diseases

Disease / Technique	Alzheimer	Cancer	Kidney Failure	HIV/AIDS
Regression	[3], [5], [6], [18]	[2], [19], [20], [30], [35]	[31], [32], [36]	[33]
Decision Trees	[3]	[2], [20], [21], [22], [30],	[23], [26], [31]	[4]
Neural Networks	[3]	[2], [20], [21], [22], [30]	[25], [29], [31],	[34]
Bayesian Classifiers		[2], [22], [24],	[28], [37]	
SVM	[3]	[20], [27], [30]		

represent ordered measurements at regular temporal intervals [9], which may be *uni* or *multivariated*, representing a single or multiple co-occurring variables.

Prediction of time series is a research field with a long history, with stock markets and signal processing the most paradigmatic cases [13]. Along the time, medical time series have also been studied, with ECG and EEC the most addressed. The techniques used vary from regression models like ARMA to recurrent neural networks [14], trying to foresee the next outcome. More recently, hidden Markov models become to be applied with considerable success on omic data, and are currently be adjusted to contemplate multivariate series [15]. Along with these approaches, dynamic Bayesian networks have been proposed, but training algorithms are just being defined [16].

3 Prognosis Based on Temporal Dependencies

The traditional classification approach has been applied to prognosis with modest success, as seen above. In all described cases, the evolution of single variables was not explored, and actually, the different time instances of their values were addressed separately, ignoring any possible hidden structure, in the majority of approaches. On the other hand, the analysis of time series is applied to predict the next outcome of a single variable.

By recognizing that estimation may be used to fill unseen variable outcomes, which in turn may be used to improve classifiers accuracy, as in asap classifiers [17], we propose to transform the prognosis into a diagnosis task, by estimating the values of the variables that constitute the snapshot in the future point in time.

Formally, let $\mathcal{A}$ be a set of attributes, C be a set of possible classes and n be the number of observations. Let the tth observation, described by m variables from $\mathcal{A}$, be the pair given by $(\bar{x}_i^t, y_i^t) = (x_{i1}^t, \ldots, x_{im}^t, y_i^t)$ that says that at observation t the instance is described by x_i^t (the *observable values*) and classified as $y_i^t \in$ C (the *predicted value*). Given an instance described by an ordered set of n observations, the goal is to predict the $n+1^{th}$ observation, as in equation (3).

$$(\bar{x}_i^1, y_i^1) \dots (\bar{x}_i^n, y_i^n) \rightarrow (\bar{x}_i^{n+1}, y_i^{n+1}) \quad (3)$$

The difference to the definition (2) is the need to predict the entire $n+1^{th}$ observation, not only the predicted value y_i^{n+1}. Indeed, if there is a model M_D, that from observable values is able to determine the predicted value, it is enough to estimate the observable values in the $n+1^{th}$ observation, and from them to predict the predicted value. This model M_D is just a simple diagnosis model as in equation (1).

According to this formulation, a prognosis model, M_P, is then the composition of several models: one estimation model M_{Ek} per each observable variable X_k and a diagnosis model M_D able to predict the class given an observation, as in equation (4), where n corresponds to the number of available observations and m the number of variables for describing each observation.

$$M_P\left((\bar{x}_i^1, y_i^1) \dots (\bar{x}_i^n, y_i^n)\right) = M_D\left(M_{E1}(\bar{x}_i^1 \dots \bar{x}_i^n) \dots M_{Em}(\bar{x}_i^1 \dots \bar{x}_i^n)\right) \quad (4)$$

By transforming the prognosis problem into a diagnosis task, the challenge becomes to be able to estimate the observation in the time point to predict, which translates into the definition of the estimation models per each observable variable.

As stated above, the art of prognosis is based on the analysis of the evolution of the different variables along time. Therefore, estimation models should be able to recognize verified evolution trends in the estimation of future values.

In this manner, we propose that an estimation model for a single variable X_k, say M_{Ek} should be a function from a sequence of the observed values to an X_k value. In particular, we propose two different approaches: the *univariate-based* and the *multivariate-based estimations*.

A *univariate-based model* for variable X_k (*UvE*) is a function from a sequence of n values of X_k to its next value, x_k^{n+1}, as in equation (5), where Dom_{Xk} represents the domain of variable X_k. These models only explore the individual values of a variable, ignoring any influence from other variables.

$$M_{UvEk}: [Dom_{Xk}]^n \rightarrow Dom_{Xk} \quad (5)$$

$$M_{UvEk}(x_k^1 \dots x_k^n) = x_k^{n+1}$$

On the counterpart, a *multivariate-based model* for variable X_k (*MvE*) is a function from a sequence of n vectors of m variables, including X_k, to its next value, x_k^{n+1} – see equation (6).

$$M_{MvEk}: [Dom_{X1} \times \dots \times Dom_{Xm}]^n \rightarrow Dom_{Xk}$$
$$M_{MvEk}(\bar{x}^1 \dots \bar{x}^n) = x_k^{n+1} \quad (6)$$

By receiving a sequence of multi-values, recorded along n observations, multivariate estimator is able to contemplate the interdependencies among the different values, and having more informed inputs, is expected to output better estimations.

3.1 Algorithm

From the previous formulation, the algorithm required to train the new classifier is simple, and is similar for both estimation models.

```
UnivariateEstimation(Dataset D, Function alg_class, int ρ)
    D - the training dataset with
                 D = {(x̄_i^t, y_i^t): ∀i,t: 1 ≤ i ≤ |D| ∧ 1 ≤ t ≤ n}
    alg_class - the training algorithm
    ρ - the number of observations to use
1  𝒜←{the set of attributes describing D}
2  // Training each estimation model
3  for each variable X_k in 𝒜
4      D_k ← π_Xk(D) = {(x_ik^(n-ρ), ..., x_ik^n): ∀x̄_i ∈ D}
5          M_Ek ← alg_class(D_k)
6  // Estimating n+1 snapshot
7  for each x_i in D
8      for each variable X_k in A
9              x_ik^(n+1) ← M_Ek(x_ik^(n-ρ), ..., x_ik^n)
10         D^(n+1) ← D^(n+1) ∪ {(x_i1^(n+1), ..., x_im^(n+1), y_i^(n+1))}
11 // Train the diagnosis model
12 M_D ← alg_class(D^(n+1))
13 // Output the composition of models
14 return M_D ∘ (M_E1, ..., M_Ek)
```

Fig. 1. Pseudocode for Univariate Estimation training

Note, that the dataset has to be composed of records containing *n* snapshots, as described before, and ρ has to be less or equal to *n*. In terms of the classification training algorithm, it should be any tabular one, like a decision tree learner, an algorithm for training neural networks or just naïve Bayes.

The difference between the models is on the creation of the estimation models (line 4), in particular on the creation of the training dataset for each variable. While for univariate model, it consists on the projection of *D* in relation to each X_k, the multivariate model uses the entire set of variables. In both cases, ρ corresponds to the number of snapshots to keep in the dataset. Since, it is usual that the instants more significant for determining the next value are the previous ones, only the last ρ snapshots are used.

After training the estimation model for each variable, the diagnosis model is learnt from the estimated snapshot for instant *n*+1 and the known class label. Then, the algorithm outputs the model resulting from the composition of the different estimators and the diagnosis model learnt from the estimated values (line 14).

4 Experimental Results

In order to validate our proposal, we used two different real datasets from the healthcare field: the ALS and the Hepatitis datasets.

The ALS dataset (https://nctu.partners.org/ProACT/) includes information from over 8500 ALS patients who participated in industry clinical trials. The data include demographic, family and medical history, the patient's history in terms of ALS symptoms, clinical and some laboratorial data. From these, we used a subset composed by the patients that had demographic data, had performed Slow Vital

Capacity exams, as well as measurements of their vitals, counting 13 variables: gender, age, height, percentage of normal, subject liters (trial 1, 2 and 3), blood pressure (systolic and diastolic), pulse, respiratory rate, temperature and Weight.

The outcome is a score that evaluates the state of the disease between 0 (severe) and 48 (normal), discretized into 4 classes (aggregations of 12 points). The subset contains 578 patients, with 5% for the 1st class, 22.3% for the 2nd, 29.1% for the 3rd and 42.7% for the 4th, and 0.88% non-classified.

The Hepatitis dataset was made available as part of the ECML/PKDD 2005 Discovery Challenge (http://lisp.vse.cz/challenge/CURRENT/), it contains information about 771 patients, and more than 2 million examinations between 1982 and 2001. Based on the work of (Watanabe et al.,2003) the data was reduced to the most significant exams. In the end 17 variables were used: gender, age, birthdate, birth decade,11 of the most significant exams (GOT, GPT, ZTT, TTT, T-BIL, D-BIL, I-BIL, ALB, CHE, T-CHO and TP) and the results from the active biopsies at the time of the exams (type, activity and fibrosis).

Fibrosis is the objective class and it is described by integer values between 0 (no-fibrosis) and 4 (most severe). The subset contains 488 patients and the following distribution of classes: 2.05% of 0, 45.9% for 1, 21.35% for 2, 15.19% for 3 and 15.40% for 4.

4.1 Diagnosis Models

As a baseline for comparison with the proposed approaches we used two models. BaselineSingleObservation is a diagnostic model where a single observation in time is used to perform the prognosis. In other words, the state of a patient at instant *n* is used to predict his class at instant *n+1*. On the other hand, BaselineMultipleObservation instead of using a single observation, uses multiple observations: all information is used here to predict the class at instant *n+1*.

A collection of techniques were used with these models, with both achieving similar results: the precision ranged between 40% and 55%, depending on the dataset, technique and number of time points used, as seen in **Fig. 2** and **Fig. 3**.

4.2 Estimation Models

Before, assessing the results of our prognosis approach, we evaluate the impact of the number of observations used, on the quality of the estimations made through the two estimation models proposed.

Since ALS observations are described by numerical variables and Hepatitis by nominal variables, we measured the prediction error (the distance from the predicted to the target value) and the number of correct predictions, respectively.

Fig. 4 and **Fig. 5** show the results with univariate and multivariate estimation models, respectively. Both estimation models were applied using a different number of observations, and the previous.

Both models reach similar levels of accuracy, with quite good results for the majority of the Hepatitis variables (above 80%) and prediction error below 20% for the majority of ALS variables. It is interesting that there is a slight trend to increase the accuracy as the number of observations get higher.

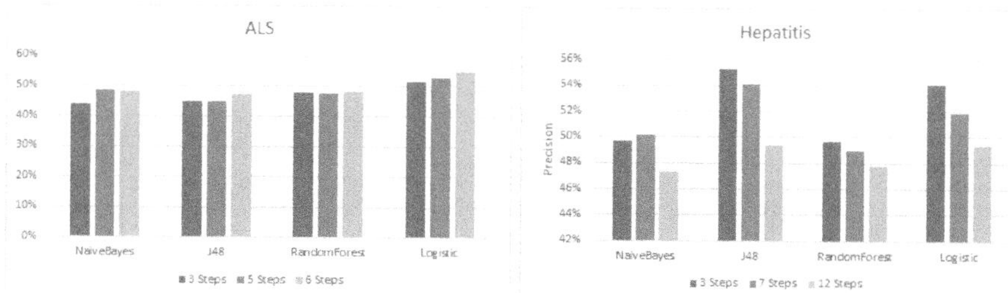

Fig. 2. BaselineSingleObs precision (several classifiers and number of observations)

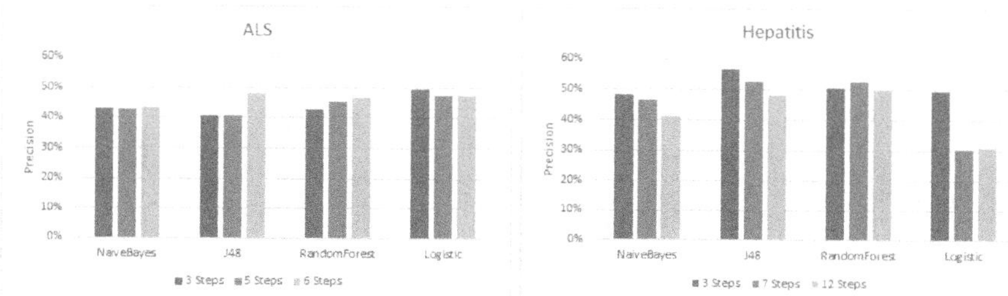

Fig. 3. BaselineMultipleObs precision (several classifiers and number of observations)

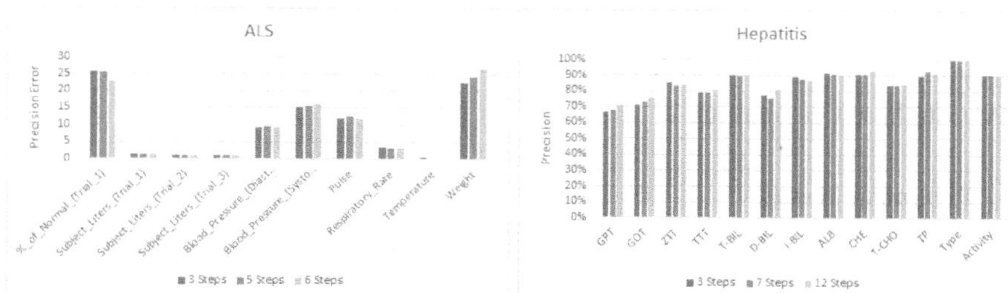

Fig. 4. Impact of the number of observations on the precision of estimation models for each variable with univariate-based estimation

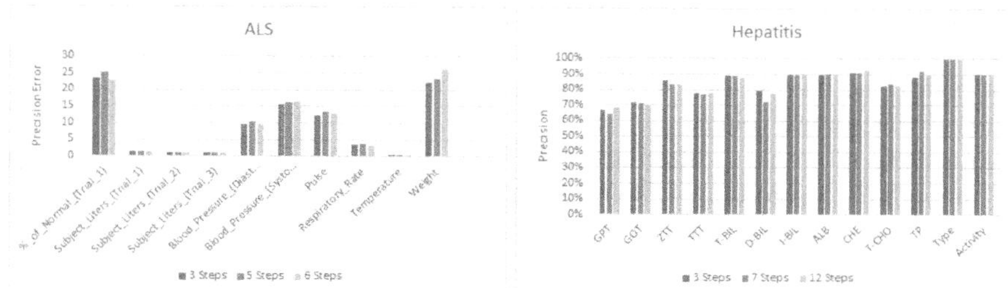

Fig. 5. Impact of the number of observations on the precision of estimation models for each variable with multivariate-based estimation

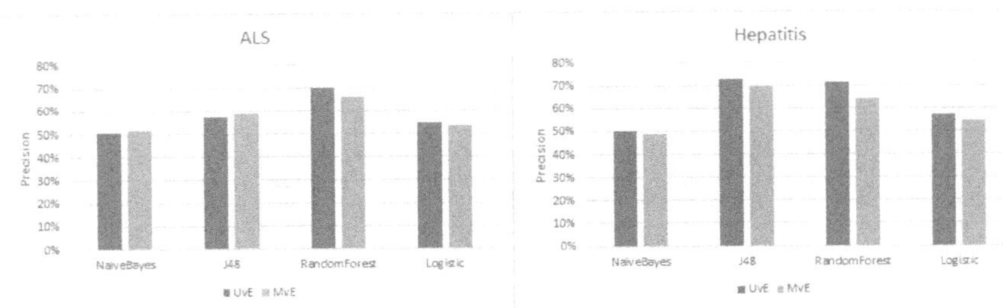

Fig. 6. Precision of different models

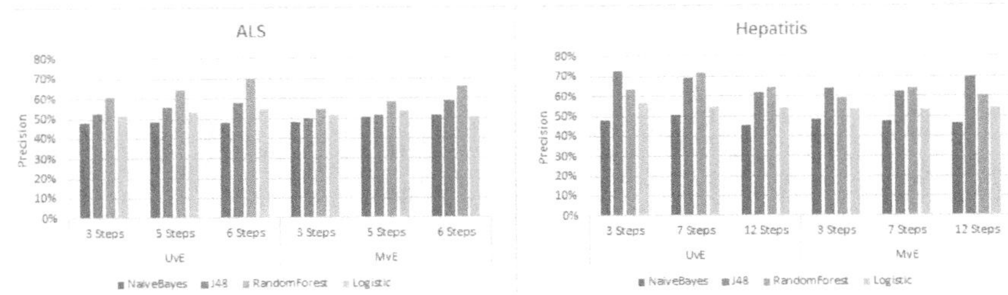

Fig. 7. Impact of the number of observation on prognosis models

Despite our expectations, it seems that there is no improvement on using multivariate-based estimation.

4.3 Prognosis Results

The overall prognosis precision achieved by using different techniques on our approaches can be seen in **Fig. 6**. The improvements on the precision of our approach are always present when compared to the ones achieved by baseline models (see **Fig. 2** and **Fig. 3**). In Hepatitis dataset the improvements round about 20%.

Fig. 7 shows the relation between the number of observations and the final precision of the prognosis, using both, *UvE* and *MvE* estimation models, and a variety of techniques. It is interesting to note that the higher number of observations become prejudicial to the UvE model, which means that the values from the long past do not help to estimate future values.

Again there is no clear difference between both estimation models, but decision trees (through C4.5 algorithm – J48) always perform better than the other models.

5 Discussion

Currently, medical practice is helped by a variety of computer-aided tools, dedicated to help physicians taking the most appropriate decisions. However, despite the importance of prognosis, it did not deserved dedicated tools, and in the majority of situations, it has been addressed as a simple diagnosis problem, without exploring the temporality involved

In order to mimic physicians practice, computer-aided prognosis should take into attention patients' evolution, considering the different observations made along time. In this paper, we formalize both diagnosis and prognosis problems, making clear the differences between them, and propose a method to transform the prognosis into a diagnosis task, based on the composition of classification over the estimation of observation values. As described above, what distinguishes this approach, from what is found in the literature, is the use of temporal dependencies of the data in order to estimate the future values of every feature and with those values perform a diagnostic in the future.

From the experimental comparison of the different approaches, over two distinct datasets (with different data characteristics, either from the medical and the data points of view), it is clear an improvement trend when using the temporal informed methods proposed. The shallow differences between the results of the estimation models, need to be deeply study, and other techniques (like HMMs or Dynamic Bayesian networks) should be explored to enrich the estimation process. In either cases, the temporality of this kind of data should be considered as a core aspect of the prognosis.

Acknowledgments. This work is partially supported by FCT Fundação para a Ciência e a Tecnologia, under research project D2PM (PTDC/EIA-EIA/110074/2009).

References

1. Sackett, D., Rosenberg, W., Muir Gray, J., Haynes, R., Richardson, W.: Evidence based medicine: what it is and what it isn't. BMJ 312(7023), 71–72 (1996)
2. Endo, A., Shibata, T., Tanaka, H.: Comparison of Seven Algorithms to Predict Breast Cancer Survival. Biomedical Soft Computing and Human Sciences 13(2), 11–16 (2008)
3. Maroco, J., Silva, D., Rodrigues, A., Guerreiro, M., Santana, I., Mendonça, A.: Data mining methods in the prediction of dementia: A real-data comparison of the accuracy, sensitivity and specificity of linear discriminant analysis, logistic regression, neural networks, support vector machines, classification trees and random forests. BMC Research Notes 4, 299 (2011)
4. Abdul-Kareem, S., Raviraja, S., Awadh, N., Kamaruzaman, A., Kajindran, A.: Classification and regression tree in prediction of survival of AIDS patients. Malaysian Journal of Computer Science 23(3), 153–165 (2010)
5. Paradise, M., Walker, Z., Cooper, C., Blizard, R., Regan, C.: Prediction of survival in Alzheimer's disease – The LASER-AD longitudinal study. Int'l Journal of Geriatic Psychiatry 24(7), 739–747 (2009)
6. Zhou, J., Yuan, L., Liu, J., Ye, J.: A Multi-Task Learning Formulation for Predicting Disease Progression. In: ACM SIGKDD Int'l Conf. Knowledge Discovery and Data Mining, pp. 814–822 (2011)
7. Steyerberg, E., Homs, M., Stokvis, A., Essink-Bot, M., Siersema, P., Study, G.: Stent placement or brachytherapy for palliation of dysphagia from esophageal cancer: A prognostic model to guide treatment selection. Gastrointestinal Endoscopy 62(3), 333–340 (2005)

8. Kharya, S.: Using data mining techniques for diagnosis and prognosis of cancer disease. Int'l Journal of Computer Science, Engineering and Information Technology (IJCSEIT) 2(2), 55–66 (2012)
9. Mitsa, T.: Temporal Data Mining. Chapman & Hall / CRC (2010)
10. Hendriksen, J., Geersing, G., Moons, K., Groot, J.: Diagnostic and prognostic prediction models. Journal of Thrombosis and Haemostasis 11(1), 129–141 (2013)
11. Nash, C., Jones, S., Moon, T., Davis, S., Salmon, S.: Prediction of outcome in metastatic breast cancer treated with adriamycin combination chemotherapy. Cancer 46(11), 2380–2388 (1980)
12. Cox, D.: Regression Models and Life-Tables. Journal of the Royal Statistical Society, Series B 34(2), 187–220 (1972)
13. Antunes, C., Oliveira, A.: Temporal Data Mining: An overview. In: 1st Workshop on Temporal Data Mining at ACM SIGKDD Int'l Conf. on Knowledge Discovery and Data Mining, San Francisco, USA (2001)
14. Palit, A.K., Popovic, D.: Computational Intelligence in Time Series Forecasting: Theory and engineering applications. Springer (2005)
15. Henriques, R., Antunes, C.: Learning Predictive Models from Integrated Healthcare Data: Extending Pattern-based and Generative Models to Capture Temporal and Cross-Attribute Dependencies. In: Hawaii Int'l Conf. System Sciences, BigIsland, Hawaii, USA (2014)
16. Murphy, K.: Dynamic Bayesian Networks: Representation, Inference and Learning. UC Berkeley, Computer Science Division (2002)
17. Antunes, C.: Anticipating student's failure as soon as possible. In: Romero, C., Ventura, S., Pechenizkiy, M., Baker, R. (eds.) Handbook for Educational Data Mining, pp. 353–363. CRC Press, New York (October 2010)
18. Ewers, M., Walsh, C., Trojanowski, J., Shaw, L., Petersen, R., Jack, C., Feldman, H., Bokde, A., Alexander, G., Scheltens, P., Vellas, B., Dubois, B., Weiner, M., Hampel, H.: Prediction of conversion from mild cognitive impairment to Alzheimer's disease dementia based upon biomarkers and neuropsychological test performance. Neurobiology of Aging 33(7), 1203–1214 (2012)
19. Lundin, M., Lundin, J., Burke, H., Toikkanen, S., Pylkkänen, L., Joensuu, H.: Artificial Neural Networks Applied to Survival Prediction in Breast Cancer. Oncology 57, 281–286 (1999)
20. Lakshmi, K.R., Krishna, M., Kumar, S.: Performance comparison of data mining techniques for prediction and diagnosis of breast cancer disease survivability. Asian Journal of Computer Science and Information Technology 3(5), 81–87 (2013)
21. Bellaachia, A., Guven, E.: Predicting Breast Cancer Survivability using Data Mining Techniques (April 2006)
22. Delen, D.: Analysis of cancer data: A data mining approach. Expert Systems 26(1), 100–112 (2009)
23. Kusiak, A., Dixon, B., Shaha, S.: Predicting survival time for kidney dialysis patients: A data mining approach. Computers in Biology and Medicine 35(4), 311–327 (2005)
24. Choi, J., Han, T., Park, R.: A Hybrid Bayesian Network Model for Predicting Breast Cancer Prognosis. Journal of Korean Society of Medical 15(1), 49–57 (2009)
25. Petrovsky, N., Tam, S., Brusic, V., Russ, G., Socha, L., Bajic, V.: Use of Artificial Neural Networks in Improving Renal Transplantation Outcomes. Graft 5(1), 6–13 (2002)
26. Osofisan, A., Adeyemo, O., Sawyerr, B., Eweje, O.: Prediction of Kidney Failure Using Artificial Neural Networks. European Journal of Scientific Research 61(4), 487 (2011)

27. Sun, B.-Y., Zhu, Z.-H., Li, J., Linghu, B.: Combined Feature Selection and Cancer Prognosis Using Support Vector Machine. IEEE/ACM Transactions on Computational Biology and Bioinformatics 8(6), 1671–1677 (2011)
28. Li, J., Serpen, G., Selman, S., Franchetti, M., Riesen, M., Schneider, C.: Bayes Net Classifiers for Prediction of Renal Graft Status and Survival Period. Int'l Journal of Medicine and Medical Sciences 1(4), 215–221 (2010)
29. Shadabi, F., Cox, R., Sharma, D., Petrovsky, N.: Use of Artificial Neural Networks in the Prediction of Kidney Transplant Outcomes. In: Negoita, M.G., Howlett, R.J., Jain, L.C. (eds.) KES 2004. LNCS (LNAI), vol. 3215, pp. 566–572. Springer, Heidelberg (2004)
30. Delen, D., Walker, G., Kadam, A.: Predicting breast cancer survivability: A comparison of three data mining methods. Artificial Intelligence in Medicine 34(2), 113–127 (2005)
31. Oztekin, A., Delen, D., Kong, Z.: Predicting the graft survival for heart–lung transplantation patients: An integrated data mining methodology. Int'l Journal of Medical Informatics 78(12), 84–96 (2009)
32. Ataide, E.C., Garcia, M., Mattosinho, T.J.A.P., Almeida, J.R.S., Escanhoela, C.A.F., Boin, I.F.S.F.: Predicting Survival After Liver Transplantation Using Up-to-Seven Criteria in Patients with Hepatocellular Carcinoma. Transplantation Proceedings 44(8), 2438–2440 (2012)
33. Egger, M., May, M., Chêne, G., Phillips, A., Ledergerber, B., Dabis, F., Costagliola, D., Monforte, A., Wolf, F., Reiss, P., Lundgren, J., Justice, A., Staszewski, S., et al.: Prognosis of HIV-1-infected patients starting highly active antiretroviral therapy: A collaborative analysis of prospective studies. The Lancet 360(9327), 119–129 (2002)
34. Dom, R., Kareem, S., Abidin, B., Kamaruzaman, A., Kajindran, A.: The Prediction of AIDS Survival: A Data Mining Approach. In: WSEAS Int'l Conf. Multivariate Analysis and its Application in Science and Engineering, pp. 48–53 (2009)
35. Wang, K.-M., Makond, B., Wu, W.-L., Wang, K.-J., Lin, Y.: Optimal Data Mining Method For Predicting Breast Cancer Survivability. Int'l Journal of Innovative Management, Information & Production 3(2), 28–33 (2012)
36. Hong, Z., Wu, J., Smart, G., Kaita, K., Wen, S.W., Paton, S., Dawood, M.: Survival Analysis of Liver Transplant Patients in Canada 1997–2002. Transplantation Proceedings 38(9), 2951–2956 (2006)
37. Ahn, J., Kwon, J., Lee, Y.: Prediction of 1-year Graft Survival Rates in Kidney Transplantation: A Bayesian Network Model. In: INFORMS & KORMS, pp. 505–513 (2000)

Ryry: A Real-Time Score-Following Automatic Accompaniment Playback System Capable of Real Performances with Errors, Repeats and Jumps

Shinji Sako, Ryuichi Yamamoto, and Tadashi Kitamura

Nagoya Institute of Technology, Gokiso-cho, Showa-ku, Nagoya, 466-8555, Japan
{s.sako,kitamura}@nitech.ac.jp

Abstract. In this work, we propose an automatic accompaniment playback system called Ryry, which follows human performance and plays a corresponding accompaniment automatically, in an attempt to realize human-computer concerts. Recognizing and anticipating the score position in real-time, known as score following, by a computer is difficult. The proposed system is based on a robust on-line algorithm for real-time audio-to-score alignment. The algorithm is devised using a delayed-decision and anticipation framework by modeling real-time music performance that includes uncertainties such as tempo fluctuation and mistakes. We developed an automatic accompaniment system that is capable of generating polyphonic music signals.

Keywords: Score following, Automatic accompaniment, Segmental Conditional Random Fields (SCRFs), Linear Dynamic System (LDS), Chord transition.

A concert is a live performance where each human performer can play a musical instrument in correspondence with others, even if the performance includes uncertainties such as tempo fluctuation and mismatch between the score. It is inadvertently or intentionally made by musicians. Real-time alignment of audio-to-score involves synchronizing an audio performance and its symbolic musical score, a process known as score following. It can be used in a wide range of real-time applications, such as the synchronization of live sounds and automatic accompaniment of human soloists or singers [1, 2]. Audio-to-score alignment can be considered as either an *off-line* or *on-line* problem. In an off-line scenario, the entire information of the given audio signal can be used in the alignment process. In contrast, in an on-line scenario, anticipated score information cannot be used because of the real-time constraint. Therefore, score following is more difficult in on-line scenarios compared to off-line scenarios.

For both on-line and off-line settings, there exist many approaches for solving the audio-to-score alignment problem by using dynamic programming methods based on hidden Markov models (HMMs) or dynamic time warping (DTW). In the off-line setting, the most likely alignment can be found using a dynamic

D. Ślęzak et al. (Eds.): AMT 2014, LNCS 8610, pp. 134–145, 2014.

programming technique, given the entire input audio signal. Even in the case of the on-line setting, the decoding algorithm works theoretically off-line; therefore, some approximation is required for on-line cases. For example, in some previous works [3–5] , an on-line greedy approximation technique was applied to the Viterbi algorithm and dynamic programming. In the case of polyphonic music, however, estimation errors occur due to uncertainties in pitch and onset, and they increase in proportion to the complexity of the input audio. Thus, the greedy approximation solution may not always be suitable.

In contrast to the dynamic programming approach, filtering methods based on state-space models have been proposed [6–8]. Although these models allow simultaneous estimation of the score position and tempo, they are prone to errors and fail to recover if the position is lost. To maintain the robustness of the solution for polyphonic music signals, it is helpful to use the tempo information to predict score position. In [9], Raphael used hybrid graphical models for estimating the score position and tempo, but this technique works only off-line.

In [10], Cont used duration-focused models consisting of hidden Markov/semi-Markov Models with an explicit tempo model. In this successful work, the greedy approximation in the Viterbi algorithm may cause estimation errors in the case of highly polyphonic signals. In [11], Arzt reported a sophisticated on-line algorithm, utilizing a forward-backward strategy that re-computes the past-determined forward path, although without an explicit tempo model.

In this work, we introduce a robust on-line algorithm for polyphonic music signals based on a delayed decision and anticipation framework, and develop a real-time automatic accompaniment playback system (Fig. 1) The advantages of our approach are that a delayed decision approximation for the Viterbi algorithm can determine highly reliable past positions utilizing future information and future position can be anticipated using an adaptively estimated tempo. In addition, we employ the state-of-the-art segmental conditional random fields (SCRFs) proposed in [12] (with a few modifications) and an explicit tempo model based on the linear dynamical system (LDS).

1 Chord Transitions Based on Segmental Conditional Random Fields

1.1 1.1 Score Alignment Problem Formulation

We first describe the audio-to-score alignment problem in the off-line situation because score following can be approximated from its on-line extension.

Given the audio music signal and its symbolic score, we address the score alignment problem as the segmentation of the audio to the chord sequence on the score (Fig. 2), where a chord is a set of concurrent notes on the score. In our approach, the transition of chords is modeled by SCRFs. SCRFs are an extension of CRFs in which the Markovian assumption is relaxed to allow a segment level that is separate from the frame level. Fig. 3 shows the transition topology of the SCRFs we used. The topology includes self-loop (repeat the same note), skip

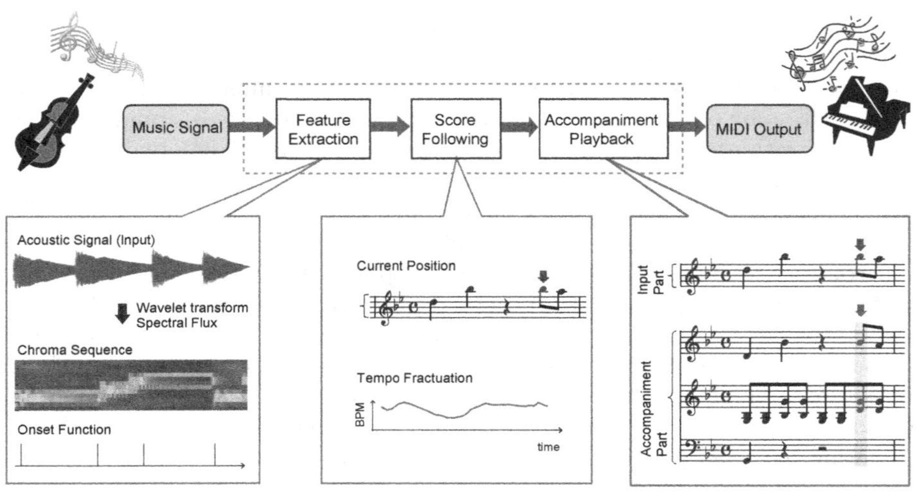

Fig. 1. Block diagram of audio-based automatic accompaniment system

(omit some notes), and jump (change the position of performance) to follow unexpected note events during a real performance.

CRFs and SCRFs were first applied to the audio-to-score alignment problem by Joder [13, 12]. They allow more flexible feature design than conventional methods such as HMMs. In particular, SCRFs can be incorporated into both frame-level and segment-level features. In contrast to Joderfs study, we model time-varying tempo as a continuous process rather than as a discrete process, as described in Sec. 2.

Let $\mathbf{o} = \{o_t\}_t$ be the observation sequence extracted from the input audio signal, where t is the frame index, and let $\mathbf{q} = \{q_n\}_n$ be the segmentation of $\mathbf{o}$, where n is the segment index. The segment $q_n = (t_n^s, t_n^e, s_n)$ consists of the start frame t_n^s , the end frame t_n^e, and the chord label s_n. The segmentation problem is formulated as

$$\hat{\mathbf{q}} = \arg\max_{\mathbf{q} \in \mathcal{Q}} p(\mathbf{q}|\mathbf{o}), \tag{1}$$

where $\mathcal{Q}$ is the set of possible segmentations. The conditional probability of a given observation sequence is defined as

$$p(\mathbf{q}|\mathbf{o}) = \frac{1}{Z(\mathbf{o})} \Psi(q_1) \prod_{n=2}^{N} \Psi(q_{n-1}, q_n) \prod_{n=1}^{N} \Phi(q_n, \mathbf{o}), \tag{2}$$

where N is the number of segments, $Z(\mathbf{o})$ is a normalization factor,$\Psi(q_{n-1}, q_n)$ are the transition functions, and $\Phi(q_n, \mathbf{o})$ are the observation functions. N is a random variable. The most likely segmentation can be calculated using the Viterbi algorithm in the off-line setting thanks to the segment-level Markovian assumption.

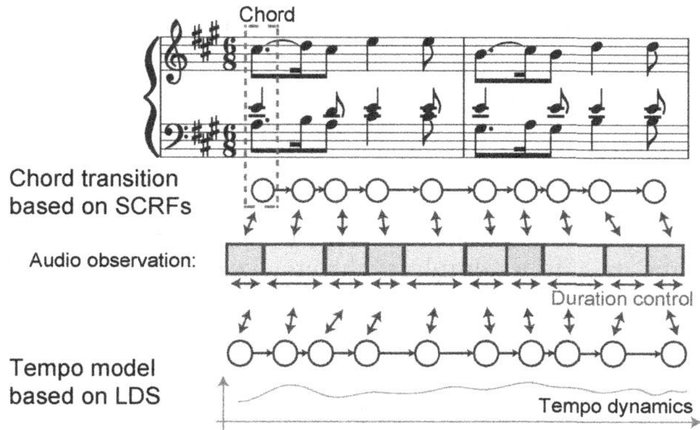

Fig. 2. Audio-to-score alignment as segmentation into a chord sequence from the feature vectors that are extracted by the input audio signal

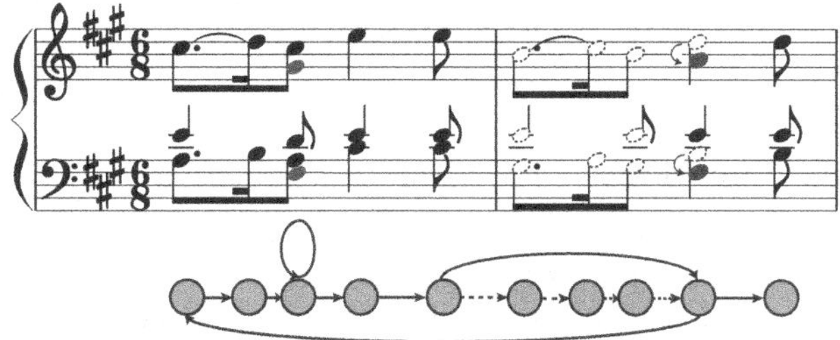

Fig. 3. Transition topology capable of real performances with errors, repeats and jumps

1.2 Observation Functions

For deriving the observation functions, which represent the relation between observations and chords, we used two acoustic features: chroma and onset features. The chroma feature is calculated on the basis of a constant-Q transform to utilize the pitch content of musical performance. The onset feature is also calculated on the basis of spectral flux to consider the burst-of-note onset. An extensive study on acoustic features in score alignment can be found in [14], which is followed in this work.

Let $\boldsymbol{o_t} = \{\boldsymbol{v_t}, f_t\}$ be the observation extracted from the audio signal, where $\boldsymbol{v_t}$ is the chroma vector and f_t is the result of the spectral-flux based onset detection method proposed in [15]. We assume that the observation function can be decomposed as

$$\Phi(q_n, \mathbf{o}) = \phi_c(q_n, \boldsymbol{v}_{t_n^s:t_n^e})\phi_a(q_n, f_{t_n^s:t_n^e}). \tag{3}$$

a) *Chroma feature*: For each segment, a chroma feature is calculated using eq. 4 given below:

$$\phi_c(q_n, \boldsymbol{v}_{t_n^s:t_n^e}) = \exp\Big\{-\lambda^c \sum_{t=t_n^s}^{t_n^e} D^{\mathrm{KL}}(\boldsymbol{v}_t || \boldsymbol{u}_{s_n})\Big\}, \tag{4}$$

where λ^c is a weighting parameter, $D^{\mathrm{KL}}(\cdot||\cdot)$ is the Kullback-Leibler (KL) divergence, and $\boldsymbol{u}_{s_n}$ is a template chroma vector built from the score for each chord in the same manner as described in [16].

b) *Onset feature*: An onset feature is defined as eq. 5 as follows:

$$\phi_a(q_n, f_{t_n^s:t_n^e}) = \exp\Big\{\sum_{g=0}^{1} \lambda_g^a \delta_{\{f_{t_n^s}, g\}} + \sum_h \mu_h^a \delta_{\{m,h\}}\Big\}, \tag{5}$$

where g and h are the indexes of the number of onsets, λ_g^a, μ_h^a are the weighting parameters, m is the number of onsets detected in the segment, and $\delta_{\{\cdot,\cdot\}}$ is Kronecker's delta function. Because of the binary representation of onset, we can take account of intuitive features such as whether the top of a segment is an onset, and the number of onsets in a segment can be detected.

1.3 Transition Functions

In our model, the duration and transition probabilities of hidden semi-Markov models (HSMMs) are incorporated in the transition functions. Let $d_n = t_n^e - t_n^s$ be the segment duration (s), r_n be the local tempo (s / beat), that is assumed to be constant in the segment, and let l_n be the chord length (beat) denoted in the score. The transition function is written as

$$\Psi(q_{n-1}, q_n) = \mathcal{N}(d_n; r_n l_n, \sigma^2) p_{s_{n-1}, s_n}, \tag{6}$$

where $\mathcal{N}$ represents a Gaussian distribution function with a mean of the expected duration $r_n l_n$ and variance of σ^2, and p_{s_{n-1}, s_n} are the HSMM transition probabilities. Note that the tempo is assumed to be constant here for allowing the inference using the Viterbi algorithm. However, the tempo is estimated adaptively during the alignment process, and it controls the transition function dynamically.

2 Linear Dynamical System for Tempo Fluctuation

To anticipate the future score position, we introduce a simple tempo model based on LDS, which is similar to an existing tempo model [9]. While the tempo can fluctuate during a human performance, in general, it does not change considerably over a short period of time. Here, we assume that tempo is constant locally. Thus, the tempo model is defined as

$$r_n = r_{n-1} + w_n, \tag{7}$$

$$d_n = r_n l_n + v_n, \tag{8}$$

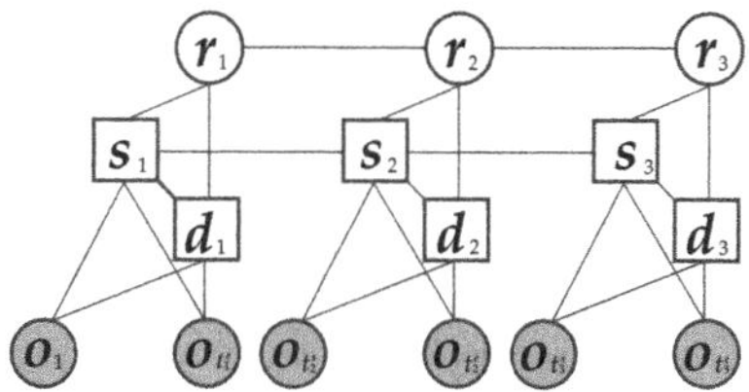

Fig. 4. Graphical representation of the SCRF and tempo model

where

$$w_n \sim \mathcal{N}(0, Q), v_n \sim \mathcal{N}(0, R), \quad (9)$$

for variance parameters Q and R. Eq. (7) and eq. (8) represent local tempo fluctuation and the observation process at inter-onset-intervals (IOI), respectively. The simple linear model allows an efficient real-time decoding using a Kalman filter, which consists of prediction and correction steps, given the result of chord segmentation, (the coupled sequence of IOI and the chord length).

Our model is a combined model of SCRFs (without a tempo variable) and a tempo model based on LDS. We assume that the chord and tempo variables have Markov chains and the duration variables among segments are mutually independent. The graphical representation of the model is shown in Fig. 4.

3 Decoding Algorithm

3.1 On-line Approximation

Time-varying tempo can be estimated using a Kalman filter in an on-line manner. However, in our approach using SCRFs, the most likely segmentation can be calculated using the Viterbi algorithm given the entire input audio. In score following, the input audio signal is fed sequentially; therefore, we need to detect the score position in a real-time using some on-line approximations. In [5, 4, 3], a greedy approximation is applied to determine the most probable current score position. However, there is a risk of estimation errors, particularly in polyphonic cases. To avoid this problem, we use a delayed-decision Viterbi algorithm for an on-line approximation that finds the most probable α-frame past score position. Fig. 5 shows an outline of the delayed-decision Viterbi algorithm. By utilization the predicted future information, highly reliable score position can be estimated. Although the idea is similar to [11], the novelty of our method is the adoption of a backward strategy in the Viterbi algorithm and future anticipation using our explicit tempo model.

3.2 Delayed Decision and Future Anticipation

We now describe our future anticipation method. Let $\{\hat{s}_1, \ldots, \hat{s}_{t-\alpha}, \ldots, \hat{s}_t\}$ be the result of chord segmentation at frame t, $\{\hat{r}_1^{-1}, \ldots, \hat{r}_{t-\alpha}^{-1}, \ldots, \hat{r}_t^{-1}\}$ be the

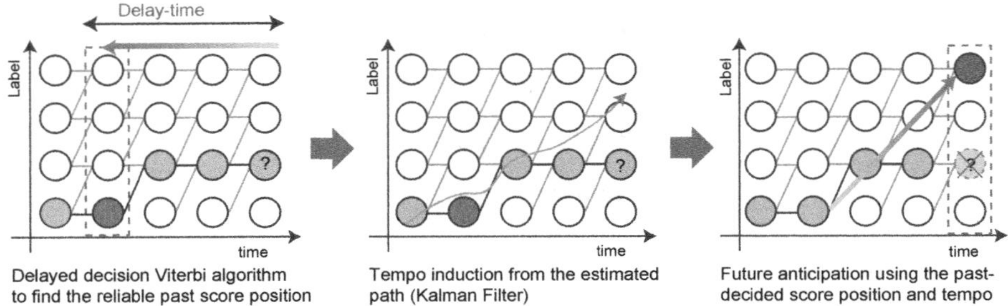

Fig. 5. Outline of the delayed-decision Viterbi algorithm

reciprocal of the tempo estimation (beats / s), and $\{b_1, \ldots, b_{t-\alpha}, \ldots, b_t\}$ be the sequence of score positions (beats) corresponding to the estimated chord sequence. The current or future score position is anticipated using eq. 10 as follows:

$$\hat{b}_t = b_{t-\alpha} + \int_{t-\alpha}^{t} \hat{r}_\tau^{-1} d\tau. \tag{10}$$

Here, if we assume that local tempo is constant from $t - \alpha$ to t, the eq. 10 can be approximated as eq. 11

$$\hat{b}_t = b_{t-\alpha} + \hat{r}_{t-\alpha}^{-1} \alpha. \tag{11}$$

The delay time α is an important parameter in our decoding algorithm. However, the determination of this parameter is not straightforward. We test various values in our experiments. Our real-time score following algorithm based on this delayed decision and anticipation framework is summarized below. The process is repeated for each frame.

Step 1: Chord segmentation using a delayed-decision Viterbi algorithm for the input observation sequence

Step 2: Tempo estimation using a Kalman filter on the chord segmentation result

Step 3: Future score position anticipation using the results of Step 1 and Step 2.

4 Experiments

4.1 Datasets

We evaluated the robustness of our delayed-decision and anticipation algorithm for polyphonic music signals using two datasets. The first dataset contained 60 classical pieces (approx. 4 hours) with perfectly aligned MIDI data from the MAPS database [17]. These recordings were real data played on a YAMAHA

Table 1. Model parameters for SCRF and LDS

SCRF	Chroma feature weight	$\lambda^c = 0.1$
	Onset feature weight	$\lambda_0^a = -0.3, \lambda_1^a = -0.01, \mu_0^a = -0.3, \mu_1^a = -0.01, \mu_h^a = -0.15h\ (h \geq 2)$
	Transition probabilities	$p_{i,j} = 1$ only if $j = i + 1$ otherwise 0
	Duration variance	$\sigma^2 = 0.18\ (\mathrm{s}^2)$
LDS	Tempo variance	$Q = 0.08\ (\mathrm{s}^2/\mathrm{beat}^2)$
	IOI variance	$R = 0.3\ (\mathrm{s}^2)$

Disklavier piano (an acoustic piano equipped with MIDI input and output interfaces), which do not contain any tempo changes but are highly polyphonic. The second dataset contained 50 pieces (approx. 3 hours) from the RWC Jazz database [18]. In contrast to MAPS database, these recordings contained many tempo changes. In almost all the recordings, nearly 15 kind of instruments (piano, guitar, etc.), including percussion, were used. The ground truth is given by manually aligned MIDI files.

4.2 Experimental Settings

For practical reasons, we only used these MIDI files as the manual annotations might have a slight gap compared to the ground truth. To evaluate our algorithm correctly, we prepare perfectly aligned recordings by synthesizing these MIDI files with a YAMAHA XG WDM SoftSynthesizer, retaining all tempo changes. All recordings are re-sampled to 44.1 kHz and 16 bit monaural data and analyzed with a 10 ms hop-size.

The algorithm was evaluated using three statistical measures: the precision, recall, and F-measure of the note onset recognition (same as in [15]). In these experiments, we reported onsets if $\hat{b}_t$ reaches theoretical onset positions in the score. The onsets detected within a tolerance threshold corresponding to the reference onset time were accepted. The error tolerance was variously set to 100, 300 and 500 ms. These parameter, listed in Table 1, were tuned by a grid search.

4.3 Results and Discussion

Table 2 shows the results of onset detection for two datasets with various delay times. In terms of the F-measure, the small delay time of 0.5 s resulted in the highest result for the 100 ms tolerance threshold, showing an increase of over 30% in both datasets compared to no delay time. With a tolerance of 300 ms, which is based on the Real-time audio-to-score alignment task in the Music Information Retrieval Evaluation eXchange (MIREX) contest [19], the results showed an improvement of 11% and 19% in F-measure for the MAPS and RWC Jazz datasets, respectively. The smaller tolerance threshold, the greater is the

Table 2. Onset detection results (%) for MAPS dataset (top) and RWC (bottom)

dataset	measure	Precision				Recall				*F*-measure			
	α	0.0	0.5 s	1.0 s	1.5 s	0.0	0.5 s	1.0 s	1.5 s	0.0	0.5 s	1.0 s	1.5 s
MAPS	100 ms	51.6	**78.9**	73.1	65.6	42.9	**78.7**	73.3	65.9	46.9	**78.8**	73.2	65.8
	300 ms	89.0	91.8	**92.7**	92.3	73.1	91.5	**93.0**	92.7	80.3	91.7	**92.8**	92.5
	500 ms	94.6	93.9	95.0	**95.3**	77.5	93.5	95.2	**95.6**	85.2	93.7	95.1	**95.4**
RWC	100 ms	37.2	**60.4**	54.7	47.6	25.4	**60.0**	54.8	48.1	30.2	**60.2**	54.7	47.8
	300 ms	74.7	79.1	**80.3**	79.1	49.4	78.2	**80.3**	80.0	59.5	78.7	**80.3**	79.6
	500 ms	85.0	84.0	85.8	**85.9**	55.4	83.1	85.9	**86.9**	67.0	83.5	85.8	**86.4**

delay time to obtain the results, which indicates that the delay time should be set according to the requirement of its application.

In case of large delay time (greater than 1.0 s), it caused the results to worsen in the small tolerance of 100 ms. This situation arises from the trade-off between the delayed decision and the future anticipation accuracy. We might expect that the large delay time would enable higher accuracy in results, because of the availability of more future information about the input signal. However, the large delay time may cause anticipation errors. There are two reasons for this: the effect of tempo estimation errors, and the assumption that the tempo within the delay time is the same as the current tempo. The tempo estimation results are sometimes not reliable in these experiments. The accuracy is about 60% with 4% tolerance in both datasets. Even if there are slight errors in the estimated tempo, the anticipation errors would increase in proportion to the delay time. However, it is worth mentioning that the results with small delay times resulted in accurate results.

In the RWC Jazz database, our method obtained less accurate results than with the MAPS database. This is because the RWC recordings have more complexity than those in the MAPS, as mentioned in Sec. 4.1. For both databases, the recall tends to be lower than the precision with no delay time, particularly so for RWC Jazz. This can be explained by the fact that highly polyphonic music signals sometimes cause instabilities in the algorithm. However, the recall is particularly improved using our delayed-decision and anticipation algorithm. These results establish the high robustness of our method for highly polyphonic music signals.

5 Implementation of Automatic Accompaniment System (Ryry)

We developed an application named Ryry for facilitating audio-based automatic accompaniment playback. Ryry can follow real performance with errors including repeats and jumps by using our robust score-following algorithm described above. Ryry generates the MIDI signal of the accompaniment part synchronized to the live performers. The sound of the accompaniment part is produced through the

MIDI sequencer. Ryry was written in C++ on the cross-platform application Qt. Ryry runs equally well on Windows, Mac OS X and Linux operating systems. The requirements for Ryry are MIDI file to be performed, a microphone and a speaker, and a software or hardware MIDI sound module for the accompaniment part playback. Ryry can operate with any music pieces if it has been described in the SMF (standard MIDI files) format 1 (multiple tracks). The user can choose the part to play from multiple tracks in MIDI data. Because Ryry can follow any jump of performance position, the user is able to start playing from the free position such as the beginning or middle of the music.

We confirmed that Ryry does not require the use of a specific type of microphone. To avoid any error of score alignment caused by environmental noises, the use of contact microphone[1] would be an effective solution in practice. The design of feature extraction and spectral template was not tuned for a particular instrument. We confirmed that Ryry accepts violin, piano, guitar and human singing voice without any problems.

At Interaction 2013 held in Japan, we successfully demonstrated Ryry. The demonstration of automatic accompaniment to follow the actual playing of the violin was performed in a noisy environment. We used a laptop computer an Intel Core i7 (1.7 GHz) dual-core CPU, 4 GB RAM that ran Mac OS X. In order to record input audio signal, one directional dynamic microphone (SHURE BETA 57A) was connected to a USB audio interface. See our demonstration video clips at `http://www.youtube.com/channel/UCVTHspYues91QWgu2cGvlkA`.

6 Conclusion

In this work, we devised a robust on-line score following algorithm for polyphonic music signals based on a delayed-decision and anticipation framework. We also implemented a real-time automatic accompaniment playback system. The key feature of the study is the delayed-decision Viterbi algorithm, which determines highly reliable past positions utilizing future signal information, and that the future position can be anticipated using an adaptively estimated local tempo thanks to our explicit tempo model.

Result of experiments performed on polyphonic music databases showed significant improvements in alignment accuracy, even for highly polyphonic cases with tempo changes. It is worth mentioning that the delayed-decision and anticipation framework can be used in existing dynamic programming-based score followers with an explicit tempo model. In future work, we intend to determine the delay time adaptively during a musical performance by considering the tradeoff between the delayed decision and anticipation accuracy. It is also important to train the model from the real-data.

Acknowledgments. This research was supported in part by the JST Adaptable and Seamless Technology Transfer Program through Target-driven R&D

[1] A contact microphone, otherwise known as a pickup or a piezo, is a form of microphone designed to sense vibrations through instruments.

and the JSPS KAKENHI (Grant-in-Aid for Scientific Research) Grant Number 26730182, and the Telecommunications Advancement Foundation (TAF).

References

1. Dannenberg, R.B., Raphael, C.: Music score alignment and computer accompaniment. Communications of the ACM 49(8), 38–43 (2006)
2. Cont, A.: Antescofo: Anticipatory synchronization and control of interactive parameters in computer music. In: Proc. of International Computer Music Conference (ICMC) (2008)
3. Cano, P., Loscos, A., Bonada, J.: Score-performance matching using hmms. In: Proc. of International Computer Music Conference (ICMC), pp. 441–444 (1999)
4. Orio, N., Dechelle, F.: Score following using spectral analysis and hidden markov models. In: Proc. of International Computer Music Conference (ICMC), pp. 151–154 (2001)
5. Dixon, S.: Live tracking of musical performances using on–line time warping. In: Proc. of International Conference on Digital Audio Effects (DAFx), pp. 92–97 (2005)
6. Otsuka, T., Nakadai, K., Takahashi, T., Ogata, T., Okuno, H.G.: Real-time audio-to-score alignment using particle filter for coplayer music robots. EURASIP Journal on Advances in Signal Processing, 1–13 (2011)
7. Duan, Z., Pardo, B.: A state space model for online polyphonic audio-score alignment. In: Proc. of International Conference on Acoustics, Speech and Signal Processing (ICASSP), pp. 197–200 (2011)
8. Montecchio, N., Cont, A.: A unified approach to real time audio-to-score and audio-to-audio alignment using sequential montecarlo inference techniques. In: Proc. of International Conference on Acoustics, Speech and Signal Processing (ICASSP), pp. 193–196 (2011)
9. Raphael, C.: Aligning music audio with symbolic scores using a hybrid graphical model. Machine Learning Journal 65(2-3), 389–409 (2006)
10. Cont, A.: A coupled duration-focused architecture for realtime music to score alignment. IEEE Transactions on Pattern Analysis and Machine Intelligence 32(6), 974–987 (2010)
11. Arzt, A., Widmer, G., Dixon, S.: Automatic page turning for musicians via real-time machine listening. In: Proc. of European Conference on Artificial Intelligence (ECAI), pp. 241–245 (2008)
12. Joder, C., Essid, S., Richard, G.: A conditional random field framework for robust and scalable audio-to-score matching. IEEE Transactions on Audio, Speech and Language Processing 19(8), 2385–2397 (2011)
13. Joder, C., Essid, S., Richard, G.: A conditional random field viewpoint of symbolic audio-to-score matching. In: Proc. of ACM Multimedia, pp. 871–874 (2010)
14. Joder, C., Essid, S., Richard, G.: A comparative study of tonal acoustic features for a symbolic level music-to-score alignment. In: Proc. of International Conference on Acoustics, Speech and Signal Processing (ICASSP), pp. 409–412 (2010)
15. Dixon, S.: Onset detection revisited. In: Proc. of International Conference on Digital Audio Effects (DAFx), pp. 133–137 (2006)
16. Hu, N., Dannenberg, R.B., Tzanetakis, G.: Polyphonic audio matching and alignment for music retrieval. In: Proc. of IEEE Workshop on Applications of Signal Processing to Audio and Acoustics (WASPAA), pp. 185–188 (2003)

17. Emiya, V., Badeau, R., David, B.: Multipitch estimation of piano sounds using a new probabilistic spectral smoothness principle. IEEE Transactions on Audio, Speech and Language Processing 18(6), 1643–1654 (2010)
18. Goto, M., Hashiguchi, H., Nishimura, T., Oka, R.: RWC music database: Popular, classical, and jazz music databases. In: Proc. of International Conference on Music Information Retrieval (ISMIR), pp. 287–288 (2002)
19. (IMIRSEL), T.I.M.I.R.S.E.L.: Real-time audio to score alignment (a.k.a score following), `http://www.music-ir.org/mirex/wiki/Real-time_Audio_to_Score_Alignment_(a.k.a_Score_Following)` (accessed May 27, 2014)

Towards Modular, Notification-Centric and Ambient Mobile Communication

User Study Supporting a New Interaction Model for Mobile Computing

Jonas Elslander and Katsumi Tanaka

Department of Social Informatics
Graduate School of Informatics, Kyoto University
Kyoto, Japan
{jonas,tanaka}@dl.kuis.kyoto-u.ac.jp

Abstract. In this paper, we synthesize the results of a qualitative user study supporting our proposed modular, notification-centric and ambient interaction model for mobile computing. Here, a context-dependent and extended implementation of notifications is introduced. For a range of personalized situations, individually tailored conceptual interfaces were compared against a baseline conform present-day mobile operating systems and the desired usage of suggested solutions as well as the perceived shortcomings of current offerings were investigated. We conclude and demonstrate the user preference for communication patterns introduced by our interaction model for mobile computing.

Keywords: HCI, interface, interaction, user study, model, context-awareness, context-dependency, notification, ambient, push, pull, modularity, visualization, design.

1 Introduction

Since the introduction of the smartphone, we have gained easy mobile access to vast amounts of information on the Internet. Adoption rates of smart devices are going strong and we're at the advent of the next information access boom, wearable computing. We can divide human-computer information transfer into two categories: user-initiated and system-initiated. In the present mobile computing environment, the former is often linked to pull-style interaction while the latter to push-style interaction. In reality however the overlap between both classifications isn't necessarily exact, so we base ourselves on the communication initiator and determine the preferable interaction later. The scope of this paper focuses on the system-initiated communication, in particular mobile notifications. We define all system-initiated information that contains data valuable to the user as a notification. Nowadays, these mainly take the form of interruptive push notifications that serve as a direct, unfiltered and context-unaware communication that looks and behaves in a similar way regardless of its relevance or inherent social importance. In a preceding user study [1], we have shown that this behavior is indeed not desirable. Applications and services are getting ever

D. Ślęzak et al. (Eds.): AMT 2014, LNCS 8610, pp. 146–159, 2014.

more capable at retrieving information that relates to our mobile lives, but the method of notification thereof is dated. A virtual layer that interrupts a mobile device user's life whenever it sees fit is doubtless not the optimal solution.

Subsequently, we have devised and proposed an adapted interaction model for mobile system-initiated communication [2]. With ever more data is becoming user accessible, there is the clear need for a structure that supports a new interaction framework and interface environment facilitating system-initiated communication in a manner more advantageous than the current approach. This publication expounds on an extensive qualitative user study that probes the usage and perception of mobile computing environments created based upon this interaction model.

The remainder of this paper is structured as follows. Section 2 explicates the proposed modular, notification-centric and ambient (MNA) mobile interaction model. In section 3, the survey methods and objective of the conducted user study are detailed. In section 4, our analyzed findings are highlighted and expounded. Section 5 synthesizes the conclusions in relation to the MNA model and future work is proposed.

2 MNA Model

2.1 Model Layers

In order to ameliorate the performance and user experience of system-initiated communication, we constructed a layered model representing the information flow for present-day mobile operating systems. Represented by Fig. 1, this consists of four layers: data, service, interface and interaction. The data layer consists of databases, both locally and through remote access, which are being consulted by mobile applications in the service layer in order for them to provide relevant information to the user. On a system level both springboard (the grid-like structure of icons that provides the device user access to all installed applications) and notification center form the starting point to retrieve (pull) and display (push) this information.

Fig. 2 shows our proposed model in an analogue fashion. The MNA model adds an extra stream layer in between interface and interaction layers. As detailed further in this paper, modularity of notifications can be achieved this way. We also suggest a change in relative value of the interface layer entities by placing the notification central as an aggregation point for a whole new set of information communicated by applications. Lastly, we introduce a new interaction paradigm labeled ambient interaction.

In the following section we elucidate the proposed model's main advances. The theory is subsequently supported by the user study described in the following sections.

2.2 M for Modular

Novel to our proposed interaction model for the mobile computing environment is the inclusion of a notification stream, an intermediary layer that handles the chronological data flow between the applications in the service layer and the notification center

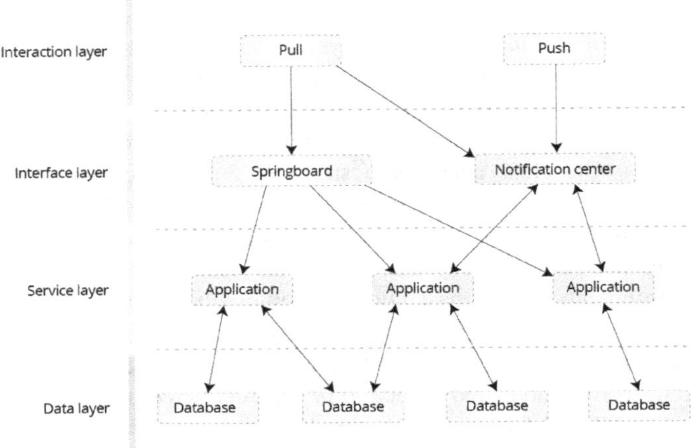

Fig. 1. Current mobile interaction model

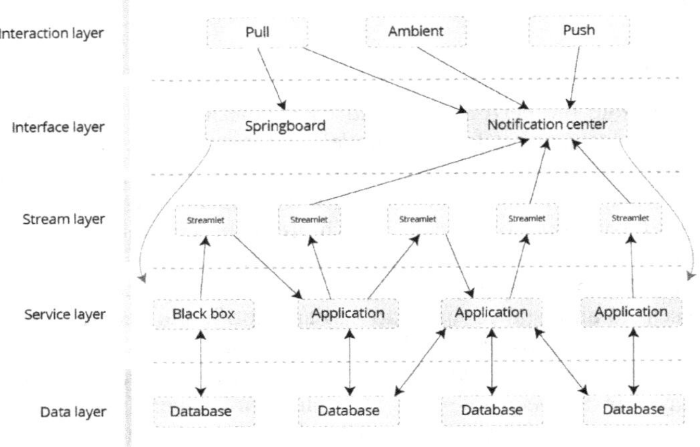

Fig. 2. Proposed mobile interaction model

in the interface layer. Preceding research conclusions have shown that the users of current-generation mobile operating systems perceive smartphone notifications as too many and of low importance/relevance. The root of the problems surrounding the current notification-handling framework can be traced back to two concepts inherent to its present design: same weight push notifications and a direct connection between applications and the device's user. Each individual application developer has ultimate control over how many notifications are being delivered to the user, while the user can only regulate which applications he accepts to get notified by. It is assumed that all notifications of the same application hold a similar value, regardless of intrinsic attributes such as the user's context or the social connections it refers to. Developers have no means to assign different notification weights. As previous work [1] has shown, often the notification centers of the present-day mobile operating systems do not suffice in this regard.

Additionally, all notification-based communication is outlined by a direct connection between user and application. There is no system-level solution for applications to automatically communicate with each other or to provide information for future delivery to the user. This direct connection prohibits notifications from being selected and delivered based on the user's context rather than application-specific triggers, which are limited by the scope of the application and thus the resources of its developer. In view of these shortcomings, we propose an intermediary stream layer in mobile operating system that serves as a structured timeline to which all notifications – called streamlets – can be written and can be subsequently read by all the system services and applications on every connected device. The notification center interfaces of all these devices can display information pulled from this stream in accordance to their specific usage pattern, the user's context and the streamlet's importance.

Both system and user could also mine this stream for search or analytical purposes, leading to better system personalization and improved insight into the cloud of data surrounding a user. Currently, mobile inter-device search is very limited in extent as the entities in the data layer are well guarded off from the system indexation by the applications as they control access to the filtered information. When allowed to output stored notifications (thus without direct transmission to the user), these can become valuable resources for search and analysis.

> The creation of an ***intermediary stream layer*** enables the storage and communication of notifications with a ***variable importance*** and ***without an immediate delivery*** to the user.

2.3 N for Notification-Centric

In contrast to the desktop computing environment that is characterized by a document-centric approach, mobile computing is application-centric. Programs control the access to information serving a specific purpose and perform their tasks independent of each other. For example, a weather application will update user on the forecast while a GPS app might give her or him the fastest route to a given destination. However, getting navigated to the closest by sunny destination would require a lot of user interaction and cognitive effort, which could be avoided if both applications involved could communicate and cooperate. We strive to turn notifications from interruptions into recommendations. Enabling the generation of this kind of insightful notifications can be accomplished through our modular approach.

We propose a notification-centric system interface approach built on top of the application-centric framework. Currently, system-initiated information can be viewed in either the application-specific interfaces or in the operating system's notification center. The latter, consisting of a list of time stamped snippets linking to their respective application interfaces, however often simply serves as a shortcut to access applications that actually contain the full information. We advocate two changes: a *redistribution* of the interaction balance between the application-specific and notification center interfaces in favor of the latter, in combination with a redesign of its *visual embodiment*. We do this by leveraging the stream layer in combination with a shift from textual towards photographic communication. Our eyes are designed to recognize familiar faces, situations, concepts and emotions in a split second. It is thus advisable to represent notifications by images familiar to the user and let him intuitively browse through the information on offer whenever feasible. A higher information density is also preferable in time-constrained situations where the attention to a mobile device should be kept at a minimum (e.g. at work or while driving) or when the screen surface is small (e.g. wearable computing).

> We accomplish a notification-centric mobile operating system interface by changing the objective of notifications from interruptive shortcuts to **recommendations** and by presenting them in a **photographic manner**.

2.4 A for Ambient

We define pull-style interaction as interaction that requires both the user's intent of accessing predefined information, as well as a set of physical actions towards achieving this goal (e.g. looking up today's recommended recipe by navigating through a cooking application prior to starting dinner preparations). Push-style interaction on the other hand is not characterized by user intent and does not have to go paired with a user's physical actions in order to obtain the information (e.g. being audibly notified of a missed text message). Both interaction styles hold certain drawbacks: push-style is interruptive – a problem whose impact can only be partly solved by making use of a context-aware delivery model – while pull-style implies preceding insight. A user either needs to know to certain extent what he is looking for or he needs to perform a lot of manual probing in order to retrieve desired information. To naturally discover contextually relevant information in a non-interruptive way, we define *ambient-style* interaction. This encompasses the intent of looking for upfront unspecified but relevant information, without the component of performing physical action in order to obtain it. In present-day mobile operating systems, the lock screen is the best example. Here, push notifications and data from a few widgets (e.g. weather, stock information, news headlines…) get aggregated. It is key that we use a similar interaction style in delivering modular notification-centric information to mobile users.

> We propose ambient-style interaction, characterized by ***intent of retrieving information*** with the ***lack of physical actions*** required hereto, as a natural way of delivering notifications.

3 Survey Methods

In order to obtain insight into the desirable ways of implementing the proposed changes, we designed two online surveys that were completed by a group of 20 respondents, randomly selected from a group of acquainted smartphone owners. The questionnaires were administered online and consisted of questions probing the personal perception and expectations regarding the problems and improvements indicated in the previous section of this paper. The results were analyzed using statistical software. Given the number of participants, the scope of the results is of a qualitative and preliminary order, which the questionnaires were drafted for and the results were weighted against.

The first survey consisted of 8 questions inquiring the personal life of the respondents. These were used to create a distinct profile for every respondent. Based upon this document, the second and 24 question counting survey has been designed on an individual basis for every single participant. In order to compare the proposed interaction model against the current mobile computing user experience, we simulated context and social awareness through the creation of life-like personal situations. By means of the first questionnaire, we informed ourselves about the most used and most important mobile applications, further referred to favorite applications, as well as the relation and names of the respondents' friends, including five persons defined as most important to each respondent. We next acquainted ourselves with the usage and workings of all applications and built a profile for all social connections, using social media websites as an information source. We also solicited the participants' favorite hobbies and a daily job descriptions including duration, usual tasks, location and means of transportation. We used this data to construct highly personalized situations in light of which we created mock-up interfaces and tailored questions. These form a life-like testing environment, as a qualitative and time-efficient alternative to developing a customized operating system that would have to be implemented in the daily life of the survey participants in order to yield similar results.

As to compare the individual respondents' answers, we followed two predefined conceptual structures. For analyzing personal situations, we invariably presented them several of the following situations. One situation was characterized by a high main task intensity and low freedom of task choice. This is often common in a work setting. Another was defined by lower main task intensity and an intermediate freedom of task choice (i.e. pastime setting). The third one was with low or no main task intensity and ample task choice (i.e. free time setting). We call the above-described situations work time, pastime and free time, respectively. For comparing notifications, we defined three categories: those communicating information originating from a single application (as is the case with all present-day mobile notifications), those combining the information of multiple applications into a single notification and those who link application-based information with one or more social connections. The latter two categories are inspired by the notion of modularity, as proposed in the MNA model. In this survey, notification-centricity and ambient interaction were also studied and compared to their respective current alternatives.

4 Analysis

4.1 Supporting Modularity

In order to validate the need for a framework that allows applications to send out notifications that will only be communicated to the user given the right conditions, we combined three individually designed situations (one describing a work related, one a pastime related and one a free time related context) with the respondents' 15 favorite applications. We asked respondents which applications they prefer to receive notifications from, given the current push-style approach and typical communicated contents, in these situations and studied the discrepancies. As we can see in Table 1, respondents prefer to receive push notifications from only one out of three favorite applications in work time or pastime situations. Only when being free of tasks, on average six out of ten applications are favored to engage in system-initiated communication. This while the information contained by all favorite applications is generally perceived as important. We also computed the situational difference: the probability that an application is preferred (not) to push information to a user in both of two given situations. About half (between 42% and 53%) of the preference sentiments towards push notifications prove to change with the situation. When we look at the summed difference – the number of applications that is preferred (not) to push information in any situation – we conclude that only one in four favorite applications is characterized by situation-independent system-initiated communication.

Table 1. Push notification preference

	Situations		
	Work time	***Pastime***	***Free time***
Preference	34%	33%	59%
Situational difference	53%		
		52%	
	42%		42%
Difference sum	25%		

Except for one survey participant, all respondents prefer to get push-notified by less than half of their favorite applications in a work or pastime related situation, with respectively 70% and 60% stating this being true for less than one in three favorite applications. Not a single participant wishes to get informed by half or more of his favorite applications in all situations. Combined, the results of notification preference and situational difference data underscore the premise of system-initiated communication being context-dependent and in need of being transmitted to the user on a situational basis. When reviewing all applications defined by the survey respondents as their favorites, we conclude four prevalent application categories: communication (e.g. mail, chat and voice call), media (e.g. photo, video and audio access), navigation (e.g. mapping and geoposition based applications) and time management (e.g. calendars, reminders and alarms). With respectively 34%, 21%, 16% and 11%, they make up for a combined 82% of all favorite applications.

Next, we probed the survey participants about their interest in search and analytics based on past notifications, both acknowledged and disregarded. The proposed aggregation of notification data from multiple sources, as described by the MNA model, was not disclosed and all answers are based on the respondents' perception of current (and thus directly communicated) notifications from their personal lives. When querying analytics informative to the user, we examined their preference for reports based on the notification data of a single application as well as those based on the correlation between notification data originating from multiple applications. We found that on average, users are interested in single application analytics for 30% of their top 15 favorite applications. Half (51%) of those are applications in the communication category. Media, navigation and time management applications are far less likely candidates, respectively seizing only 19%, 14% and 5% of the total demand for reports based on single application notification data. One in five respondents is not interested in single application analytics at all, while all survey participants are keen on reports detailing aggregated notification data. All except for one respondent created the maximum of three non-compulsory lists of applications they would like to see notifications from combined in an analytical report. On average, navigation and communication applications score best, being included in 87% and 80% of any given report based on multiple application notification data, while media and time management applications are both present in 40% of those reports. In half (50%) of all analytical reports based on multiple application notification data, respondents wish to correlate the data obtained from applications with their own social graph.

We also suggested the respondents to come up with multiple open questions they would like their smartphone to be able to answer. The results – averaging 1,6 questions per respondent per situation – were cross-referenced with their individual top 15 favorite applications clustered by application category. We did this in order to discover the most contributing categories in providing answers based on past notifications. Table 2 shows the probability of applications belonging to a specific category in supplying information to the notification center that supports the user in his information need given a certain context. During working situations, the probability of any given question being at least partially solvable with information contained by system-initiated communication originating from applications within the time management category is as high as 66% (roughly three times higher than in pastime or free time situations). In pastime situations, the combined applications in the navigation (51%) and media (40%) categories contribute answer data to as much as nine out of ten registered questions. Regarding free time situations, the probability of partial question solvability by information contained by notifications originating from applications belonging to the media category reaches 74%. This is almost twice as high as in the pastime context and three times higher than during work situations. The combined partial solvability gives us an insight into the complexity of the recorded questions per situation. Within the context of free time, we note an increase of roughly 25% compared to pastime and work time situations. This points to more questions being partially solvable by system-initiated information originating from multiple application categories.

Table 2. Probability of question solving

	Situations		
	Work time	*Pastime*	*Free time*
Communication	21%	23%	42%
Media	24%	40%	74%
Navigation	32%	51%	43%
Time management	66%	26%	20%
Complexity	1,43	1,40	1,79

As described earlier, the inclusion of a notification stream in the MNA interaction model allows for modular notifications that contain information from either a single application, multiple applications or that link application-based information to the user's social graph. Based on their favorite – and thus highly familiar – applications, we created a participant-specific list of nine notifications placed in random order in a mock-up interface resembling the notification center of current mobile operating systems (removing distracting visuals as soft buttons, timestamps and taskbars), as shown in Fig. 3a. In every mock-up interface, three notifications communicate information originating from a single application, three combine information from multiple applications and three include one or more social connections of the participant. Subsequently, the respondents were asked to closely observe and rank all given notifications in order of their expected importance, imagining a relevant corresponding situation if needed. We then could determine the presence of all aforementioned categories in all top 3 and top 5 results. We learn that notifications receiving information from multiple applications are 14% more likely to appear in the top 3 notifications perceived as most important by the respondents. When analyzing the top 5 results, no significant (more than 10% difference from the mean value) preferences are observed between the three notification categories. It is noteworthy that currently no notifications source information from multiple applications or the user's social graph. Yet, they are well received by the participants of our survey. Whereas information sourced from multiple applications is highly appreciated by users for analytical use, it does not enjoy a preferential treatment over information sourced from a single application for notification purpose.

4.2 Supporting Notification-Centricity

Within the breadth of this survey we also compared the current snippet-based visual embodiment of the notification center (Fig. 3a) as a baseline to an exemplary mock-up of the proposed photographic notification center interface (Fig. 3b). We propose the use of images for displaying recommendation notifications as it corresponds best to our brain's natural information processing capabilities. We filter visual data rich in detail fast on sight. To analyze if this also applies to contents typical to mobile notifications, we studied the information recall for both textual (present-day) and photographic (proposed) interface mock-ups containing similar notifications.

First off, we randomly selected eight applications out of each respondent's 15 favorite applications and incorporated these in two mock-up notification center interfaces, each representing one of the visualization styles. We individually showed both interfaces twice to every respondent: at first for a duration of 5 seconds, subsequently for another 10 seconds. After each showing, we urged them to list everything they could remember about the communicated information. We ranked the results for every displayed notification on a three-point scale: no recall, partial recall of information and full recall of information. For the textual interface, we only accounted for recall of the information contained in the notification title. As shown in Table 3, for every timespan information communicated through a photographic notification center interface scores significantly better at being partially and fully recalled. When comparing the partial recall rate of both interfaces, we see the proposed photographic interface outperforming textual interface with a factor of two: respondents recall twice as much information across all notifications for both timespans. Despite 30% of all notifications being partially recalled after a combined 15 seconds for textual interfaces, a full recall is near to non-existent within this short timeframe. However, when displaying the photographic interface mock-up, one in five notifications can be fully recalled after 5 seconds and two in five notifications after a combined 15 seconds.

Table 3. Information recall rate

	Textual		**Photographic**	
	5 sec.	*10 sec.*	*5 sec.*	*10 sec.*
Partial recall	21%	29%	38%	58%
Full recall	2%	7%	21%	38%

(a)

(b)

Fig. 3. (a) Mock-up of textual notification center.(b) Mock-up of photographic notification center.

When asking the respondents to rate their performance in recalling the most important information, while showing both interfaces to them for an extended time, six out of ten answered positive regarding the textual interface (some citing a better recall for the notifications that were displayed towards the top of the mock-up) while 85% said so concerning the photographic interface. It is our belief that photographic interfaces can reduce task distraction – with far reaching positive consequences as we also use mobile computing in potentially hazardous situations – and make preferred information stand out easier through the display of familiar (and thus arguably more important) concepts.

After becoming acquainted with both interface styles, we probed the respondents about their preference for each in nine individually created situations, three for every situation category (work time, pastime and free time). We also inquired their context-independent preference for both interfaces. The results, displayed in Table 4, show a strong situation dependency of the interface preferences. In general, a textual notification center interface is slightly (75% versus 65%) better received, while the photographic interface is somewhat (20% versus 10%) more undesired than the textual interface. In a work context however, we see a very strong preference for the textual interface, while in free time and – even more so – in pastime situations a photographic interface is both more (two to three times) preferred and less (two times) undesired. Surprisingly, the presently de facto textual notification center interface is opposed by almost half of its users in a context of leisure (i.e. in both pastime and free time situations).

Table 4. Situation-interface dependency

	Textual		**Photographic**	
	-	+	-	+
Work time	0%	100%	50%	35%
Pastime	45%	20%	25%	70%
Free time	40%	45%	20%	80%
General	10%	75%	20%	65%

4.3 Supporting Ambience

In order to determine the optimal interaction with information following a notification, we presented the survey participants six communication methods: three user-to-device and three device-to-user, each group covering tangible (T), visible (V) and audible (A) communication. Tangible and audible user-to-device communication relate to pull-style interaction, through touch interfaces and voice commands respectively. Tangible and audible device-to-user interaction can refer to push-style interaction through vibrations and sound notifications, but in this context we portend it as feedback communication as we focus on interaction following the notification. The respondents were asked to state their preference for all communication methods in three personal situations for every situation category, plus a general context-independent preference. From the quantified results, displayed in Table 5, we remark an aversion towards audible communication in both directions and towards device-to-user physical contact for work situations. Three out of four respondents approve these communication styles in general or in a leisure context. During free time situations, users are significantly keener on interacting using non touch-based gestures (user-to-device visible communication) than during other situations or in general.

Table 5. Interaction (after notification) preference

	User-to-device			Device-to-user		
	T	*V*	*A*	*T*	*V*	*A*
Work time	91%	53%	30%	52%	74%	42%
Pastime	73%	44%	82%	80%	65%	76%
Free time	92%	76%	72%	83%	68%	71%
General	90%	40%	75%	80%	70%	75%

As the MNA interaction model is designed with multiple-device and wearable mobile computing in mind, we also probed the desirability of the aforementioned communication methods for two wearable computing device categories: wrist-worn devices (e.g. smart watches) and face-worn devices (e.g. augmented reality glasses). As shown in Table 6, we see a small drop of 20% (compared to the context-independent interaction with portable mobile devices as shown in Table 5) in preferred usage of user-to-device touch interaction for wrist-worn devices. A more significant 40% drop (same comparison) of device-to-user audible communication can be explained by the troublesome physical action of bringing such device close to the user's ear. Quite different are the results regarding face-worn devices. Here, we profess an aversion towards tangible communication in both directions, in particular user-to-device being welcomed by only one in three users. As much as 85% of the respondents would like to interact with a face-worn device using non touch-based gestures, twice as much as for wrist-worn or portable devices for general usage.

Table 6. Interaction (wearable computing) preference

	User-to-device			Device-to-user		
	T	*V*	*A*	*T*	*V*	*A*
Wrist-worn	70%	45%	75%	90%	65%	30%
Face-worn	30%	85%	80%	50%	90%	65%

Because our proposed interaction model introduces a novel way of communicating notifications through ambient interaction, we asked the respondents to assert their preference for push, pull and ambient interaction in three personal situations for every situation category, as well as in general (context-independent) and for wearable devices. The results are presented in Table 7. Earlier we discovered only a selection of even the most important applications are favored to send us push notifications. Again, we see an overall disliking of push-style interaction, especially when the context is characterized by a to be executed main task (work time and pastime situations). Compared to work and free time situations, all forms of interaction are less preferred during pastimes. Also except in this situation category, ambient interaction is received well by nine out of ten users. Wearable devices yield a significantly different interaction preference for system-initiated communication. Only one in three respondents is fond of pull-style interaction to access notifications on wrist-worn devices, while almost half prefers push notifications on face-worn devices. Ambient interaction is perceived equally well on wearable devices.

Table 7. Interaction (notification) preference

	Interaction style		
	Push	***Pull***	***Ambient***
Work time	22%	98%	91%
Pastime	24%	74%	63%
Free time	37%	82%	92%
General	10%	85%	90%
Wrist-worn	30%	35%	90%
Face-worn	45%	70%	95%

5 Conclusions and Future Work

The modular, notification-centric and ambient interaction model for mobile computing was conceived based on the hypothesis that present-day system-initiated communication is ineptly limited and uniform. The findings of this research support the claim that notification preferences regarding interaction, visualization and information origins differ greatly based on the user's context. We have also shown that notifications originating from different application categories and from one or multiple sources contribute to a different usage perception. We determined a user need for search and analytics based on a modular notification stream as proposed by our model, as well as the preference for a photographic notification interface in the leisure context. This photographic visualization style also yields a significantly higher information recall rate and a better natural highlighting in time-constrained situations. Finally, we support the preference for ambient interaction over push-style interaction for system-initiated communication on portable and wearable devices.

In future work we aim to develop and test a functional notification center prototype based on our proposed interaction model and the conclusions of this survey. Consequently, we can thoroughly study the influence of different design approaches regarding the realization of the notification stream, information visualization and ambient interaction.

Acknowledgements. The authors would like to thank every contributor to the research work discussed in this paper; in specific all survey participants and Prof. Adam Jatowt. This work was supported in part by the following projects: Grants-in-Aid for Scientific Research (Nos. 24240013) from MEXT of Japan.

References

1. Elslander, J., Tanaka, K.: A Notification-Centric Mobile Interaction Survey and Framework. In: Jatowt, A., et al. (eds.) SocInfo 2013. LNCS, vol. 8238, pp. 443–456. Springer, Heidelberg (2013)
2. Elslander, J., Tanaka, K.: Designing an Ambient Interaction Model for Mobile Computing. In: Streitz, N., Markopoulos, P. (eds.) DAPI 2014. LNCS, vol. 8530, pp. 3–14. Springer, Heidelberg (2014)
3. Scott McCrickard, D., Chewar, C.M.: Attuning notification design to user goals and attention costs. Commun. ACM 46(3), 67–72 (2003)
4. Maglio, P.P., Campbell, C.S.: Tradeoffs in displaying peripheral information. In: Proceedings of the SIGCHI Conference on Human Factors in Computing Systems (CHI 2000), pp. 241–248. ACM, New York (2000)
5. Maglio, P.P., Barrett, R., Campbell, C.S., Selker, T.: SUITOR: An attentive information system. In: Proceedings of the 5th International Conference on Intelligent User Interfaces, IUI 2000 (2000)
6. Booker, J.E., Chewar, C.M., McCrickard, D.S.: Usability testing of notification interfaces: Are we focused on the best metrics? In: Proceedings of the 42nd Annual Southeast Regional Conference (ACM-SE 42), pp. 128–133. ACM, New York (2004)
7. Grudin, J.: Partitioning digital worlds: Focal and peripheral awareness in multiple monitor use. In: Proceedings of the SIGCHI Conference on Human Factors in Computing Systems (CHI 2001), pp. 458–465. ACM, New York (2001)
8. Erickson, T., Kellogg, W.A.: Social translucence: An approach to designing systems that support social processes. ACM Trans. Comput.-Hum. Interact. 7(1), 59–83 (2000)
9. Saket, B., Prasojo, C., Huang, Y., Zhao, S.: Designing an effective vibration-based notification interface for mobile phones. In: Proceedings of the 2013 Conference on Computer Supported Cooperative Work (CSCW 2013), pp. 149–1504. ACM, New York (2013)
10. Hazlewood, W.R., Stolterman, E., Connelly, K.: Issues in evaluating ambient displays in the wild: two case studies. In: Proceedings of the SIGCHI Conference on Human Factors in Computing Systems (CHI 2011), pp. 877–886. ACM, New York (2011)
11. Messeter, J., Molenaar, D.: Evaluating ambient displays in the wild: Highlighting social aspects of use in public settings. In: Proceedings of the Designing Interactive Systems Conference (DIS 2012), pp. 478–481. ACM, New York (2012)
12. Mankoff, J., Dey, A.K., Hsieh, G., Kientz, J., Lederer, S., Ames, M.: Heuristic evaluation of ambient displays. In: Proceedings of the SIGCHI Conference on Human Factors in Computing Systems (CHI 2003), pp. 169–176. ACM, New York (2003)
13. Kim, T., Hong, H., Magerko, B.: Design requirements for ambient display that supports sustainable lifestyle. In: Proceedings of the 8th ACM Conference on Designing Interactive Systems (DIS 2010), pp. 103–112. ACM, New York (2010)
14. Ryu, H.-S., Yoon, Y.-J., Lim, M.-E., Park, C.-Y., Park, S.-J., Choi, S.-M.: Picture navigation using an ambient display and implicit interactions. In: Proceedings of the 19th Australasian Conference on Computer-Human Interaction: Entertaining User Interfaces, OZCHI 2007, pp. 223–226. ACM, New York (2007)

Body Posture Recognition as a Discovery Problem: A Semantic-Based Framework*

Michele Ruta[1], Floriano Scioscia[1], Maria di Summa[2], Saverio Ieva[1], Eugenio Di Sciascio[1], and Marco Sacco[3]

[1] Politecnico di Bari, Bari, Italy
{michele.ruta,floriano.scioscia,saverio.ieva,eugenio.disciascio}@poliba.it
[2] Consiglio Nazionale delle Ricerche, Bari, Italy
maria.disumma@itia.cnr.it
[3] Consiglio Nazionale delle Ricerche, Milano, Italy
marco.sacco@itia.cnr.it

Abstract. The automatic detection of human activities requires large computational resources to increase recognition performances and sophisticated capturing devices to produce accurate results. Anyway, often innovative analysis methods applied to data extracted by off-the-shelf detection peripherals can return acceptable outcomes. In this paper a framework is proposed for automated posture recognition, exploiting depth data provided by a commercial tracking device. The detection problem is handled as a semantic-based resource discovery. A simple yet general data model and a corresponding ontology create the needed terminological substratum for an automatic posture annotation via standard Semantic Web languages. Hence, a logic-based matchmaking allows to compare retrieved annotations with standard posture descriptions stored as individuals in a proper Knowledge Base. Finally, non-standard inferences and a similarity-based ranking support the discovery of the best matching posture. This framework has been implemented in a prototypical tool and preliminary experimental tests have been carried out w.r.t. a reference dataset.

Keywords: Action recognition, Resource Discovery, Semantic-based matchmaking, Ubiquitous Computing.

1 Introduction

The detection of articulate activities has been studied for a long time, mainly focusing research on video analysis. Nevertheless, recent technological enhancements opened the way for novel possibilities. Infrared depth sensors allow to discern three-dimensional shapes in an environment, a kind of information which is often hard to derive from standard video data. Unfortunately, until latest

* The authors wish to acknowledge support from National Operative Program project *Res Novae* (Grid, Building and Road Objectives for Environment and Energy).

D. Ślęzak et al. (Eds.): AMT 2014, LNCS 8610, pp. 160–173, 2014.

years depth sensors were very expensive and therefore they were used in limited applications and circumstances. More recently, following some product and process evolutions, several low-cost multi-sensor devices become commercially available, as for example *Microsoft Kinect*.[1] It is equipped with a standard RGB video camera, a microphone and an infrared depth sensor with resolution and accuracy enough for several practical applications. It must be also considered that deficiencies in capture precision (particularly in general-purpose use cases, where performance decreases due to variety and generality of the input), could be counterbalanced by novel software-side analyses often profiting from the large availability of data corpuses.

In this paper a framework is proposed for an automated posture detection, exploiting depth data provided by the *Microsoft Kinect* tracking device. A recognition problem is handled as a resource discovery one, grounded on a semantic-based matchmaking [1]. The needed terminology (*a.k.a.*, ontology) for geometry-based semantic descriptions of postures has been encapsulated in a Knowledge Base (KB) also including several instances representing poses templates to be detected. Skeleton model data retrieved by the Kinect are preprocessed on-the-fly to identify *key postures*, *i.e.*, unambiguous and not transient body positions (which typically correspond to the early or the final state of a gesture). Each key posture is then annotated adopting standard Semantic Web languages based on the Description Logics (DL) formalism [2]. Hence, non-standard inferences allows to compare the retrieved annotations with templates populating the KB and a similarity-based ranking supports the discovery of the best matching posture. The theoretical framework has been implemented in a prototype and several experiments have been carried out w.r.t. a public dataset [3]. Preliminary results report a satisfactory recognition precision for various kinds of postures, validating the feasibility and effectiveness of the proposed approach.

The remainder of the paper is organized as follows. Most relevant related work is surveyed in Section 2, the theoretical framework and the proposed approach are presented in Section 3 while details about designed prototype along with the related evaluation are in Section 4. Conclusion and perspectives in Section 5 terminate the paper.

2 Related Work

After the needed preparatory steps on data extracted by capturing devices, specific recognition algorithms can be divided in *machine learning* and *ontology* based ones. Approaches based on machine learning theory can be either supervised or unsupervised. In *supervised* techniques, collected data is divided into a training set and a test set, in order to train the recognition algorithm on the former and evaluate its performance on the latter. A limit of that kind of approaches is that they require a relatively large corpus of labeled data to be built for training, usually by hand. Furthermore, the resulting models achieve

[1] `http://www.microsoft.com/en-us/kinectforwindows/`

good accuracy only for the specific scenarios they are thought for. They are not reusable and scalable when individual behavior or environmental conditions change. Hence, the recognition of a large diversity of activities in real-world application scenarios could be deemed as impractical. *Unsupervised* methods try to construct recognition models directly from unlabeled data, by manually assigning a probability to each possible activity and using a graph-based, algebraic or probabilistic model. The limit of the unsupervised learning methods lies in the assignment of probabilistic parameters. *Semi-supervised* learning [4] has recently received significant attention as a technique to balance system accuracy and required human and computational effort. It combines small-scale expert labeled data and large-scale unlabeled data based on certain assumptions.

Ontology-based activity recognition follows a completely different approach. It exploits a logic-based knowledge representation for activity and sensor data modeling, and logical reasoning to perform activity detection. Such approaches: (i) use a semantically rich formalism to explicitly define a library of models for all possible instances in a domain; (ii) aggregate and translate sensed data in logical formulae grounded on the above terminological box; (iii) perform reasoning to infer a minimal model based on the set of observed actions. Ontology-based approaches bridge the semantic gap between low-level observations and high-level detected phenomena. In order to exploit this benefit, in [5] a video movement ontology was engineered to allow automatic annotation of human movements in the classic Benesh notation. A standard ontology-based framework for video annotation is presented in [6], allowing a hierarchical representation of events, by means of Video Event Representation Language (VERL) and Video Event Markup Language (VEML). The description of complex events is built by aggregating elementary concepts relating them by means of temporal relationships. However, VERL is rather complex and verbose, so that exhaustive definition of recognition rules is not practical for large sets without domain-specific customizations and/or user-friendly tools. Automated analysis of surveillance video is one of the most frequent applications of ontology-based activity recognition and annotation [6,7,8,9]. Chen and Nugent [10] proposed an ontology-based approach which is more similar to the one described here: an Activities of Daily Living (ADL) DL ontology was produced for activity modeling and reasoning in the context of smart homes. Subsumption is used to enable a flexible activity recognition at different levels of detail, depending on the amount of knowledge acquired from the environment. However, as pointed out in [11], classical DL inference services are not enough in these cases, since recognition/interpretation tasks cannot be trivially considered as *classification* ones, but they are more similar to *model construction*. The main weakness of logic-based approaches is their general inability to represent vagueness and uncertainty. This issue could be partially solved by exploiting fuzzy DLs. Furthermore, most of ontology-based approaches offer no mechanism for deciding whether one particular model is more effective than another. Finally, they have adopted a top-down approach so far, focusing only on high-level activities and events.

3 Framework and Approach

The framework proposed here can be considered as a part of the gesture recognition technique based on key posture detection [12]. Each recognition process occurs in three steps: (i) *posture description*, which provides posture annotations; (ii) *posture detection*, which sequentially identifies a few reference poses, namely *key postures*; (iii) *gesture identification*, which labels recognized gestures from sequences of key postures. The present work focuses on the first two stages. Data capture is provided by the popular Kinect platform. Particularly, the NUI (Natural User Interface) API, provided in the Kinect for Windows SDK,[2] uses the depth data stream to detect human presence in the infrared sensor range: at most two people can be recognized and tracked simultaneously. For each of them, the NUI produces a human body model, named *skeleton*, composed by 20 joints. Each joint point is defined by its *(x, y, z)* coordinates, expressed w.r.t. a Cartesian spatial reference system whose origin is located on the depth sensor itself. NUI can also mark a point as "inferred" if its coordinates are not directly detected by the sensor (*e.g.*, the body part is occluded by another object), but are estimated via proprietary algorithms by the processing unit embedded in the Kinect device. Inferred joint data are commonly affected by significant noise.

3.1 Architecture

The architecture of the proposed system is depicted in Fig. 1. It is based on three main components:
1. **Posture annotator**, which exploits skeleton tracking capabilities, in order to give a description of body pose with unambiguous semantics. A proper domain ontology has been developed to this aim, as described in the next section.
2. **Posture repository**, storing key postures to be recognized as instances in a Knowledge Base –expressed w.r.t. the shared reference ontology;
3. **Semantic matchmaking engine**, exploiting non-standard logic-based reasoning to support approximated matches, key posture ranking and explanation of outcomes.

3.2 Skeleton Representation

Due to software-side correction effort, a considerable accuracy in detection is not needed –if compared to on-screen rendering– hence a straightforward joint-angle skeleton model is adopted. It provides invariance to sensor orientation and skeleton variations among different individuals. A less approximated representation could improve the posture labeling process, but it would introduce not negligible technical issues: (i) heavier processing; (ii) lack of robustness of the annotation procedure w.r.t. detection errors from joint position data obtained by Kinect sensor; (iii) more complex semantic matchmaking, adversely affecting precision and recall of key posture recognition. The body postures are

[2] `http://www.microsoft.com/en-us/kinectforwindows/develop/`

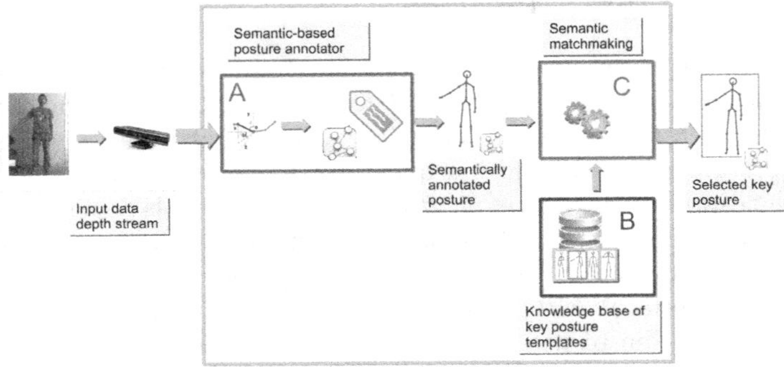

Fig. 1. Architectural block diagram of the proposed framework

basically determined by the mutual position of bones (for example a leg is bent or stretched depending on status of femur and tibia). The proposed framework adopts a posture description model similar to the one in [13], by converting each joint position, defined w.r.t. the Cartesian coordinate system of the NUI API, to a local spherical system. The new reference system keeps x and y axes parallel to the former ones, while z (depth) axis is opposed, *i.e.*, it points toward the Kinect sensor, as illustrated in Fig. 2a. The origin is locally and progressively translated to the "parent" joint, according to the hierarchical order defined by the Kinect NUI API. Following this model, each skeleton segment is represented by zenith and azimuth angles $\{\theta, \varphi\}$; the radius is ignored because the length of each bone is fixed for a given subject. The proposed model omits to annotate body extremities (feet and hands) because they are often inferred by the NUI API, so the posture detection process could be affected by some inaccuracy. In spite of its simplistic nature, this model has been chosen because it allows to represent a broad variety of human postures, also keeping under control the complexity of both recognition and annotation automatic procedures. Such raw angular information are labeled using the Cone-Shaped Directional (CSD) logic framework [14] as formal reference. Particularly, in the proposed model a set of labeled directions is used for given θ and φ values between each parent-child joint couple. This results in a series of cone-shaped 3D regions, as illustrated in Fig. 2b. Regions are defined to conform to qualitative intuition or common-sense knowledge, *e.g.*, for φ the back and forward regions are wider than the ones on the sides.

3.3 Semantic-Based Posture Annotation

In order to enable a fully automated posture annotation as well as the further matchmaking for recognition, the skeleton representation model described above must be translated using an ontology language grounded on a given logic and provided with a proper semantics. A prototypical ontology modeling the domain

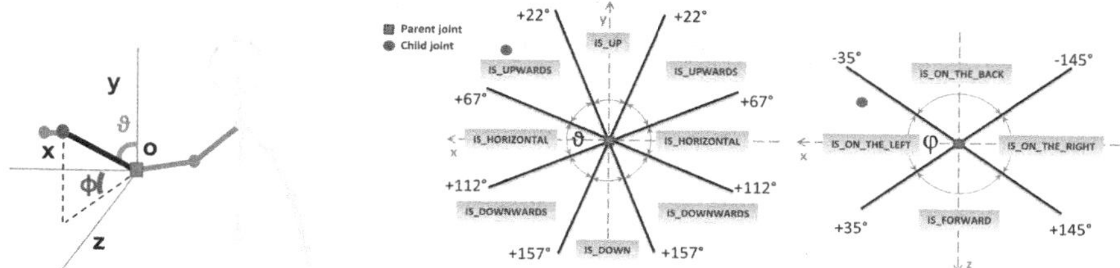

(a) Spherical coordinates reference system

(b) The set of Cone Shaped Directional relations adopted for θ and φ in the proposed model

Fig. 2. Proposed model

of interest has been defined, using a subset of *OWL 2*[3] elements corresponding to the $\mathcal{ALN}$ (Attributive Language with unqualified Number restrictions) formal language of DLs family. An ontology is a formal conceptualization of the problem domain. Relevant entities like the human body parts and the steps of the recognition process (posture, gesture, action) can be modeled explicitly with unambiguous meaning. Furthermore, this knowledge can be shared among researchers, developers and practitioners, allowing the latter to extend the core model in order to meet requirements of their specific use cases. Basically, an ontology is composed of: (i) *classes* (*a.k.a.*, concepts), denoting types of objects; (ii) *properties* (*a.k.a.*, roles), representing relationships either between pairs of objects as classes instances (*object properties*) or between class instances and data-oriented attributes for which a data type is provided (*datatype properties*, *a.k.a.*, features on concrete domains). These basic elements are used to build concept expressions by exploiting logical constructors; each language of the DLs family is characterized by a given set of allowed constructors, which affects algorithmic complexity of inference procedures. Main patterns of the $\mathcal{ALN}$ ontology designed for the purposes of this work are reported hereafter, adopting the example in Fig. 3.

– Joints are modeled as subclasses of the `SkeletonJoint` class. Likewise, skeleton segments are expressed as subclasses of `SkeletonSegment`. Each segment is related to the joints at its extremities through `hasParentJoint` and `hasChildJoint` properties.

– Skeleton body part positions are expressed by means of subclasses of the `SkeletonBodyPart` element, modeling common body part poses (*e.g.*, `RightArmRaised`). Each configuration is related to a subclass of `SkeletonSegment` through azimuth and zenith properties. The mapping between $\{\theta, \varphi\}$ values and the object properties is achieved via the CSD framework, as described above.

– A set of classes representing pre-defined body postures have been modeled as subclasses of `BodyPosture`. They are related to the above-mentioned body part position classes through the `hasPosition` property. As an example, Fig. 3

[3] OWL 2 Web Ontology Language Document Overview (Second Edition), W3C Recommendation, 11 December 2012, `http://www.w3.org/TR/owl2-overview/`

depicts only a portion of `StandingArmsRaisedPosture` definition, which describes the right arm; remaining body parts are described following the same modeling pattern.

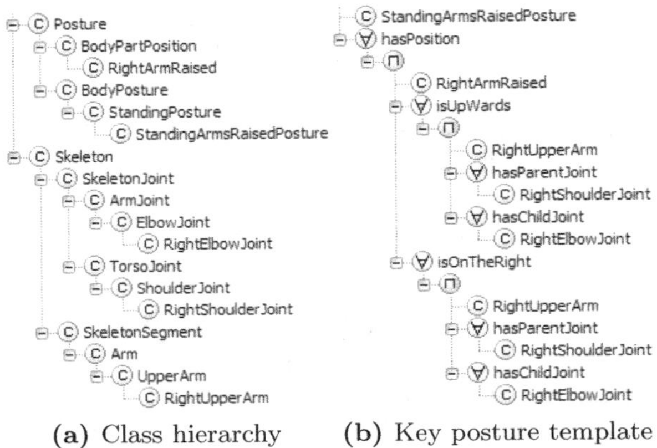

(a) Class hierarchy **(b)** Key posture template

Fig. 3. Posture ontology model

The proposed approach offers many benefits. Practitioners (*e.g.*, physiotherapists) will be able to extend easily the provided ontology with their domain-specific knowledge, also in a collaborative way. The high level of abstraction allows changing the assumptions underlying an implementation if the knowledge about the domain changes. Hard-coding assumptions in programming languages makes them not only hard to find and understand but also hard to change, in particular for someone without programming expertise. In addition, explicit specifications of domain knowledge are useful for new users who must learn the meaning of the terms in the domain. Possible extensions of the recognition framework to include gestures and actions detection simply lead to an incremental expansion of the domain ontology accordingly.

3.4 Semantic-Based Key Posture Recognition

The proposed approach is devoted to use technologies borrowed from the Semantic Web initiative and deductive inferences to solve the posture detection problem: (i) by exploiting machine-understandable annotated descriptions; (ii) by reasoning on obtained expressions inferring implicit knowledge from annotations; (iii) by adopting the Open World Assumption (OWA), saying that the lack of a feature in a resource representation should not be necessarily interpreted as a constraint of absence. Such a general modeling framework allows users to create Knowledge Bases of annotated posture templates, suitable for different scenarios. Then, the key posture identification is managed as a semantic-based resource discovery on the KB, where logic-based inference services provide a similarity ranking which can be seen as a confidence degree in the identification.

Also a detailed semantic-based explanation of results is returned as useful outcome. *Semantic matchmaking* can be defined as the process of finding the best matches among n resources $S_i(i = 1, \ldots n)$ w.r.t. a given request R, where both request and resources are annotated w.r.t. a common reference ontology [1]. In the proposed approach, S_i are the key posture templates in the KB, while R is the current annotated posture. Most reasoners usually provide two standard *Satisfiability* and *Subsumption* inference services for matchmaking. In particular, *Subsumption* returns *true* iff all features requested in R are provided by S_i, but full matches are infrequent in practical scenarios, so it usually gives hopeless 'no match' results. *Concept Abduction* (CA) non-standard inference service, originally formalized and applied in e-commerce scenarios [1], is adopted in this work, allowing to: (i) provide an outcomes explanation beyond the trivial "yes/no answer of subsumption tests and (ii) enable a logic-based relevance ranking of a set of available resources w.r.t. a given query. If R and S_i are compatible –*i.e.*, not contradictory– but S_i does not fully satisfy R, CA allows to determine what is missing in S_i in order to completely satisfy R. The solution H (for *Hypothesis*) to CA represents "why" the subsumption relation does not hold. H can be interpreted as *what is requested in R and not specified in* S_i. In this way, it is possible to support non-exact matches and to define metrics upon H to compute logic-based ranking of resources best approximating the request [1]. Given S and R in Conjunctive Normal Form [1], Algorithm 1 (reported later on) finds a minimal solution for CA in $\mathcal{ALN}$ DL w.r.t. the number of conjuncts in H and computes the corresponding *penalty function* [1] for S w.r.t. R. A toy example should clarify the above process. A semantic-based request of a *person standing up with straight and parallel legs, left arm straight along left side, right arm pointing downward to the right, head up* can be formally expressed as:

(**R**)**DetectedPosture** $\equiv$ $Skeleton \sqcap \forall\ hasPosition.(\forall\ isUp.(Head \sqcap SpinalColumnSegment) \sqcap \forall\ isDownWards.(RightUpperArm \sqcap RightLowerArm) \sqcap \forall\ isOnTheRight.(RightUpperArm \sqcap RightLowerArm) \sqcap \forall\ isDown.(LeftUpperArm \sqcap LeftLowerArm \sqcap LeftUpperLeg \sqcap LeftLowerLeg \sqcap RightUpperLeg \sqcap RightLowerLeg))$.

Let us consider the following resources in the key posture templates KB.

Standup with right arm outstretched (S_1)*: person standing up with straight and parallel legs, left arm straight along left side and right arm outstretched, head up looking straight ahead.* W.r.t. domain ontology, it is expressed as:

(**S₁**) $\equiv$ $StandupPosture \sqcap \forall\ hasPosition.(HeadUp \sqcap LeftArmAlongSide \sqcap RightArmOutstretched))$.

Standup with raised arms on side (S_2)*: person standing up with straight and parallel legs, left arm raised on left side and right arm raised on right side, head up looking straight ahead.* In DL notation:

($\mathbf{S_2}$) $\equiv StandupPosture \sqcap \forall\ hasPosition.(HeadUp \sqcap LeftArmRaised \sqcap RightArmRaised))$.

It can be noticed that the structure of the proposed ontology allows to keep posture annotations short and easy to understand, because details are encapsulated in the definition of referenced classes. **StandupPosture** $\equiv BodyPosture \sqcap \forall\ hasPosition.(RightLegStraight \sqcap LeftLegStraight)$.

RightLegStraight $\equiv \forall\ isDown.(RightUpperLeg \sqcap RightLowerLeg)$.

LeftLegStraight $\equiv \forall\ isDown.(LeftUpperLeg \sqcap LeftLowerLeg)$.

RightArmOutstretched $\equiv$ $\forall$ $isHorizontal.(RightLowerArm)$ $\sqcap$ $\forall$ $isDownwards.(RightUpperArm)$ $\sqcap$ $\forall$ $isOnTheRight.(RightUpperArm \sqcap RightLowerArm)$.
RightArmRaised $\equiv$ $\forall$ $isUpwards.(RightLowerArm)$ $\sqcap$ $\forall$ $isHorizontal.(RightUpperArm)$ $\sqcap$ $\forall$ $isOnTheRight.(RightUpperArm \sqcap RightLowerArm)$.
LeftArmRaised $\equiv$ $\forall$ $isUpwards.(LeftLowerArm)$ $\sqcap$ $\forall$ $isHorizontal.(LeftUpperArm)$ $\sqcap$ $\forall$ $isOnTheLeft.(LeftLowerArm \sqcap LeftUpperArm)$.
LeftArmAlongSide $\equiv$ $\forall$ $isDown.(LeftUpperArm \sqcap LeftLowerArm)$.
RightLowerLeg $\sqsubseteq$ $\forall$ $hasParentJoint.(RightKneeJoint)$ $\sqcap$ $\forall$ $hasChildJoint.(RightAnkleJoint)$.
LeftUpperArm $\sqsubseteq$ $\forall$ $hasParentJoint.(LeftShoulderJoint)$ $\sqcap$ $\forall$ $hasChildjoint(LeftElbowJoint)$.
RightUpperArm $\sqsubseteq$ $\forall$ $hasParentJoint.(RightShoulderJoint)$ $\sqcap$ $\forall$ $hasChildjoint(RightElbowJoint)$.

When an annotated posture (request) is received, the following processing steps are performed.
1. Stored key postures (resources) are extracted from the repository.
2. The reasoning engine computes CA with Algorithm 1 between request and each resource.
3. Results of semantic matchmaking are transferred to the utility function calculation module, which computes the final ranking according to the scoring function reported afterwards.
4. Finally, the ranked list of best resource records is returned. Furthermore, a similarity threshold is introduced: resources key posture templates having an overall score w.r.t. the request worse than this threshold will not give back. Otherwise, detected posture(s) are provided along with their scores and H values, which justify the outcome. By solving the Concept Abduction Problem it is possible to compute the missing features of the key posture templates S_i, needed to reach a full match with R. In particular, in S_1 *RightLowerArm* should be in horizontal position. On the other hand, in S_2 *RightUpperArm* should be slightly lowered and $LeftLower\ Arm$ and $LeftUpperArm$ should be down. In formulae:
$\mathbf{H_{R,S_1}} \equiv \forall\ isDownwards.(RightLowerArm)$.
$\mathbf{H_{R,S_2}} \equiv \forall\ isDown.(LeftUpperArm \sqcap LeftLowerArm) \sqcap \forall\ isDownwards.(RightUpperArm)$.
In the proposed approach, semantic similarity is computed via the following *utility function*:

$$f(R, S) = 100 * \left[1 - \frac{penalty(R, S)}{penalty(R, \top)}\right] \tag{1}$$

where *penalty* measures the CA-induced semantic distance between real-time annotation R and key posture template instance S; this value is normalized dividing by the distance between R and the universal concept –*Top* or *Thing*– which depends only on axioms in the ontology and is the maximum possible value. In the example, the overall similarity score is 91% for S_1 and 60% for S_2.

4 Prototype and Experiments

As a proof of concept, the proposed framework has been implemented in a software prototype, extending the existing *Kinect Toolbox*.[4] Thanks to the GUI in Fig. 4, the tool enables users to compose semantic annotations for body postures,

[4] Kinect Toolbox, `http://kinecttoolbox.codeplex.com/`

Algorithm: *abduce* $(\langle \mathcal{L}, D, S, \mathcal{T} \rangle)$

```
Require: ⟨L, R, S, T⟩ with L = ALN, acyclic T
Ensure: ⟨H, penalty⟩ with penalty ≥ 0 and H ∈ ALN
H := ⊤;
penalty := 0;
for all concept name A in R do
    if no B in S exists s.t. B ⊑ A then
        H := H ⊓ A;
        penalty := penalty + 1;
    end if
end for
for all concept (≥ x P) in R do
    if (≥ y P) exists in S and y < x then
        H := H ⊓ (≥ x R);
        penalty := penalty + (x−y)/x;
    else if no (≥ y P) exists in S then
        H := H ⊓ (≥ x P);
        penalty := penalty + 1;
    end if
end for
for all concept (≤ x P) in R do
    if (≤ y P) exists in S and x < y then
        H := H ⊓ (≤ x P);
        penalty := penalty + (y−x)/x;
    else if no (≤ y P) exists in S then
        H := H ⊓ (≤ x P);
        penalty := penalty + 1;
    end if
end for
for all concept ∀P.E in D do
    if ∀P.F exists in S then
        ⟨H', penalty'⟩ := abduce (⟨L, E, F, T⟩);
        H := H ⊓ ∀P.H';
        penalty := penalty + penalty';
    else
        H := H ⊓ ∀P.E;
        penalty := penalty + 1;
    end if
end for
return ⟨H, penalty⟩
```

Algorithm 1. Concept Abduction in $\mathcal{ALN}$ DL

without requiring specific knowledge of Semantic Web languages and underlying logic-based formalisms. In literature many tools aim to support developers, *e.g.*, *KINA* toolkit [15] and *DejaVu* [16]. Conversely, the goal of the proposed system is to allow users, who are typically practitioners and not necessarily developers, to build a set of representative postures for a given domain by composing visually a high-level description. A typical usage experience consists of the following interaction steps:

– **Data capture:** input streams include depth and RGB data provided by Kinect in real-time. When a subject is facing the Kinect sensor, her/his movements are tracked and skeleton data are retrieved and displayed on the panel (A). The automatic annotation engine described in the previous section calculates body segment angles and builds the corresponding semantic description. The system allows also to process pre-recorded data, in the form of skeleton frame sequences.

– **Annotation:** if the user presses the 'Annotate Posture' button on panel (D), it is shown the annotation of the just captured posture. Panel (B) provides an intuitive tree-like graphical representation. The above described ontology is loaded and its elements populate the upper part of the panel: (i) classes (*e.g.*, `SkeletonSegment`, `BodyPart Position`, etc.); (ii) object properties (*e.g.*, *has-Position* links a `Body Posture` to a `BodyPartPosition`); (iii) datatype properties.[5] The current posture annotation is displayed in the lower portion of panel (B). The user can edit it through drag-and-drop of classes and properties from the ontology: context menus appear whenever additional information should be specified. Then s/he can save it in the KB by clicking on the 'Apply' button just below the annotation.

– **Semantic matchmaking:** the current posture is detected as a key posture only if it lasts for a tunable *motion sensitivity* parameter. In that case, it is automatically processed for recognition. An embedded lightweight reasoner [17]

[5] This kind of constructors is not used in the current version of the ontology, but may be used in future extensions to deal with gestures and actions.

Fig. 4. Prototype tool screenshot

is exploited to perform semantic matchmaking between the real-time annotated posture and each key posture instance stored in the KB. The result is a list of recognized key postures ordered w.r.t. the overall compatibility score. The recognized key posture is the one that best matches the real-time annotation, and it is added to the list (C).

Basic key postures can be slightly modified or even enriched to help domain experts in fitting requirements of a specific scenario. Expert personnel can customize posture descriptions through the above interface to improve the recognition capabilities. . For example, in telerehabilitation scenarios, a physiotherapist records for a patient the postures to avoid in front of the Kinect sensor. Then he loads the data, annotates the postures and saves them in the KB. From then on, every time the patient takes one of these bad postures, the system automatically recognizes and registers it and the physical therapist can figure out the patients compliance remotely. An experimental campaign was carried out to obtain a preliminary performance evaluation of the proposed approach. Postures were selected from a subset of gestures collected in [3]. Each gesture was repeated at least 10 times consecutively. The sequences were recorded after giving instructions to the subjects using different formats (images, video and/or text). Preliminarily, by exploiting the developed tool, a set of annotated key postures were defined. Then the prototype tool was tuned to a motion sensitivity of 0.3 sec -equivalent to 9 frames at the default NUI API sampling frequency of 30 frames/sec- and a precision threshold of 60%. Results are reported in Table 1 and refer to seven key postures belonging to a subset of five gestures taken from the dataset. For each key posture the following data were measured: (i) relative frequency of detected occurrences N, divided by the total number of real executions observed in each sequence n; (ii) average semantic score $\overline{f(R,S)}$; (iii) standard deviation $\sigma_{f(R,S)}$. Outcomes show that the proposed method is reliable, having 88.0% rate of correct posture recognition, particularly if the subject

Table 1. Experiments result

KP	Sequence1			Sequence2			Sequence3			Sequence4			Sequence5		
	N/n	$\overline{f(R,S)}$	$\sigma_{f(R,S)}$	**N/n**	$\overline{f(R,S)}$	$\sigma_{f(R,S)}$	**N/n**	$\overline{f(R,S)}$	$\sigma_{f(R,S)}$	**N/n**	$\overline{f(R,S)}$	$\sigma_{f(R,S)}$	**N/n**	$\overline{f(R,S)}$	$\sigma_{f(R,S)}$
Start music/raise volume															
	11/11	100	0	13/13	100	0	12/12	100	0	11/11	100	0	12/12	100	0
	8/10	96.14	6,91	6/11	98,53	2,65	11/11	97,61	5,05	10/10	86,11	3,08	11/11	98,71	1,26
Crouch or hide															
	10/10	80,43	2,46	10/10	73,89	4,57	10/10	76,14	5,36	10/10	80,97	3,10	10/10	84,60	2,25
Navigate to next menu															
	6/10	90,41	4,83	6/12	93,05	3,58	14/16	87,39	0	12/12	88,17	2,58	4/10	67,26	3,62
	6/10	94,46	2,49	7/11	94,94	3,24	16/16	94,92	2,52	11/11	94,87	1,50	0/10	-	-
Put on night vision goggles to change the game mode															
	10/10	91,59	2,08	9/10	73,87	11,72	10/10	90,15	1,17	11/11	93,79	2,09	11/11	89,88	1,97
Take a bow															
	11/11	72,84	2,59	10/10	71,73	0,52	10/10	69,66	0,74	8/10	71,67	1,75	10/10	69,50	3,61

is scrupulous in the execution of the task. Looking in greater detail, the following remarks can be made:
– *Start music/raise volume*: results are good except for the *Sequence2*, where the achieved recognition rate is lower because the subject did not raise arms enough in some cases.
– *Crouch or hide* and *Take a bow*: even if key postures are always identified correctly, low average scores were obtained, probably because of skeleton tracking instability and occlusion issues in the input data.
– *Navigate to next menu*: recognition frequencies are not very high for some sequences, while semantic scores are always high, except for *Sequence5*. That case is anomalous because the gesture was performed with the wrong hand; in fact, the system detected the symmetric posture in some cases.
– *Put on night vision goggles*: obtained results are very good in terms of frequency and semantic score; only *Sequence2* did not achieve a perfect recognition rate, because of some skeleton tracking instability in the input data.

5 Conclusion

The paper introduced a general-purpose framework and approach for semantic-based posture annotation and recognition. It exploits 3D skeleton joint position data provided by a Microsoft Kinect device as input. A general model was devised to characterize body parts and most common poses, and an ontology was designed to give them the needed formal terminology through standard Semantic Web languages. The posture recognition problem has been basically handled as resource discovery via semantic matchmaking, exploiting non-standard inference services from managing approximate matching. The theoretical framework has been implemented in a prototypical tool devoted to prove the effectiveness of the proposed approach: results obtained w.r.t. a reference dataset provide a promising proof of concept.
Future work aims to enhance the presented framework toward gesture recognition. This goal will require an extension of both data model and domain ontology allowing to annotate a gesture as ordered sequence of postures. Also the semantic matchmaking framework will be extended to support a more articulate recognition. Finally, a broader experimentation and comparison with state-of-the-art approaches is planned.

References

1. Colucci, S., Di Noia, T., Pinto, A., Ragone, A., Ruta, M., Tinelli, E.: A Non-Monotonic Approach to Semantic Matchmaking and Request Refinement in E-Marketplaces. International Journal of Electronic Commerce 12(2), 127–154 (2007)
2. Baader, F., Calvanese, D., Mc Guinness, D., Nardi, D., Patel-Schneider, P.: The Description Logic Handbook. Cambridge University Press (2002)
3. Fothergill, S., Mentis, H., Kohli, P., Nowozin, S.: Instructing people for training gestural interactive systems. In: Proceedings of the SIGCHI Conference on Human Factors in Computing Systems, pp. 1737–1746. ACM, New York (2012)

4. Zhang, T., Xu, C., Zhu, G., Liu, S., Lu, H.: A generic framework for video annotation via semi-supervised learning. IEEE Transactions on Multimedia 14(4), 1206–1219 (August)
5. Saad, S., De Beul, D., Mahmoudi, S., Manneback, P.: An ontology for video human movement representation based on benesh notation. In: 2012 International Conference on Multimedia Computing and Systems (ICMCS), pp. 77–82. IEEE (2012)
6. François, A.R.J., Nevatia, R., Hobbs, J., Bolles, R.C., Smith, J.R.: VERL: an ontology framework for representing and annotating video events. IEEE MultiMedia 12(4), 76–86 (2005)
7. Vrusias, B., Makris, D., Renno, J.-P., Newbold, N., Ahmad, K., Jones, G.: A framework for ontology enriched semantic annotation of CCTV video. In: Eighth International Workshop on Image Analysis for Multimedia Interactive Services, WIAMIS 2007, p. 5. IEEE (2007)
8. Akdemir, U., Turaga, P., Chellappa, R.: An ontology based approach for activity recognition from video. In: Proceedings of the 16th ACM International Conference on Multimedia, MM 2008, pp. 709–712. ACM, New York (2008)
9. SanMiguel, J.C., Martinez, J.M., Garcia, A.: An ontology for event detection and its application in surveillance video. In: Sixth IEEE International Conference on Advanced Video and Signal Based Surveillance, pp. 220–225. IEEE (2009)
10. Chen, L., Nugent, C.: Ontology-based activity recognition in intelligent pervasive environments. International Journal of Web Information Systems 5(4), 410–430 (2009)
11. Gómez-Romero, J., Patricio, M.A., García, J., Molina, J.M.: Ontology-based context representation and reasoning for object tracking and scene interpretation in video. Expert Syst. Appl. 38(6), 7494–7510 (2011)
12. Miranda, L., Vieira, T., Martinez, D., Lewiner, T.: Real-time gesture recognition from depth data through key poses learning and decision forests. In: Sibgrapi 2012 (XXV Conference on Graphics, Patterns and Images), Ouro Preto, MG. IEEE (August 2012)
13. Raptis, M., Kirovski, D., Hoppe, H.: Real-time classification of dance gestures from skeleton animation. In: Proceedings of the 2011 ACM SIGGRAPH/Eurographics Symposium on Computer Animation, pp. 147–156. ACM, New York (2011)
14. Renz, J., Mitra, D.: Qualitative direction calculi with arbitrary granularity. In: Zhang, C., Guesgen, H.W., Yeap, W.-K. (eds.) PRICAI 2004. LNCS (LNAI), vol. 3157, pp. 65–74. Springer, Heidelberg (2004)
15. Reis, B., Teixeira, J.M., Breyer, F., Vasconcelos, L.A., Cavalcanti, A., Ferreira, A., Kelner, J.: Increasing Kinect application development productivity by an enhanced hardware abstraction. In: 4th ACM SIGCHI Symposium on Engineering Interactive Computing Systems, pp. 5–14. ACM, New York (2012)
16. Kato, J., McDirmid, S., Cao, X.: Dejavu: Integrated support for developing interactive camera-based programs. In: 25th Annual ACM Symposium on User Interface Software and Technology, pp. 189–196. ACM, New York (2012)
17. Ruta, M., Scioscia, F., Di Sciascio, E., Gramegna, F., Loseto, G.: Mini-ME: the Mini Matchmaking Engine. In: Horrocks, I., Yatskevich, M., Jimenez-Ruiz, E. (eds.) OWL Reasoner Evaluation Workshop (ORE 2012). CEUR Workshop Proceedings, vol. 858, pp. 52–63. CEUR-WS (2012)

Literal Node Matching Based on Image Features toward Linked Data Integration

Takahiro Kawamura[1,2], Shinichi Nagano[1], and Akihiko Ohsuga[2]

[1] Corporate Research & Development Center, Toshiba Corp.
[2] Graduate School of Information Systems, University of Electro-Communications, Japan

Abstract. Linked Open Data (LOD) has a graph structure in which nodes are represented by Uniform Resource Identifiers (URIs), and thus LOD sets are connected and searched through different domains. In fact, however, 5% of the values are literal (string without URI) even in DBpedia, which is a *de facto* hub of LOD. Since the literal becomes a terminal node, and we need to rely on regular expression matching, we cannot trace the links in the LOD graphs during searches. Therefore, this paper proposes a method of identifying and aggregating literal nodes that have the same meaning in order to facilitate cross-domain search through links in LOD. The novelty of our method is that part of the LOD graph structure is regarded as a block image, and then image features of LOD are extracted. In experiments, we created about 30,000 literal pairs from a Japanese music category of DBpedia Japanese and Freebase, and confirmed that the proposed method correctly determines literal identity with F-measure of 99%.

1 Introduction

The goal of this paper is to connect literals (string values without URI) of triples $< Resource, Property, Value >$ as much as possible, and to create "linked" data of the original meaning. DBpedia, which represents part of Wikipedia, is currently a *de facto* hub of LOD in the world, but according to our research approx. 5% of the values in DBpedia are literals. Since the literal becomes a terminal node, and we need to rely on regular expression matching, we cannot trace the links in the LOD graphs during searches. In addition, we are now working on LOD generation from Web information and social media, but at least in the initial stage of the generation many literal values are created, since the original data are sentences. In the near future, this will become a major issue, when LOD sets of different domains are connected and merged in the LOD cloud in the world [1].

Therefore, in order to determine the identity of literal values and support data linkage, we propose a method for matching literal values using LOD graph structures. These initiatives will be applied for cross-domain search through links in LOD. Literal matching corresponds to "name identification," which is a traditional but important problem in system integration (SI) projects. It is

D. Ślęzak et al. (Eds.): AMT 2014, LNCS 8610, pp. 174–186, 2014.

also similar to instance matching in ontology alignment, although the matching target is not an instance (resource in LOD) but a value.

The novelty of this paper is that the proposed method regards the target literal and surrounding information in LOD as a block image, and extracts image features around the literal. Then, it determines the identity of the literals through similarity discrimination of two images. This method is inspired by recent computer shogi, in which records of games are regarded as figures, and a game is played to make a good figure in the record [2,3]. Thus, the contribution of this paper is the introduction of a new feature in Linked Data integration.

The rest of this paper is organized as follows. Section 2 presents related work, and Section 3 proposes image feature extraction from LOD. Then, Section 4 evaluates the matching results of the literals using ensemble learning, comparing with simple string matching. Finally, Section 5, refers to future work and concludes this paper.

2 Related Work

In regard to the literal matching proposed in this paper, we first introduce research on instance matching.

A well-known tool for instance matching is SILK [4,5]. With SILK, a user can select features to identify the similarity and define the link specifications for each dataset. Also, Maali et al.[9] implemented four kinds of extensions for instance matching in Google Refine [8], that is a data cleansing tool provided by Google: (1)SPARQL query, (2)SPARQL with full-text search, (3)SILK Server API, (4)Keyword-based Sindice search API [7] with (2) (since Sindice result is a list of document URIs containing matching RDF data, and not a list of the actual matching resources). Then, they measured accuracy and computational performance for each extension. Although we can apply the approaches similar to the extensions to literal matching, all of them have a problem that search queries and translation rules cannot be constructed without knowing structures of two dataset to be compared in advance. Thus, research on automatic rule generation based on Genetic Programming [6] is also conducted. A similar approach can be found in [10], in which matching rules for each dataset are acquired based on semi-supervised learning and are repeatedly refined. Other research can be divided into approaches depending on domain (dataset) knowledge to raise accuracy, and domain-independent approaches that dispense with the need to prepare training data for each domain.

The former research is Rong et al.[11], in which instance matching is defined as a binary classification problem and determined by a learning method. However, they also utilize a transfer learning method, TrAdaBoost [12], to reduce the training data depending on a domain. ObjectCoref [13] introduces a self-learning framework that repeatedly finds pairs of specific properties and values for training data reduction. This research also provides useful guidance for literal matching.

The latter research is RiMom [14]. In regard to domain-independent approaches, most research combines several features, for instance string similarity such as Jaccard coefficient, cosine, TF/IDF and Levenshtein distance, and semantic similarity such as WordNet and inverse functional property, and also structural similarity without any learning method. RiMom can suggest the proper combination of features according to the dataset characteristics. In addition, [15] is a useful reference for surveying instance matching.

Next, we introduce Apache Solr [16] that is used for name identification in SI projects, since we could not find research similar to literal matching of LOD. Narrowly conceived, name identification means confirming the consistency of customers' information for financial institutions and administrative organizations, whereas broadly conceived it usually refers to data cleansing. Frameworks such as Solr calculate word similarity using TF/IDF etc., and then suggest words that might have the same meaning. Developers visually confirm the words and register them in a dictionary. In the early stage of the Semantic Web, freedom from word ambiguity was considered advantageous, but nowadays if users search for "dentist" on the web, "dental clinic" is also included in the results owing to such frameworks. However, synonyms with totally different notations are not suggested by using word similarity. Moreover, homonyms with the same notations cannot be distinguished by dictionary registration.

Literal matching is a problem similar to instance matching, but an independent problem. Literal matching matches several nodes that are not instances (resources in LOD), if they indicate the same entity in the real world. This problem remains even after performing instance matching and property matching [17]. For example, even if movie resources A and B are matched by instance matching and 'author' property of A and 'writer' property of B are matched by property matching, in the case that there are two author/writer in the movie the problem that which node indicates which author/writer remains. There is also a case that 'director' property of a movie resource and 'writer' property of a book resource indicate the same person. In this case, the nodes of the same person cannot be found after instance matching. In regard to technical aspect, although instance matching can utilize semantic and structural similarity of properties and nodes connected to the instance as features, literal matching is for a terminal node that only connects to a property, and then needs to define a new feature. Figure 1 indicates the relationship of the above-mentioned three kinds of matching.

As LOD has become widely available, those matching techniques have been implemented in commercial applications. Fujitsu [18] released a data matching solution that is included in its open data platform. However, to the best of our knowledge from [19,20], the dataset is not in a linked data format. For the purpose of the classification of microblogs, advertisement and product recommendation for senders and readers, they combine search results from Google and Baidu, lexical matching and popularity information, and then try to match terms in a microblog with corresponding entities in their knowledge base, that contains vast amount of name variations of entities such as acronyms, confusable names, spelling variations, and nick names etc.

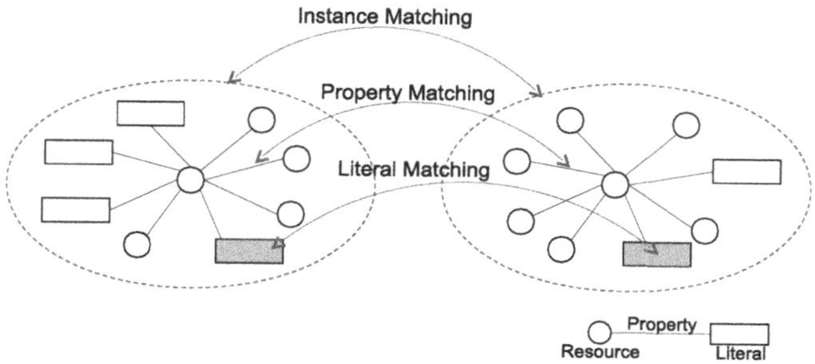

Fig. 1. Relationship to other works

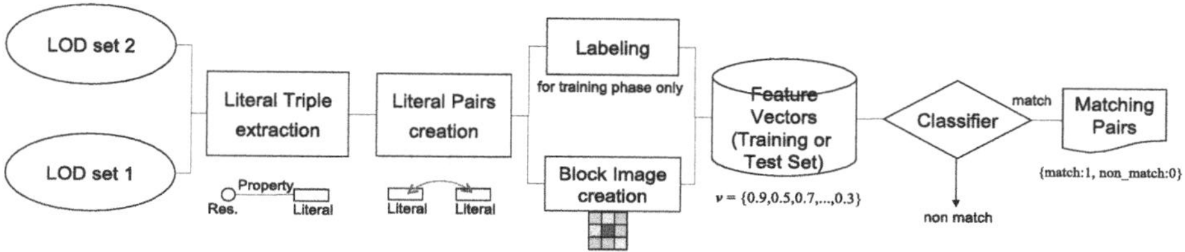

Fig. 2. Workflow of literal matching

3 Extraction of Image Features from Graph Structure

The workflow of literal matching is shown in Fig.2.

First, we construct feature vectors as input for the classifier. In this paper, we propose a method to construct the feature vector based on Scale-Invariant Feature Transform (SIFT) [21], which is a well-known method in computer vision. SIFT extracts local features for each key point in an image. The area surrounding a key point is divided into 4×4 blocks, with each block providing illumination changes (gradient) of 8 orientations per 45 deg., and then a vector with 128 dimensions is created in SIFT algorithm. As the feature vectors in Computer Vision, there are also SURF (Speeded Up Robust Features) that is a high speed version of SIFT, haar-likeCand HOG (Histogram of oriented Gradient), each of which has limitations according to recognition objects. The reason why we selected SIFT is that it extracts features for each key point in an image. By regarding the target literal as a key point, we considered that a graph structure can be mapped to SIFT algorithm. In contrast, haar-like feature is for each image, and HOG is for a pixel.

Fig.3 illustrates a method of generating a block image from two LOD graphs. We select properties and resources connected from two literal values for comparison, and another two properties and values connected from the above resources, and then generate a grayscale image of 3×3 blocks by calculating each similarity Sim_l, Sim_p, Sim_r. Each block has a value $[0, 1]$, where 1 represents black. Regarding the selection of another two properties and values, if we find that

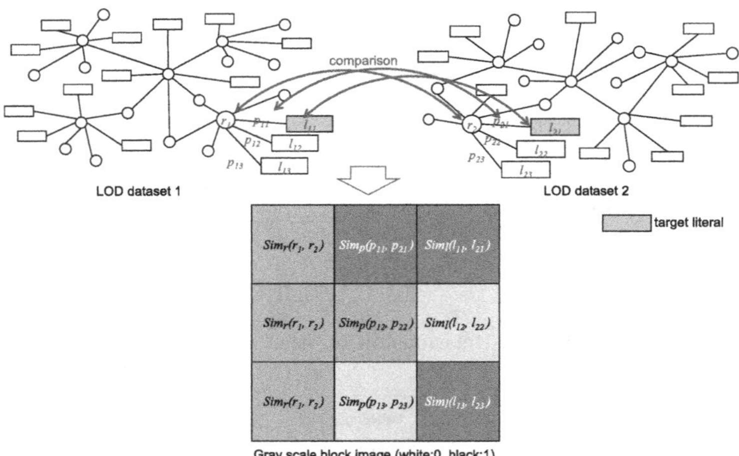

Fig. 3. Image generation from LOD

the above two resources have common properties, we alphabetically select two properties and their values as the second and the third row. If they don't have common properties, we select the two most similar properties and their values. Thus, we attempt to generate an image with specified regularity. The similarities Sim_l, Sim_p, Sim_r are composed of string similarity and semantic similarity and defined as follows:

$$Sim_l(l_{1i}, l_{2j}) = \alpha\ StringSim(l_{1i}, l_{2j}) + \beta\ SemanticSim(l_{1i}, l_{2j}) \tag{1}$$

$$Sim_p(p_{1i}, p_{2j}) = \gamma\ StringSim(p_{1i}, p_{2j}) + \delta\ SemanticSim(p_{1i}, p_{2j}) \tag{2}$$

$$Sim_r(r_1, r_2) = \epsilon\ StringSim(r_1, r_2) + \zeta\ SemanticSim(r_1, r_2) \tag{3}$$

$$StringSim(str1, str2) = \frac{|str1 \cap str2|}{|str1 \cup str2|} \tag{4}$$

$$SemanticSim(str1, str2) = \begin{cases} 1.0 & \text{if str1 = str2} \\ 0.75 & \text{else if str1 is } \textit{Synonym} \text{ of str2} \\ 0.5 & \text{else if str1 is } \textit{Hypernym} \text{ or } \textit{Hyponym} \text{ of str2} \\ 0.25 & \text{else if str1 has same } \texttt{namespace} \text{ as str2} \\ 0.0 & \text{else} \end{cases} \tag{5}$$

l_{ij}, p_{ij}, r_{ij} mean identifiers of a literal node, a property and a resource, respectively. Also, parameters $\alpha, \beta, \gamma, \delta, \epsilon, \zeta$ satisfy $0 \leqq \alpha, \beta, \gamma, \delta, \epsilon, \zeta, \alpha+\beta, \gamma+\delta, \epsilon+\zeta \leqq 1$. These parameters depend on problems to be solved, but we set 0.5 for each in the next section. *StringSim* is a Jaccard coefficient, and *SemanticSim* is based on the semantic similarity measurement between two words according to path lengths between two corresponding classes in ontology [22]. We then simplified the calculation to reduce the processing time. Also, $str1, str2$ are strings, and are regarded as sets of characters in *StringSim*. To determine *SynonymCHypernymCHyponym*, we refer to the external ontologies, and we use WordNet [23] and Japanese WordNet [24] in the next section.

The intuitive meaning of this image is representation of contextual information on the graph. That is, the image represents the connection of semantic contents on the graph. For example, if the literals are synonyms with different notations and their semantic similarity cannot be determined by ontology, then the image that has high similarities in each block except for a block of the literals like Fig.4(a) can be expected. In contrast, the block of the literals will have high similarities, but most of the other blocks will have low similarities in the case of homonyms with the same notation like Fig.4(b). In this manner, we represent the similarities of the target nodes and surrounding links and nodes as an image, and characterize the connection of meaning in the graph. This method can be regarded as a variation of comparison of tree-like structures in instance and/or ontology matching. As explained above, however, a literal node is a terminal and cannot be seen as a tree on the graph. Therefore, we generated the block image from common and similar properties and their values.

Finally, we regard a target literal value as a key point in the image, and construct the feature vector from the similarity changes of the adjacent 9 blocks. Unlike SIFT, no orientation is set and an absolute value of the similarity at the key point is included as a basis. The feature vector $\boldsymbol{v}$ is represented by the following equation. Also, each of the parameters η, θ, λ is set to 1.0 in the next section.

$$\begin{aligned}
\boldsymbol{v} =& Sim_l(l_{11}, l_{21}) + \Delta_{p1} + \Delta_{r1} \\
& \Delta_{l2} + \Delta_{p2} + \Delta_{r2} \\
& \Delta_{l3} + \Delta_{p3} + \Delta_{r3} \\
\Delta_{li} =& \eta \left(Sim_l(l_{1i}, l_{2i}) - Sim_l(l_{1(i-1)}, l_{2(i-1)})\right) \\
\Delta_{pi} =& \theta \left(Sim_p(p_{1i}, p_{2i}) - Sim_l(l_{1i}, l_{2i})\right) \\
\Delta_{ri} =& \lambda \left(Sim_r(r_1, r_2) - Sim_p(p_{1i}, p_{2i})\right) \quad (i = 1, 2, 3)
\end{aligned} \tag{6}$$

In SIFT, the gradient is taken as the feature in order to absorb the difference in illuminance values by lighting etc. In the comparison of LOD sets, if both sets are written in the same schema (property definitions), the similarities of the links and nodes surrounding the matching literals are expected to change at a high level, although they are expected to change at a lower level as a whole in the case of sets with different schemas as illustrated in Fig.5. Therefore, in order to

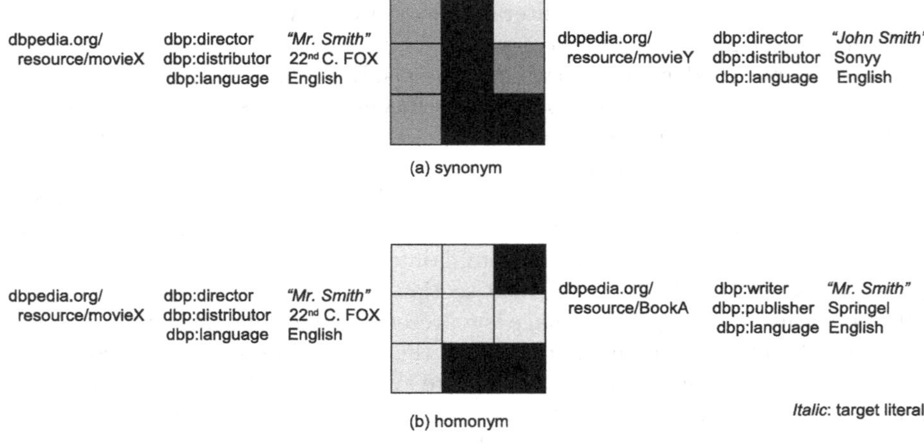

Fig. 4. Example features of synonym and homonym

absorb the difference in the relative level of the similarity between the LOD sets to be compared, the proposed method does not use the absolute values of Sim, but the differences from adjacent blocks (similarity gradient) as the training data for generic features.

In the next section, we present the results of our experiments using a learning method as the classifier.

4 Classification Experiment on Literal Pairs

This section presents the experimental results for evaluating the effectiveness of the proposed feature. Unlike in the case of instance matching, we could not find any public research tool to perform a comparison for literal matching on LOD. We thus compared the proposed method with a simple string matching of literal values as a baseline method, and the matching results of the same schema sets with different schema sets.

4.1 Experimental Setting

For literal matching between LOD sets with the same schema, we first extracted triples ($N = 8,504$) that have literal values, from the resources in the "JPOP" (Japanese pop music) category of DBpedia Japanese [25], and created literal pairs ($(N^2 - N)/2 = 36,154,756$). We then selected 17,391 pairs, excluding the pairs obviously not matching such as time and location, and manually marked the literal matching to them TRUE or FALSE. The literal pairs regarded as matching include: inconsistent spellings of values for properties such as `label` and `alias`, inconsistent spellings and different notations of album titles and artists' names

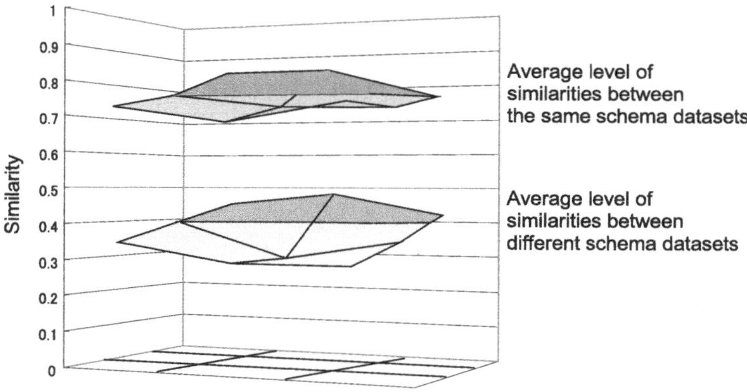

Fig. 5. Difference in the relative level of the similarity between the LOD sets

that have not become resources, and award record such as "Oscar nomination". Some descriptions of addresses and dates have also not become resources.

For matching between LOD sets with different schemas, we extracted triples $M = 2,879$ that have literal values, from the resources in the "J-POP" category of Freebase LOD [26], and created literal pairs $N^2 = 24,483,016$ combining the above triples $N = 8,504$ extracted from DBpedia. We then selected 13,702 pairs, excluding the pairs obviously not matching as mentioned above, and manually marked the literal matching to them TRUE or FALSE.

Next, we calculated the image feature described in the previous section for each of the above-mentioned pairs. We used an ensemble learning method, Random Forest [27], as the classifier, and then evaluated with 10-fold cross-validation. Ensemble methods use multiple weak learners to obtain better predictive performance, and algorithms such as Bagging, Boosting, and Random Forest exist. Random Forest operates by constructing a multitude of decision trees at training time and outputting the result. Random Forest is advantageous in that it gives estimates of what features are important in the classification, and has an effective method for estimating missing data and maintains accuracy when a proportion of the data is missing. It also runs efficiently on large datasets and is faster than, for example, Support Vector Machine (SVM). Other research on instance matching [11] indicated that Random Forest surpasses other methods owing to the nature of the problem. In the following experiments, we also confirmed that the precision can be easily raised, but the recall tend to fall in the case of SVM, since accuracy of SVM is largely dependent on training dataset. There are 10 decision trees, each of which randomly takes 4 features and unlimited depth of the tree. The algorithm of each tree is an implementation (modified REPTree) based on C4.5.

4.2 Comparison of the Baseline Method and the Proposed Method

We first show the results of the baseline method and the proposed method for the same schema sets. As a baseline method, we used string comparison of literal

Table 1. Classification result of string comparison (%)

	$jcc > 0.7$		$jcc > 0.9$	
	Precision	Recall	Precision	Recall
TRUE	61.4	58.0	82.8	54.7
FALSE	98.8	99.0	98.7	99.7

values with Jaccard coefficient jcc. Table 1 shows the matching results, in which TRUE and FALSE mean matching pairs and non-matching pairs, respectively.

As a result, FALSE indicates that both precision and recall have high scores. Since the matching pairs are only about 3% of all the pairs, if we strictly determine the identities of the pairs, and raise the accuracy of the majority FALSE, then we can easily raise the total accuracy. Moreover, comparison of the results of TRUE in the case of $jcc > 0.7$ and $jcc > 0.9$ indicates that if we strictly determine the string matching, we can also easily raise the precision of TRUE. However, the recall of TRUE decreased, and we lost many matching pairs as a result. In Solr described in section 2, matching is based on a dictionary registered by the string similarity, and thus abbreviations and synonyms that are not suggested by the similarity have not been covered. The above results are considered to correspond to the current situation using Solr. Therefore, we can understand that solving the problem of literal matching will involve raising the recall of TRUE without decreasing the precision by separately looking at the results of TRUE and FALSE.

Next, in Table 2 we present the result using the proposed method.

Table 2. Classification Result of image features (for the same schema sets) (%)

	Precision	Recall	F-Measure
TRUE	87.5	82.6	85.0
FALSE	99.5	99.7	99.6
Weighted Avg.	99.2	99.2	99.2

As a result, we found that the proposed method improved the recall of TRUE 25 points compared with the case of $jcc > 0.7$, improving the precision 5 points compared with the case of $jcc > 0.9$. In addition, the weighted average according to the number of pairs including FALSE resulted in over 99% precision, recall and F-measure. As mentioned before, however, the weighted average is highly raised by a large number of obvious FALSE pairs. The out-of-bag (OOB) error was about 1%. Thus, the effectiveness of the proposed method was confirmed.

4.3 Comparison of the Same Schema Sets and Different Schema Sets

Next, we show the result of the proposed method for different schema sets. Table 3 shows the matching results.

Table 3. Classification Result of image features (for different schema sets) (%)

	Precision	Recall	F-Measure
TRUE	83.6	69.8	76.1
FALSE	99.5	99.8	99.7
Weighted Avg.	99.3	99.3	99.3

As a result, we found that the recall of TRUE fell 13 points compared with the result of the same schema sets, although the precision remained 4 points fall. The reason for less recall than for the same schema sets is considered to be the difference in the properties. Within the datasets we collected for this experiment, the common property between DBpedia Japanese and Freebase is only `http://www.w3.org/2000/01/rdf-schema#label`. Even if a property has the same meaning, Freebase defines the original property name. Also, DBpedia has many kinds of properties, whereas Freebase does not necessarily have properties corresponding to them. Because of these differences, discrete patterns of the properties could not be learned. We intend to increase the size of the dataset and try to improve the recall.

4.4 Comparison of Similarity Gradient and Absolute Values

Moreover, we conducted the evaluation using the absolute values of Sim, not the differences from adjacent blocks (similarity gradient), as the feature vector. In order to confirm that similarity gradient can be more generic features than the absolute values, we conducted the evaluation on about 30 thousand pairs that are composed of the same schema dataset and difference schema dataset, which are different in the relative level of the similarity. As a result, we confirmed that the gradient slightly improved the precision of TRUE, by 1 point (84.7% $\rightarrow$ 85.5%). There was no difference in the recall of TRUE (approx. 77%) and the accuracy of FALSE. As to the reason that the results for two kinds of feature vectors were similar, since the above $StringSim$ and $SemanticSim$ are normalized to $[0, 1]$, the absolute values did not differ markedly in terms of the relative level of similarity. In fact, the average of similarities between the same schema datasets (DBpedia) in Fig.5 was 0.40, and the average of similarities between different schema datasets (DBpedia and Freebase) was 0.27.

Furthermore, since the proposed features of similarity gradient include an absolute value of similarity between target literals as a basis, there is a possibility that the classification is largely determined by the absolute value. Therefore, we

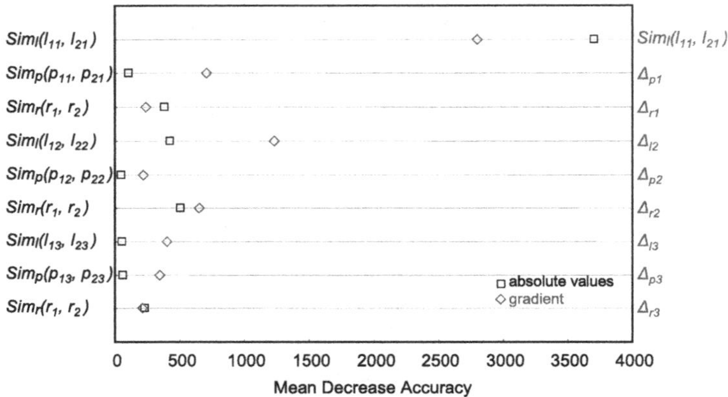

Fig. 6. Comparison of predictor importance

calculated predictor importance (Mean Decrease Accuracy in Random Forests) in both kinds of feature vectors, which is shown in Fig.6. As a result, although the importance of the basis was rather high, we confirmed that the importance decreased and importance of surrounding nodes and links increased in the features of the gradient, comparing with the features of absolutes values. Thus, the features of the gradient can be regarded as less dependence on a particular predictor, and effective when the training data is missing and changing.

5 Conclusion and Future Work

This paper addressed literal matching to determine identities of literal values in LOD, in order to give a URI to the literals, which exist in LOD in a certain proportion, and allow search through different LOD sets. Then, we proposed a novel method to extract image features from LOD, inspired by computer shogi. In an experiment, we extracted literal pairs from Japanese music category of DPpedia Japanese and Freebase, and determined the identities of the pairs using the ensemble learning that is a well-known classifier in computer vision. We then confirmed that it successfully classified the pairs with F-measure of 99%.

Future works include performance evaluations. Matching between datasets is not performed at users access, but at the dataset creation, and partially performed again at the update of the dataset. Therefore, considering the matching is determined for each category, the processing time is not a critical issue. In the future, however, the relationship between data size and the processing time needs to be investigated for making a plan to open the dataset on the web.

This method in principle does not depend on the music domain, and we intend to apply it to other domains. In addition, we intend to make a module of the classifier and publish it as a web service that dynamically determines the matching of literals. We expect these initiatives to facilitate cross-domain search through links in LOD.

References

1. Bizer, C., Heath, T., Berners-Lee, T.: Linked Data - The Story So Far. International Journal on Semantic Web and Information Systems (IJSWIS) 5(3), 1–22 (2009)
2. Obata, T., Sugiyama, T., Hoki, K., Ito, T.: Consultation Algorithm for Computer Shogi: Move Decisions by Majority. In: van den Herik, H.J., Iida, H., Plaat, A. (eds.) CG 2010. LNCS, vol. 6515, pp. 156–165. Springer, Heidelberg (2011)
3. Hoki, K., Kaneko, T.: The Global Landscape of Objective Functions for the Optimization of Shogi Piece Values with a Game-Tree Search. In: van den Herik, H.J., Plaat, A. (eds.) ACG 2011. LNCS, vol. 7168, pp. 184–195. Springer, Heidelberg (2012)
4. Volz, J., Bizer, C., Gaedke, M., Kobilarov, G.: Silk - A link discovery framework for the web of data. In: Proc. of 2nd Linked Data on the Web Workshop (LDOW). CEUR Workshop Proceedings, vol. 538 (2009)
5. Volz, J., Bizer, C., Gaedke, M., Kobilarov, G.: Discovering and maintaining links on the web of data. In: Bernstein, A., Karger, D.R., Heath, T., Feigenbaum, L., Maynard, D., Motta, E., Thirunarayan, K. (eds.) ISWC 2009. LNCS, vol. 5823, pp. 650–665. Springer, Heidelberg (2009)
6. Isele, R., Bizer, C.: Learning linkage rules using genetic programming. In: Proc. of 6th International Workshop on Ontology Matching (OM). CEUR Workshop Proceedings, vol. 814 (2011)
7. Sindice - The Semantic Web Index (2014), `http://sindice.com/`
8. Google Refine (in transition over to Open Refine) (2014), `https://github.com/OpenRefine`
9. Maali, F., Cyganiak, R., Peristeras, V.: Re-using Cool URIs: Entity Reconciliation Against LOD Hubs. In: Proc of 4th Linked Data on the Web Workshop (LDOW 2011). CEUR Workshop Proceedings, vol. 813 (2011)
10. Niu, X., Rong, S., Wang, H., Yu, Y.: An Effective Rule Miner for Instance Matching in a Web of Data. In: Proc. of 21st ACM International Conference on Information and Knowledge Management (CIKM), pp. 1085–1094 (2012)
11. Rong, S., Niu, X., Xiang, E.W., Wang, H., Yang, Q., Yu, Y.: A Machine Learning Approach for Instance Matching Based on Similarity Metrics. In: Cudré-Mauroux, P., Heflin, J., Sirin, E., Tudorache, T., Euzenat, J., Hauswirth, M., Parreira, J.X., Hendler, J., Schreiber, G., Bernstein, A., Blomqvist, E. (eds.) ISWC 2012, Part I. LNCS, vol. 7649, pp. 460–475. Springer, Heidelberg (2012)
12. Dai, W., Yang, Q., Xue, G., Yu, Y.: Boosting for transfer learning. In: Proc. of 24th International Conference on Machine Learning (ICML), pp. 193–200 (2007)
13. Hu, W., Chen, J., Cheng, G., Qu, Y.: Objectcoref & falcon-ao: results for oaei2010. In: Proc. of 5th International Workshop on Ontology Matching (OM). CEUR Workshop Proceedings, vol. 689 (2010)
14. Li, J., Tang, J., Li, Y., Luo, Q.: Rimom: A dynamic multistrategy ontology alignment framework. IEEE Transactions on Knowledge and Data Engineering 21(8), 1218–1232 (2009)
15. Castano, S., Ferrara, A., Montanelli, S., Varese, G.: Ontology and Instance Matching. In: Knowledge-Driven Multimedia Information Extraction and Ontology Evolution, pp. 167–195 (2011)
16. Apache Solr (2014), `http://lucene.apache.org/solr/`
17. Gunaratna, K., Thirunarayan, K., Jain, P., Sheth, A., Wijeratne, S.: A Statistical and Schema Independent Approach to Identify Equivalent Properties on Linked Data. In: Proc. of 9th International Conference on Semantic Systems (I-SEMANTICS), pp. 33–40 (2013)

18. Fujitsu Laboratories of Europe Limited: "Fujitsu Laboratories Develops Technology for Automatically Linking with Open Data throughout the World" (2014), http://www.fujitsu.com/global/news/pr/archives/month/2014/20140116-01.html
19. TAC KBP 2013 Entity Linking Track (2014), http://www.nist.gov/tac/2013/KBP/EntityLinking/index.html
20. Miao, Q., Lu, H., Zhang, S., Meng, Y.: Simple Yet Effective Method for Entity Linking in Microblog-Genre Text. In: Zhou, G., Li, J., Zhao, D., Feng, Y. (eds.) NLPCC 2013. CCIS, vol. 400, pp. 440–447. Springer, Heidelberg (2013)
21. Lowe, D.G.: Object recognition from local scale-invariant features. In: Proc. of 7th International Conference on Computer Vision (ICCV), vol. 2, pp. 1150–1157 (1999)
22. Li, Y., Bandar, Z.A., McLean, D.: An Approach for Measuring Semantic Similarity between Words Using Multiple Information Sources. IEEE Transactions on Knowledge and Data Engineering 15(4), 871–882 (2003)
23. Princeton University: WordNet: A lexical database for English (2014), http://wordnetweb.princeton.edu/perl/webwn
24. National Institute of Information and Communications Technology: Japanese WordNet (2014), http://nlpwww.nict.go.jp/wn-ja
25. National Institute of Informatics: DBpedia Japanese (2014), http://ja.dbpedia.org/sparql
26. OpenLink Software: OpenSearch (2014), http://lod.openlinksw.com/sparql/
27. Breiman, L.: Random Forests. Machine Learning 45(1), 5–32 (2001)

Audio Features in Music Information Retrieval

Daniel Grzywczak and Grzegorz Gwardys

Warsaw University of Technology
Institute of Radioelectronics
Warsaw, Poland
{D.Grzywczak,G.Gwardys}@ire.pw.edu.pl

Abstract. The rapid increase in the amount of on-line music services creates a need for automatic extraction of information from songs, music indexing and recommendation. But Music Information Retrieval is not an easy task and is still under development. This review paper focuses on features in Music Information Retrieval. We present audio engineering features and feature learning approach.

1 Introduction

Today, on-line music services are more and more popular in the whole world. Therefore, dealing with a large amount of music data is becoming a challenge. International Federation of the Phonographic Industry (IFPI[1]) reports a constantly decreasing trend in global revenues from physical records and increasing in digital records. Physical records revenues in 2012 were only $8.7 billion with 56% of market, in 2013 revenues were $7.7 billion. In 2012 digital records were 35% of market with $5.6 billion of revenues, in 2013 it was 39% of market with $5.9 billion of revenues. One of the significant factors of this change, was an increasing impact of on-line music services. IFPI reports that revenues from subscription streams increased by 51% and were $1.1 billion and the number of paying service subscribers increased by 40% to 28 million. These kind of services have many advantages over classic CD's. For instance, one can buy the desired songs only, instead of the whole album. Moreover, you pay only once for an access to numerous music tracks which are available everywhere, where the internet connection is available. One of the most popular online music service is Spotify[2]. This service has about 24 million users, 6 millions whom pay $10 monthly to get an unlimited access on all devices to over 20 million songs database without any advertisements. To keep up to .date Spotify adds 20 000 songs per day to the database [3]. Dealing with such a huge amount of audio files requires music resources to be indexed and classified.

Music Information Retrieval (MIR) use audio signal to describe and classify songs. The biggest advantage of this kind of systems is that, they do not need any

[1] http://ifpi.org
[2] http://spotify.com
[3] http://press.spotify.com/us/information/

D. Ślęzak et al. (Eds.): AMT 2014, LNCS 8610, pp. 187–199, 2014.

human effort like rating or any additional data like systems based on metadata. Even modern recommendation systems, based on an user‘s history, their ratings and correlation between them (Collaborative Filtering Approach [1]), should provide feature extraction module to prevent from the cold start effect or the lack of user‘s songs rating.

MIR topic appeared in the 60‘s of 20th century [2] but its significant development started at the beginning of the 21st century. MIR covers a variety of topics, such as user recommendations, music similarity, mood and genre classification, artists identification, tempo estimation and beat tracking, melody extraction and others. MIR is an interdisciplinary research area that involves not only many parts of computer science, signal processing and information theory, but also musicology, music theory and psychology. International Society for Music Information Retrieval Conference (ISMIR) is an annual event, established in 2000, which connects MIR researches from different areas [3]. To support MIR algorithms development, Music Information Retrieval Evaluation eXchange (MIREX[4]) contest is organized since 2005. In 2014 MIREX evaluates MIR algorithms in 17 different categories such as cover song identification, melody extraction and query by singing, humming and tapping [1, 3–5].

This paper focuses on feature extraction from audio signals. In Section 2 a general MIR system architecture is introduced. The Section 3 concentrates on the most popular features extracted in many MIR systems, whereas the Section 4 focuses on learning features approach.

2 System Architecture

General scheme of MIR system is presented in Figure 1. Each audio signal is represented by an audio features, extracted from this signal. Audio features can be learnt from training data (Feature Learning Approach - see Section 4), or extracted directly from a signal (Audio Features - see Section 3). Feature extraction can by preceded by segmentation and preprocessing step. Extracted features from collection of audio signals form a feature set, which can be used to build a data model. Data model can be constructed using approaches known from data analysis and machine learning such as k-means, Support Vector Machines etc. There can be many uses of data model - music track genre recognition, music similarity, efficient music retrieval.

3 Audio Features

3.1 Low-Level Signal Features

Low-level Signal features form a large set of spectral and temporal parameters of audio signal. They are not perceptually motivated and describe the specificity of

[4] `http://www.music-ir.org/mirex/wiki/MIREX_HOME`

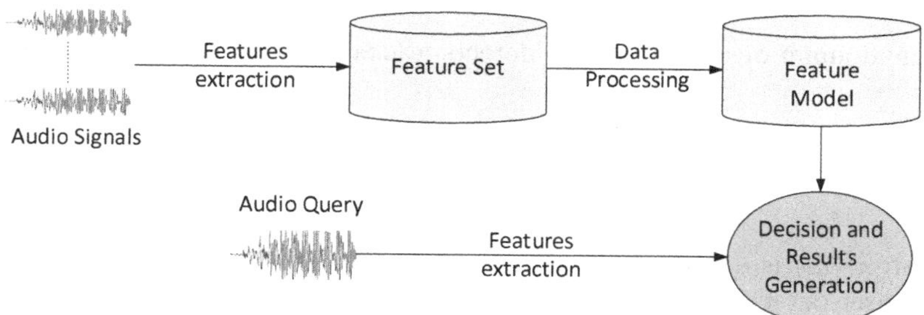

Fig. 1. General MIR System Architecture

signal on time or frequency domain. Because music usually varies significantly in time, therefore feature extraction of physical features is performed in short overlapping windows. In frequency domain, for the same reason, Short Time Fourier Transform (STFT) is used, which is Fourier Transform performed on short overlapping windows. Most popular physical feature in time domain is *Zero Crossing Rate* which describes the number of zero crossing in time domain during the windows analysis. The most popular features in frequency domain are [6]: *Spectral Centroid* which describes the center of spectrum gravity, *Spectral Rolloff* a frequency below which 85% of the spectrum magnitude is concentrated, *Spectral Flux* which describes the difference between the consecutive frames energies, *Short-Time Energy* and *Band-Level Energy* which describe the whole energy in a window and energy in each subband in window respectively , *Fundamental Frequency* which is first harmonic of audio signal. Also all parameters used in statistics as variance or skewness can be used as a low level feature for audio signal classification. There are several tools for feature extraction like JAudio [7] or Marsyas [8]. Good example of MIR system based on physical features is CUIDADO [9].

3.2 Tempo Estimation

The goal of tempo estimation is to detect particular beats in a music track and estimate the tempo of a song. Beats detection and tempo estimation is not always unambiguous, for example in classic music with rubato tempo, in jazz and blues compositions, during live performance artists have place for their interpretation.

Tempo estimation can be divided into two stages. First is the onset detection which goal is to detect the beginnings of notes. This phase is often preceded by some preprocessing. This stage usually emphasises strong transients, loud and bass notes and wide spectrum events. The objective of the second stage is to look for periodicity in the detected onset.

Onset Detection. There are several methods for onset detection [10] [11]:

- Time domain onset detection - detects a local increase of signal envelope

$$E_0 = \frac{1}{N} \sum_{m=-\frac{N}{2}}^{\frac{N}{2}-1} |x(n+m)|w(m) \tag{1}$$

  where $w(m)$ is a smoothing kernel. Also an increase in energy can be used for onset detection

$$E = \frac{1}{N} \sum_{m=-\frac{N}{2}}^{\frac{N}{2}-1} [x(n+m)]^2 w(m) \tag{2}$$

  Onset detection in time domain can be unreliable for music with loud background.
- Frequency domain onset detection - focuses on detecting increase in energy in frequency domain

$$SF(n) = \frac{1}{N} \sum_{k=-\frac{N}{2}}^{\frac{N}{2}-1} H(|X(n,k)| - |X(n-1,k)|) \tag{3}$$

  where $H(x) = \frac{x+|x|}{2}$ is a half-wave rectifier, and $|X(n,k)|$ spectral amplitude in n-th frame and k -th frequency bin.
- Phase onset detection - focuses on detecting phase deviation in frequency spectrum in adjacent windows

$$PD(n) = \frac{1}{N} \sum_{k=-\frac{N}{2}}^{\frac{N}{2}-1} |\psi^{''}(n,k)| \tag{4}$$

  where $\psi(n,k)$ is a phase of $X(n,k)$, $\psi(n,k)^{'} = \psi(n,k) - \psi(n-1,k)$ and $\psi(n,k)^{''} = \psi^{'}(n,k) - \psi^{'}(n-1,k)$.
- Complex Domain onset detection - detects difference between complex vector in frequency domain, so it combines phase and magnitude informations

$$CD(n) = \sum_{k=-\frac{N}{2}}^{\frac{N}{2}-1} |X(n,k) - X_T(n,k)| \tag{5}$$

  where $X_T = X(n-1,k)e^{\psi(n-1,k)+\psi^{'}(n-1,k)}$.

Onset detection can also be used as a segmentation method, applied before feature extraction stage.

Periodicity Functions. After onset detection phase, periodicity is detected. There are several methods used in beat tracking [12]:

- *Autocorrelation* - Describes how a function is similar to itself.

$$AC(\tau) = \sum_{n=0}^{N-\tau-1} x(n)x(n+\tau) \tag{6}$$

  This approach can be used is some extensions like Autocorrelation Phase Matrix [13]
- *Comb Filters Bank* - set of filters where each one is tuned to a particular frequency. The construction of filters is crucial for periodicity detection. General equation for a filter is [14]:

$$y_\tau(t) = \alpha_\tau y_\tau(t-\tau) + (1-\alpha_\tau)x(t) \tag{7}$$

  where α_τ is a gain for given time shift τ
- *Inter-Onset Intervals* - intervals between detected onsets are grouped into the clusters of intervals of the same length. Intervals may partially or completely overlap on each other. After clustering, groups are ranked which gives the most probable beat hypothesis. Inter-Onset Intervals approach was used in the BeatRoot System [11]

3.3 Mel Frequency Cepstrum Coefficients

MFCC are well known features in speech recognition domain but were also adapted to Music Information Retrieval domain [15], despite the fact that it discards pitch information, a valuable feature in MIR. MFCC provide perceptual representation of short-time spectrum envelope.

In Figure 2 it can be seen that MFCC are much more discriminative than spectrograms which may contain a lot of redundant linearly-spaced frequency channels.

To compute MFCC first STFT is computed. Then, Mel filter-bank is applied by summing squares of the spectral magnitudes within each band. After Mel sub bands division, logarithm of each sub band is computed and finally Discrete Cosine Transform (DCT) is computed for the whole spectrum. Usually only some of the first MFCC are used (10 - 15 coefficients).

3.4 Daubechies Wavelet Coefficient Histograms

Wavelet Transform, similarly to Fourier Transform is used for signal decomposition into different components. However, unlike Fourier Transform, Wavelet Transfom decomposes a signal into components called wavelets, which frequency and amplitude vary in time. To obtain Daubechies Wavelet Coefficient Histograms (DWCH) we need to first compute db_8 wavelet decomposition, where db_8 is a Daubechies wavelet with 8 vanishing moments. Then we construct a histogram and for each subband we compute the first three moments and energy, which give us 28 features. The wavelet approach is also used in image processing and data mining [16].

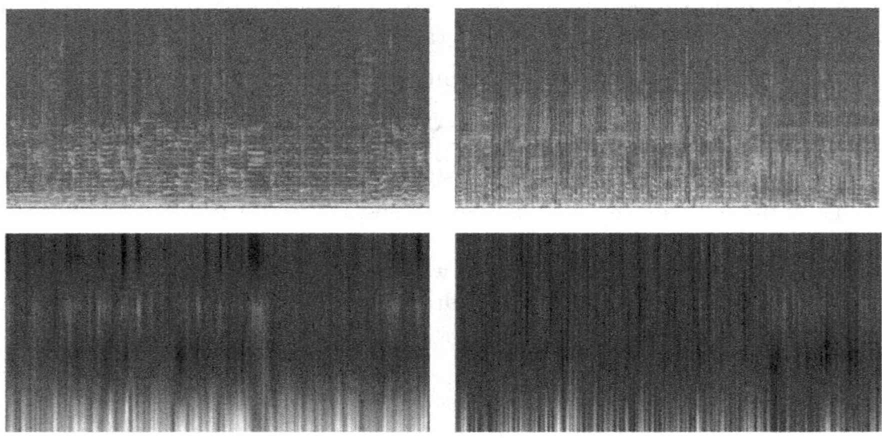

Fig. 2. top left: spectrogram of country song, top right: spectrogram of hiphop song, bottom left: MFCC of country song, bottom right: MFCC of hiphop song

3.5 Chroma Based Feature

Chromagram, also known as Pitch Class Profile or Harmonic Pitch Class Profile, is a type of frequency folding, where energy levels are related to pitch classes in twelfth-octave bands[5]. Example of Chromagram is presented in Figure 3.

Chroma features are powerful representation for music audio - it is known from music theory that notes at a distance of one octave are perceived as particularly similar. Therefore, knowing only the distribution of Chroma, without the absolute frequency, can provide meaningful music information.

Generally, Chroma features cope well with noise such as ambient noise or percussive sounds. What is more they are independent of timbre, loudness and dynamics [18]. In some MIR tasks, these qualities might make them lead to better results than popular MFCCs - for example in cover song identification [18] or in music similarity [19]. Chromagrams are used in many other MIR tasks such as Key Extraction [20] or Automatic Chord Recognition [21, 22].

4 Feature Learning Approaches

The purpose of feature learning or representation learning [23] is to determine such a transformation of an input, that its result (representation) could be effectively used in another task, for example classification. There is a whole set of feature learning algorithms in machine learning: autoencoders [24], dictionary learning, matrix factorization [25], Restricted Boltzmann Machines (RBM) [24]

[5] however there are solutions with 24, 36 or 48 bands in case of finer resolution of pitch information [17].

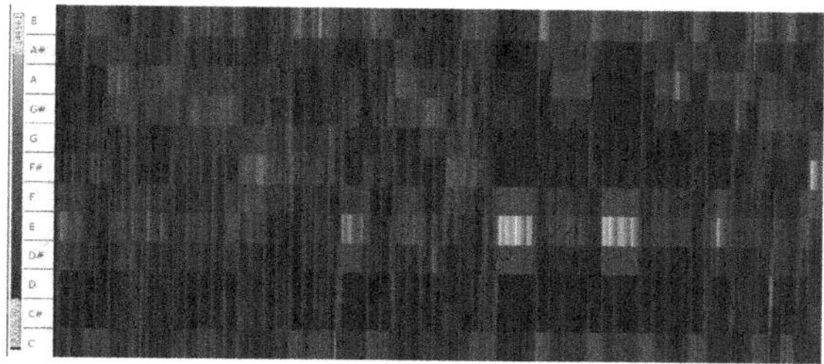

Fig. 3. Chromagram - extracted a series of tonal chroma vectors from the audio

or even well known k-means clustering [24] is classified in this group. The way of learning representation consists of multiple non-linear transformations of the data (for example multilayer neural networks) then we call such a technique a deep learning one [23].

In computer vision, speech recognition and natural language processing (NLP) domains the feature learning is a leading approach - it improves the previous state of the art by a large margin [26–28].

Therefore it is not surprising that Humphrey et al. [29] recommend the use of deep architectures to solve MIR problems.

4.1 Bag of Words Model

Bag of words (BOW) model is a representation that assumes no order between words, which originates from the the NLP domain. It became popular in computer vision applications [30], as well as MIR tasks [31–33].

To construct a BOW representation, we need features - in NLP these are simply words, however in Computer Vision or MIR we must somehow define them. It is achieved by choosing some low-level features and then clustering them into codewords. While in Computer Vision Scale Invariant Feature Transform (SIFT) [34] or Speeded Up Robust Features (SURF) [35] descriptors are common choices for these low-level features, in MIR it is popular to choose MFCCs or Chromagrams. Once the codebook has been constructed on a large corpus of low-level features, we can represent any song in a bag of words model, by performing vector quantization (VQ).

Once we get a BOW representation, we can benefit from some well known techniques from NLP such as Topic Modelling. For example, Latent Semantic Indexing technique can be used to represent a whole audio clip in the latent perceptual space [31].

4.2 Restricted Boltzmann Machines and Deep Belief Networks

RBM is a generative stochastic neural network, originally invented under the name Harmonium by P. Smolensky in 1986 [36], however the RBMs gained enormous attention after G. Hinton and his colleagues presented the fast learning algorithms in the mid-2000s. RBMs were successfully used in dimensionality reduction [37], classification [38], feature learning [24] and topic modelling [39].

Deep Belief Networks can be viewed as a composition of RBMs (see Figure 4), that can be optionally fine-tuned with gradient descent and backpropagation[6].

RBMs and DBNs are also well known in machine listening community. Lee et al. [40] used DBNs to acoustic signals noticing improvement over MFCCs for speaker, gender, and phoneme detection. P. Hamel and D. Eck also demonstrate an increase in performance after adopting DBNs to the musical genre identification and autotagging tasks [41,42]. DBNs and RBMs were also used in rythm and melody recognition [43,44] or in modelling drum patterns [45].

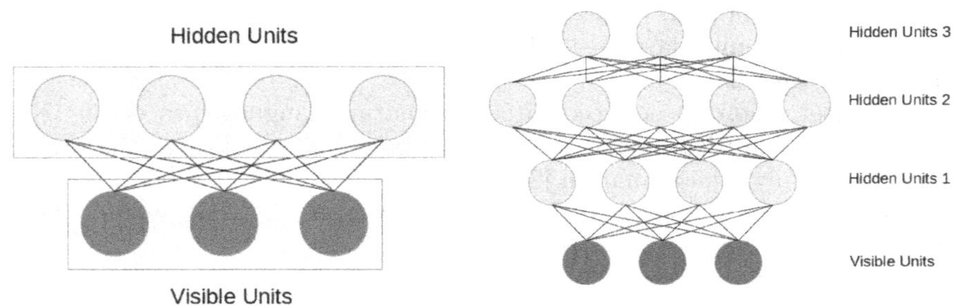

Fig. 4. On left: Restricted Boltzman Machine with 3 visible units and 4 hidden units; On right: Deep Belief Network with 3 hidden layers

The RBM can be seen as an undirected graphical model consisting of two fully inter-connected layers, one consisting of hidden units and the other one of visible units. "Restriction assumption" states that there is no connection between nodes from the same layer. This allows for more efficient training algorithms like gradient-based contrastive divergence algorithm.

The conventional type of RBM has binary-valued hidden and visible units. Connections between hidden and visible units are represented by matrix of weights $W = (w_{i,j})$.There ale also bias weights - a_i for the visible units v and b_j for the hidden units . Energy of configuration is defined as:

$$E(v,h) = -\sum_i a_i v_i - \sum_j b_j h_j - \sum_i \sum_j h_j w_{i,j} v_i = -a^{\mathrm{T}} v - b^{\mathrm{T}} h - h^{\mathrm{T}} W v \quad (8)$$

In terms of energy, we can define joint probability distribution:

$$P(v,h) = \frac{1}{Z} e^{-E(v,h)} \quad (9)$$

[6] http://www.scholarpedia.org/article/Deep_belief_neworks

Thanks to lack of intra-layer connections, we can define conditional probabilities as:

$$P(h|v) = \prod_{j=1}^{n} P(h_j|v)$$
$$P(v|h) = \prod_{i=1}^{m} P(v_i|h) \tag{10}$$

The activation probabilities of the units are defined as:

$$P(h_j = 1|v) = \sigma(b_j + \sum_{i=1}^{m} w_{i,j}v_i)$$
$$P(v_i = 1|h) = \sigma(a_i + \sum_{j=1}^{n} w_{i,j}h_j) \tag{11}$$

where σ denotes a logistic sigmoid.

RBM is trained to maximize probability or equivalently expected log probability of visible units configuration:

$$\arg\max_{W} \prod_{v \in V} P(v)$$
$$\arg\max_{W} E[\sum_{v \in V} \log P(v)] \tag{12}$$

4.3 Multiscale Approach

S. Dieleman and B. Schrauwen present in [46] a shallow feature learning approach which is a multiscale one. One of the main blocks of feature learning pipeline (see Figure 5) is a spherical K-means algorithm [47]. [24] presents results showing that shallow feature learning techniques such as the K-means algorithm can exceed deep learning techniques like RBMs, autoencoders and sparse coding.

The proposed pipeline is applied to three multiscale architectures: multiresolution spectrograms, Gaussian pyramid and Laplacian pyramid. The multiresolution spectrograms is the simplest approach - like in [42] mel-spectrograms are constructed with different window sizes [7]. Gaussian pyramid is obtained by smoothing and sub-sampling operations performed on subsequent levels [46]. The Laplacian pyramid can be derived from Gaussian pyramid as it is a similar concept - each level of the Laplacian pyramid is the difference between two adjacent levels of the Gaussian one.

[7] in [46] spectrogram window size was doubled at each level.

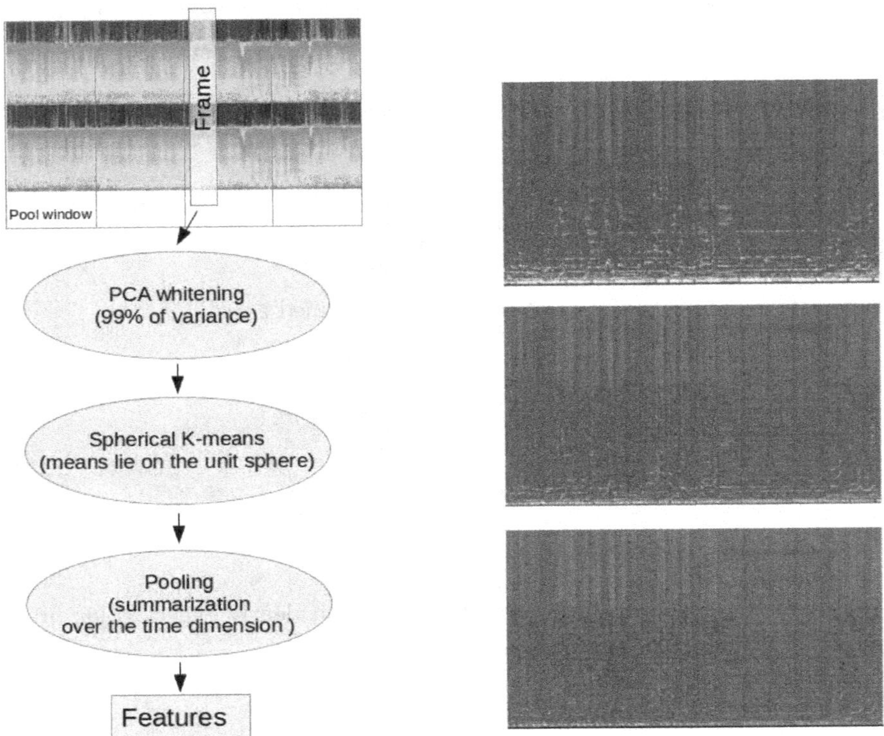

Fig. 5. Feature learning pipeline (left) and multiscale archictecture - multiresolution spectrograms (right)

The proposed pipeline is applied to each architecture to many levels separately. The feature vectors are constructed then by concatenation. The authors of [46] do not choose one winner among the proposed multiscale architectures, however learning features at multiple timescales outperform the single-timescale approaches.

5 Conclusions

In this paper, we presented a wide range of music features, which are the main components of content-based approach in MIR. Paper was divided into two parts. In the first part we focused on automatic extracted audio features. Here were shown features such as MFCCs and Chromagrams, Tempo estimation or Daubechies Wavelet Coefficient Histograms. These features are easy in extraction and do not require additional training data.

In the second part we presented a feature learning approach, a natural successor of engineered audio features. Feature learning, in contrast to audio features approach, needs training data and obtained features are related to data statistics. We gave an overview of both deep and shallow learning techniques as well mentioned a multiscale approach.

Both approaches are used in MIR tasks. In latest MIREX Mood and Genre Classification the winning solution was to perform Gabor filtering on spectrograms with additional beat tracking information (respectively 68% and 76% of accuracy) [48]. On the other hand, the use of convolutional neural network (CNN), which are popular in computer vision domain [27], allowed to get first place in Audio Onset Detection task (87% of avg F-measure) [49].

It must be said, that MIREX datasets can be to small for feature learning techniques to get winning results, nevertheless the interest in this approach is very high in machine listening community. We believe, that this kind of techniques, as in computer vision or NLP, it will gain more and more attention, especially in the appearance of new annotated data sets.

References

1. Song, Y., Dixon, S., Pearce, M.: A survey of music recommendation systems and future perspectives. In: 9th International Symposium on Computer Music Modeling and Retrieval (2012)
2. Kassler, M.: Toward Musical Information. Perspectives of New Music 4(2), 59–66 (1966)
3. Futrelle, J., Downie, J.S.: Interdisciplinary Communities and Research Issues in Music Information Retrieval. In: Library and Information Science, pp. 215–221 (2002)
4. Downie, J.S., West, K., Ehmann, A.F., Vincent, E.: The 2005 Music Information Retrieval Evaluation Exchange (MIREX 2005): Preliminary Overview. In: International Conference on Music Information Retrieval, pp. 320–323 (2005)
5. Downie, J.S., Ehmann, A.F., Bay, M., Jones, M.C.: The music information retrieval evaluation exchange: Some observations and insights. In: Ras, Z.W., Wieczorkowska, A. (eds.) Adv. in Music Inform. Retrieval. SCI, vol. 274, pp. 93–115. Springer, Heidelberg (2010)
6. Rao, P.: Audio signal processing. In: Prasad, B., Prasanna, S.R.M. (eds.) Audio Signal Processing. SCI, vol. 83, pp. 169–189. Springer, Heidelberg (2008)
7. Mcennis, D., Mckay, C., Fujinaga, I.: Jaudio: A feature extraction library. In: International Conference on Music Information Retrieval (2005)
8. Tzanetakis, G., Cook, P.: Marsyas: A framework for audio analysis. Org. Sound 4, 169–175 (1999)
9. Peeters, G.: A large set of audio features for sound description (similarity and classification) in the CUIDADO project. Technical report, Icram (2004)
10. Bello, J.P., Daudet, L., Abdallah, S., Duxbury, C., Davies, M., Sandler, M.B.: A tutorial on onset detection in music signals. IEEE Transactions on Speech and Audio Processing 13, 1035–1047 (2005)
11. Dixon, S.: Onset Detection Revisited. In: Proc. of the Int. Conf. on Digital Audio Effects (DAFx 2006), Montreal, Quebec, Canada, pp. 133–137 (2006)
12. Klapuri, A., Davy, M.: Signal Processing Methods for Music Transcription. Springer, New York (2006)

13. Eck, D.: Beat tracking using an autocorrelation phase matrix. In: Proceedings of the 2007 International Conference on Acoustics, Speech and Signal Processing (ICASSP), pp. 1313–1316. IEEE Signal Processing Society (2007)
14. Robertson, A., Stark, A.M., Plumbley, M.D.: Real-time visual beat tracking using a comb filter matrix. In: Proceedings of the International Computer Music Conference (2011)
15. Logan, B.: Mel frequency cepstral coefficients for music modeling. In: International Symposium on Music Information Retrieval (2000)
16. Ogihara, M.: Toward intelligent music information retrieval. IEEE Transactions on Multimedia 8, 564–574 (2006)
17. Casey, M.A., Veltkamp, R., Goto, M., Leman, M., Rhodes, C., Slaney, M.: Content-based music information retrieval: Current directions and future challenges. Proceedings of the IEEE 96(4), 668–696 (2008)
18. Serra, J., Gómez, E., Herrera, P., Serra, X.: Chroma binary similarity and local alignment applied to cover song identification. IEEE Transactions on Audio, Speech, and Language Processing 16, 1138–1151 (2008)
19. Yu, X., Zhang, J., Liu, J., Wan, W., Yang, W.: An audio retrieval method based on chromagram and distance metrics. In: 2010 International Conference on Audio Language and Image Processing (ICALIP), pp. 425–428. IEEE (2010)
20. Van De Par, S., McKinney, M.F., Redert, A.: Musical key extraction from audio using profile training. In: ISMIR, pp. 328–329 (2006)
21. Lee, K.: Automatic chord recognition from audio using enhanced pitch class profile. In: Proc. of the International Computer Music Conference (2006)
22. Harte, C., Sandler, M.: Automatic chord identifcation using a quantised chromagram. In: Audio Engineering Society Convention 118. Audio Engineering Society (2005)
23. Bengio, Y., Courville, A., Vincent, P.: Representation learning: A review and new perspectives (2013)
24. Coates, A., Ng, A.Y., Lee, H.: An analysis of single-layer networks in unsupervised feature learning. In: International Conference on Artificial Intelligence and Statistics, pp. 215–223 (2011)
25. Srebro, N., Rennie, J.D., Jaakkola, T.: Maximum-margin matrix factorization. In: NIPS, vol. 17, pp. 1329–1336 (2004)
26. Dahl, G.E., Yu, D., Deng, L., Acero, A.: Context-dependent pre-trained deep neural networks for large-vocabulary speech recognition. IEEE Transactions on Audio, Speech, and Language Processing 20, 30–42 (2012)
27. Krizhevsky, A., Sutskever, I., Hinton, G.E.: Imagenet classification with deep convolutional neural networks. In: NIPS, vol. 1, p. 4 (2012)
28. Socher, R., Perelygin, A., Wu, J.Y., Chuang, J., Manning, C.D., Ng, A.Y., Potts, C.: Recursive deep models for semantic compositionality over a sentiment treebank
29. Humphrey, E.J., Bello, J.P., LeCun, Y.: Moving beyond feature design: Deep architectures and automatic feature learning in music informatics. In: ISMIR, pp. 403–408 (2012)
30. Sivic, J., Zisserman, A.: Efficient visual search of videos cast as text retrieval. IEEE Transactions on Pattern Analysis and Machine Intelligence 31, 591–606 (2009)
31. Sundaram, S., Narayanan, S.: Audio retrieval by latent perceptual indexing. In: IEEE International Conference on Acoustics, Speech and Signal Processing, ICASSP 2008, pp. 49–52. IEEE (2008)
32. Seyerlehner, K., Widmer, G., Knees, P.: Frame level audio similarity-a codebook approach. In: Proceedings of the 11th International Conference on Digital Audio Effects (DAFx 2008), Espoo, Finland (2008)

33. Hoffman, M.D., Blei, D.M., Cook, P.R.: Easy as cba: A simple probabilistic model for tagging music. In: ISMIR, vol. 9, pp. 369–374 (2009)
34. Lowe, D.G.: Object recognition from local scale-invariant features. In: The Proceedings of the Seventh IEEE International Conference on Computer Vision, vol. 2, pp. 1150–1157. IEEE (1999)
35. Bay, H., Tuytelaars, T., Van Gool, L.: SURF: Speeded up robust features. In: Leonardis, A., Bischof, H., Pinz, A. (eds.) ECCV 2006, Part I. LNCS, vol. 3951, pp. 404–417. Springer, Heidelberg (2006)
36. Smolensky, P.: Parallel distributed processing: Explorations in the microstructure of cognition, vol. 1, pp. 194–281. MIT Press, Cambridge (1986)
37. Hinton, G.E., Salakhutdinov, R.R.: Reducing the dimensionality of data with neural networks. Science 313(5786), 504–507 (2006)
38. Larochelle, H., Bengio, Y.: Classification using discriminative restricted boltzmann machines. In: Proceedings of the 25th International Conference on Machine Learning, pp. 536–543. ACM (2008)
39. Salakhutdinov, R., Hinton, G.E.: Replicated softmax: an undirected topic model. In: NIPS, vol. 22, pp. 1607–1614 (2009)
40. Lee, H., Pham, P.T., Largman, Y., Ng, A.Y.: Unsupervised feature learning for audio classification using convolutional deep belief networks. In: NIPS, vol. 9, pp. 1096–1104 (2009)
41. Hamel, P., Lemieux, S., Bengio, Y., Eck, D.: Temporal pooling and multiscale learning for automatic annotation and ranking of music audio. In: ISMIR, pp. 729–734 (2011)
42. Hamel, P., Bengio, Y., Eck, D.: Building musically-relevant audio features through multiple timescale representations. In: ISMIR, pp. 553–558 (2012)
43. Schmidt, E.M., Kim, Y.E.: Learning rhythm and melody features with deep belief networks. In: ISMIR (2013)
44. Cherla, S., Weyde, T., d'Avila Garcez, A.S., Pearce, M.: A distributed model for multiple-viewpoint melodic prediction. In: de Souza Britto Jr., A., Gouyon, F., Dixon, S. (eds.) ISMIR, pp. 15–20 (2013)
45. Battenberg, E., Wessel, D.: Analyzing drum patterns using conditional deep belief networks. In: ISMIR, pp. 37–42 (2012)
46. Dieleman, S., Schrauwen, B.: Multiscale approaches to music audio feature learning (November 4-8, 2013)
47. Coates, A., Ng, A.Y.: Learning feature representations with k-means. In: Montavon, G., Orr, G.B., Müller, K.-R. (eds.) Neural Networks: Tricks of the Trade, 2nd edn. LNCS, vol. 7700, pp. 561–580. Springer, Heidelberg (2012)
48. Wu, M.J.: Mirex 2013 submissions for train/test tasks (draft) (2013)
49. Schlüter, J., Böck, S.: Cnn-based audio onset detection mirex submission, http://www.music-ir.org/mirex/abstracts/2013/SB1.pdf

Model of Auditory Filters and MPEG-7 Descriptors in Sound Recognition

Aneta Świercz and Jan Żera

Institute of Radioelectronics, Warsaw University of Technology
ul. Nowowiejska 15/19, 00-665 Warsaw, Poland
{a.swiercz,j.zera}@ire.pw.edu.pl

Abstract. It was examined whether applying a model of human auditory filter could improve the quality of sound recognition with the use of MPEG-7 standard audio descriptors. Modeling of filtering in the auditory system was with a bank of 38 gammatone filters closely spaced across the audible frequency range. The bank of filters was implemented as a low-level audio descriptor to replace the short-term Fourier transform (STFT) MPEG-7 audio descriptor. Sound recognition tests were conducted on a large set of sounds of nine musical instruments and speech of twelve speakers. The results showed that the proposed descriptor employing a bank of gammatone filters led to improved recognition of musical instruments and speakers as compared to the STFT-based original low-level MPEG-7 audio descriptor.

Keywords: MPEG-7 audio descriptors, gammatone filter, sound recognition, instrument recognition, speaker recognition.

1 Introduction

Although most Internet searches are still done using text-based information, there are many kinds of non-text data available from many sources. However, the possibility of finding information by using data such as audio signal analysis is a very promising prospect. Unified standard rules for sound recognition and indexing have been described in the ISO/IEC MPEG-7 standard [7,8]. The first step of the sound recognition process is the extraction of sound features from sound material. Sound features described by the low-level descriptors of the MPEG-7 standard are extracted using purely statistical magnitudes mostly derived using short-term Fourier transform (STFT) of audio signal, with no consideration of properties of the human hearing system. There are many examples, however, of existing models of the auditory system based on banks of gammatone filters which successfully represent various aspects of perception. The motivation for this study was to propose a descriptor applying a bank of gammatone filters that models filtering in the auditory system. New descriptor design is consistent with the MPEG-7 standard. It was examined whether the use of the gammatone filter improved recognition of instruments and speakers.

D. Ślęzak et al. (Eds.): AMT 2014, LNCS 8610, pp. 200–211, 2014.

2 MPEG-7 Audio Descriptors

The MPEG-7 standard [7, 8] attempts to describe sound features in the form of audio low-level descriptors and uses high-level tools for sound applications. Low-level descriptors are calculated as statistical quantities derived from audio signals such as: audio signal waveform envelope, spectrum envelope based on the STFT, spectrum reduced by using singular value decomposition (SVD), spectrum centroid, spectrum spread, spectrum flatness, and harmonicity. Audio high-level tools make use of the output obtained from the low-level descriptors and are used by algorithms that are designed, among others, for sound recognition and indexing, instrument recognition or melody recognition. Description of sound obtained by using MPEG-7 audio low-level descriptors does not represent directly auditory sensory features and specifically the filtering that takes place in the auditory system.

The MPEG-7 standard high-level tools for general sound recognition and indexing use hidden Markov models (HMMs) as a classifier. These tools use STFT sound spectrum (AudioSpectrumEnvelope) reduced with the SVD algorithm. The AudioSpectrumEnvelope descriptor preserves information on sound spectrum envelope averaged from the STFT coefficients over 30-ms time frames.

3 Gammatone Filter

Typical numerical modeling of auditory system is based upon a bank of bandpass filters followed by some model of neural activity, and a decision statistics [3, 4, 9, 12]. Bandpass properties of hearing system are most commonly modeled with the use of gammatone or more advanced gammachirp filters [5, 6, 13]. These filters allow for time varying analysis of sound in frequency bands similar to that determined in masking procedures for the hearing system. In this study, the gammatone filter was used which has symmetric frequency response sufficient for analysis of sound at moderate sound pressure levels (not exceeding 80-85 dB SPL). The impulse response of the gammatone filter is given by equation:

$$gammatone(t) = at^{n-1}e^{-2\pi bERB(f_c)t}cos(2\pi f_c t + \Phi), t > 0,$$

where: f_c – filter center frequency, $ERB(f_c)$ – equivalent rectangular bandwidth of the filter, t – time, n – rank order of gamma function, a, b – constants.

4 The GT and MP7 Applications

An AudioSpectrumEnvelopeGT (GT for GammaTone) descriptor was designed and implemented as compatible with the MPEG-7 standard. This descriptor filtered sound with the use of gammatone filters. Output of the AudioSpectrumEnvelopeGT descriptor provided averaged energy in the 38 consecutive ERB bands in a format compatible with low-level MPEG-7 AudioSpectrumEnvelope descriptor. The entire sound recognition process made use of audio high-level tools for general sound recognition and indexing, and the AudioSpectrumBasis and AudioSpectrumProjection low-level descriptors. Sound information obtained from low-level AudioSpectrumEnvelope descriptor was reduced using the SVD and stored in the AudioSpectrumBasis and AudioSpectrumProjection.

In this study, the AudioSpectrumEnvelopeGT replaced the original AudioSpectrumEnvelope MPEG-7 descriptor. Therefore, it allowed us to compare the sound recognition with the use of the gammatone filters (AudioSpectrumEnvelopeGT descriptor) with the sound recognition of STFT analysis (original AudioSpectrumEnvelope MPEG-7 descriptor). The MPEG-7 audio descriptors used in this study followed the implementation by [2].

5 Tests

Experiments dealt with the recognition of musical instruments and speakers (male and female voices). The first experiment examined recognition among nine orchestral instruments. The second experiment was designed to show how using entirely different recordings for training and testing processes influenced sound recognition. The third experiment explored recognition of twelve male and female speakers. Finally, the fourth experiment was dedicated to joint recognition of instrumental and speech sounds.

5.1 Test Sounds and Method

In the experiments, a set of 3087 sounds were used representing nine orchestral instruments (woodwinds, brass and strings): flute (fl), oboe (ob), clarinet (cl), trumpet (tr), trombone (trb), French horn (cor), violin (vn), viola (vla), and cello (vc). Samples comprised simple isolated sounds and those of more complex articulation (0.2-15.5 s duration). Instrumental sounds were derived from two entirely different recording sources. The first source represented sounds of the highest quality digital studio recording [11]. These sounds provided complete representation of instruments in terms of pitch scale and various manners of articulation. The second source was a set of 660 sounds of six instruments (flute, clarinet, trumpet, trombone, French horn, and violin) recorded at the Fryderyk Chopin University of Music (FCUM) in Warsaw. Sounds were played by students who were asked to freely articulate isolated random sounds across the entire instrument pitch scale with unrestricted articulation. Recordings were made with the use of Schoeps MK21 and MK2S microphones and an audio interface RME Fireface 800.

Speech sounds used in the experiments comprised 4680 words of Polish language recorded by 12 speakers, six female (K1, ..., K6) and six male (M7, ..., M12) voices. Two sets of words were used: set A included 240 monosyllabic words of 10 phonetically balanced lists [10], set B contained 150 mono- and polysyllabic words randomly selected from a book [1]. Speech sounds were recorded in sound-proof room with the use of a DAT recorder Tascam DA-P1 and a Neuman TLM103 microphone.

There were two rules applied for splitting sounds into training and testing sets: fixed and random. In fixed selection, training sets consisted either of instrument sounds of simplest possible articulation (long and short, soft and loud isolated sounds) or for the recognition of speakers set A of words. Testing sets included instrument sounds of more complex articulation (e.g.: grace notes, mordents, tremolo, glissando, chords on strings), or for speakers set B of words. In random selection, 70% of all sounds used in a given test were randomly selected for training sets. Remaining 30% of sounds comprised the testing sets. Both for fixed and random selection no item of training sets was included in testing sets. The reason for using fixed selection for training and testing sets was to examine whether training carried out on simple objects (simplest articulation instrumental sounds or monosyllabic words) could be effective for the recognition of more complex structures. Within this context, the random selection, often used in various studies served as a reference.

Each result presented in this study either for fixed or random selection and the GT or MP7 application is an average of 100 simulations. In all figures, additional vertical lines indicate ±one standard deviation around the mean value. In the fixed selection, 100 simulations differed in initial randomization seed. In the random selection, training and testing sets were randomly drawn 10 times, and 10 simulations were conducted for each selection with different randomization seed (10x10 equals a total of 100 simulations). The use of 10 different training and testing sets provided additional randomization, which allowed us to avoid bias resulting from specific selection of both sets.

5.2 Recognition of Nine Orchestral Instruments

Recognition tests of nine instruments were conducted on a number of samples in fixed and random selection listed in Table 1. Test results are shown in Fig. 1 (see also main diagonals of matrices in Table 2). These results showed that use of the AudioSpectrumEnvelopeGT descriptor in the GT application generally improved instrument recognition as compared to the original MPEG-7 AudioSpectrumEnvelope descriptor of MP7 application. In most cases, in both fixed and random split of the data, results obtained using gammatone filters had a higher percentage of correct instrument recognitions. Only in the case of French horn in fixed selection, the MP7 application achieved higher recognition rate by 12 pp (percentage points) than the GT application. The largest differences between both applications occurred for the flute and amounted to 45 and 35 pp in fixed and random selections, respectively. In both cases, correct recognition was well below 50%, which can-not be considered satisfactory for identifying any object. Similar but smaller difference of 41 pp between the applications was observed for the violin in fixed selection. This difference, however, was significant in terms of the quality of recognition. The violin was not an instrument well recognized by the MP7 application (43%), but was relatively well recognized by the GT application (84%). The smallest differences between the GT and MP7 applications were 6 pp in fixed and 3 pp in random selections for recognition of the trombone. In all cases, the advantage of using gammatone filters in GT application over the original MPEG-7 descriptor was larger in fixed than random selection of training and testing sets. For all instruments (except French horn), lower recognition rate obtained either in fixed or random selection by the GT application was higher than the recognition by the MP7 application in any of selection conditions. All discussed differences between the two applications were statistically significant, as was shown by the ANOVA and Student's t-test (p=0.01).

Table 1. Numbers of sounds in training and testing sets for recognition of 9 instruments

	Fixed selection									Random selection								
	Woodwinds			Brass			Strings			Woodwinds			Brass			Strings		
Set	fl	ob	cl	tr	trb	cor	vn	vla	vc	fl	ob	cl	tr	trb	cor	vn	vla	vc
Training	112	139	176	120	185	118	178	142	185	186	180	282	161	176	124	205	177	209
Testing	153	118	227	110	66	59	115	111	113	79	77	121	69	75	53	88	76	89

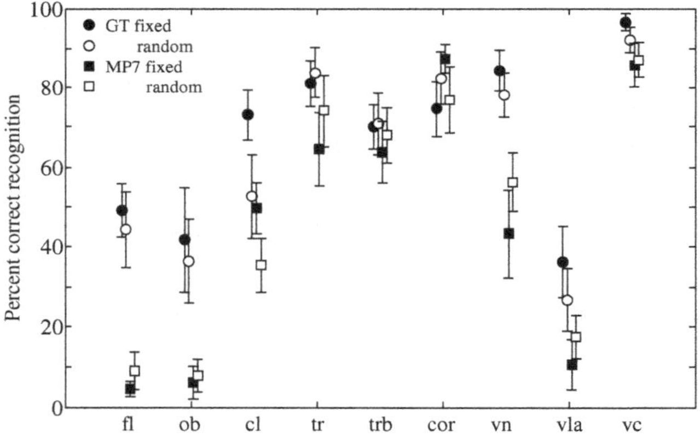

Fig. 1. Recognition of 9 instruments by the GT and MP7 applications

Confusion matrices (Table 2) show less incorrect classifications by the GT than MP7 application. The GT application classifies erroneously less than 33% and 37% of instrumental sounds for fixed and random selections, respectively. The MP7 application incorrectly classifies more than 52% for fixed, and 54% for random selections. It may be noted that for the GT application correct classifications of instruments are dominant except for viola sounds in random selection. In the case of MP7 application, sounds of three instruments (flute, oboe, viola) were predominantly classified as other instruments, and their classification to

Table 2. Confusion matrices for recognition of 9 instruments

Classifi-cation	Fixed selection Instruments									Random selection Instruments								
	fl	ob	cl	tr	trb	cor	vn	vla	vc	fl	ob	cl	tr	trb	cor	vn	vla	vc
	GT application																	
fl	49	1	-	-	-	-	-	-	-	44	3	-	-	-	-	-	-	-
ob	1	42	-	-	-	-	-	-	-	-	36	1	-	-	-	-	-	-
cl	1	12	73	3	-	1	-	-	-	-	4	53	2	-	-	-	-	-
tr	4	13	6	81	4	5	1	-	-	10	22	13	84	1	1	3	-	-
trb	21	29	4	12	70	17	5	12	1	13	25	6	11	71	16	5	7	1
cor	11	3	15	2	25	75	6	8	-	17	4	23	2	27	82	9	15	2
vn	7	1	1	1	-	2	84	31	-	9	5	1	1	-	-	78	30	2
vla	-	-	-	-	-	-	3	36	2	1	-	-	-	-	-	2	27	3
vc	4	-	1	-	-	-	1	12	97	6	1	5	-	1	-	2	22	92
	MP7 application																	
fl	5	1	-	-	-	-	1	-	-	9	2	-	-	-	-	1	-	-
ob	3	6	-	-	-	-	1	2	1	2	8	-	-	-	-	1	1	-
cl	18	14	50	1	1	-	4	6	1	5	5	35	-	-	-	1	1	-
tr	1	7	4	64	2	-	1	-	-	3	12	9	74	2	-	7	1	-
trb	6	9	8	10	64	11	21	25	3	9	10	12	11	68	19	17	14	1
cor	56	59	36	22	32	87	27	30	6	47	42	35	11	27	77	14	22	4
vn	4	4	1	2	-	-	43	6	1	12	13	2	2	1	-	56	9	1
vla	2	-	-	-	-	-	1	11	3	2	2	1	-	-	-	1	18	7
vc	7	1	1	-	-	-	1	19	86	10	5	5	1	3	3	1	35	87

respective classes were on long positions. In both fixed and random selections, when majority of testing sounds of all instruments was incorrectly classified by the GT and MP7 applications the correct recognition rate was below 27%.

5.3 Recognition of Sounds from Different Recording Sources

Tests were carried out on sets of sounds comprising samples of six instruments: flute, clarinet, trumpet, trombone, French horn and violin. In the main test, the training sets were the same as those used for recognition of nine instruments [11] (Table 1). The testing sets comprised 110 sound samples per instrument from the second source i.e. recorded at the FCUM. The main test was followed by the reference test in which both training and testing sets were the same as those used for recognition of nine instruments [11].

The use of entirely different sources for sound samples in training and testing sets had a significant impact on instrument recognition by both the GT and MP7 applications (Fig. 2 and Fig. 3). The recognition rate of most of the instruments was reduced in the main test as compared to the reference test. Differences were greater for the performance of the MP7 application. For the MP7 application,

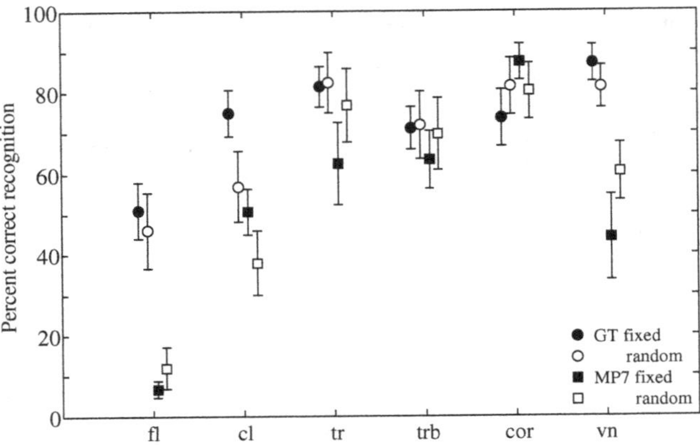

Fig. 2. Recognition of 6 instruments in main test (training and testing sets from two different sources)

the difference between main and reference tests ranged from 2 pp (improved recognition of the clarinet in fixed selection) to 61 pp (reduced recognition of the trumpet in random selection). For the GT application, the decrease in correct recognition ranged from 0.3 pp (flute) to 41 pp (trumpet), in fixed selection of training and testing sets.

Regardless of different sources used for training and testing sets, the GT application preserved higher correct recognition rate than the MP7 application (Fig. 2). There are certain cases, however, in which the MP7 application proved more correct in identifying instruments than the GT application in the main test as compared to the reference test (Fig. 3): recognition of French horn in both selections, and recognition of the clarinet in random selection. For the clarinet, trumpet and trombone, using training and testing sets from different sources largely altered differences in recognition by the GT and MP7 applications as compared to the reference test. Changes were associated with clear increase or decrease in differences between GT and MP7 applications. The differences between GT and MP7 applications were statistically significant (p=0.01) as showed by ANOVA and Student's t-test (with exception for the trombone and French horn in the reference test in random selection, and clarinet in the main test in fixed selection).

5.4 Recognition of Speakers

The use of the GT application resulted in better recognition of twelve speakers by the MP7 application in fixed selection of training and testing sets (Fig. 4) except speakers K6, M8 and M9. Speaker M9 was correctly recognized by both applications in 100% of cases. The largest difference in favor of the GT

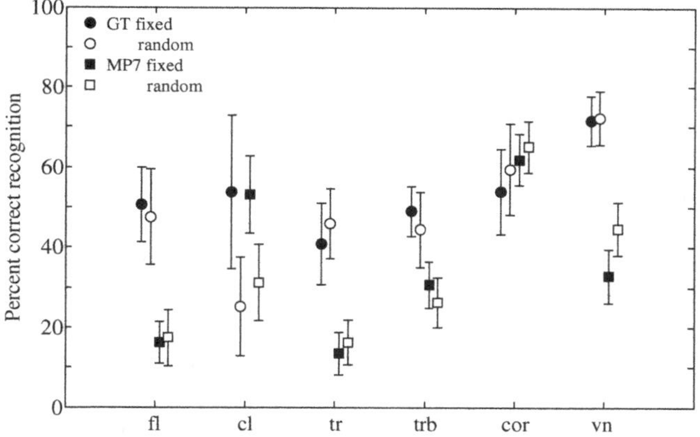

Fig. 3. Recognition of 6 instruments in reference test (training and testing sets from single source)

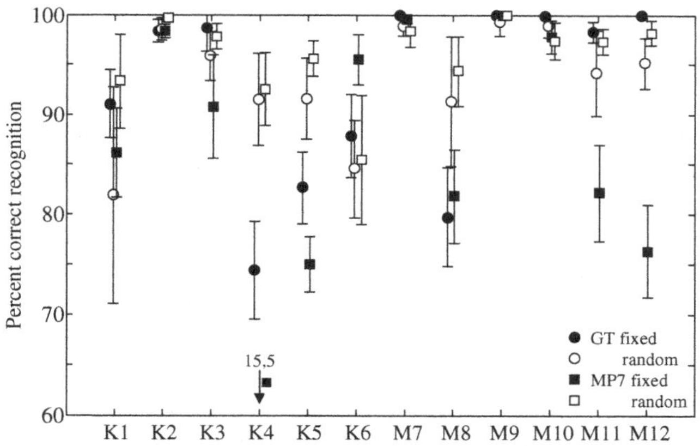

Fig. 4. Recognition of 12 speakers by the GT and MP7 applications

application was 59 pp (K4). In random selection, slightly higher recognition rate was obtained for the MP7 than GT application (except for M7 and M10). The differences between the GT and MP7 applications were, however, smaller than 4 pp with the exception of 11-pp difference for speaker K1. Therefore, the results can be considered comparable. For comparison, correct recognition of five speakers was more than 7 pp higher for the GT than MP7 application in fixed selection. All discussed differences were statistically significant (ANOVA test and Student's t-test, p=0.01) except for speakers K2 and K4 in fixed selection and speaker K6 in random selection of training and testing sets.

For the GT application, there were fewer incorrect classifications of speakers than for the MP7 application (GT 7.0%, MP7 10.5%; Table 3). In fixed selection, the MP7 application had more than twice as many erroneous classifications as the GT application (GT 7.4%, MP7 16.8%). In random selection, the GT application showed minimally more erroneous classifications (GT 6.5%, MP7 4.2%). The GT application assigned most testing words of all speakers to the respective classes. This same was true for the MP7 application but only in random selection. In the case of fixed selection, speaker K4 was predominantly classified erroneously as speaker K6. Very exceptionally male voices were erroneously classified as female voices or vice versa. Most of such cases (24%) occurred for the MP7 application in fixed selection (speaker M12). For the GT application and fixed selection, there was the largest (7%) number of misclassifications of female voices (K1, K5, K6) as male voices.

5.5 Recognition within Joint Set of Instrumental and Speech Sounds

Recognition tests within a joint set of instrumental and speech sounds was conducted on three instruments (clarinet, French horn, cello: 878 sounds), and voices of two male (M7, M9) and two female (K2, K3) speakers (1560 word samples). Instruments selected were best recognizable instruments as averaged over the GT and MP7 applications and fixed and random selection of training and testing sets in the first experiment (Sec. 5.2), within families of woodwinds, brass and strings. Selected speakers showed the highest recognition among male and female voices (Sec. 5.4). Training and testing sets for each object were the same as for recognition test within the set of nine instruments (Sec. 5.2), and recognition of twelve speakers (Sec. 5.4).

Recognition scores shown in Fig. 5 provide no explicit evidence that any of the GT or MP7 applications showed a higher efficiency in recognition. Speakers were recognized by both applications at a similar level of more than 96.5% correct recognitions. Differences between the GT and MP7 applications were usually not exceeding 1 pp. Only speaker K3 in random selection was better recognized by the MP7 than GT application by 2 pp. In the case of instruments, the rate of correct recognitions was much lower and showed larger differences (GT from 51% to 98%, MP7 from 40% to 95%). A considerable difference of 13 or 12 pp in favor of the GT application occurred for the recognition of the clarinet in both fixed and random selections. A difference of 17 pp in favor of the MP7 application occurred for fixed selection in identifying French horn. All differences between the GT and MP7 applications seen in Fig. 5 were statistically significant (ANOVA and Student's t-test, p=0.01) except for M7 in fixed and the cello in random selection of training and testing sets.

Table 3. Confusion matrices for recognition of 12 speakers

Classifi-cation	Fixed selection Speakers												Random selection Speakers											
	Female (K)						Male (M)						Female (K)						Male (M)					
	1	2	3	4	5	6	7	8	9	10	11	12	1	2	3	4	5	6	7	8	9	10	11	12
	GT application																							
K1	91	-	-	3	5	2	-	-	-	-	-	-	82	-	-	2	-	4	-	-	-	-	-	-
K2	1	98	1	-	3	3	-	-	-	-	-	-	4	99	3	1	4	3	-	-	-	-	-	-
K3	-	1	99	-	3	1	-	-	-	-	-	-	2	-	96	-	2	1	-	-	-	-	-	-
K4	1	-	-	74	2	-	-	-	-	-	-	-	3	-	-	92	2	4	-	-	-	-	-	1
K5	-	-	-	-	83	-	-	-	-	-	-	-	4	-	1	-	92	1	-	-	-	-	-	-
K6	7	-	-	22	4	88	-	-	-	-	-	-	3	-	-	4	-	85	-	-	-	-	-	-
M7	-	-	-	1	1	4	100	17	-	-	-	-	1	-	-	1	-	2	99	5	1	-	1	2
M8	-	-	-	-	-	-	-	80	-	-	-	-	-	-	-	-	-	-	1	91	-	-	-	-
M9	-	-	-	-	-	1	-	1	100	-	-	-	-	-	-	-	-	-	-	1	99	-	1	-
M10	-	-	-	-	-	-	-	1	-	100	1	-	-	-	-	-	-	-	-	-	-	99	4	-
M11	-	-	-	-	-	-	-	1	-	-	98	-	1	-	-	1	-	-	-	2	-	-	94	2
M12	-	-	-	-	-	-	-	-	-	-	-	100	-	-	-	-	-	-	-	-	-	-	-	95
	MP7 application																							
K1	86	-	-	-	2	-	-	-	-	-	-	3	93	-	-	-	-	-	-	-	-	-	-	-
K2	-	98	-	-	1	-	-	-	-	-	-	-	-	100	1	-	-	-	-	-	-	-	-	-
K3	-	1	91	-	2	-	-	-	-	-	-	-	-	-	98	-	2	-	-	-	-	-	-	-
K4	-	-	-	15	-	-	-	-	-	-	-	6	-	-	-	93	-	11	-	-	-	-	-	-
K5	1	-	-	-	75	-	-	-	-	-	-	-	4	-	1	-	96	1	-	-	-	-	-	-
K6	12	1	9	85	19	95	-	-	-	-	-	15	2	-	-	7	2	85	-	-	-	-	-	-
M7	-	-	-	-	-	3	100	17	-	-	16	-	-	-	-	-	-	1	98	4	-	-	1	-
M8	-	-	-	-	-	-	-	82	-	-	-	-	-	-	-	-	-	-	1	94	-	-	-	-
M9	-	-	-	-	-	1	-	-	100	2	1	-	-	-	-	-	-	-	-	-	100	2	1	-
M10	-	-	-	-	-	-	-	-	-	98	1	-	-	-	-	-	-	-	-	-	-	97	1	-
M11	-	-	-	-	-	-	-	1	-	-	82	-	-	-	-	-	-	-	-	1	-	-	97	2
M12	-	-	-	-	-	-	-	-	-	-	-	76	-	-	-	-	-	-	-	-	-	-	-	98

Confusion matrices (not included here) showed that there were very few errors in speaker recognition by both applications. No speaker sound was misclassified as that of an instrument. In contrast, erroneous classification of instruments as speakers was relatively frequent. In the case of the MP7 application only, most of sounds of an object (clarinet) were incorrectly classified (random selection).

Small and inconsistent differences between the GT and MP7 applications can likely be attributed to selection of sounds best recognized in earlier tests. It is likely that if more diverse objects from previous tests were selected for the present test, the differences between the GT and MP7 applications would turn out to be larger and indicate the dominance of one application over the other.

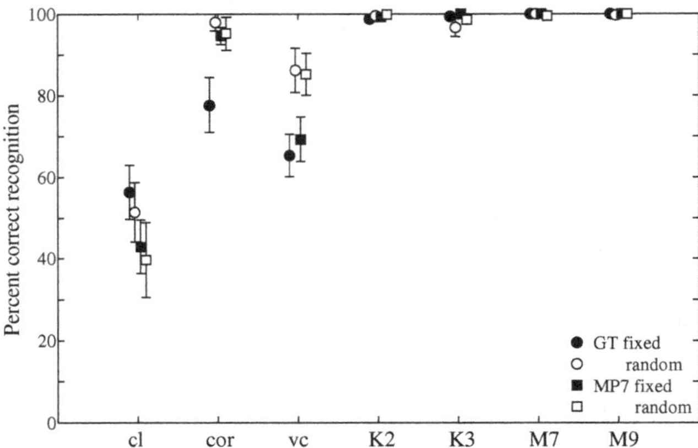

Fig. 5. Recognition of 3 instruments and 4 speakers by the GT and MP7 applications

6 Conclusions

The study demonstrated that replacing the STFT by gammatone filters in sound analysis led to improved instrument and speaker recognition by high-level tools of MPEG-7 audio. For most cases in which the differences in the correct recognition exceeded 5 percentage points, correct recognition by the GT was higher than by the MP7 application. Clear advantage of the GT application was observed in recognition of musical instruments. For speech sounds, the GT application demonstrated a higher recognition rate of speakers in fixed selection of training and testing sets. In random selection of sets, differences between the gammatone and STFT disappeared, being in most cases smaller than 4 percentage points. A similar result of no clear advantage of any application, regardless of the selection of training and testing sets, with differences below 4 percentage points, occurred when the set of recognized objects included both instruments and speakers. The present study showed the merit of undertaking studies of sound recognition with the use of gammatone filters. Possible modifications may include change of gammatone filter parameters and more complex processing of signal following the filtering in subsequent frequency bands. The AudioSpectrumEnvelopeGT descriptor can be also used for calculating various other MPEG-7 audio descriptors.

Acknowledgments. This research was supported by grant NN516377634 founded by the Polish National Science Centre, statutory grant 504P0413 from the Ministry of Science and Higher Education, and the Foundation for the Development of Radiocommunication and Multimedia Technologies.

References

1. Buchner, E.: Linia rozwoju życia, PWM (1994) (in Polish)
2. Casey, M.: General sound classification and similarity in MPEG-7. Organised Sound 6(2), 153–164 (2001)
3. Dau, T., Puschel, D., Kohlrausch, A.: A quantitative model of the "effective" signal processing in the auditory system, I. Model structure. J. Acoust. Soc. Am. 99(6), 3615–3622 (1996)
4. Dau, T., Puschel, D., Kohlrausch, A.: A quantitative model of the "effective" signal processing in the auditory system. II. Simulations and measurements. J. Acoust. Soc. Am. 99(6), 3623–3631 (1996)
5. Hartmann, W.M.: Signals, Sound, and Sensation, AIP Series in modern acoustics and signal processing. Springer, New York (2000)
6. Irino, T., Patterson, R.D.: A time-domain. level-dependent auditory filter: the gammachirp. J. Acoust. Soc. Am. 101, 412–419 (1997)
7. ISO/IEC FDIS 15938-4:2001(E), Information Technology - Multimedia Content Description Interface - Part 4: Audio (2001)
8. Manjunath, B.S., Salembier, P., Sikora, T.: Introduction to MPEG-7 Multimedia Content Description Interface. Wiley (2002)
9. Patterson, R.D., Allerhand, M., Giguere, C.: Time-domain modelling of peripheral auditory processing: A modular architecture and a software platform. J. Acoust. Soc. Am. 98, 1890–1894 (1995)
10. Pruszewicz, A., Demenko, G., Richter, A., Wika, T.: Nowe listy artykulacyjne do badań audiometrycznych. Logopedia 20, 139 (1993) (in Polish)
11. Siedlaczek, P.: Peter Siedlaczek's Advanced Orchestra, Recordings CD Vol. 1-5
12. Tchorz, J., Kollmeier, B.: A model of auditory perception as front end for automatic speech recognition, J. Acoust. Soc. Am. 106(4), 2040–2050 (1999)
13. Unoki, M., Irino, T., Glasberg, B., Moore, B.C.J., Patterson, R.D.: Comparison of the roex and gammachirp filters as representation of the auditory filter. J. Acoust. Soc. Am. 120, 1474–1492 (2006)

A Scheme for Robust Biometric Watermarking in Web Databases for Ownership Proof with Identification

Vidhi Khanduja[1], Shampa Chakraverty[1], Om Prakash Verma[2], and Neha Singh[1]

[1] Department of Computer Engineering, Netaji Subhas Institute of Technology, Delhi, India
{vidhikhanduja9,apmahs.nsit}@gmail.com, nehalohchubh@yahoo.com
[2] Department of Information Technology, Delhi Technological University, Delhi, India
opverma.dce@gmail.com

Abstract. We propose a robust technique for watermarking relational databases using voice as a biometric identifier of ownership. The choice of voice as the distinguishing factor is influenced by its uniqueness, stability, universality, and ease of use. The watermark is generated by creating a statistical model of the features extracted from the owner's voice. This biometric watermark is then securely embedded into selected positions of the fractional parts of selected numeric attributes using a reversible bit encoding technique. In case of a dispute regarding true ownership, the relative scores of the extracted watermark are generated by comparing features of the disputed voices with the extracted one. We experimentally demonstrate the robustness of the proposed technique against various attacks. Results show that watermark is extracted with 100% accuracy even when 97% tuples are added or when 90% tuples are deleted or when 80% tuples are altered. More significantly, biometric identification helped identify the correct owner even when only 60% of the watermark could be extracted.

Keywords: Relational Databases, Ownership Protection, Robust Watermarking, Voice Biometrics.

1 Introduction

Digital convergence technologies have enabled the seamless integration of human activities with powerful machine capabilities and thereby helped humanity achieve new milestones in almost every aspect of life. Ubiquitous intelligent interfaces perceive snippets of information continually from numerous environmental sources. The data collected are then transferred through the internet to build up huge compendiums of information. As a result, the web is today populated by large collections of digital databases. Unfortunately, the menace of piracy and rights infringement has also reached dangerous proportions by feeding upon these very technological advancements. Malicious or benign attacks pose continuous threats to the owners of digital and multimedia content. This has stimulated the need for powerful combative measures which can protect the ownership and integrity of databases efficiently.

Technological Protective Measures (TPM) backed by anti-circumvention laws, have been adopted by many nations to deter such threats [1]. Digital watermarking is

D. Ślęzak et al. (Eds.): AMT 2014, LNCS 8610, pp. 212–225, 2014.

a cost-effective and widely used self help technological measure to protect multimedia content and, in recent years, digital databases [2]. In this paper, we focus on watermarking relational databases augmented by biometrics as a means to safeguard the interests of owners against illegal claims of ownership. It is well known that biometrics provide a promising technique to uniquely identify an individual based on his biological features and gives little scope for duplications. Contrary to traditional watermarking techniques which entail any random pattern as a watermark to be embedded into the database, we generate a meaningful sequence using the voice of the owner. The choice of voice is influenced by the fact that voice biometric is unique to an individual, is difficult to forge or fake, is universally available and is easy to capture through cheap audio sensors in a man-machine environment. The proposed technique is robust and reversible and can protect the ownership rights of the owners of web databases with a high degree of accuracy.

The rest of this paper is organized as follows. Section 2 gives a glimpse of the work done in the domain concerned. Section 3 describes the implementation of the proposed watermarking technique. Section 4 presents the experiments performed to check the robustness achieved in a biometrically watermarked database. Section 5 concludes our paper.

2 Related Work

One of the earliest works that aimed at protecting ownership rights of relational databases was proposed in [3]. The authors proposed a bit level watermarking technique that effectively marks numeric attributes by modifying the least significant bit (lsb) of selected numeric attributes. Later, in a bid to enhance security, Xinchun et al. proposed a modification to the simple bit level watermarking technique by introducing the concept of weighted attributes wherein all candidate attributes for marking are given different weights by the owner in order to model the marking rate of each attribute as per her preference [4]. However, these lsb based techniques are not resilient to simple bit based attacks such as random bit flipping which can degrade the watermark.

Some watermarking techniques have been proposed that optimize the statistical properties of watermarked attributes in order to maximize discernability while respecting usability constraints [5, 6, 7]. The technique proposed by M.Shehab et al. is primary key dependent and computationally inefficient [5]. In [6] the authors proposed a primary-key independent, optimization based watermarking technique using Bacterial Foraging Algorithm with increased robustness and efficiency. Recently, Kamran and Farooq proposed an information-preserving watermarking technique on numeric attributes, highlighting the need of rights protection in Electronic Medical Records (EMR) systems [7]. Some researchers have also exploited non-numeric and categorical attributes for rights protection [8, 9, 10].

Scant attention has been paid towards investigating the potential of biometrics to prove the ownership of digital databases. In [11], Haiqing Wang proposed a technique for embedding the owner's voice to establish ownership rights. There are some serious shortcomings in their approach. Firstly, it is not reversible and hence any changes made to embed the watermark are permanent and cannot be undone. Secondly, the authors have not outlined any specific approach to reach a decision about the ownership of the data in question, which is the primary objective of the whole process. In this paper, we too adopt voice as a biometric source of watermark to establish ownership with identification. However, in sharp contrast with [11] we adopt a reversible technique which not only reverses all modifications made to the database during the process of watermark insertion but also enables a higher degree of robustness. Furthermore, we demonstrate systematically how ownership conflicts can be resolved by identifying the contending voices against the extracted watermark.

3 Proposed Technique

Fig. 1 gives the overall architecture of the proposed voice biometric based watermarking technique for a relational database. It consists of the sub-modules: *Watermark Preparer*, *Watermark Embedder* and *Watermark Extractor* and *Decision Generator*. We now describe each of these modules in the following sub-sections.

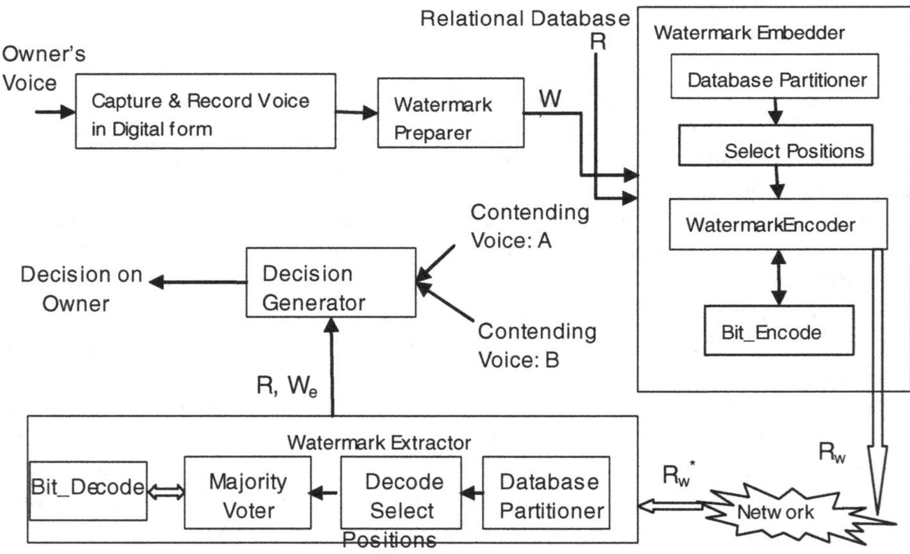

Fig. 1. Architecture of the biometric watermarking system

3.1 The Data Model

The database R is built upon a relational model. It includes the primary key attribute P_k and η number of float type numeric attributes which are candidates for embedding the watermark bits. Additionally, it may have non-numeric attributes, but they are not used for watermarking. The owner supplies a set of secret parameters:

Security Key. A key K is used for generating virtual partitions in a secure manner as well as for embedding the watermark bits into specific positions in the database.

Tuple Selection Parameter. The parameter Υ determines the fraction of all tuples in a database that can be utilized for embedding watermark bits.

Bit Position Parameter. The parameter Ω defines the total number of potential locations within the fractional part of float numeric attributes where a watermark bit can be inserted.

3.2 Watermark Preparer

The foremost step is to generate the watermark bit sequence using the database owner's biometric characteristic. The process Watermark Preparer does this by using the owner's voice as the source to construct the watermark.

The process starts by capturing and digitally recording the owner's voice sample using audio sensors, analog to digital converters and voice recording software. The digital representation of voice data is used as a training sample to extract its features using Mel Frequency Cepstral Coefficients (MFCC). This involves a series of processing steps which ensure that the final compressed signal adequately represents what the human auditory system can truly discern. These include, performing pre-emphasis of the voice data, framing and windowing, generating the Fast Fourier Transform (FFT) coefficients of the signal that represent its complex frequency spectrum, generating the power spectrum from the complex spectrum, applying Mel filtering to the power spectrum using Mel filter banks, taking logarithms of the Mel filter bank energies, performing Discrete Cosine Transform (DCT) on the log Mel filter bank energies, selecting appropriate DCT coefficients and appending delta energy features [12].

After extracting the MFCC feature vectors χ , a statistical model of the data is generated using Gaussian Mixture Model (GMM). The goal is to estimate the parameters of the GMM that best matches the distribution of MFCC training feature vectors χ [13]. This is achieved by employing the maximum log likelihood estimation technique which returns (i) the mean matrix (ii) the nodal diagonal covariance matrix and (iii) the matrix of mixture weights. These three matrices define the parameter of the GMM model λ. The individual elements of the matrices are then rounded off and concatenated to yield a bit sequence $W[1:L]$ of length L to be used as the watermark.

The use of voice as a biometric identifier strengthens proof of ownership. Instead of using any random pattern that has no intrinsic correlation with the owner of the watermark, the use of biometrics provides a meaningful sequence that can be deciphered to identify the owner. The owner need not pre-register her watermark along with the associated registration costs in order to resolve possible ownership disputes in a court of law. The extracted watermark automatically identifies the real owner. Thus, it provides a truly self-help technological measure for database protection.

3.3 Watermark Embedder

The block *Watermark Embedder* in Fig. 1 comprises all the processes that are needed to embed the watermark sequence $W[1:L]$ in the relational database R. The watermark is embedded into specific bits of the fractional portions of selected numeric attributes in target tuples. The following processes are carried by the Watermark Embedder using the secret parameters K, Υ , and Ω.

Create Virtual Partitions. Firstly, subroutine DB_Patition(.) shown in Fig. 2, carries out the task of partitioning R into N_p virtual partitions $S_0,\ldots,S_{N_p-1}$ which are initially empty. Next, taking each tuple in turn, the process concatenates the secret key K at both ends of the unique primary key $t.P_k$ of a tuple t and generates the hash $H_p(t)$ for secure partitioning by applying the MD5 hash function [14]. Finally it returns the partition index $i(t)$ for tuple t, which is equal to $H_p(t) \bmod N_p$. The virtual partition S_i thus gets its new member t.

The purpose of partitioning is to boost the resilience of the watermark against random tuple reordering attacks. As the assignment of each tuple to its virtual partition is a function of its primary key value rather than its spatial position in the database, reordering does not change its membership. The partitioning process uses the secret key K known only to the owner. Further it employs a randomizing hash function, thus making it almost impossible for an attacker to guess the partition assigned to any tuple.

Select Secret Embedding Positions. The next step is to select potential locations for embedding the L bits of the watermark $W[1:L]$. For each partition S_i the process Select_EmbedPositions(.) shown in Fig. 3, calculates the following parameters:

- *Watermark Bit Index.* A specific watermark bit index z_i within the L-bit watermark W is selected to be embedded in partition S_i (line 2). Actually, z_i as well as the next bit position z_i+1 are embedded in partition S_i.
- *Candidate Attributes.* Two among η candidate numeric attributes A_{i1} and A_{i2}, into which the selected bit will be embedded are selected (line 3). Readers may

note that more than two attributes per tuple can be selected for embedding the watermark bits in order to increase the level of redundancy. However, we have selected only two watermarking attributes as an illustration for introducing redundancy.

- *Target Tuples.* Specific tuples which are earmarked for embedding (line 5-6) are selected. To decide which tuples in a partition will participate in the embedding process, the secret key K is concatenated with the tuple's primary key and its hash $H_m(t)$ is generated for secure embedding. If the value of $(H_m(t) \bmod \Upsilon)$ is 0, then the tuple t is selected for watermarking (line 6). In effect, $1/\Upsilon$ fraction of the total number of tuples are thus earmarked. This secure hashing mechanism serves to ward off attackers who may venture to guess the watermarked tuples.
- *Embed Position.* A specific bit position $b(t)$ within the fractional parts of the selected attributes A_{i1} and A_{i2} is chosen to store the watermark bits z_i and $z_i + 1$ respectively (line 7). The hash $H_m(t)$ is used to choose $b(t)$.

Encode Watermark Bit. For each selected tuple t in a virtual partition S_i two consecutive watermark bits $W[Z_i]$ and $W[Z_i + 1]$ are embedded at the chosen bit position $b(t)$ of selected attributes A_{i1} and A_{i2} respectively. Fig. 4 shows the reversible bit encoding sub-routine Bit-Encode(.) that inserts a watermark bit w into the bth bit of a numeric attribute A of tuple t. The process Watermark_Encoder(.) in Fig. 1, calls Bit_Encode(.) twice for each selected tuple t. The first call Bit_Encode $(H_m(t), A_{i1}, b(t), W[z_i])$ embeds watermark $W[i]$ into bit position b of attribute A_{i1} and the second call Bit_Encode $(H_m(t), A_{i2}, b(t), W[z_i + 1])$ embeds watermark $W[z_i + 1]$ into bit position b of attribute A_{i2}.
Alterations in the fractional portion of numeric attributes ensure minimal change to their values. While inserting the watermark bits, the system verifies that the predefined usability constraints are respected. If the alterations violate any of these usability constraints, then the attribute value is not changed.

The reversible encoding process is based on the technique proposed in [15]. The fractional part of each of the two selected numeric attributes is first separated from its integer part, subtracted from the tuple's hash and converted to its binary form *Bin* (lines 1-3). Next, its binary value $Bin(b)$ at position b and the chosen watermark bit w are re-arranged so that b now contains w while $Bin(b)$ is appended at the end of the fractional part (lines 4-6). The modified fraction is converted back to decimal, subtracted again from the tuple's hash and added to the integer part I (lines 7-9).

The above secure encoding process serves two functions. Firstly, the secure hash function introduces a random perturbation in each attribute' value, thus making it impossible for an attacker to detect any pattern that can correlate the original and new values of the watermarked attributes without knowledge of the secret parameters. Secondly, the attribute can revert back to its original value after extracting the watermark bit. Fig. 6 illustrates this reversibility with the help of an example. Fig. 6a illustrates the watermark bit encoding procedure by tracing through the steps of Bit_Encode(.) and Fig. 6b illustrates the reverse procedure to restore the original attribute value.

The watermarked database R_w is now prepared and can be transferred through a network where it may be exposed to different kinds of attacks.

3.4 Watermark Extractor

The block Watermark Extractor in Fig. 1 contains all the processes needed for extracting the embedded watermark sequence from a suspected database R_w^* using a majority voting technique. The process also restores back the altered values of the database to its original form by exploiting the reversible nature of the embedding technique. The process of watermark extraction starts by repeating the first two steps of watermark embedding. It calls DB_Paritions(.) to generate the virtual partitions and Select_EmbedPositions(.) to select the watermarked tuples, attributes and bit positions.

***DB_Partition*(.)**	***Select_EmbedPositions*(.)**
1. Initialize virtual partitions $S_0, S_1 \ldots S_{Np-1}$ to ϕ	1. **For each** virtual partition S_i {
2. **For each** tuple t in R {	2. Calculate watermark bit index, $z_i = I \bmod L$.
3. Calculate partition hash $H_p(t) = hash\ (K // t.Pk // K)$	3. Calculate $A_{i1} = (i \bmod \eta)$ and $A_{i2} = (i \bmod \eta\text{-}1)$
4. Calculate index $i(t) = (H_p(t) \bmod N_p)$	4. **For each** tuple t in S_i {
5. Assign tuple t to the ith partition, $S_i = S_i \cup t$	5. Calculate $H_m(t) = hash\ (K // t.Pk)$
6. }	6. **If** $((H_m(t) \bmod \Upsilon) == 0)$ {
	7. Calculate $b(t) = ((H_m(t) \bmod \Omega)+1)$
	8. }
	9. }
	10. }

Fig. 2. Algorithm for generating virtual partitions of the database

Fig. 3. Algorithm for selecting positions for embedding watermark bits

Next, it calls the process Majority_Voter(.) described by the pseudo-code in Fig. 5, to reconstruct the watermark. Majority_Voter(.) takes each virtual partition in turn and initializes all elements of the $Count[1:L][0:1]$ matrix to zero. An element $Count[x][0]$ counts the number of times $W[x]$ is extracted as zero and $Count[x][1]$ counts the number of times a $W[x]$ bit is extracted as one. The sub-routine Bit_Decode(.) given in Fig. 7, retrieves an embedded watermark bit value from a watermarked tuple t from its possibly compromised attribute A' from bit position b. Majority_Voter(.) iteratively calls Bit_Decode(.) for each watermarked tuple and uses the retrieved values corresponding to the z and $(z+1)th$ bit of the watermark to update the corresponding elements of the $Count[.][.]$. Finally, the full watermark sequence is reconstructed using majority voting on the pair $Count[x][0]$ and $Count[x][1]$ for each watermark bit $x \in 1..L$.

Fig. 6b illustrates how the modification done in an attribute's value due to the encoding carried out by the example in Fig. 6a is reversed by tracing through the steps of Bit_Decode(.).

3.5 Decision Generator

The Majority_Voter(.) hands over the extracted bit sequence We to the process Decision_Generator(.), described in Fig. 8. Decision_Generator(.) first converts back We into the mean matrix, the nodal diagonal covariance matrix and the mixture weight matrix as parameters of the prior trained GMM. Let us denote this model as λ. Next, Decision_Generator(.) inputs voice samples A and B of two contenders for ownership, presumably one of whom is the genuine owner, and extracts their respective MFCC feature vectors. These serve as test observation sequences χ_A and χ_B. For testing, the multigaussian log-likelihood is computed for each observation feature vector using the trained model λ as the reference to yield the respective identification scores for A and B [13]. These comparative scores conclusively identify the owner of the database; the speaker with the maximum score is declared the true owner.

4 Experimental Results and Discussion

We used a 2.2 GHz Intel i5 core processor with 4 GB RAM Windows 7 OS and the MATLAB computing environment for our experiments. We applied the proposed voice-enabled watermarking algorithm on a national geochemical survey database of the US [16]. This database comprises a complete geochemical coverage of the United States and has been used for a wide variety of studies in geological and environmental sciences. For our experiments, we used a database of 76071 records. The number of virtual partitions was fixed to be 732.

Bit_Encode(H(t), A, b, w)

1. Segregate integral part I and fractional part F of attribute A
2. Calculate $P_e = H_m(t) - F$
3. Convert P_e to binary, $Bin = binary(P_e)$.
4. $Temp = Bin[b]$
5. $Bin[b] = w$
6. Concatenate *Temp* at the end of *Bin*
7. Convert *Bin* to decimal, $P_e' = decimal(Bin)$
8. Calculate $F' = H_m(t) - P_e'$
9. Calculate updated attribute value $A_j' = I + F'$

Fig. 4. Algorithm for embedding watermark bits

Majority_Voter(.)

1. Assign $Count[1:L][0] = Count[1:L][1] = 0$
2. **For each** partition S_i^* {
3. **For each** tuple t in S_i^* {
4. **If** $((H_m(t) \bmod \Upsilon) == 0)$ {
5. $B_1^* = Bit\_Decode\ (H_m t), A_{i1}, b(t))$
6. $B_2^* = Bit\_Decode\ (H_m(t), A_{i2}, b(t))$
7. Assign $Count[z_i][B_1^*] += 1$ and $Count[z_{i+1}][B_2^*] += 1$
8. }}}
9. **For** g =1 to L{
10. **If** $(Count[g][1] > Count[g][0])$
11. $W^*[g] = 1$
12. **Else**
13. $W^*[g] = 0$ }
14. $Generate\_Decision(W^*[1:L])$

Fig. 5. Algorithm for extracting the watermark

a. Bit_Encode(.)	***b. Bit_Decode(.)***
Let, $A_i = 1.9800$, $H_m(t) = 177$, $W[i] = 0$ and position $b(t)=3$.	*Let*, $A_i = 1.2600$, $H_m(t) = 177$, $W[i] = 0$ and position $b(t)=3$.
1. I=1, F=98	1. $I'=1, F'=26$
2. $P_e = H_m(t) - F = 177-98=79$	2. $P_e' = H_m(t) - F'=151$
3. $Bin = binary(P_e) = 1001111$	3. $Bin = binary\ (P_e') = 10010111$
4. $Temp = Bin[b] \Rightarrow Temp = 1$	4. $Temp = lsb(B_{in}) = 1$ and delete from *Bin* $\Rightarrow Bin = 1001011$
5. $Bin[b] = W[i] = 0 \Rightarrow Bin[b] = 0$	5. $W[i]$ as b^{th} lsb of $Bin \Rightarrow W[i] = 0$
6. Concatenate *Temp* , $Bin = 10010111$	6. $Bin[b] = Temp \Rightarrow Bin = 1001111$
7. $P_e' = decimal(Bin) = 151$	7. $P_e = decimal(Bin) = 79$
8. $F' = H_m(t) - P_e' = 26$	8. $F = H_m(t) - P_e = 98$
9. $A_j' = I + F' = 1.2600$	9. $A_j = I + F = 1.98$

Fig. 6. Illustration of reversible watermark bit encoding

Bit_ Decode($H_m(t)$, A', b)	***Decision_Generator(.)***
1. Separate A' into fractional part F'& integer part I' 2. Calculate $P_e' = H_m(t) - F'$ 3. Convert P_e' to binary; $Bin= binary(P_e')$. 4. Store *lsb* of B_{in} in *Temp* & delete from *Bin* 5. Extract *w* as *Bin*[*b*] 6. Assign *Bin*[*b*] = *Temp.* 7. Convert B_{in} to decimal; $P_e = decimal(Bin)$ 8. Calculate $F = H_m(t) - P_e$ 9. Calculate original attribute value $A=I+F$	1. Decode the extracted watermark W_e into mean, variance and weight matrices as parameters of the trained GMM λ (refer 3.1). 2. Input the voice samples *A* and *B* of contenders. 3. Extract MFCC feature vectors χ_A and χ_B for *A* and *B*. 4. Calculate identification scores using log-likelihood of χ_A and χ_B with respect to model λ. 5. The speaker who has obtained maximum score is declared as the owner of the database.

Fig. 7. Algorithm for decoding watermark bits

Fig. 8. Algorithm for determination of database owner

4.1 Robustness Analysis

An attacker Mallory may try to execute malicious attacks or benign updates on a watermarked database with the intention of degrading its contents or destroying the embedded watermark. It is also possible that an authorized receiver / user of the database may decide to extract a portion of the database or even alter some portions of it for subsequent use. The watermark must be robust enough to persist despite such malicious or allowable alterations. The resilience of any watermark scheme against different kinds of modifications is a measure of its usability and efficiency. We now illustrate the robustness of our proposed technique against various subset attacks and draw a comparison with [11].

Subset Addition Attack. In this attack, Mallory inserts records from other sources into the watermarked relation. The unmarked tuples fall under different partitions randomly and contribute to noise. However the extraction process utilizes the repetitive embedding of each bit of the watermark in numerous tuples and attributes by conducting a majority voting to reconstruct the original watermark.

We simulated the subset addition attack by adding different quantities of random tuples to the database and tested what percentage of the watermark could still be recovered as a result. The results are plotted in Fig. 9 alongside the results of subset addition attack for the other technique as reported in [11]. On adding a subset of size 97% of the original database, the full watermark could be fully recovered in our case, whereas in [11] with the watermark could not be recovered at all under the same situation, dropping sharply to 85% on addition of 60% extra tuples and then degrading beyond recognition. This clearly indicates the superior robustness of our technique and justifies the selection of multiple attributes for embedding against the technique in [11], where the watermark is embedded into a single attribute.

Subset Alteration Attack. In this attack Mallory alters certain values in the database in order to distort the watermark. But in doing so, Mallory is faced with the dilemma of compromising the usability of the database. Therefore, Mallory can only resort to randomly altering certain values or altering the least significant bits of certain attributes. But no matter what she chooses, our technique is highly resilient against this class of attacks by enforcing secure mechanisms of partitioning, selecting tuples and attributes and deciding locations for embedding.

We simulated the subset alteration attack by altering some fraction of the all tuples in the database randomly. We plotted the results and compared the results reported in the work of [11], as shown in Fig. 10. Till the point where 80% of the tuples were altered, we could correctly extract the watermark and on 97% tuple alteration, 60% of the watermark could be extracted. In contrast, the results for subset alteration attacks as reported in [11], show that their technique failed to extract the watermark on 90% alteration. This clearly establishes the superior performance of our reversible and multi-attribute insertion technique.

Subset Deletion Attack. Now Mallory tries to delete randomly selected subsets of tuples in the watermarked database so as to obliterate the watermark. There is a dilemma for her again as excessive deletions runs the risk of rendering the database useless. Mallory may delete tuples randomly which will result in the loss of some watermarked tuples from all partitions. The lost bit can be recovered from other positions and decoded using majority voting. Thus, after deletion even if one marked tuple is found from a partition, it helps in extracting the watermark sequence.

We simulated the subset deletion attack by deleting randomly selected tuples of the database. The watermark extracted was recorded as shown in Fig. 11. Experimental result reveals that even when 90% of tuples were deleted, 100% of watermark could be extracted correctly. In contrast, the results for subset deletion attack as reported in [11] show that their technique failed to extract the watermark on 90% deletion and in fact, steadily degrades from 20% tuple deletion onwards.

The experimental results also show that even if a small subset of the database is retained for further use by an end user, it still remains protected.

4.2 Score Based Identification

Earlier experiments revealed that due to the redundancy introduced in embedding the watermark coupled with majority voting, we were able to extract it fully against all subset attacks till 80% modifications were introduced. Beyond that level of noise, the watermark does suffer degradation. We now show that the use of biometrics gives the added advantage of correctly identifying the owner despite the degraded watermark.

Fig. 12 shows the variation of relative scores of the owner's and Mallory's identification when compared with altered voice extracted from the database. We can infer that even when only 60% of the original watermark is detected, the relative score of owner's voice against the features recovered from the extracted watermark is always higher than that of Mallory.

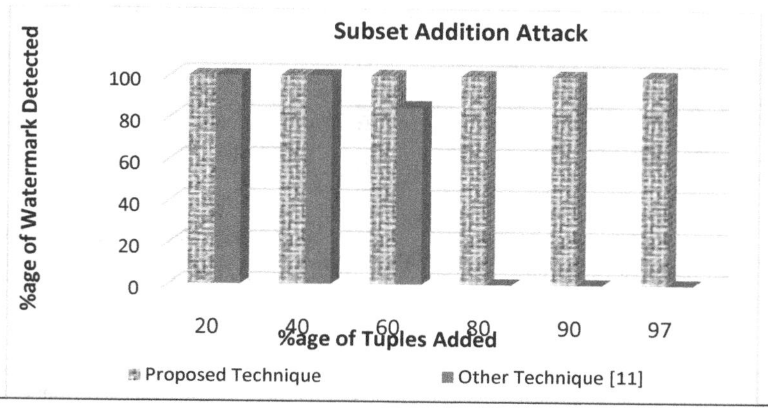

Fig. 9. Watermark degradation by subset addition attack

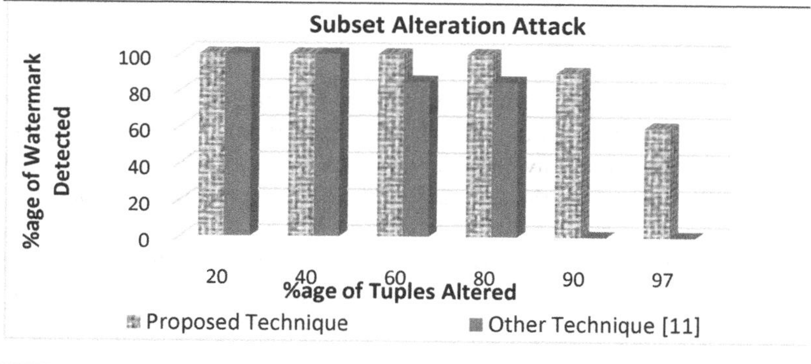

Fig. 10. Watermark degradation by subset alteration attack

In [11], the authors have completely ignored the issue of monitoring the degree to which the voice can be detected when the extracted watermark suffers degradation, thereby suggesting that a perfect match is needed. On the other hand, we have adopted a comparative score based detection of voice identification. Even when there is an alteration in the extracted voice sequence, the owner's voice scores higher than the attacking contender's. Thus, there is always a sense of clarity regarding the decision which makes the final conclusion more trustworthy. Further, it is clear from Fig. 12 that the watermark degrades even for a mere deletion of 20% tuples in case of [11] and eventually gets completely destroyed. The detection mechanism would obviously fail under such high levels of watermark degradation.

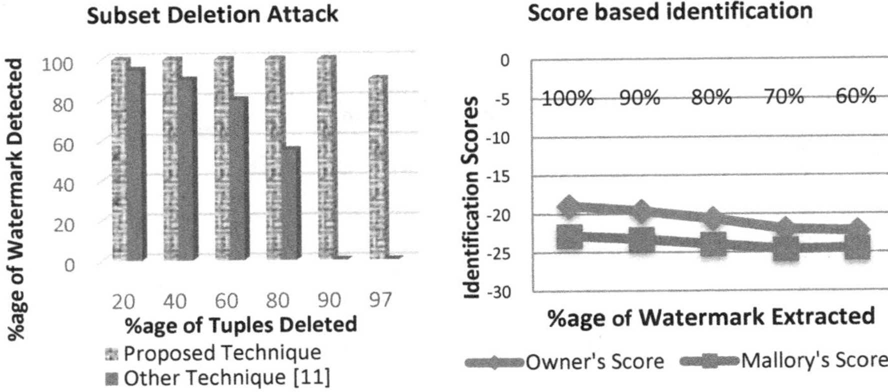

Fig. 11. Water degradation by subset deletion attack

Fig. 12. Variation in scores of owner & Mallory after altering the watermark

5 Conclusion

At the core of our proposed watermarking scheme lies the principle of using a biometric feature for watermarking a digital database. Specifically, we have used the owner's voice as a stable, readily available and reliable mark of identification to prepare the watermark. This increases the accuracy of the decision regarding ownership of the database.

The encoding technique is reversible. This ensures that changes made during the embedding process are fully reversed. Hence, the quality of information is not compromised for the sake of ensuring ownership protection. The fact that our technique is completely reversible allows us to introduce a greater degree of redundancy in embedding the watermark, thereby increasing robustness significantly. This is borne out by our experimental results. We were able to extract the watermark with 100% accuracy despite 97% subset addition or 90% subset deletion or 80% tuple alteration attacks.

Instead of applying absolute voice recognition to identify the owner in case of conflicts, we used comparative scores of the voice samples of the contenders. This relative evaluation approach allows us to correctly identify the owner for a large range of degradation up to 40% in the quality of the watermark that is extracted from a compromised database. We are currently investigating the potential of multiple biometrics for improving the reliability of ownership proof in noisy environments.

References

1. Chakraverty, P., Khanduja, V.: Digital Database Protection-A Technico-legal Perspective. In: 7th National Conference on Computing for Nation Development, New Delhi, India, pp. 361–364 (2013)
2. Kerr, I.R., Maurushat, A., Christian, S.T.: Technical Protection Measures: Tilting At Copyright's Windmill. Ottawa Law Review 34, 6–82 (2003)
3. Agrawal, R., Kiernan, J.: Watermarking relational databases. In: 28th Very Large Data Bases Conference (VLDB), vol. 28, pp. 155–166 (2002)
4. Xinchun, C., Xiaolin, Q., Gang, S.: A Weighted Algorithm for Watermarking Relational Databases. Wuhan University J. of Natural Sciences 12(1), 79–82 (2007)
5. Shehab, M., Bertino, M., Ghafoor, A.: Watermarking Relational Databases Using Optimization-Based Techniques. IEEE Transactions on Knowledge and Data Engineering 20(1), 116–129 (2008)
6. Khanduja, V., Verma, O.P., Chakraverty, S.: Watermarking Relational databases using Bacterial Foraging Algorithm. In: Multimedia Tools and Applications. Springer, US (2013), doi:10.1007/s11042-013-1700-9
7. Kamran, M., Farooq, M.: An information Preserving Watermarking Scheme for right protection of EMR systems. IEEE Transactions on Knowledge and Data Engineering 24(11), 1950–1962 (2012)
8. Khanduja, V., Khandelwal, A., Madharaia, A., Saraf, D., Kumar, T.: A Robust Watermarking Approach for Non-Numeric Relational Database. In: IEEE International Conference on Communication, Information & Computing Technology, ICCICT-2012, pp. 1–5 (2012)
9. Sion, R., Atallah, M., Prabhakar, S.: Rights Protection of Categorical Data. IEEE Transactions on Knowledge and Data Engineering 17(7), 912–926 (2005)
10. Khanduja, V., Chakraverty, S., Verma, O.P.: Robust Watermarking for Categorical data. In: IEEE International Conference on Control, Computing, Communication and Materials, ICCCCM-2013, Allahabad, India, pp. 174–178 (2013)
11. Wang, H., Cui, X., Cao, Z.: A Speech Based Algorithm for Watermarking Relational Databases. In: International Symposium on Information Processing, pp. 603–606 (2008)
12. Klautau, A.: The MFCC, `http://www.cic.unb.br/~lamar/te073/Aulas/mfcc.pdf`
13. Reynolds, D., Rose, R.C.: Speaker Identification using Gaussian Mixture Speaker Models. IEEE Transactions on Speech and Audio Processing 3(1), 72–83 (1995)
14. Schneier, B.: Applied cryptography Protocolss, Algorithms and Source in C. Wiley, India (1996)
15. Mahmoud, E.F., Horng, S.-J., Lai, J.-L., Run, R.-S., Chen, R.-J., Khan, M.K.: A Blind Reversible Method for Watermarking Relational Databases Based on a Time-stamping Protocol. Expert Systems with Applications 39(3), 3185–3196 (2012)
16. National Geochemical Database, `http://mrdata.usgs.gov/geochem`

An Intelligent Multi-Agent Based Detection Framework for Classification of Android Malware

Mohammed Alam and Son Thanh Vuong

Department of Computer Science
University of British Columbia
Vancouver, Canada
{malam,vuong}@cs.ubc.ca

Abstract. Android is currently the most popular operating system for smartphone devices with over 900 million installations until 2013. It is also the most vulnerable platform due to allowing of software downloads from 3rd party sites, loading additional code at runtime, and lack of frequent updates to known vulnerabilities. Securing such devices from malware that targets users is paramount. In this paper, we present a Jade agent based framework targeted towards protecting Android devices. We also focus on scenarios of use where such agents can be dynamically launched. We believe, a detection technique has to be intelligent due to limited battery constraints of these devices. Moreover, battery utilization might become secondary in certain settings where detection accuracy is given a higher preference. In this framework, the expensive analysis components utilizing machine-learning algorithms are pushed to server side, while agents on the Android client are used mainly for intelligent feature gathering.

Keywords: Multi-agent systems, Android, Machine Learning, Security in Ubiquitous environment, Intelligent Agents.

1 Introduction

Monitoring sleep patterns using wearable devices such as Fitbit; paying for purchases using mobile banking using Paypal; accessing and authorizing 2-factor authentication devices for checking email or Amazon Web Services using Google authenticator; are just some of the tasks performed on a daily basis on smartphones. Over 900 million devices based on the Google Android platform have been deployed since 2008. The platform allows software downloads from third party sites; allows loading additional code at runtime [1], and does not support patching vulnerabilities in older Android devices. As such, malware authors have predominantly focused on the Android platform. Today, it is the most vulnerable mobile platform with over ninety nine percent of malware built for it. Devices based on the Android platform can be infected to send SMS to premium rate numbers [2] to cause financial harm to users, used as part of botnets [3] to cause

D. Ślęzak et al. (Eds.): AMT 2014, LNCS 8610, pp. 226–237, 2014.

distributed denial of service attacks, and steal banking credentials when these devices are used for 2-factor authentication. A broad characterization of Android malware was presented by Zhou and Jiang [4] that provides the observed trends of malware for Android. They presented over 1260 malware samples in 49 malware families with a best-case detection rate of 79.6 percent and a low of 20.2 percent by commercial antivirus companies. Many of the malware samples used drive-by-download and update attacks similar to personal computing devices.

We believe that many new malware can be detected by monitoring features allowed to be traced by Android phones. Based on the subset of features observed, multiple machine learning classifier models can be used to analyze the probability of a device exhibiting malware like behavior. In this paper, we use a multi-agent framework to gather such data.

The main contributions of this paper are (1) Provide a multi-agent architecture to collect features to apply machine learning classifiers on Android features, (2) Provide intelligence built into some of the agents on the device to dynamically modify its behavior based on battery, location, and device contexts, (3) Discuss issues with the current Agent based framework, (4) Provide initial experimental data of network behavior that can be modeled using the ticker behavior of Jade agents. The rest of this paper is organized as follows: Section 2 presents related work in the area of agent use, and android malware, Section 3 presents our architecture and data gathering strategies, Section 4 provides implementation and experimental data obtained from our tests, Section 5 discussion challenges with the current solution, and finally Section 6 summarizes our paper.

2 Related Work

The research work presented in this paper extends our previous work [5], [6] and covers multiple areas of research, including but not limited to the use of agent technology and the use of machine learning algorithms. Use of agent technology can further be applied specifically to mobile devices for detecting malware. Similarly, machine learning classifiers could be applied specifically to analyze Android related features.

2.1 Agent Systems and Their Advantages

According to the Artificial Intelligence community, an agent is defined to be a program that behaves autonomously to perform tasks dynamically based on observed data. A multi-agent system (MAS) is composed of multiple agent types that interact in a given environment. In our system architecture, we use MAS, some of which run on the Android device, and some higher CPU and memory requiring agents that are placed in servers. Some of the advantages of MAS include [7]: *asynchronous* interaction between agents allowing to better handle network disconnections; *easy software upgrades* that allows to use only relevant and newer agent features; and the ability to function in *heterogeneous platforms* by using

FIPA [1] compliant Agent Communication Language (ACL) to transfer data in case agents cannot migrate their state between (inter platform) or within a platform (intra platform). Agent based systems have been previously researched in the area of intrusion detection. Kruegel et al. proposed the Sparta system [8], [9] to detect intrusions and security violations using an event query language. Crosbie and Spafford [10] proposed using genetic modelling using automatically defined functions (ADF) to detect intrusions.

2.2 The Need for Machine Learning and the Random Forest Classifier

With the vast amount of data generated by sensors on Android devices, it would be difficult for a human to compute relevant features that could be used to classify malware. This task is best done by automation by applying machine learning algorithms to generate models from the collected features. The Random Forest Algorithm by Breiman [11] is one such algorithm that can be used. It is comparatively more robust to the space-time complexity dependent upon the size of the features and the number of data points collected. The hyper parameters of this algorithm that can be modified include the number of decision trees to be used, the depth of each tree, and the number of random features that are to be selected at every decision node. Choosing log m features, where m represents the number of available features, yields optimal results in most cases. Our previous work in [5] provides the performance of the Random Forest algorithm on an Android feature dataset.

2.3 Android Features to Observe for Learning

Applying machine-learning algorithms first requires feature vectors to be generated by the agent system. The set of features that can be accessed from Android systems depends on the Android version in use, and the device root status i.e. if the device allows for installation of additional software to monitor system behavior using administrative privileges. Android uses the Linux kernel as the bottom layer [12]. All layers on top of the kernel layer run without privileged mode. Applications are prohibited from accessing other application data without explicit permissions. If features of Android are collected in unrooted mode, then system information made available by Android API only can be used. Some of them include IP addresses being communicated with, active connections, connection states, binder inter process communication calls, cpu usage, and system memory utilization. On the other hand, having a rooted device allows one to install system tools that could gather features from underlying host and network behavior. Example of features that can be obtained from rooted devices include: using tcpdump to collect data being sent by applications, using strace to track the system calls being invoked, etc. The related work in the area of machine learning approaches to Android data either deal with feature collection from unrooted devices [13], [14], or rooted devices [15], [16], [17].

[1] `www.fipa.org`

2.4 Machine Learning Classification of Android Malware

Malware detection can be classified into two categories: static detection or dynamic detection. Static detection uses signature matching in order to make decisions. Dynamic detection techniques monitor anomaly in the behavior of running software. They are prone to higher false positives compared to static detection if the classifier is not adequate trained. This technique however, can detect new or slightly modified malware samples, which is not possible by signature matching in static detection techniques. As our approach uses machine-learning classification in the context of Android operating system, the following related papers fall in the same domain.

Amos [13] created an automated system to analyze Android malware samples using adb scripts and using the adb-monkey tool to simulate user action. They observed 42 features at 5 second intervals on 1330 malicious and 407 benign samples. For our study in [5], we used a modified data set obtained from this work that takes into account class imbalance. The authors have provided the largest dataset available for the application of machine learning classifiers on Android. The author compared results from using the following classifiers without any hyper parameter modification from the weka library: Bayes Net, J48 decision tree, Logistic Regression, Multilayer Perceptron, Naive Bayes and Random Forest. As shown in our previous work [5], hyper parameters of an algorithm play a big role in classification accuracies.

Shabtai et al. [14] monitored 88 features on an unrooted device. They tested their approach on 16 benign applications and 4 self-generated malware by monitoring 88 features at 2-second intervals. The authors compared the performance of k-Means, Logistic Regression, Histograms, Decision Tree, Bayesian Networks and Naive Bayes. Their solution yielded a true positive rate of 80 percent and a false positive rate of 12 percent.

Kim et al. [15] performed an automatic feature extraction using JavaScript for static detection on 1003 Android applications. The features used include the total number of permissions requested by the application; phone management API invoked from code; counting API calls related to phone control and privacy in code. The authors used J48 Decision Tree classifier from the Weka library. They observed a TPR of 82.7 percent and false negative rate of 17.3 percent.

Burguera et al. [16] used strace to capture number of system calls made by an Android application on a rooted Android device. Each feature vector was formed from 250 linux 2.6.23 system call counts for the applications. They used 2-means clustering algorithm to distinguish between benign and malware versions of an application.

Dini et al. [17] used a rooted device to design a host based real time anomaly detector based on 1-Nearest Neighbour classifier. They monitored 13 features, 2 at user-level (phone active/inactive; SMS sent when the phone is inactive) and the rest at kernel level to observe system calls. Their TPR was 93 percent.

3 System Design

3.1 Architecture

The proposed multi-agent framework has a client-server architecture. This was done to move the computationally intensive tasks to a server since Android devices are battery constraint. For example, the analysis agents in the system are placed on the server as they are responsible for generating the machine learning classifier models and for comparing the feature vectors generated. In our system, the Android devices are each considered as clients.

Based on JADE specifications for mobile devices, in order to consume fewer resources, the containers for hosting the agents on Android devices follow a split-container model. In this method a slim front-end client runs on the actual device whereas a back-end is maintained on the server [2]. A dedicated connection is used to communicate between the two components. It is robust to communication disconnections [18]. Each Android device runs an Android application that is wrapped using a JADE runtime service split-container.

3.2 Agent Types and Their Functionality

We designed our agent types based on the procedures outlined in [19]. Fig. 1 shows the agent types in our system. We now provide a description of their functionality.

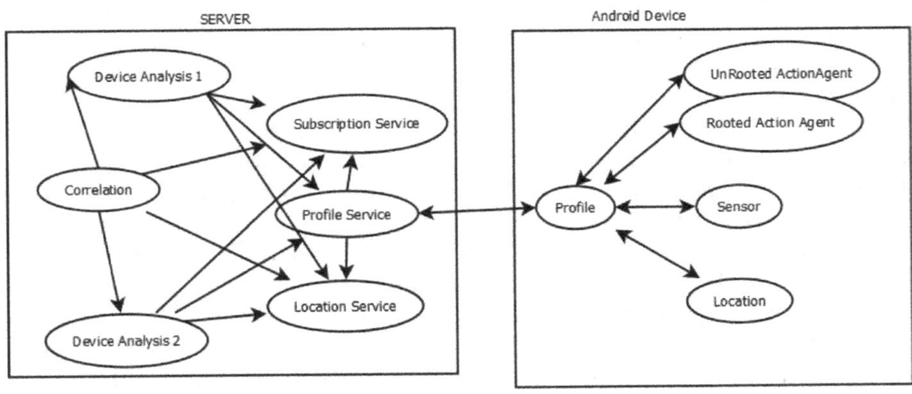

Fig. 1. Agent Interaction

The following four agent types reside on an Android device. Some of them are active all the time, whereas others wake up periodically to update information depending on if a cyclical behavior is required.

[2] A full container service is allowed to be run on an Android device, but is generally not used to avoid high consumption of system resources.

- *Sensor agents:* These periodically track device and network related information. Some of them include IP address assigned to the device, current Android version of the device, its Linux kernel version, and the network type it is connected to. This information is relayed to the profile agent. This information is required as some of the action agents are launched only if the Android version is compatible with an action agent.
- *Location agents:* These agents use Android intent messages to subscribe to changes in GPS location of the device. This information is communicated to the profile agent on the device. The profile agent periodically reports the location to the Profile Service agent on the server. Moreover, the Profile agent on the device launches Action Agents if the location reported by the agent matches any rules to trigger Action Agents.
- *Profile agents:* These agents perform vital checks on the Android device, and are always in a running state. This agent communicates with the profile service agent on the server to communicate any required behavioral changes, such as raising alert levels, or launching new agents based on requests sent by the Profile service agent. This agent also maintains the threshold level to keep track of resource consumption over a period of time to avoid running too many agents on a resource constraint device. This agent uses the information provided by the location and sensor agents on the device to launch appropriate action agents. The features collected by the Action agents for machine learning classification are reported to the profile service agent on the server. The system administrator can program the frequency of reporting when the client registers with the services on the server.
- *Action agents:* These agents are dynamically launched by the Profile Agent based on set rules. Their primary task is to collect features to form feature vectors based on the model of the features required. Some of the tasks programmed include monitoring application installations, memory and CPU usage, android binder API calls, IP addresses communicated with, network transmission amounts. If a device is rooted, and applications such as tcpdump for ARM processors has been installed, many more information such as network packet content can be inspected. In order to work properly, Action agent has to maintain the Android API-level. This requirement is needed as certain functionalities are only allowed under certain versions of Android. Google periodically adds public API to access information based on changing privacy or feature needs. For example, Android 4.3 introduced the ability for applications to access Android notifications by creating a notifications listener. Each Action agent also stores its resource utilization level in terms of memory, CPU, disk and network utilization. Currently it is set manually based on approximation of tasks performed by it, as there is no methodology that exists on Android devices to track individual agent resource consumption. We foresee agents being able to automatically compute these values if Android provides public API to access resource consumptions on a device per process, or per thread. Action agents are launched if and only if the resource constraints are met.

The following five service agents are maintained on the server side. It is important to note that not all the server side agents have to be maintained on a single server. The agents can be distributed among multiple machines as the agents in the system communicate with each other using Agent Communication Language (ACL) using an application specific ontology.

- *Profile Service agent:* This agent maintains a list of all the Android devices that have connected to the system. This agent communicates with the Profile agent on the Android device when the device is connected, to send it rules regarding launching of Action agents and frequency of data collection based on information collected from sensors on the device. If the device Analysis agent requests a change to the behavior of data collection on the Android device, the Profile Service agent relays the information to the Profile agent using agent communication language (ACL) messages.
- *Subscription Service agent:* This agent maintains historical information of Android devices such as uptime of devices, and the current connection state of Android devices connected to the system.
- *Location Service agent:* This agent maintains the location history of all Android devices in the system. Location update information sent by the Profile agent to the Profile Service agent on the server is relayed to this agent. This agent can be utilized to query devices in the vicinity of a given device if correlation of information among devices is required.
- *Device Analysis agent:* The Device Analysis agent performs the machine learning tasks using the Weka library. There exists a device Analysis agent for each device on the system. Its task is to use information obtained from Location, Subscription and Profile service agents to detect patterns. This agent is launched by an administrator based on the models of the features required. Table 1 lists some of the features that our system currently collects.
- *Correlation agent:* The Correlation agent is currently being investigating by us. These agents cluster devices based on device type, Android API level, or similar infection patterns. These agents could interact with Analysis, Subscription and Location Service agents to gather required temporal and spatial data to cluster devices together. This would allow informing currently benign devices that could have been infected by an Android device infected by malware.

4 Experiment and Evaluation

4.1 Implementation

We built our prototype system using JADE with the Leap plugin for version 4.3.2 and created a split-container micro-runtime service on Android devices. The ontology [20] for ACL communication between agents was developed using Protege [3] with the OntologyBeanGenerator [4] plugin to generate the required

[3] `http://protege.standford.edu`

[4] `http://protege.cim3.net/cgi-bin/wiki.pl?OntologyBeanGenerator`

Table 1. This table provides some of the features collected by our system [5]

Category	Features observed per category
Binder	Transactions, Reply, Acquire, Release, ActiveNodes, TotalNodes, ActiveRef,TotalRef, ActiveDeath, TotalDeath, ActiveTransactions, TotalTransactions, ActiveTransactionsComplete, TotalTransactionsComplete, TotalNodesDifference, TotalReferenceDifference, TotalDeathDifference, TotalTransactionDifference, TotalTransactionCompleteDifference
Memory	memActive,memInactive, memMapped, memFreePages, memAnonPages,memFilePages, memDirtyPages, memWritebackPages
Network	TotalTXPackets,TotalTXBytes, TotalRXPackets, TXPacketsDifference TXBytesDifference, RXPacketsDifference, RXBytesDifference
CPU	CPU Usage
Battery	OnCharging, Voltage, Temperature, Battery Level, BatteryLevelChange
Permission	Number of Permissions requested
Location	Number of Location Changes, Devices In Vicinity

class files. We extended the various agent behaviors available in the Jade platform to perform various tasks. Some of the behaviors used include *one-shot behavior* to gather static information on Android devices such as kernel and Android API; *ticker behaviors* with battery level as a parameter to estimate the frequency with which information is queried by the Action agent, etc. GPS locations were retrieved by the Location agent by receiving updates from Android *LocationManager*. We used *RootTools* [5] to monitor if a device is in a rooted state. The Android API level of a mobile device is retrieved by accessing the *android.os.Build.VERSION.SDK_INT* value of the system. In order to evaluate the performance of our prototype system, multiple experiments were conducted to test various portions of our system. We discuss some of them next.

4.2 Experiment 1: Rooted Device Data Retrieval Test

This experiment was run on a rooted Android Virtual Device (AVD). All the packages mentioned next were installed using command line using Android Debug Bridge (*adb*). To access the root (*su*) shell in the emulator, we installed *superuser* using command line tools. We then installed *Busybox*[6] and *tcpdump*[7] to check if our agents can retrieve secured information. As mentioned previously,

[5] `https://github.com/Stericson/RootTools`
[6] `http://www.busybox.net/`
[7] `http://tcpdump.org`

the agents detect if the device is rooted using *RootTools*. This test passed, as our action agent was able to retrieve network data gathered by tcpdump for ARM processors if the device has been rooted.

4.3 Experiment 2: Feature Collection Using Action Agent

In this experiment, we generate feature vectors from the features mentioned in Table 1. Our Action agents are launched with a parameter that sets the frequency with which the features mentioned in Table 1 are to be retrieved. This decision is made by the rules stored on the Profile agent. The Profile agent launches Action agent based on battery life to compute the said features at 5 second intervals or at 1 minute intervals. For our initial test, we set a 50 percent battery requirement to test one or the other. For brevity, following is an example feature vector in CSV of some of the values obtained in no particular order.

```
0,0,36,0,7289,5063,1141,252,418,623,712,1141,168,294,2,12348,
0,0,30,53,7,993,0,98.747765,168340,115344,54996,53874,32473,
38440,4,0,-1,-1,-1,-1,0,0,0,0,1094,3,104,4,0,true
```

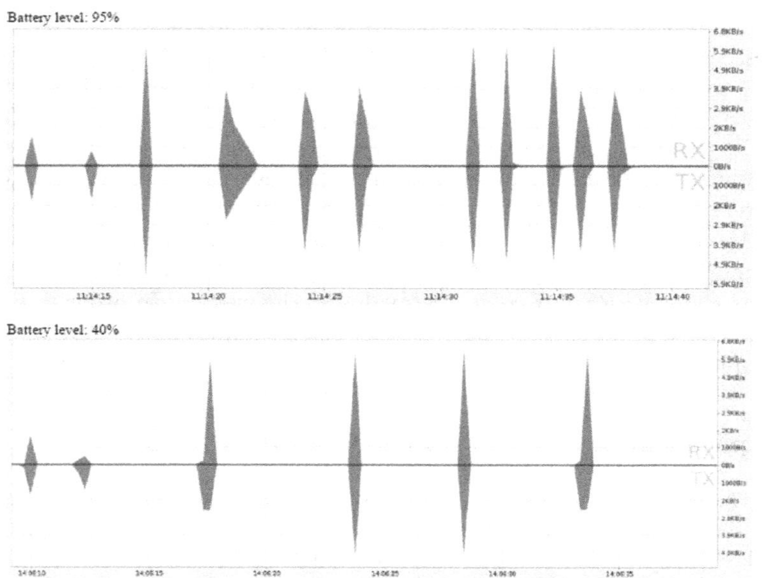

Fig. 2. Network usage change based on battery context measurements [6]

4.4 Experiment 3: Dynamic Network Transmission Test

As shown in Fig. 2, we performed a test to change the pattern of data transmission from the profile agent on an android device to the Profile service agent on the server side based on battery context [6]. In this test, the Profile agent collects and sends features at a rate dependent on the battery level. For this we make use of the *ticker behavior* class of Jade and set a parameter based on reading the battery level. Thus, there will be a gradual drop in the rate of reporting data if the battery level is less.

5 Discussion

Based on our initial experimentation, we list some of our findings and limitations.

5.1 Feature Collection

The classifiers used can only be as good as the features collected. The features that can be obtained in Android depends on if the Android device is rooted or not. As mentioned previously in related work, a rooted device allows access to substantial amount of information such as system calls invoked, content of data packets, etc. Unrooted devices are limited to access allowed by the Android API. Moreover, the amount of noise in the features would reduce the quality of features. i.e. if the features measured (such as network transmission rate of the device) are global in nature, it would be difficult to detect the specific application that behaves maliciously.

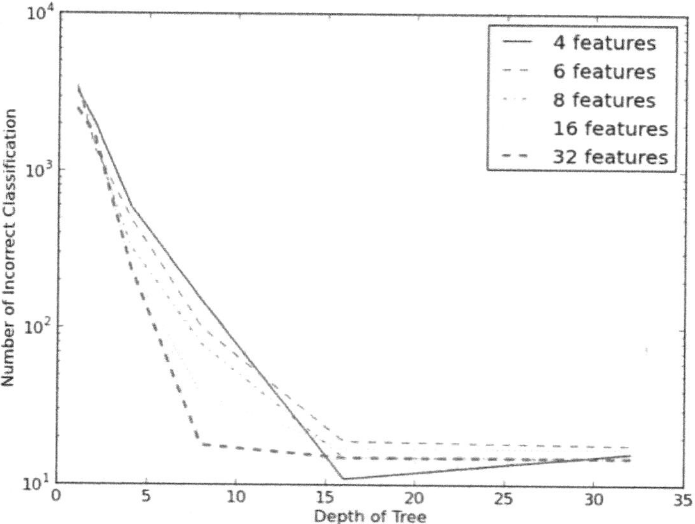

Fig. 3. Performance of Random Forest algorithm on a subset of featues of Table 1 [5]

5.2 Machine Learning Classifier

In our previous work [5], we worked with the Random Forest classifier as it is robust to inter dependence between the features collected. As Fig. 3 shows, the algorithm performed well on a subset of data features of Table 1. This however might not be the case if the data features are noisy.

5.3 Per Agent Resource Utilization

Currently, there is no mechanism in Android to measure the battery resource cost to run each agent. This would cause an issue in detecting the amount of workload that a mobile device has to endure in order to run the JADE agent system on a per-agent basis.

6 Conclusion and Future Work

In this paper we have provided an initial multi agent based framework based on the JADE agent platform on Android that could be used to gather features to apply machine learning classifiers to generate or verify feature vectors. These feature vectors could be used to detect malware if appropriate malware models exist with the Analysis agents on the server side. The intelligence of our system is managed by the Profile agent on the Android device based on communicating with the Profile service agent on the server side. We intelligently launch Action agents based on the Location or Network that the device is connected to. We appropriately change the frequency of data collected by the Action agents based on these contexts.

In the future, we plan on testing state-only mobility of Jade-Android agents using the new feature of JADE-Android. We also plan on testing the detection accuracy of malware of our classifiers based on various features collected on a real device running malware.

Acknowledgments. We thank Zhiyong Cheng for his contributions to the initial development of the related work to this project [6] on ACL generation and preliminary experimentation.

References

1. Poeplau, S., Fratantonio, Y., Bianchi, A., Kruegel, C., Vigna, G.: Excute This! Analyzing Unsafe and Malicious Dynamic Code Loading in Android Applications. In: NDSS (2014)
2. TrustGo Security, New Virus SMSZombie.A Discovered by TrustGO Security Labs, `http://blog.trustgo.com/SMSZombie/`
3. Xiang, C., Binxing, F., Lihua, Y., Xiaoyi, L., Tianning, Z.: Andbot:towards advanced mobile botnets. In: Proceedings of the 4th USENIX Conference on Large-scale Exploits and Emergent Threats (LEET 2011). USENIX Association, Berkeley (2011)

4. Zhou, Y., Jiang, X.: Dissecting Android Malware: Characterization and Evolution. In: Proceedings of 2012 IEEE Symposium on Security and Privacy (2012)
5. Alam, M., Vuong, S.: Random Forest Classification for Android Malware. In: Proceedings of IEEE International Conference on Internet of Things (2013)
6. Alam, M., Cheng, Z., Vuong, S.: Context-aware multi-agent based framework for securing Android. In: Proceedings: The 4th International Conference on Multimedia Computing and Systems, ICMCS (2014)
7. Bieszczad, A., White, T., Pagurek, B.: Mobile Agents for Network Management. Proceedings of IEEE Communicaations Surveys (1998)
8. Krugel, C., Toth, T., Kirda, E.: SPARTA, a Mobile Agent Based Intrusion Detection System. In: Proceedings of the First Annual Working Conference on Network Security: Advances in Network and Distributed Systems Security, November 26 - 27. IFIP Conference Proceedings, vol. 206, pp. 187–200. Kluwer B.V, Deventer (2001)
9. Krugel, C., Toth, T.: Flexible, Mobile Agent based Intrusion Detection for Dynamic Network. In: Proceedings of the European Wireless (2002)
10. Crosbie, M., Spafford, G.: Defending a Computer System using Autonomous Agents. In: Proceedings of the 8th National Information Systems Security Conference (1995)
11. Breiman, L.: Random Forests. Machine Learning 45(1), 5–32 (2001), doi:10.1023/A:1010933404324
12. Google. Android Security Overview, http://source.android.com/tech/security
13. Amos, B.: Antimalware, https://github.com/VT-Magnum-Research/antimalware
14. Shabtai, A., Kanonov, U., Elovici, Y., Glezer, C., Weisee, Y.: Andromaly: a behavioral malware detection framework for android devices. Proceedings: Journal Intelligent Systems 38, 161–190 (2012)
15. Kim, D., Kim, J., Kim, S.: A Malicious Application Detection Framework using Automatic Feature Extraction Tool on Android Market. In: Proceedings:3rd International Conference on Computer Science and Information Technology (ICCSIT 2013), January 4-5 (2013)
16. Burguera, I., Zurutuza, U., Nadjm-Tehrani, S.: Crowdroid: Behavior-Based Malware Detection System for Android. In: Proceedings: SPSM (2011)
17. Dini, G., Martinelli, F., Saracino, A., Sgandurra, D.: MADAM: A multi-level anomaly detector for android malware. In: Kotenko, I., Skormin, V. (eds.) MMM-ACNS 2012. LNCS, vol. 7531, pp. 240–253. Springer, Heidelberg (2012)
18. Bellifemine, F., Caire, G., Greenwood, D.: Developing multi-agent systems with JADE. John Wiley and Sons (2007)
19. Nikraz, M., Caire, G., Bahri, P.A.: A methodology for the development of multiagent systems using the JADE platform. Proceedings: International Journal of Computer Systems Science and Engineering 21(2), 99–116 (2006)
20. Cheng, Z.: A Multi-Agent Security System for Android Platform. Masters Thesis, Dept. Comp. Sci., University of British Columbia, Vancouver, BC (2012)

Outlier Analysis Using Lattice of Contiguous Subspaces

Vineet Joshi and Raj Bhatnagar

Dept. of Electrical Engineering and Computing Systems
University of Cincinnati, Cincinnati, OH, USA
raj.bhatnagar@uc.edu
joshivt@mail.uc.edu

Abstract. Many anomaly detection techniques consider all the data-space dimensions when looking for outliers, and some others consider only specific subspaces, in isolation from other subspaces. However, interesting information about anomalous data points is embedded in the inter-relationships of the subspaces within which the data points appear to be outliers. Important characteristics of a dataset can be revealed by looking at these inter-relationships among subspaces. We describe a methodology for searching for outliers within the context of contiguous subspaces in the subspace lattice of a domain. We demonstrate additional insights about the outliers gained from this approach compared to finding the outliers in only specific subspaces or in the complete data-space. This additional information points an analyst to peculiar sets of subspaces to investigate further the underlying structure of the data space and also of the anomalous nature of the data points.

Keywords: Anomaly detection, outliers, subspace lattice, causality analysis, subspaces, intentional knowledge.

1 Introduction

Anomaly detection aims at identifying those data points that in some sense stand-out from the rest of the population. It has applications in domains such as credit card fraud detection, network intrusion detection, finding mal-functioning components in a complex machinery etc. Most of the existing research on this topic has focused on algorithms and discriminating functions to identify outliers, either in the whole attribute space, or in subspaces of attributes. Nevertheless, another important facet of anomaly detection is the identification of subspaces within which outliers are detected. An analyst may get important clues about the underlying processes that lead to outlying data points from an analysis of the set of attributes that, in their subspace, reveal specific outliers.

As an example, consider a dataset D with four attributes, 1, 2, 3 and 4. Fig. 1 displays the subspace lattice for such a dataset. In this figure, each node represents a subspace. Let us say that we checked for outliers in all these subspaces and found that some data point P is identified as anomalous in the subspaces 1,

D. Ślęzak et al. (Eds.): AMT 2014, LNCS 8610, pp. 238–250, 2014.

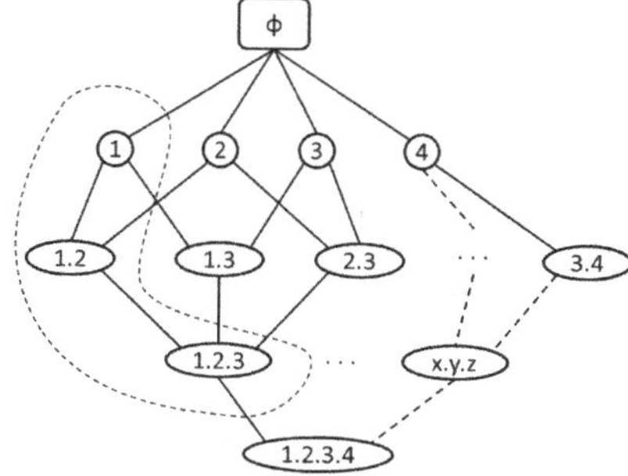

Fig. 1. Subspace Lattice with a Contiguous Region of Subspaces

1.2 and 1.2.3, and is not seen as outlier in the remaining subspaces. Note that these subspaces are connected to each other through paths that only include these subspaces in the lattice and are encapsulated in the dotted region. This is a *contiguous region* of the subspace lattice within which the same point P has displayed anomalous behavior. Such contiguous regions can provide important insights into the processes that give rise to outliers.

For example, in the above mentioned dataset, an analyst can get a hint that there is something peculiar about attribute 1 because even in presence of attribute 2 in subspace 1.2 or both 2 and 3 in subspace 1.2.3, the point P continues to be an anomaly. At the same time, when only the attributes 1 and 3 are present in subspace 1.3, the point comes across as normal even though P was anomalous in the subspace above and below the subspace 1.3. Further the data point also comes across as normal in subspaces containing only attribute 2 or attribute 3.

We believe that valuable insights can be gained not only by knowing about subspace 1 which is the *topmost* subspace in this contiguous region, but also by knowing the *bottom-most* subspaces such as 1.2.3 of such a region and also the other lattice nodes that lie inside the region. When we have a larger lattice with more subspaces below 1.2.3, such as 1.2.3.4 in Fig. 1, we can infer the peculiarity of 1.2.3 because it is the lower-most subspace in which P was anomalous. All such subspaces identify the sets of attributes and their specific combinations under which P turns out to be anomalous.

We call subspaces such as 1.3 in which P appears normal but which lies between subspaces such as 1 (top) and 1.2.3 (bottom) where P again becomes an anomaly as the *absent* subspaces.

Taken together, such topmost, bottom-most and absent subspaces along with the aggregate of subspaces inside the contiguous lattice region can provide hints about the source attributes due to which anomalies are arising in a domain. We believe that having such insights in addition to knowing the list of anomalous data points is important for discovery of the sources of outliers in datasets.

2 Related Work

The research on outlier detection can be divided into two broad categories. The first category of research is about developing efficient methods for finding outliers. Examples of this kind of research include finding new distance metrics, new definitions of outliers and better algorithms of searching for outliers in datasets of large sizes and dimensions etc. The second category of research involves trying to understand or explain the anomalies detected using the methods in the first category.

Our work provides a framework to address the issue of trying to look for the reasons for observed outliers. The methodology we propose here can utilize any criterion for identifying outliers in the context of subspace, and use it to provide details about the subspace structure in the lattice of subspaces within which the outliers are being generated. Hence our work belongs to the second broad direction of research on outlier detection that involves understanding and explaining some insights about the detected outliers. A representative work of this type was by Knorr et al.[1] where the goal was to find 'intensional knowledge' about outliers by determining the minimal subspaces in which anomalous behavior was detected along with the detected outliers. The authors claimed that these minimal subspaces, called 'strongest outlying subspaces' provide a kind of explanation of the outlying behavior.

While our work belongs to the same category of research as that of Knorr mentioned above, we aim at not only finding the minimal subspaces, but the whole multiple disjoint contiguous subspace regions of subspace lattice where the same data point exhibits anomalous behavior. Further we also consider those exception subspaces that lie between the topmost and the bottom-most subspaces of such regions, and in which the data point appears normal. Hence we provide a more comprehensive treatment of the problem of finding insights about the possible causal factors of outlier behavior of data points.

Keller et al. have recently proposed a data pre-processing method wherein the goal is to determine the subspaces where interesting outliers may be found [2]. The analyst can then specifically apply outlier detection algorithms in these subspaces. However this method is limited to working with pre-selected subspaces and does not look at the relationships between various contiguous subspaces in which interesting outliers can be found.

Zhang et al. have proposed an algorithm called HOS-Miner (High-dimensional Outlier Subspace Miner)[3]. The goal of this algorithm is to detect some subspaces within which a given data point is an outlier. However, this method again did not look at the lattice structure of the subspaces to extract deeper insights from them. Further, this method compares distances across different subspaces without normalizing these distances with respect to the number of attributes in the subspace. This yields closure properties that allow pruning of the set of subspaces to be searched. Further, as noted by Aggarwal[4], while this pruning can make the method efficient, it can miss out the large chunks of the subspace lattice in which the outlier-containing subspaces lie between two subspaces that themselves don't have outliers.

3 Our Approach

Our methodology is independent of specific distance metrics and definitions of outliers. It builds upon the existing outlier detection algorithms. It consists of the following distinct steps:

- Systematically explore the subspace lattice and identify the outliers in various subspaces.
- Summarize the collected information for each outlier and identify the subspaces in which a given outlier is detected.
- Finally, analyze the subspaces computed in previous step and report the contiguous subspace regions, topmost subspaces, bottom-most subspaces and absent subspaces.

We briefly explain the three phases now.

3.1 Subspace Lattice Exploration

As explained earlier, our methods builds upon existing techniques and definitions of outliers and similarity measures. Hence we can pick any existing method and utilize it for further outlier analysis. For the sake of discussion in this paper, we have adopted the following two definitions as examples: *a)* Z-Scores of distances to 'k' nearest neighbors, and *b)* *Local Outlier Factor(LOF)*[5]

Knorr and Ng[1] proposed one of the earliest definitions of distance based outliers as:

Definition 1. *An object O in a dataset T is a $DB(p, D) -$ outlier if at least fraction 'p' of the objects in T lies at greater than distance D from O.*

Angiulli and Pizutti[6] presented a related and widely used definition of outliers, described by Nguyen[7] as:

Definition 2. *The dissimilarity of a point p with respect to its k nearest neighbors is known by its cumulative neighborhood distance, defined as :*

$$F_{out}(p) = \sum_{m \in kNN_p} D(p, m)$$

We adapt a variation of these two definitions as our definition of outliers as :

Definition 3. *Outliers are the data points whose $z-$scores of distances from their k-nearest neighbors, which a subspace, are greater than some threshold t.*

In most of our tests, we have used value of 3.0 for t. Let us call this definition of outliers '$Z-Dist$'. With this definition, the only other parameter that needs to be specified for identifying outliers is the value of 'k', i.e. the number of nearest neighbors to consider. This definition is not a new contribution. It has been used in the field of geographic datasets[8].

Furthermore, we also demonstrate our analysis using another definition of outliers to exemplify that our proposed methodology can plug-in any definition of outliers. This second definition of outliers is the widely used *Local Outlier Factor* proposed by Breunig et al.[5].

The pseudo-code for algorithm to explore the outliers in subspace lattice is in the listing of algorithm 1.

Algorithm 1. Lattice Exploration Using Z-Dist

```
Input:
    k: number of nearest neighbors
    L: subspace lattice
    DS: data set

Output: Subspace lattice L containing the detected outliers in each
        subspace
for each subspace S ∈ L do
    knn_dist ← allocate array of length |DS|
    i ← 1
    for each data point P ∈ DS do
        knn_dist[i] ← avg. dist. to k nearest neighborhood of P
        i ← i + 1
    end
    z_score ← calculate z − score of distances in knn_dist
    Select points P whose z−score is greater than 3 and record them in S
end
return L
```

This procedure performs the following functions: *a*) walks through the subspaces in the lattice, *b*) computes the outliers in each subspace, *c*) stores the outliers detected in a given subspace within the data structure representing the subspace.

If the number of subspaces is m and number of data points is n, the time complexity of this procedure is $O(mn^2)$. This is because we iterate through each subspace and calculate distance matrix for the n data points. A similar procedure to detect outliers using LOF can be designed. We are omitting that to save space.

The information about outliers contained within each subspace is then seeded into the next step below.

3.2 Summarization of Subspaces for Outliers

Algorithm 1 collects information about the outliers identified within the various subspaces in subspace lattice. Using this, we next find out the subspaces within which a given outlier is identified. The method is specified in Algorithm 2

Let us suppose that on average, the count of outliers in each subspace is q. The average time complexity of algorithm 2 is $O(mq)$ where, as before, m is the number of subspaces. Consider a case in which no outlier was found in any subspace. In this case we will execute the outer for-loop m times. Hence the lower bound of the time complexity is $O(m)$. Similarly, consider the other extreme case where all n points get flagged as outlier. This is not a realistic scenario; however this is the extreme case that the algorithm may theoretically encounter. The time complexity in this case has the bound $O(mn)$.

The set of outliers obtained in the second phase is fed into the third phase.

Algorithm 2. Subspace Summarization for Outliers

Input:
 L: subspace lattice containing the outliers identified in each subspace of the lattice
 DS: data set

Output: Set S_{out} of outliers, with each outlier data structure containing the set of subspaces within which this outlier has been identified.

```
S_out ← allocate new set of outliers
for each subspace s ∈ L do
    for each outlier o ∈ s do
        if o ∈ S_out then
            o.subspace_set.append(s)
        end
        else
            o ← allocate new outlier object
            o.subspace_set.append(s)
            S_out.add(o)
        end
    end
end
return S_out
```

3.3 Determination of Contiguous Subspaces

The goal of this phase is to determine the set of contiguous subspaces for each identified outlier. A subspace lattice can be modeled analogous to a graph a follows:

Definition 4. *A lattice L is a tuple:*

$$L = (S, E, \phi, s_{last})$$

where:
S is a set of independent sets of subspaces
E is a set of edges between subspaces in S
ϕ is the NULL subspace that contains no attributes
s_{last} is the last subspace that contains all attributes

Note that as S is a set of independent sets, the edges in E can only consist of subspaces in two different independent sets in S.

Further, if two subspaces p and q are connected by an edge, then p has exactly one more (or less) attribute than q.

Definition 5. *If an edge exists between two subspaces p and q in L, then if p has one more attribute than q, it is called a child of q. Otherwise p is called a parent of q.*

Definition 6. *A path between two subspaces p and q consists of the sequence of subspaces $(s_1, s_2 ... s_{n-1}, s_n)$ such that:*

$$for\ 1 \leq i < n \text{ there is an edge between } s_i\ and\ s_{i+1},$$

$$\text{and } s_1 = p \text{ and } s_n = q$$

The subspaces $s_1, s_2, ... s_{n-2}, s_{n-1}$ in the path from p to q are called intermediary subspaces.

We now define a set of contiguous subspaces. Informally, a contiguous region of a subspace lattice is the set of subspaces such that we can start from any subspace in this region and reach any other subspace in this same region without needing to visit a subspace outside the region. A formal definition of contiguous subspaces will now follow.

Definition 7. *Two subspaces p and q are said to cohabit a region R if:*

- *Either p and q are connected by an edge*
- *Or, there exists a path between p and q with intermediary nodes $s_1, s_2 ... s_{n-1}, s_n$ such that for $1 \leq i < n$, s_i and s_{i+1} also cohabit the regions R.*

Definition 8. *The 'cohabit' relationship as described above defines an equivalence relation on subspaces in S that can be used to divide S into disjoint sets. Each such disjoint set is called a 'contiguous region' in the given subspace set S.*

Definition 9. *For a contiguous subspace lattice region R, all nodes p such that no parent(child) of p exists within R are called 'topmost nodes' ('bottom-most nodes) in region R.*

Definition 10. *Subspace q is descendant (ancestor) of subspace p if : (a) Either q is child (parent) subspace of p, (b) Or q is descendant (ancestor) of a child (parent) subspace of p.*

The above definition recursively defines the concept of descendant subspaces in subspace lattice. It is analogous to the recursive definition of binary trees.

The next step is to identify contiguous regions among the set of subspaces in which the same outlier has been identified. The procedure described in algorithm 3 picks up a subspace in which this given outlier has been identified. Then it keeps exploring those child and parent nodes of this subspace in which the same outlier is identified. Each completion of this step yields one contiguous region of subspaces in which the outlier is identified. This step is continued till all the subspaces in which the given outlier is identified have been explored, potentially generating more than one contiguous region. The procedure repeats this computation for all the outliers that have been identified.

Let us denote the number of outliers by q and the number of subspaces by m. In the worst case, all q data points will be outliers in all the m subspaces. Hence the worst case complexity of the above procedure is $O(mq)$. However, in real datasets it is unlikely that a data point will appear anomalous in all subspaces and this can happen only for pathological cases in very simple datasets. Hence this worst case is extremely unlikely to occur. The average behavior depends on the average occurrence of outliers. Since this is difficult to quantify a-priori what the average number of outliers and their distribution of various subspaces can be, the complexity of algorithm in average case is difficult to specify because of lack of information about the average input data.

Next, we determine the topmost subspaces in any given contiguous region. Algorithm 4 iterates through all the subspaces in a given contiguous region and identifies and returns those subspaces whose parents don't exist in this contiguous region.

The worst case complexity for this procedure is $O(q)$, where we assume that all q subspaces in the lattice belong to a single contiguous region.

A procedure analogous to algorithm 4 can be written for determining the bottom-most subspaces in a contiguous region by considering the child subspaces instead of the parent subspaces of the subspace s in the *for* loop.

Next, we determine the absent subspaces for a given contiguous region. The procedure starts from the topmost subspaces and places them in a queue. For each subspace in the queue, it determines whether the bottom-most subspaces are descendants of this subspace, and also whether this subspace exists in the region. If this subspace does not exist in the region, it is considered absent subspace. If it does exist in the region, its child nodes are placed in the queue for similar examination.

In the worst case, we will end up iterating through all the subspaces in the lattice. Hence the worst case complexity is $O(m)$ where m is the number of subspaces in the lattice.

As as result of execution of these algorithms, we obtain the following:

Algorithm 3. Contiguous Region Computation

Input:
S_{out}: set of outliers identified. Outliers include the subspaces in which they were identified.

Output: Collection of contiguous regions of subspaces.

```
for each outlier o ∈ S_out do
    C ← allocate a collection of sets of subspaces
    subspaces_unexplored ← o.subspace_set
    while subspaces_unexplored is not empty do
        R ← allocate new subspace set
        s ← get a subspace from subspaces_unexplored
        list_subspace ← allocate a new list of subspaces
        Add s to list_subspace
        while list_subspace is not empty do
            s_tmp ← get a subspace from list_subspace
            Add s_tmp to R
            Add to list_subspace all parents of s_tmp that
              exist in subspace_unexplored and have not yet
              been explored, and remove them from
              subspace_unexplored
            Add to list_subspace all children of s_tmp that
              exist in subspace_unexplored and have not yet
              been explored, and remove them from
              subspace_unexplored
        end
        add R to C
    end
end
return C
```

Algorithm 4. Computation of Topmost Subspaces

Input:
R: A contiguous region of subspaces

Output: Set S_{out} of outliers, with each outlier data structure containing the set of subspaces within which this outlier has been identified.

```
S ← allocate a set of subspaces
for each subspace s ∈ R do
    if R does not contain any parent of s then
        Add s to S
    end
end
return S
```

Algorithm 5. Computation of Absent Subspaces

Input:
R: A contiguous region of subspaces
$S_{topmost}$: topmost subspaces of R
$S_{bottom-most}$: bottom-most subspaces of R

Output: Set of subspaces that exist between $S_{topmost}$ and $S_{bottom-most}$ in the contiguous subspaces lattice, but do not lie within the contiguous region R.

```
list_absent-subspaces ← allocate list of subspaces
list_subspaces ← allocate list of subspaces
Add all subspaces in S_topmost to list_subspaces
while list_subspaces is not empty do
    s ← get a subspace from list_subspaces
    if S_bottom-most contains descendants of s then
        Add children of s to list_subspaces
        if R does not contain s then
            Add s to list_absent-subspaces
        end
    end
end
return list_absent-subspaces
```

- The set of contiguous regions within which a given data point displays anomalous behavior
- The topmost and bottom-most subspaces in the aforementioned contiguous regions.
- The subspaces between the topmost and bottom-most subspaces within which the data points appear normal.

This information is instrumental in providing insights into the anomalous behavior of data points.

4 Results

We executed the procedure mentioned in the previous section on the following datasets obtained from the UCI Machine Learning Repository[9]: *a*) Iris dataset, and *b*) Seeds dataset

We used the two definitions of outliers mentioned before as drivers in our analysis of outliers over various subspaces.

4.1 Iris Dataset

Let us first consider the analysis of Iris dataset using *ZDist* that we defined earlier. Iris is well-studied dataset with 150 data points and 4 attributes that represent properties of the Iris flower. The attributes are: sepal length, sepal width, petal length and petal width.

All measurements are in centimeters. We represent these attributes with numerals 1, 2, 3 and 4 respectively. We represent a subspace as a dot-separated sequence of attributes that constitute this subspaces. For example, the sequence 1.3.4 represents the subspace comprising of attributes 1, 3 and 4.

If we consider the whole attribute space, only three data points, namely 132, 118 and 119 are identified as outlier using $ZDist$, with a neighborhood of 30 points. However, when we search through subspaces, 18 data points including the above mentioned 3 data points are identified as outliers across various subspaces.

Further, we gain interesting insights such as, data point 123 is identified as outlier in the contiguous region of subspaces $\{1, 3, 1.3, 3.4, 1.2.3, 1.3.4\}$. The topmost subspaces of this region are 1 and 3. The bottom-most subspaces are $\{1.2.3\}$ and $\{1.3.4\}$. Interestingly, the subspaces 1.2, 1.4 and 2.3 that lie on the subspace lattice along the paths from the topmost subspaces to bottom-most subspaces do not identify the data point 123 as an outlier. This indicates a peculiarity in these subspaces because the same data point appears anomalous in the subspaces immediately above and below these subspaces, but not within them. Note also that this data point is not identified as an outlier in the complete attribute space. It is anomalous only within specific subspaces.

Table 1 summarizes a few examples obtained from the analysis of Iris dataset using $Z - Dist$. To save space, we have included only a few representative cases (out of the 18 found) in this table.

Table 1. Outlier Analysis on Iris Dataset using Z-Dist

Data Point	Is Outlier in Complete Attribute Space	Contiguous Regions	Topmost Subspaces	Bottom-most Subspaces	Absent Subspaces
132	Yes	{1, 1.2, 1.3, 1.4, 2.3, 2.4, 1.2.3, 1.2.4, 1.3.4, 1.2.3.4}	1, 2.3, 2.4	1.2.3.4	2.3.4
118	Yes	{1, 3, 1.2, 1.3, 2.3, 2.4, 3.4, 1.2.3, 1.2.4, 1.3.4, 2.3.4, 1.2.3,4}	1, 3, 2.4	1.2.3.4	1.4
119	Yes	{1, 3, 1.2, 1.3, 2.3, 3.4, 1.2.3, 1.2.4, 1.3.4, 2.3.4, 1.2.3,4}	1, 3	1.2.3.4	1.4
106	No	{3}	3	3	None

Table 2. Outlier Analysis on Seeds Dataset using LOF

Data Point	Is Outlier in Complete Attribute Space	Contiguous Regions	Topmost Subspaces	Bottom-most Subspaces	Absent Subspaces	Region Size
150	No	{1, 1.2, ..., 1.3.4.5, 1.2.3.4.5}	1, 2.4, 3.5	1.2.3.4.5	2.3.5, 2.4.5, 3.4.5, 2.3.4.5	10
84	No	{1, 1.3, ..., 1.4.5, 1.3.4.5}	1	1.3.7, 1.3.4.5	Empty	10
84	No	{4.6, 6.7, ... 4.5.6.7, 3.4.5.6.7}	4.6, 6.7	3.4.5.6.7	4.6.7	11
175	No	{1, 1.3, ..., 1.2.4.5, 1.2.3.4.5}	1, 3.5	1.2.3.4.5	1.2, 1.2.3, 1.4.5, 2.3.5, 3.4.5, 1.3.4.5	13

4.2 Seeds Dataset

The seeds dataset consists of 210 data points and 7 attributes. The lattice generated from this dataset consists of 127 subspaces. Being a larger lattice, this yields richer set of information than the Iris dataset.

Next, we discuss the analysis of this dataset using LOF as the definition of outliers. When we consider all the attributes, 10 outliers were identified. However, on searching across various subspaces, 89 outliers including the aforementioned 10 data points were identified as outliers. Data point 125 is an example of the identified outliers in the contiguous region $\{2.4, 4.5, 2.3.4, 2.4.5, 4.5.7, 2.3.4.5, 2.4.5.7, 3.4.5.7, 2.3.4.5.7\}$. The topmost subspaces for this regions are 2.4 and 4.5. The bottom-most subspace is 2.3.4.5.7. The subspaces 2.4.7, 3.4.5 and 2.3.4.7 lie between the topmost and bottom-most subspaces, but are absent from the contiguous region. The size of this contiguous region is 9. We show a few more representative results in table 2.

5 Conclusion

In this paper we have presented a methodology that aims at providing greater insights into outlier behavior instead of just reporting the outliers and the subspaces with which they are identified. We compute the *contiguous regions* of the subspace lattice within which the same data point is identified as outlier. We also compute *topmost and bottom-most subspaces* in this region. These subspaces are important points of inflection along the subspace lattice where specific data points either start appearing as anomalies, or stop being anomalies. Finally, we compute subspaces that lie within the topmost and bottom-most subspaces mentioned above, and in which the data point in question does not appear anomalous.

These subspaces introduce attributes under the influence of which the data point again appears normal. We term these subspaces the *absent subspaces.*

Taken together, all this information provides deeper insight to the analyst about causal factors (in terms of responsible attributes) of anomalies. This can help to either mitigate the reasons that give rise to anomalies, or anticipate anomalies and prepare to reduce their impact.

Our methodology can work in conjunction with any definition of outliers. We have demonstrated the insights gained from our methodology and its operability with multiple definitions of outliers by showing examples of results of analysis over the Iris and seeds datasets, using $Z - Dist$ and LOF as two competing definitions of outliers.

References

1. Knorr, E.M., Ng, R.T.: Finding Intensional Knowledge of Distance-Based Outliers. In: Proceedings of 25th International Conference on Very Large Data Bases VLDB 1999, Edinburgh, Scotland, UK, September 7-10, pp. 211–222. Morgan Kaufmann (1999)
2. Keller, F., Muller, E.B.K.: HiCS: High Contrast Subspaces for Density-Based Outlier Ranking Data Engineering (ICDE). In: 2012 IEEE 28th International Conference on, pp. 1037–1048 (2012)
3. Zhang, J., Lou, M., Ling, T.W., Wang, H.: HOS-Miner: A System for Detecting Outlying Subspaces of High-dimensional Data. In: Proceedings 2004 VLDB Conference, pp. 1265–1268. Morgan Kaufmann (2004)
4. Aggarwal, C.C.: Outlier analysis. Springer (2013)
5. Breunig, M., Kriegel, H.-P., Ng, R.T., Sander, J.: LOF: identifying density-based local outliers. SIGMOD Rec. 29(2), 93–104 (2000), `http://doi.acm.org/10.1145/335191.335388`, doi:10.1145/335191.335388
6. Angiulli, F., Pizzuti, C.: Outlier mining in large high-dimensional data sets. IEEE Transactions on Knowledge and Data Engineering 17, 203–215 (2005)
7. Nguyen, H.V., Gopalkrishnan, V., Assent, I.: An unbiased distance-based outlier detection approach for high-dimensional data. In: Yu, J.X., Kim, M.H., Unland, R. (eds.) DASFAA 2011, Part I. LNCS, vol. 6587, pp. 138–152. Springer, Heidelberg (2011)
8. Ebdon, D.: Statistics in geography. John Wiley and Sons (1985)
9. Bache, K., Lichman, M.: UCI Machine Learning Repository. University of California, School of Information and Computer Science, Irvine, CA (2013), `http://archive.ics.uci.edu/ml`

eSelect: Effective Subspace Selection for Detection of Anomalies

Vineet Joshi and Raj Bhatnagar

Dept. of Electrical Engineering and Computing Systems
University of Cincinnati, Cincinnati, OH, USA
joshivt@mail.uc.edu
raj.bhatnagar@uc.edu

Abstract. Anomaly detection is used for many applications such as detection of credit card fraud, medical diagnosis and computer system intrusion detection. Many interesting real-world data sets are high dimensional. Detection of anomalies in such datasets is hampered due to the curse of dimensionality. In such datasets the anomalies are hidden in smaller subspaces of attributes. However the number of subspaces possible from a given attribute set increases in a combinatorial fashion. Consequently an exhaustive search through all possible subspaces for anomalies is not computationally feasible. In this paper we propose a method for exploring the subspaces in a high dimensional data set in an effective and organized way to detect anomalies within them while avoiding an exhaustive search over all possible subspaces. The new method is called *eSelect*. Through extensive experimentation we compare *eSelect* to a well-established subspace selection method and demonstrate that our newly proposed method attains marked improvements.

Keywords: anomaly detection, outliers, high-dimensional data, unsupervised data mining, subspace sampling.

1 Introduction

Anomaly detection is an important problem within data mining with diverse applications such as credit card fraud detection, computer system intrusion detection and medical diagnosis [1]. There have been many approaches to anomaly detection including the statistical, distance based, information theory based etc[1].

The increase in the sizes and dimensionality of datasets has introduced new challenges about detecting anomalies. The existing methods of anomaly detection do not perform well when directly applied to high dimensional datasets because of the *curse of dimensionality*[2]. In this paper we propose a new technique for effectively searching for anomalies in high-dimensional datasets. Through extensive experiments we demonstrate that our technique performs better than other methods such as the subspace sampling method proposed by Lazarevic[3].

Our method builds upon the foundations laid by Aggarwal et al.[2] and Lazarevic et al.[3]. Our specific contributions in this paper are:

D. Ślęzak et al. (Eds.): AMT 2014, LNCS 8610, pp. 251–262, 2014.

1. A subspace sampling algorithm that selects subspaces that are effective in discovering anomalies.
2. Not require the analyst to provide input parameters that are difficult to know *a-priori.*
3. Allow the analyst to adapt the computational overhead according to the resources available, instead of simply failing to terminate in the face of very high-dimensional data.

Our method improves upon the method proposed by Aggarwal et al.[4] by not requiring the analyst to supply a parameter ϕ that specifies the number of *equi-depth* portions into which each attribute should be divided. It is difficult to know the correct value for this parameter apriori and hence eliminating the need for such a parameter is one of the goals of our method.

Further, we also aim at not basing our method on arbitrary parameters that, while reducing the attribute search space, cannot be universally justified. For example, Nguyen et al.[5] proposed an anomaly detection method in which they have implicitly assumed the value of ϕ from Aggarwal's work to be 10. Choosing a high value of this parameter helps reduce the search space. However in cases where this is not a suitable value, the search space is unjustifiably pruned. The goal of our method is to not prune the search space without justification, and instead provide the analyst with parameters that they can use to control the computational overhead to suite their computational resources.

Hence our contribution in this paper is an anomaly detection method for high-dimensional data that eliminates these shortcomings.

2 Related Work

It has been observed that anomaly detection techniques developed for low dimensional datasets lose effectiveness with high dimensional datasets because of the curse of dimensionality[2]. Statistical techniques were among the earliest attempts at detecting anomalies in datasets[1]. These techniques depend on the assumption of a statistical model to represent the complete dataset. They identify outliers as datapoints that don't conform to this assumed model for the complete dataset. These statistical techniques don't scale to high dimensional datasets because of the computational complexity associated with determining the correct statistical model and choosing its parameters to effectively model the data[1]. The proximity-based methods also tend to lose meaning because the concept of distance loses meaning in high-dimensional datasets[2].

Aggarwal et al. were among the early researchers to explore the problem of anomaly detection in very high dimensional datasets. They observed that in very high dimensional dataset all points seem almost equally distant with no clear outliers. Instead, the real outliers are embedded in subspaces formed by subsets of attributes. Hence the use of subspaces for anomaly detection has been proposed, analogous to the use of subspaces for data clustering[2].

However, the number of subspaces available for a given set of attributes is a combinatorial number and increases at a very rapid rate as the number of attributes increases. Consequently, an exhaustive search through all possible subspaces in a high-dimensional dataset is not computationally feasible. Aggarwal et al. have termed this situation as searching for needle within a haystack when the number of haystacks itself is exponential[6].

To tackle this problem of exploring a very large number of subspaces, Lazarevic et al. proposed the method of subspace sampling[3]. Here the idea is to randomly sample subspaces of sizes varying from $\lfloor d/2 \rfloor$ to $(d-1)$ where d represents the total number of attributes in the dataset. It has been noted that unlike association rule mining, the outlier-ness of a datapoint is not upward or downward closed with respect to the set of subspaces. Hence, the exploration of outlier-ness of a datapoint in a given subspaces does not provide any guidance about its outlier-ness in any other subspace in the complete lattice of subspaces[6]. The subspace sampling method proposed by Lazarevic et al. implicitly acknowledges this and resorts to randomly sampling the subspaces to search for outliers[6].

Aggarwal and Yu have investigated the question of the number of attributes of a dataset that need to be examined in order to determine the anomalies embedded in the subspaces of attributes[2][4]. They argued that for an input parameter ϕ, the number of attributes that need to be examined is $\lfloor \log_\phi \left(N/s^2+1\right) \rfloor$. Here the value of parameter ϕ needs to be judiciously chosen. This parameter represents the number of portions into which each attribute is divided. These portions are not equal-sized. Instead, the sizes of portions are chosen to each contain the same number of data points. The authors use the term *equi-depth* to describe these ranges. These regions across different attributes divide the complete data space into small cubes. The proposed method then uses evolutionary algorithms to find the cubes that have abnormally low density of data. Choosing too large a value of ϕ is counterproductive because it would cause most data cubes to contain too small a number of data points to yield a meaningful conclusion about outliers within them. Choosing a smaller value would increase the number of attributes that would need to be included in the search for sparsely populated data cubes and hence would increase the search space. Choice of the correct value would depend on the analyst's insights about the dataset.

Nguyen et al. implicitly choose the value of ϕ as 10 and explore all the subspaces of sizes from 1 to $\lfloor \log_{10} N \rfloor$ where N represents the number of records in the dataset[5]. From the context of Aggarwal's proposal, this means that each attribute is always divided into 10 equi-distant regions. however Nguyen does not provide a basis for this choice, and it seems somewhat arbitrary. Further, for large datasets, this value can cause a very large number of subspaces to be explored and the algorithm does not provide any method to control the computational overhead. For sufficiently large datasets, the method may fail to terminate in a reasonable length of time, and the analyst does not have any control over this.

These related works provide a platform for the method we propose in this paper. We are proposing a new method of sampling subspaces based on the theoretical consideration provided by Aggarwal[2]. Our subspace sampling method performs better than the Lazarevic's method of subspace sampling.

The new method is based on the hypothesis that low-dimensional subspaces are more useful than high dimensional subspaces in the search for outliers. Therefore the computational resource should be preferably spent exploring low-dimensional subspaces as compared to the high-dimensional subspaces as compared to the high-dimensional subspaces. The results of experiments conducted thoroughly vindicate our faith in this hypothesis and the new method performs substantially better than Lazarevic's subspace sampling method.

3 Our Approach

We start from Aggarwal's conclusion that the number of attributes that need to be examined to detect anomalies is $\lfloor \log_{\phi}(N/s^2+1) \rfloor$, where ϕ is an input parameter. However, we don't depend on the analyst to supply the appropriate value of ϕ. Unlike Nguyen's method[5] we set it to 2, which results in the maximum number of attributes to be searched. This avoids arbitrarily restrictions on the search space, as would happen if we choose any higher arbitrary value of ϕ. Hence we use $\lceil \log_2 N \rceil$ as the maximum size of the subspaces that need to be searched for anomalies. Then we sample subspaces from the set of subspaces of sizes from 1 and $\lceil \log_2 N \rceil$.

Further, we use the heuristic that outliers are more likely to be discovered in low-dimensional subspaces rather than in high-dimensional subspaces in the subspace lattice. Consequently, when we sample the subspaces of sizes between 1 and $\lceil \log_2 N \rceil$, we choose a greater fraction of low dimensional subspaces than the high-dimensional subspaces.

In this paper we focus on the exploration of subspaces for detection of outliers within them. Our method can accommodate any technique of outlier detection that generates outlier scores. Like Lazarevic's proposal, we collect the outlier scores for all records across various sample subspaces. Then we use the cumulative sum approach described by Lazarevic et al.[3] to draw conclusions about the outliers detected in various subspaces.

Our method of searching for outliers in different subspaces does not depend on any specific definition of outlier-ness. We only require that the definition of outlier-ness generate outlier scores that signify the extent of outlying behavior demonstrated by records in the dataset. For our discussion in this paper we have used the distance to k nearest neighbors[7] as the definition of outliers. Any other definition generating the outlier scores such as the *Local Outlier Factor* (LOF)[8] would also serve our purpose.

Once the outlier scores have been computed, the analyst can apply thresholds to the score and select the records whose cumulative outlier scores exceed the threshold. This step is similar to the application of thresholds to the LOF scores to identify the outliers.

To control the computational complexity, we introduce a user defined parameter c, which is a number between 1 and $\lceil \log_2 N \rceil$. While choosing subspaces and computing outliers within them, we terminate the process when all subspaces of sizes 1 to c have been already selected. We will use n as a short notation to represent the quantity $\lceil \log_2 N \rceil$ in our discussion from now on-wards.

Let D denote the complete dimensionality of the dataset, and N denote the number of records in the dataset.

It should be noted that the exhaustive search through all the subspaces of sizes between 1 and $\lceil \log_2 N \rceil$ can also be computationally intensive. This step can be scaled down to suit the computational resources that the analyst is able to offer. We choose the input parameter c such that instead of exploring all the subspaces of sizes between 1 and n, we terminate when all subspaces of sizes from 1 to c have been explored.

Further, note that the number of subspaces with 1 attribute will be smaller than the number of subspaces with 2 attributes, which in turn will be smaller than the number of subspaces with 3 attributes and so on. In a general case, if D represents the dimensionality of the dataset, then till $D/2$ attributes, the number of subspaces will increase as the number of attributes is increased. Consequently, when we choose subspaces by first selecting the number of attributes based on the uniformly distributed random number, we will end up exhausting all subspaces of size 1 before we exhaust all subspaces of size 2. Similarly, we will use up all subspaces of size 2 before we use up all subspaces of size 3, and so on. Hence, we will end up exhausting all subspaces of size $(d-1)$ before we exhaust all subspaces of size d.

We can use the input parameter $1 \leq c \leq n$ that we described before to indicate that we intend to stop the computation once all subspaces of size up to c have been explored. Hence the analyst can use this parameter to terminate the search for outliers if the complete search through all subspaces of size up to n is more than the computational resources available.

We would like to observe that restricting the size of subspaces searched to n is only a heuristic to make the problem of searching through all possible subspaces computationally tractable. However, it does not absolutely guarantee that all outliers have been found. In fact, such a guarantee is not possible without performing exhaustive search through the complete subspace lattice because anomalous behavior in the lattice does not demonstrate upward or downward closed properties[6].

It is possible to further restrict the computational resources required by introducing another input parameter f which represents the fraction of subspaces of sizes from 1 to c that needs to be fully explored.

Putting these restrictions in place reduces the thorough-ness of the subspace search process with a possibility that some outliers escape detection, or at least don't score as high as they should in the cumulative outlier-scores. On the other hand, they provide the analyst with control over the computational resources that the analyst will consume. This is a trade-off that the analyst has to make while analyzing very high dimensional datasets.

3.1 Computational Complexity

The number of subspaces of size from 1 to n is

$$O\left(n \times \binom{D}{n}\right)$$

Algorithm 1. Cumulative Outlier Score Computation

Input:

N: number of records in the dataset

D: number of dimensions in the dataset

Output:

$C[1..n]$: cumulative outlier scores of each record in the dataset.

```
initialize n ← ⌈log_2 N⌉
allocate array M[1..n]
allocate array P[1..n]
allocate array C[1..n]
for i ← 1 to n do
    initialize M[i] ← (D choose i)
    initialize P[i] ← 0
    initialize C[i] ← 0
end
while true do
    continue_flag ← false
    for i ← 1 to n do
        if P[i] < M[i] then
            continue_flag ← true
            break
        end
    end
    if continue_flag is true then
        r ← {uniformly distributed random number between 1 and n}
        while P[r] == M[r] do
            r ← (r + 1)
        end
        Choose a subspace S_r of size r that
          has not been chosen before.
        P[r] ← P[r] + 1
        Compute the outlier scores of all records
          in the dataset in subspace S_r.
        Add outlier scores for each record[i]
          to its cumulative outlier score in C[i].
    end
    else
        break
    end
end
```

where D is the total number of dimensions in the complete dataset. Hence, the worst case computational complexity of the process of exploring subspaces also is the same. This represents only the complexity of the search through the subspaces. Complexity of the specific definition of anomalous behavior will vary from one method to another.

4 Empirical Evaluation

We have compared our method of subspace sampling with the Lazarevic's method[3]. We have used the following datasets for classification tasks from the UCI Machine Learning repository[9] for comparison:

- Musk dataset
- Spam dataset
- Wisconsin Breast Cancer (Diagnostic) dataset

Classification datasets are not ideally suited for anomaly detection tasks. A general practice is to designate the records belonging to the smaller class as anomalies, and evaluate the anomaly detection algorithm on whether these small-class records are detected by the algorithm. However, this is not a good test because the algorithm might end up considering the distances of records of smaller class from other records in the same class, and hence might not find them unusual enough to flag them as anomalies. Thus, the algorithm will underperform because the dataset used to evaluate it does not have 'true-enough' anomalies.

A better strategy is to embed a very small number of records from the smaller class among the records of the larger class. In this case, records of the smaller class will stand-out as anomalies with respect to their neighbors because most of the neighbors will belong to the bigger (i.e. a different) class. Hence, such a dataset is a better test for anomaly detection algorithm. We follow this argument, and hence create such an augmented dataset wherein we embed a very small number of records from smaller class among the records of the large class.

Musk Dataset. The musk dataset contains records belonging to two classes, the musk and non-musk records as identified by human observers. We used a uniformly distributed random number generator to randomly pick 10 records from the non-musk records and inserted them within the records for musk data points. The final augmented dataset contained 217 data points. We generated 3 such augmented datasets and repeated our experiments with all of them.

We used $z - score$ of distances to k nearest neighbors [7] as outlier scores in our experiment. Fig. 1 contains the results for one augmented dataset for 10 nearest neighbors. Subspaces were selected till all subspaces of size 1 got picked. For each subspace, only the records for which the outlier-score exceeded a threshold of 3 were selected. This was done to prevent records with low outlier scores from obscuring the true outliers because of accumulation of scores over multiple subspaces. As is evident from the figure, the *area under curve* (AUC) for *eSelect* is 0.944 while for Lazarevic's method it is 0.798. We have presented the average AUC scores for all the 3 augmented datasets in table 1.

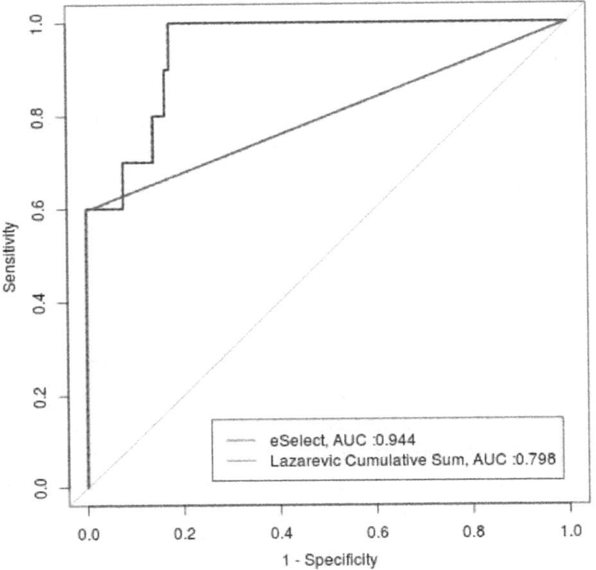

Fig. 1. Musk Dataset, $k = 10$, $attrib = 1$

Table 1. Avg. AUC values for Musk dataset

k	*eSelect*	Lazarevic's Method
5	0.932689	0.739935
10	0.929307	0.710064
15	0.924637	0.710628

Spam Dataset. The spam dataset contains two classes, one corresponding to the non-spam emails and another for spam emails. We again randomly picked 10 records from the class for spam records and inserted them within the records for non-spam records using a random number generator. The final augmented dataset contained 2798 data points. Again, we generated 3 such augmented datasets for our experiments.

Again, we used $z - score$ of distances to k nearest neighbors as outlier scores in our experiment. Fig. 2 contains the results for one augmented dataset for 5 nearest neighbors. Subspaces were selected till all subspaces of size 1 got picked. Again, only the records for which the outlier-score exceeded a threshold of 3 were selected. The *area under curve* (AUC) for *eSelect* is 0.823 while for Lazarevic's method it is 0.631. We have presented the average AUC scores for all the 3 augmented datasets in table 2.

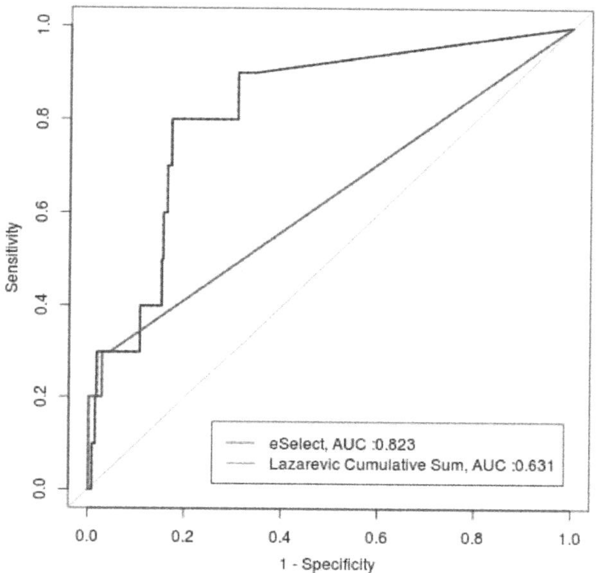

Fig. 2. Spam Dataset, $k = 5$, $attrib = 1$

Table 2. Avg. AUC values for Spam dataset

k	*eSelect*	Lazarevic's Method
5	0.868681	0.665967
10	0.728425	0.614030
15	0.755697	0.628006

Wisconsin Breast Cancer Dataset. The Wisconsin Breast Cancer dataset contains two classes, one for malignant cases and the other for benign cases. We randomly picked 10 records from the malignant class using a uniformly distributed random number generator, and inserted them within the class containing records for the benign case. The final augmented dataset contained 367 records. As with previous cases, 3 augmented datasets were created.

As before, we used $z - score$ of distances to k nearest neighbors as outlier scores in our experiment. Fig. 3 contains the results for one augmented dataset for 10 nearest neighbors. Subspaces were selected till all subspaces of size 2 got picked. Again, only the records for which the outlier-score exceeded a threshold of 3 were selected. The *area under curve* (*AUC*) for *eSelect* is 0.973 while for Lazarevic's method it is 0.844. We have presented the average *AUC* scores for all the 3 augmented datasets in table 3.

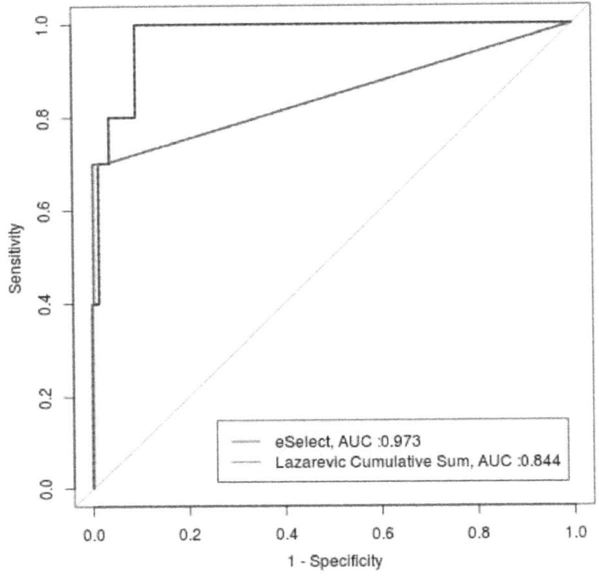

Fig. 3. Wisconsin Breast Cancer Dataset, $k = 5$, $attrib = 2$

Table 3. Avg. *AUC* values for Wisconsin Breast Cancer dataset

k	*eSelect*	Lazarevic's Method
5	0.878944	0.792950
10	0.915639	0.776143
15	0.931139	0.794864

5 Conclusion

Detection of anomalies in high-dimensional data poses unique problems because of the *curse of dimensionality*[10]. The techniques developed for low-dimensional data are not very effective when applied to the dataset with very large number of dimensions[2].

It has been suggested that in such high dimensional data the outliers are actually hidden in the low-dimensional projections of the dataset[2]. However, the number of low-dimensional projects, also called the subspace, is a combinatorial number with respect to the number of attributes and its exhaustive exploration is not possible.

We have presented in this paper a subspace sampling based method called *eSelect* which is an effective method for sifting through the search space consisting of all possible subspaces in the subspace lattice of the dataset. The new method out-performs the currently well-established methods of sampling the subspaces.

As stated earlier, our method is based on the hypothesis that low dimensional subspaces are more useful in the search for outliers as compared to high dimensional subspaces. Hence our method preferably sifts through the low dimensional subspaces over high dimensional subspaces. Validity of this assumption is the fundamental reason behind the improvements demonstrated by *eSelect*. Our dependence on this hypothesis is vindicated by the results of our experiments.

Further, unlike Aggarwal's method that requires the analyst to input a parameter to represent the number of equi-depth ranges into which the attributes need to be split, we place no such requirement upon the analyst. Unlike the method proposed by Nguyen et al.[5] which uses the arbitrary value of 10 as the number of regions into which the attributes need to be split, we use the value of 2 which does not unwarrantedly restrict the number of attributes in the subspaces that need to be explored. Further, unlike Nguyen's method, we don't exhaustively search through all the subspaces up to a certain size. Instead, we sample for subspaces among all the available subspaces up to a maximum number of attributes, as described in the description of our method. This allows us to scale our method to arbitrarily large datasets.

Finally, our method includes parameters that specify the cardinality of the subspaces that need to be completely explored as a heuristic to further control the execution of computation so that it does not exceed the computation resources available to the analyst. We aim that the search for anomalies should gracefully degrade in the face of insufficient resources as the number of dimensions grows. The search should not instead simply fail to terminate in a reasonable period of time.

Hence our method can operate on dataset with huge number of records and attributes and produce results with accuracy governed by the computational resources committed by the analyst, instead of simply failing to accommodate such huge datasets.

References

1. Chandola, V., Banerjee, A., Kumar, V.: Anomaly detection: A survey. ACM Comput. Surv. 41(3), Article 15, 58 pages (2009), `http://doi.acm.org/10.1145/1541880.1541882`, doi:10.1145/1541880.1541882
2. Aggarwal, C.C., Yu, P.S.: Outlier detection for high dimensional data. In: Sellis, T., Mehrotra, S. (eds.) Proceedings of the 2001 ACM SIGMOD International Conference on Management of Data (SIGMOD 2001), pp. 37–46. ACM, New York (2001), `http://doi.acm.org/10.1145/375663.375668`, doi:10.1145/375663.375668

3. Lazarevic, A., Kumar, V.: Feature bagging for outlier detection. In: In Proceedings of the eleventh ACM SIGKDD International Conference on Knowledge Discovery in Data Mining (KDD 2005), pp. 157–166. ACM, New York (2005), http://doi.acm.org/10.1145/1081870.1081891, doi:10.1145/1081870.1081891
4. Aggarwal, C., Yu, P.S.: An effective and efficient algorithm for high-dimensional outlier detection. The VLDB Journal 14(2), 211–221 (2005), http://dx.doi.org/10.1007/s00778-004-0125-5, doi:10.1007/s00778-004-0125-5
5. Nguyen, H.V., Gopalkrishnan, V., Assent, I.: An unbiased distance-based outlier detection approach for high-dimensional data. In: Yu, J.X., Kim, M.H., Unland, R. (eds.) DASFAA 2011, Part I. LNCS, vol. 6587, pp. 138–152. Springer, Heidelberg (2011)
6. Aggarwal, C.C.: Outlier analysis. Springer (2013)
7. Ebdon, D.: Statistics in geography. John Wiley and Sons (1985)
8. Breunig, M.M., Kriegel, H.-P., Ng, R.T., Sander, J.: LOF: identifying density-based local outliers. SIGMOD Rec. 29(2), 93–104 (2000), http://doi.acm.org/10.1145/335191.335388, doi:10.1145/335191.335388
9. Bache, K., Lichman, M.: UCI Machine Learning Repository. University of California, School of Information and Computer Science, Irvine, CA (2013), http://archive.ics.uci.edu/ml
10. Bellman, R.: Adaptive control processes: a guided tour, vol. 4. Princeton university press (1961)

Active Recommendation of Tourist Attractions Based on Visitors Interests and Semantic Relatedness

Yi Zeng, Tielin Zhang, and Hongwei Hao

Institute of Automation, Chinese Academy of Sciences, Beijing, China
{yi.zeng,zhangtielin2013,hongwei.hao}@ia.ac.cn

Abstract. Many visitors always search on tourist attractions related information on the Web so as to get more information on the places they are visiting or plan their next trips. In this study, we introduce CASIA-TAR, an active tourist attractions recommendation system, which provides relevant knowledge of specific tourist attractions and make recommendations for other relevant places to visit based on semantic relatedness among the specific tourist attraction and potentially interesting places. Two algorithms are introduced to calculate the semantic relatedness among different tourist attractions based on the tourist attraction semantic knowledge base with relevant knowledge mainly extracted from Web-based encyclopedias. As an integrated portal for tourist attraction recommendation, CASIA-TAR also provides images, news and microblog posts that are relevant to specific tourist attractions so that visitors could obtain relevant information in an integrated Web-based system.

Keywords: Semantic Relatedness, Active Recommendation, User Interests, Knowledge Base, Information Integration.

1 Introduction

The Web provides a platform for visitors to obtain tourist attractions related information. The information related to specific tourist attractions is always distributed in different Web sources. Information integration is needed to integrate various information from different sources and a unified portal is needed to provide users with useful and most updated tourist attractions information. In this paper, we designed and implemented a tourist attraction information portal, which integrate various tourist attractions information from Web-based encyclopedias, images and news search engines, as well as microblog posts.

Traditional ways of selecting interesting tourist attractions to visit are mostly based on recommendations from other people or local tourist information lookup services. The visitors need to take actions to obtain relevant information. We propose that when certain previous visits information is obtained, active recommendations of interesting tourist attractions can be made automatically based on the unified tourist attraction information portal. Semantic relatedness plays an important role for developing such kind of active recommendation algorithms.

Two strategies for tourist attraction active recommendations are introduced in Section 2. The construction of the tourist attractions knowledge base is discussed in

D. Ślęzak et al. (Eds.): AMT 2014, LNCS 8610, pp. 263–273, 2014.

Section 3. An algorithm for semantic relatedness calculation for tourist attraction recommendation is proposed in Section 4. Section 5 introduces the services provided by CASIA-TAR, the active Tourist Attractions Recommendation system designed and implemented at Institute of Automation, Chinese Academy of Sciences (CASIA).

2 Strategies for Active Recommendation Based on User Interests and Semantic Relatedness

Given the previous visits information of a specific visitor, two recommendation strategies based on semantic relatedness among tourist attractions are proposed to meet different types of needs from visitors with contrary preferences. Here we adopt Interest Logic to formally describe these two recommendation strategies [1].

Strategy 1 (Positive Relevance Recommendation): If the visitor (denoted as v) is interested in the tourist attraction denoted as t, and there is a list of candidate tourist attractions (denoted as $T=\{t_1,t_2,\ldots,t_n\}$), an ordered list of candidate tourist attractions T' will be recommended to this visitor. The recommendation sequence is positive relevant to the semantic relatedness among t and the candidates in T. For each of the candidate, the following rule holds:

$$K_r I_v t \wedge K_r(t \sim t') \rightarrow K_r I_v t \wedge K_r I_v t',$$

where r denotes the tourist recommendation system, K is the epistemic operator, I is the interest operator, and t' is a candidate tourist attraction from the ordered list T. $t \sim t'$ indicates that these two tourist attractions are semantically related. The rule indicates if r knows that v has a previous interest in visiting t, and t is semantically related to t', then r knows that v is also interested in t'. For visitors who prefer this strategy, the tourist attractions they prefer should be similar in some way.

Strategy 2 (Negative Relevance Recommendation): Suppose the visitor v is interested in the tourist attraction t, and there is a list of candidate tourist attractions ($T=\{t_1,t_2,\ldots,t_n\}$), an ordered list of tourist attractions (T^*) will be recommended to v. The recommendation sequence is negative relevant to the semantic relatedness among t and the candidates in T. Using this strategy, the recommendation system assumes that v likes to visit the tourist attractions which are not very similar to previous ones. For visitors who adopt this strategy, the following rule holds:

$$K_r I_v t \wedge K_r \dashv (t \sim t') \rightarrow K_r I_v t \wedge K_r I_v t',$$

where $\dashv$ is the negation operator, and $\dashv (t \sim t')$ denotes that these two tourist attractions are not semantically similar. For visitors who prefer this strategy, novelty of the tourist attractions compared to previous visits is very important.

3 The Tourist Attractions Knowledge Base

Knowledge bases play central roles in various intelligent systems [2]. The tourist attractions knowledge base (denoted as TAKB) is also the core for the active tourist attraction recommendation system CASIA-TAR. TAKB is a branch knowledge base of CASIA-KB, hence it follows the design principles of CASIA-KB. CASIA-KB is a general purpose semantic knowledge base. It extracts and integrates multi-modal knowledge from various resources (e.g. Web-based Encyclopedias, Microblog posts and News on the Web) [3,4]. Entities in TAKB are linked with other entities in CASIA-KB to form a unified and highly connected knowledge base.

The names for different tourist attractions are keys to gather and integrate information to build the tourist attractions knowledge base. In this study, two strategies are considered to collect these names. Firstly, the names which owns an upper concept among "Tourist Attractions, Place of Interests" in CASIA-KB are selected (These upper concepts are extracted from three different Web-based Encyclopedias, namely, Wikipedia, Baidu Encyclopedia, and Hudong Encyclopedia. They are collectively contributed by these encyclopedia users). Secondly, the names listed in the Tourist Attraction Rating Categories of China (ranging from Level A to Level AAAAA) are selected. There are 12,275 tourist attractions which are finally included in the tourist attractions knowledge base.

Several types of information are extracted from different sources to construct the TAKB. From the encyclopedia Web pages, we extract the following items: (1) Contents in information boxes (They are extracted and stored as triple form declarative knowledge about specific tourist attractions); (2) The introductory section at the beginning of each page (These paragraphs are for quick introduction of specific tourist attractions. They are stored as textual resources and are connected to specific tourist attraction entities by the binary relation "Has Introduction"); (3) Word terms under the section of "Open Classification" and "Related Items" are extracted (Terms under "Open Classification" are upper concepts for tourist attractions, while the ones under "Related Items" are considered to be related resources, such as similar tourist attractions. These two resources are of vital importance to quantitatively calculate the semantic relatedness among tourist attractions); (4) Images embedded in the tourist attractions page (This resource will provide additional visualized information for visitors). From the image search engine, we extract at least 10 pictures for each of the tourist attractions. All the resources are stored as triple form knowledge (as shown in Table 1) in the TAKB so that they can be accessed and used in a semantic approach.

Table 1. Illustrative Triple Form Knowledge Stored in Tourst Attraction Knowledge Base

Types of Knowledge	Subject	Predicate	Object
Information box contents	National Stadium	Other Name	Birds Nest
	National Stadium	Height	69 meters
Pictures	National Stadium	Has Picture	http://img1.gtimg.com/news/pics/22304/22304186.jpg
Introduction Section	National Stadium	Has Introduction	"Beijing National Stadium, located in the south central area of the Olympic…"

The knowledge triples in TAKB are not isolated. Instead, most of them are connected to each other to form a large knowledge network. Figure 1 presents the knowledge network of 297 tourist attractions in China. As shown in the figure, there are several pivotal nodes in the network, whose node degrees are much larger than the rest. We find most of them are upper concepts of various tourist attractions. They are the hubs to connect different tourist attractions and other related knowledge. They are also the keys to calculate the semantic relatedness among different tourist attractions for making recommendations.

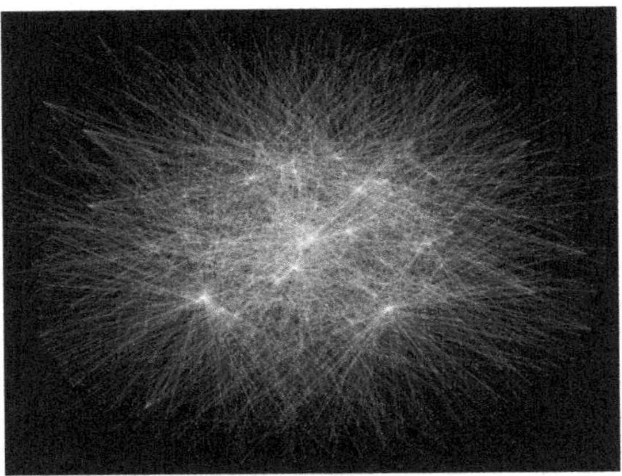

Fig. 1. The Visualized Knowledge Network of Tourist Attractions Knowledge Base

4 Semantic Relatedness Computation for Tourist Attractions

In Section 2, we proposed two strategies for active recommendation on tourist attractions. Both of them rely on the semantic relatedness of a previous visit place and the candidate places. In this section, we propose an algorithm named CIREER for computing the entity semantic relatedness based on knowledge in the tourist attraction knowledge base, as shown in the following Algorithm 1.

In CASIA-TAR, t is a previous visited tourist attraction or the one which is looked up, while $T=\{t_1,t_2,...,t_n\}$ is the list of candidate tourist attractions. Two factors serve as the key for the proposed algorithm. Firstly, as stated in Section 3, various upper concepts of different tourist attractions connect them together. Hence, the common upper concepts indicate the semantic relatedness of different tourist attractions from one perspective (For example, "the Palace Museum" and "the Ming Tombs" share upper concept "World Cultural Heritage", "Culture", "Tourism", etc.). Secondly, for each of the tourist attractions, there are several word terms which serve as "Related Items" in the TAKB (For example, "the Summer Palace" owns several "Related Items" including "the Palace Museum", "the Ming Tombs", and "the Temple of Heaven", etc.). These related items are manually tagged by encyclopedia users (Sometimes there is a negotiation process so that the listed items are agreed by most users). Hence they collectively reflect various users' opinion on whether they are related. But only with this information, it is still hard to quantitatively measure the relatedness of these related items. Hence, the second factor serves as an additional source, and these two factors need to be combined together for the semantic relatedness computation. In the upper algorithm, if a specific candidate tourist attraction t_n is listed as "Related Item" of t, then the semantic relatedness computed based on the common upper concepts will be strengthened by the weight ω.

Algorithm 1: Conceptualization Intersection and Related Entity Information based Entity Relatedness Computation (CIREER)

Input: compared entity name (t), candidate entity list ($T=\{t_1,t_2,...,t_n\}$), semantic knowledge base (TAKB)

Output: Ordered list of related entities (T') by semantic relatedness

Begin

Conceptualize the compared entity t to get all the upper concepts (the set of concepts is denoted as C_t), and conceptualize all the entities in the candidate list T to obtain $\{C_1, C_2, \dots, C_n\}$, where $C_1, C_2, \dots, C_n$ are all set of upper concepts corresponding to $\{t_1, t_2, \dots, t_n\}$.

Obtain an intersection set between each t_n in T and t by $C_t \cap C_n$.

If t_n is listed as related items of t in the knowledge base TAKB, then $sr(t, t_n) = \omega * |C_t \cap C_n|$, where ω is the weight to strengthen the relatedness ($\omega > 1$). Else, $sr(t, t_n) = |C_t \cap C_n|$.

Order T by $sr(t, t_n)$ and produce an ordered set of entities T' (If $sr(t, t_n) > sr(t, t_m)$, then t_n is ranked before t_m).

End

The left side of Table 2 shows a list of results based on the upper Algorithm 1. Here $\omega = 2$, and we restrict the tourist attractions within China. Two tourist attractions are selected to generate two lists of semantically related tourist attractions. As comparative studies, the right side of Table 2 provides a list of results, which only consider conceptualization intersection, while ignoring the impact of "Related Items" for candidates (This algorithm is denoted as CIER. In this case, CIER is a simplified version for ontology based semantic relatedness computation [5]). Namely, in the CIER algorithm, $sr(t, t_n) = |C_t \cap C_n|$ always holds.

Note that in Table 2, the results of relatedness are normalized for each column so that the differences of the relatedness values are more convenient to observe. User Evaluations of CIREER and CIER algorithm will be provided in Section 6.

For tourist attractions recommendation, if the visitors prefer to visit tourist attractions with similar types of previous visits, the order of recommendation is consistent with the semantic relatedness ranking (namely, Strategy 1 in Section 2 is adopted). If the visitors prefer novel types of tourist attractions, a reverse order of the upper list should be adopted (namely, Strategy 2 in Section 2 is adopted).

Table 2. Semantic Relatedness among Different Tourist Attractions

Tourist Attraction to Compare: The Ming Tombs					
Algrithm	Ranking	Relatedness	Algorithm	Ranking	Relatedness
CIREER	Palace Museum	1	CIER	Sanqing Mountain	1
	Great Wall at Badaling	0.714286		Palace Museum	1
	Temple of Heaven	0.714286		Fujian Tulou	1
	Jingshan Imperial Park	0.571429		Old Town of Lijiang	1
	Beihai Park	0.571429		Wudang Moutain	0.857143
	Sun Yat-Sen Park	0.571429		Chongsheng Temple	0.857143
	Sanqing Mountain	0.5		Tianzhu Moutain	0.857143
	Fujian Tulou	0.5		Big Wild Goose Pagoda	0.857143
Tourist Attraction to Compare: The Palace Museum					
Algrithm	Ranking	Relatedness	Algorithm	Ranking	Relatedness
CIREER	Ming Tombs	1	CIER	Fujian Tulou	1
	The Great Wall at Badaling	0.714286		Tanzhe Temple	0.888889
	The Temple of Heaven	0.714286		Grand View Garden	0.777778
	Fujian Tulou	0.642857		Sanqing Mountain	0.777778
	Jingshan Imperial Park	0.571429		South Lake	0.777778
	Tanzhe Temple	0.571429		Tianzhu Moutain	0.777778
	Beihai Park	0.571429		Ming Tombs	0.777778
	Sun Yat-Sen Park	0.571429		Old Town of Lijiang	0.777778

5 Active Tourist Attractions Recommendation System

CASIA-TAR is aimed at providing a unified platform for tourist attractions information lookup and recommendation. A screen shot of the system is shown in Figure 2. The system is currently available as a Web-based service platform[1]. The main functions of the system can be summarized as follows: Visitors can access CASIA-TAR on the Web. The user input for this system is a name for a tourist attraction (it can be a place for a previous visit or a place where the visitors are interested in). The output for this system is (1) related textual information and images about the tourist attraction which is queried; (2) Recommendations of possible interesting tourist attractions for next trips (based on the strategies in Section 2 and the algorithm in Section 4).

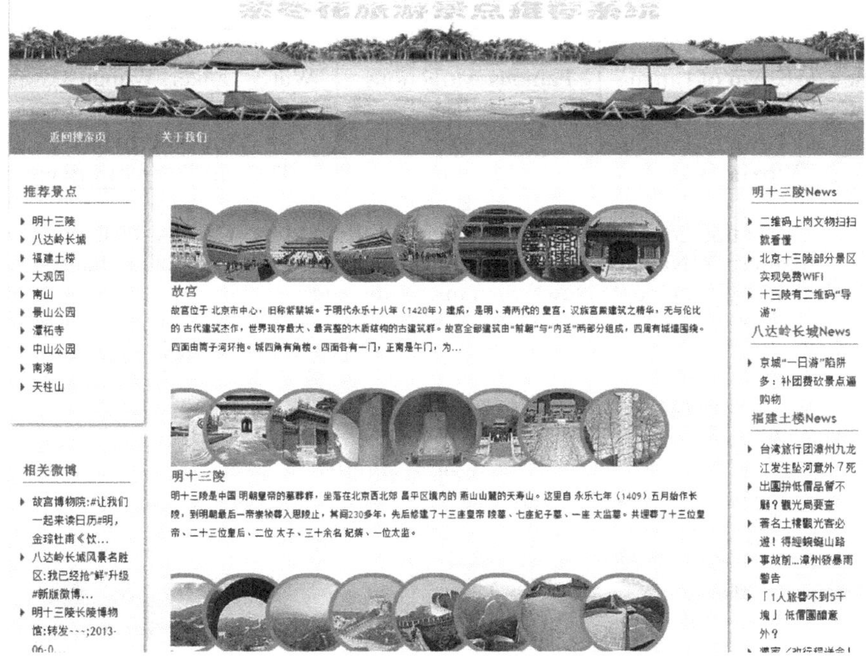

Fig. 2. A Screen Shot of the Active Tourist Attractions Recommendation System CASIA-TAR

The tourist attraction information that CASIA-TAR provides to visitors includes the following two types: (1) Not very frequently updated information (Declarative knowledge from information boxes, upper concepts, related items, and introduction section from Web-based encyclopedias. Table 3 illustrates the function which presents the declarative knowledge for a tourist attraction); (2) Very frequently updated information: news and microblog posts related to a specific tourist attraction. Note that the

---

[1] CASIA-TAR can be accessed at:
[http://www.linked-neuron-data.org/CASIA-TAR]

information which is frequently updated plays a unique role. For example, when there are special events hosted by the tourist attractions, it is probable that this kind of information appears in news Web pages and microblog posts. In addition, the most updated information might be very important for potential visitors. For example, if one gets the information that the Great Wall is full of people in these two days on the Sina microblog, chances are that he/she might reconsider the visiting plan to avoid too crowded visits.

Table 3. Declarative Knowledge on the Old Town of Lijiang presented by CASIA-TAR

Entity	Attribute	Attribute Value
The Old Town of Lijiang	Train Station	The Lijiang Train Station
The Old Town of Lijiang	Postal Code	674100
Mu's Residence	Location	The Old Town of Lijiang
The Old Town of Lijiang	Area Code	0888
The Old Town of Lijiang	Airport	The Lijiang Airport
The Old Town of Lijiang	Famous Attractions	Water Tanker, Wangu Tower
The Old Town of Lijiang	Alias	The Dayan ancient town
The Old Town of Lijiang	Location	Yunnan Province
Lijiang Woxingwosu Inn	Location	The Old Town of Lijiang

The information sources which are integrated to build the CASIA-TAR portal are shown in Figure 3 (Namely, Baidu encyclopedia, Hudong Encyclopedia, images from SoSo search engine, Google News, Sina microblog, etc.).

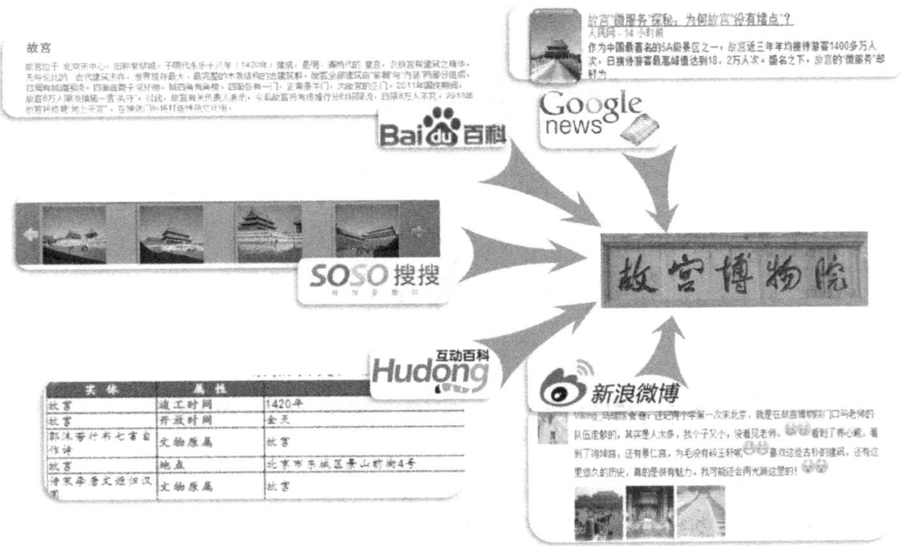

Fig. 3. Tourist Attraction Information Fusion from Various Sources for CASIA-TAR

On the left column of CASIA-TAR, a list of recommended tourist attractions is provided based on the input. When clicking on a specific recommendation, the reason for this recommendation is shown to users, as shown in Table 4.

When querying the CASIA-TAR, the system may produce several candidate tourist attractions if they share the same name. For example, if the query term is "the Palace Museum", CASIA-TAR will have five candidates, namely "the Palace Museum" in Beijing, Nanjing, Shenyang, Taipei and Seoul respectively, because they share the same label in the knowledge base and CASIA-TAR cannot identify which one the visitor is looking for.

Table 4. Recommendation Analysis Shown to the Visitors by CASIA-TAR

Tourist Attractions	The Palace Museum, The Ming Tombs
Common Tags	Art, Cultural heritage of the world, Leisure, History, Nation, Palace, Building, Construction, Culture, Cultural relic, Tourist, Tourist Attractions, Life, Documentary film, Ruins, Heritage Park

In most cases, entity disambiguation problems can be handled by considering co-occurred contextual literal information [6]. In this study, visitors only have the word term "Summer Palace" as inputs for the recommendation system. An alternative strategy needs to be proposed. In this study, we assume that whey a specific visitor is looking up relevant tourist attraction information, he/she might be in the same city with the tourist attraction. Since CASIA-TAR is a Web-based service platform, and can be accessed through mobile phones, in this case, we use the location information from the mobile phone to disambiguate the tourist attractions which share the same name. For the upper problem, if we obtain the information from the mobile that the visitor is in Beijing, the Palace Museum in Beijing will be selected and relevant information will be provided.

6 User Evaluations

The proposed strategies, algorithms and the CASIA-TAR platform were evaluated from several perspectives. Participants were asked to perform the following tasks: (1) Compare the CIREER and the CIER algorithm for relatedness between a given tourist attraction and the generated lists of candidates by these two algorithms. (2) From the recommendation perspective, compare the two lists of recommendations by the two algorithms on which list do they prefer as a possible list for future visit. (3) Decide which types of recommendations do they like, the positive relevance recommendation strategy (Strategy 1) or the negative relevance recommendation strategy (Strategy 2). Figure 4 provides a summary on the evaluation results.

User studies on 20 persons show that: (1) Judging from the relatedness perspective, the CIREER algorithm significantly outperforms the CIER algorithm, as shown in Figure 4(a). It indicates that the "related" relations from the crowds have positive contribution to the semantic relatedness computation. (2) Judging from the interestingness perspective, CIREER (55%) also outperforms CIER (45%), as shown in Figure 4(b). (3) Interestingly, although most users said that they prefer negative

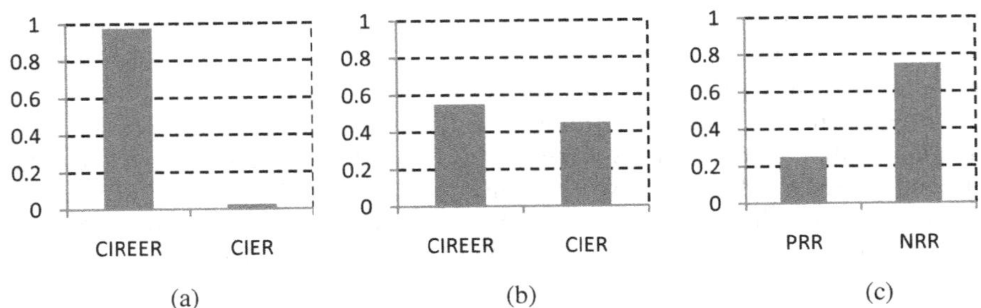

Fig. 4. Recommendation Analysis Shown to the Visitors by CASIA-TAR

relevance recommendation (NRR, as shown in Figure 4(c)), 95% of them found the recommendations by CASIA-TAR are reasonable and convenient for them, while by default, CASIA-TAR adopt the positive relevance recommendation (PRR) strategy.

7 Conclusion and Future Works

This paper introduces CASIA-TAR, an active tourist attractions recommendation system and its related core strategies and algorithms. The whole platform is based on information fusion from various Web resources such as Web-based encyclopedias, News on the Web, Microblog posts, etc., and it provides visitors a unified platform for tourist attractions information lookup and active recommendations.

The recommendation strategies proposed in this paper is based on semantic relatedness computation. The relatedness computation algorithm CIREER utilizes two sources of knowledge related to specific tourist attractions, namely the ontological structure and the related items. User studies show that the proposed CIREER algorithm outperforms the CIER algorithm which only consider ontological structures. The proposed strategies and algorithms focus on how semantics and its related computation play a role in tourist attractions recommendation, hence they can be generally considered as content based filtering techniques. More factors and methods (such as collaborative filtering, and user personalization) need to be considered and integrated for better recommendation performance.

Acknowledgement. We thank Yue Gao who participated in developing the beta version of CASIA-TAR. We thank Professor Chenglin Liu who provided helpful discussions on the design issues of the system. This study was supported by the Young Scientists Fund of the National Natural Science Foundation of China (61100128) and the Strategic Priority Research Program of the Chinese Academy of Sciences (XDA06030300).

References

1. Zeng, Y., Huang, Z., Liu, F., Ren, X., Zhong, N.: Interest Logic and Its Application on the Web. In: Xiong, H., Lee, W.B. (eds.) KSEM 2011. LNCS (LNAI), vol. 7091, pp. 12–23. Springer, Heidelberg (2011)

2. Douglas, B., Lenat, C.Y.C.: A Large-Scale Investment in Knowledge Infrastructure. Communications of the ACM 38(11), 32–38 (1995)
3. Zeng, Y., Wang, D., Zhang, T., Wang, H., Hao, H., Xu, B.: CASIA-KB: A Multi-source Chinese Semantic Knowledge Base Built from Structured and Unstructured Web Data. In: Kim, W., Ding, Y., Kim, H.-G. (eds.) JIST 2013. LNCS, vol. 8388, pp. 75–88. Springer, Heidelberg (2013)
4. Zeng, Y., Wang, H., Hao, H., Xu, B.: Statistical and Structural Analysis of Web-based Collaborative Knowledge Bases Generated from Wiki Encyclopedia. In: Proceedings of the 2012 IEEE/WIC/ACM International Conference on Web Intelligence (WI 2012), Macau, China, December 4-7, pp. 553–557 (2012)
5. Lee, W.-N., Shah, N., Sundlass, K., Musen, M.: Comparison of Ontology-based Semantic-Similarity Measures. In: Proceedings of the 2008 AMIA Annual Symposium, pp. 384–388 (2008)
6. Zeng, Y., Wang, D., Zhang, T., Wang, H., Hao, H.: Linking Entities in Short Texts based on a Chinese Semantic Knowledge Base. In: Zhou, G., Li, J., Zhao, D., Feng, Y. (eds.) NLPCC 2013. CCIS, vol. 400, pp. 266–276. Springer, Heidelberg (2013)

Twitter Sentiment Polarity Analysis: A Novel Approach for Improving the Automated Labeling in a Text Corpora

Pablo A. Tapia and Juan D. Velásquez

Department of Industrial Engineering
University of Chile
Santiago, Chile
patapia@ing.uchile.cl, jvelasqu@dii.uchile.cl

Abstract. The high penetration of Twitter in Chile has favored the use of this social network for companies and brands to get additional information on user opinions and feedback about their products and services. In recent years there have been many studies to determine the polarity of a comment on Twitter mainly considering three classes: positive, negative and neutral. One big difference inherent in the problem of Sentiment Analysis on Twitter as opposed to the Web is the ease with which you can obtain data to perform a supervised training algorithm with its API. To take advantage of this characteristic it is necessary to find a semi-automatic method for obtaining tweets to generate the corpora and avoid the traditional method of manual labeling which is very demanding in time and money. This paper goes deeper into the work of using a semi-automated generated corpora for Twitter sentiment polarity classification, introducing a novel approach of tweet selection for corpora consolidation and the addition of a fourth class of tweets that doesn't correspond to any of the above. This new class includes tweets that are irrelevant for classification and do not contain much information, a type of posts that Twitter is full of. Experimental evaluations show that the usage of the fourth class of the denominated meaningless tweets together with the tweet filter criteria for corpus generation improves the system accuracy.

Keywords: Twitter sentiment analysis, opinion mining, polarity detection.

1 Introduction

Companies have always been worried about the perception that customers have of their brands, and have conducted marketing studies such as surveys or focus groups to obtain information that supports decision making. Some time ago, when a person wanted to buy a product, he was limited to asking friends and family opinions. Recent studies have shown that with the explosive growth of social media on the Web (reviews, forums, blogs, micro-blogs, etc), individuals are

D. Ślęzak et al. (Eds.): AMT 2014, LNCS 8610, pp. 274–285, 2014.

increasingly using opinions [5] present in this media before making a purchase, and are greatly influenced by the reviews and comments that can be found on the Web [1]. It is therefore of great interest for companies to know what users of their products say, but given the vast quantity of available reviews and the specific characteristics of the Web, specialized tools are needed to extract the opinions of Websites [2]. In recent years the penetration of Twitter has been constantly increasing at an enormous rate in Chile and the world [9]. For this reason, different industries such as banks, retails and others are using this platform for customer service and to evaluate how and how much users are talking about them. Nowadays, because of the specific way that users interact in this platform, common sentiment analysis tools are not very effective at performing this task and the labeling is done by a human tagging process. This paper goes deeper into the emoticon-based approach for generating both positive and negative corpora, an automated method for collecting a neutral class corpus and the addition of a fourth class of tweets to improve the classifiers accuracy, which corresponds to tweet posts that cannot be classified into any of the previous classes because they are meaningless or lack content, a type of tweets that Twitter is full of.

The paper is organized as follows. Section 2 overviews the related work previously done for attempting the sentiment analysis task in Twitter. Section 3 presents a brief description of how the corpora were collected using previous approaches. In Section 4 the corpora are analyzed in order to evaluate and present some problems found following the accepted corpora-generation approach, some solutions to the problems found and the new corpus structure for each class applying them. Section 5 describes the classification algorithms based on the Nave Bayes classifier that are going to be used for the experiments. Section 6 examines the results of the classifiers using the proposed improvements, Section 7 includes the analysis of a Chilean presidential debate where the system was used to obtain the sentiment polarity of each candidate, and Section 8 presents some conclusions and discusses future work.

2 Related Work

The ease of collecting Twitter comments makes the problem of sentiment analysis on Twitter very different than doing the same task focused on the Web. Thanks to the Twitter API we are able to extract millions of tweets to train a supervised machine learning algorithm. The set of text posts used for this purpose is called the corpus. The collected corpora must be labeled in order to train an algorithm which can later receive a tweet as input, and is able to correctly assign a class from the ones in the training data. The traditional method of labeling the corpora is reading all training data and labeling it manually. This approach has two problems, the first related to the subjectivity present in assigning labels, this task usually being performed by various people, even on different days and at different fatigue levels, which affects the labeling criteria. The other problem is that this methodology cannot take advantage of the Twitter properties in order to get a large number of tweets for the generation of the corpora due to the large amount of time and money that is required to perform the task of manual labeling. For these reasons researchers have recently been developing studies for a methodology that allows automatic generation of a corpus, i.e. a corpus of tweets for each class with an automatic labeling approach.

Go et al. 2009 in [3] pioneered this problem using a study by Read 2005[7] that refers to the extraction of opinions in movie reviews. One of Reads conclusions is that a polar emoticon defines the polarity of texts that contain it. The main problem of this approach is that in the same review many fragments can be presented with different types of polarity, so it is a problem to determine the overall polarity of a review. For Twitter this fact does not present a big problem as a comment on the social network has a maximum length of 140 characters and they most often consist of a single idea. This is the approach taken by Go et al. to establish that a tweet with a polar emoticon presents the polarity of the emoticon. However, their work had focused on the polarity of sentiment analysis into two classes (positive and negative), and only makes a minor analysis considering three classes, including for a neutral class tweets that don't present any emoticon in their content, which is a very weak assumption. In [4] Koppel et al. show why it is important to consider the neutral class, and among its findings are that training using only positive and negative examples does not allow an accurate neutral example classification, and that classification using neutral data contributes to a better distinction between positive and negative examples. This implies that considering a neutral class corpus for training and classification significantly increases the overall system accuracy.

Since then, many other studies have been done improving sentiment polarity classification using semi-automated corpora generation, but mainly focusing on improving the training algorithm and feature extraction rather than the collection and selection of the corpora.

To use a large number of tweets for training requires then a selection approach that allows automatic labeling of neutral-type comments. To resolve this issue Pak et al. proposed in [6] to extract opinions issued by official accounts of newspapers, making the assumption that these tweets are neutral. While this approach produced good results for classification in general, fundamental differences with the structure of the obtained subjective tweets were not considered using the explained subjective and objective corpora collection methods.

In this paper we work with these differences and problems of the automatic corpora collection approach and propose solutions to improve the system classification. Also as a consequence of the proposed solutions addressed we considered a fourth kind of tweet, which is based on a classification rule, and that includes all the tweets that are not relevant to be classified in any other class because they lack content or are meaningless, comments of which Twitter is full. Assigning this type of tweet to its own class improves the overall performance of the classifier by avoiding bad training examples, noise in the other corpus classes and misclassifications.

3 Collecting the Corpora

Corpora collection is one of the most important tasks in the process of Sentiment Analysis. Considering the fact that training data should be sufficiently robust so it can generate a model for later classifying new tweets, there is an interest in collecting as many tweets as possible for the algorithm training. We also need this data labeled according to the different classes in which the tweets will later be classified. Because of these characteristics the manual labeling of tweets is not a viable option due to the large numbers of people and man hours that would be required to label all tweets used for training.

In this work a corpus of tweets was collected to build a dataset of three classes: Positive, Negative and Neutral. For the collection of positive and negative tweets the same methodology proposed by Go et al. 2009 [3] was followed, which involves collecting tweets with polar emoticons justified by the work of Read 2005 in [7] which concludes that a polar emoticon has a high correlation with the polarity of text containing it.

Using the Twitter API all the comments were collected for 2 months that come with the country code attribute referencing CHILE and that contain the following emoticons:

- Positive Emoticons: " :) ", " =) ", " ;) "
- Negative Emoticons: " :(", " =(", " ;("

To collect the neutral tweet class, the same procedure proposed by Pak & Paroubek was followed, selecting tweet posts from popular newspapers and Twitter press accounts. Some of these media are: @latercera, @emol, @lasegunda, @biobio and @cooperativa among others, which are supposed to be objective and not express any opinion about the news they broadcast.

For each one of the classes 80,000 tweets were selected according to the described rules and procedures.

4 Analyzing the Emoticon-Based Tagging Approach

In previous work, the emoticon-based labeling approach has been accepted for collecting the corpora and the efforts have focused on improving the accuracy of the classifier algorithm using text mining [10] and natural language processing techniques. However in this work we wanted to test the proposed approaches for collecting the corpora and establish how accurate these semi-automated labeling methods are.

To make a first evaluation of the described semi-automated approach, a Simple Random Sampling was performed for each of the classes of the collected corpora. For this task 100 tweets were randomly selected from each class and they were checked one by one to determine if the labels assigned by the above methodologies were correct. The following table shows the results of this experiment, obtaining the best result for the neutral class, and the worst for the positive class. The overall accuracy for these three classes combined was 81%, which is quite low for an attempt to obtain a methodology that can classify tweets according to their polarity with an accuracy of at least 85%.

Table 1. Simple Random Sampling to evaluate the semi-automated generated corpora

Class	Positive	Negative	Neutral	Total
Correct-tagged cases	75%	82%	86%	81%

Because of the results obtained by the simple random sampling experiment, each corpus was analyzed in depth in order to find any problems the proposed methodology might have. Inspecting the wrong-tagged cases, various problems were found that generate noise and, in consequence a bad tweet labeling, which are specified below:

- **Meaningless tweets:**
  Only filtering tweets by the emoticon present in their content (for the subjective classes), is a very simple rule for a tweet labeling approach. When analyzing the collected corpora, we found tweets like the following:
  1. #AvenidaBrasil :)
  2. #FelizLunes a @Brennedy2 :) #TuitUtil http://t.co/sNd2Fmtsxe

  For the first example, the tweet is made up of one hashtag and one emoticon. For the second example the tweet consists of a word, two hashtags, a mention of another user, a link and the emoticon. The problem lies in the fact that these tweets are not useful for training the algorithm because they do not add distinct information to each corpus. In addition, the number of features that can be extracted from them through processing hashtags, links and references to standard codes are small. Meaningless tweets are thus not useful for training the algorithm because they do not contribute with distinct features to each corpus.

- **Neutral tweets:** Being neutral texts obtained from newspapers and press accounts, these have a considerable number of additional words compared to positive and negative comments (as seen in table 2). This is mainly because these types of accounts have to summarize the news in a tweet and also add a link to it to a Web page, so they generally use most of the limited space available to communicate content. Also because of this, they have more features making a tweet more likely to be classified as a neutral type.

- **Retweets:** Another major problem identified in the corpus is retweets. A retweet sent by an influential person or that refers to very important news can be retweeted many times. The tweet that was shared more times in the history of Twitter was posted by Ellen Degeneres, a famous American TV host for the Oscars with more than 3.3 million retweets and 1.9 million favorites[1]. The second one was Barack Obama's tweet when he was re-elected as President of the United States with over 750 thousand retweets and 290 thousand favorites[2]. The main problem with this type of tweets is that they have exactly the same content, not contributing to the variability of the corpus, and further deepening the polar characteristic possessed by the words within that tweet. For example, the tweet "I'm bored of studying, now I'll play :)", using the semi-automatic labeling methodology assumes that each word in the tweet has a positive polarity , then because retweets are copies of the original tweet, it accentuates the positive polarity that the training algorithm extracts from the words of this particular tweet.

- **Repeated Tweets:** Another important problem found in the tweets of the collected corpora was related to marketing campaigns in Twitter. Given the high penetration of the platform and quick viralization of campaigns, companies and brands have begun to use Twitter as a means to conduct raffles and sweepstakes. In order to gain or increase their chances of winning users are able to emit dozens of equal comments by adding a number to the end of the comment in order to avoid the duplicate restriction of Twitter comments. Below there is an example of a tweet that was posted over 90 times by the same user in order to win a ticket for a concert.

  @LollapaloozaCL #ViveLollapalooza #Cocacolafmcl @CocaCola_CL Quiero ir!! asi que a poner tweets se ha dicho!! :D 92

  The problem with this kind of tweet to the conformation of the corpus following the semi-automatic method proposed, is essentially the same as the effect that is caused by retweets, where the words contained in these comments become very polarized according to the emoticon labelling criteria, and also they do not contribute to the variability of words in the Corpus because they have almost the same. The anomalous peaks that can be seen for the number of words in the corpus of the positive class in Figure 1 is actually a consequence of a local campaign to win tickets to a famous singers concert.

4.1 Proposed Solutions for the Identified Problems

- **Meaningless tweets:** To handle this kind of tweet, the approach followed is to make a filter of hashtags, RT mentions (initials that refer to retweets) and URLs. After removing these kinds of strings, and according to the number of words, the tweet is selected to be part of the corpus only if it has more than a certain number of words. This number is then defined later in the experiments. We call these kind of Twitter strings *Twitter words*

---

[1] `https://twitter.com/TheEllenShow/status/440322224407314432`

[2] `https://twitter.com/BarackObama/status/266031293945503744`

- **Retweets:** To avoid this type of tweet, they are filtered using the rt_twitter_user_id attribute and excluded if it is not null because it will be then a retweet of another users tweet. This attribute comes within the ones that Twitter API includes.

- **Repeated Tweets:** Since these tweets cause the same effect as retweets, they will also be removed. According to the general structure that they have, we queried each corpus in order to find tweets that end with correlative numbers and from users that have a high frequency of posts.

In the next table it can be observed how varied the number of words and characters after applying these filters. It reduces, on average, 3.5 words per tweet for the positive class, almost three words for the negative class and five words for the neutral class. So tweets in the corpus no longer have fewer than a certain threshold of words (excluding the strings mentioned) which is defined later after experimenting with a different number and selecting the one that presents the best results for the classification.

Table 2. Corpora structure applying the tweet filters

	Type	Words		Characters	
Class		Mean	STD	Mean	Std
Positive	Without Filter	14,01	6,58	88,35	35,75
	With Filter	10,55	6,23	59,73	31,82
Negative	Without Filter	13,54	6,75	84,33	35,58
	With Filter	10,87	6,5	61,48	33,16
Neutral	Without Filter	18,96	3,89	110,67	19,65
	With Filter	13,48	3,87	79,24	20,19

Thanks to only the use of these three filter criteria for the selection of tweets in corpora generation, the overall accuracy for a new simple random sample experiment increased by almost 5%, which is fairly good considering the simplicity of the rules applied.

Table 3. New simple random sampling experiment with filtered corpora

Class	Positive	Negative	Neutral	Total
Correct-tagged cases	86%	84%	84%	85%

One question that was caused after this experiment was what we should do with the meaningless tweets. Despite not being considered for the corpora generation, we know that Twitter is full of them and in any attempt to classify them they will be incorrectly assigned to a class making noise and decreasing accuracy. Unlike other types of filtered tweets, i.e. repeated tweets and retweets that can be well classified because original examples are considered in corpora generation, for meaningless tweets it isn't like that, and there is no tweet like this in the entire corpora. This is why we propose the use of a particular class that only meaningless tweets belong to. In this study we will use the same filtering approach to classify these tweets into the new category, but further work can be done using the filtered examples of the original corpus to train a classifier in order to learn more characteristics about meaningless tweets. The approach considered had good results improving the overall accuracy of the classifier and the system. In this work we will use unigrams to buid the classifier in order to find out if the filters criteria improves the system accuracy. The results of these experiments will be presented in the following sections.

5 Feature Extraction and Classification Algorithms

Feature Extraction:
The collected corpora are used to extract features that will be used to train the sentiment classifier. Pang et al. and Pak et al. among others have experimented with unigrams and bigrams, some obtaining better results using unigrams for movie review and tweets polarity extraction while others found that bigrams outperform unigrams. The process for obtaining the unigrams from the Twitter posts is as follows:

1. Filtering: All type of Twitter words (i.e. @mentions, #Hashtags, URLS, RTs) were removed from the tweets replacing them by a very distinct string: [@] for mentions, [#] for hashtags, [U] for Urls and RT. Emoticons were also extracted from tweets so the model does not depend on them.
2. Tokenization: The tokenization criteria consist of splitting strings by spaces and punctuation marks.
3. Stopwords removal: we use the stopwords list provided in [3] and remove the tokens matching any of those terms.
4. The same procedure was followed to extract the features from tweet POS tags using Treetagger

Classification Algorithhms:
In this work we used Nave Bayes Classifier because it was proven to have good performance in sentiment classification problems for Twitter. The Nave Bayes Classifier is based on the Bayes Theorem and for this work assigns to a tweet T, represented by a vector of features F, the class from the set of classes C which maximizes the following expression:

$$P(c|F) = \frac{P(c)(F|c)}{P(F)} \tag{1}$$

[3] Snowball Tartaurus website:
`http://snowball.tartarus.org/algorithms/spanish/stop.txtl`

The prior term *P(c)* is the same for every class in our experiments because we use an equal number of training tweets per corpus. Conditional independence between the features was also assumed. Even when it is known that a natural language text's words are dependent on each other, it is proven that the Nave Bayes classifier performs very well for text classification [] under this assumption. We can simplify the equations to:

$$P(c|F) = \frac{P(F|c)}{P(F)} \tag{2}$$

$$P(F) = \sum_{c \in C} P(F|c)P(c) \tag{3}$$

And choose then the class that maximizes: $arg \max_{c \in C} P(c|F) \sim P(F|c)$

These criteria were used to build two classifiers using both types of features explained above, unigrams and parts-of-speech (POS) tags.We developed two classification algorithms: one using a combined approach of the two features selected, proposed by Pak et al. in [6] and a two-level algorithm that first classifies a tweet into subjective or objective classes using only POS tags as features. Subjective tweets are then classified into positive and negative considering unigrams of words as features for this problem.

1. **Nave Bayes Classifier for Three Classes:**
   To use both types of feature and generate a three-class algorithm the posterior probability can be written as a combination of the posteriors for each model if we assume conditional independency between POS tags and unigrams of words:

   $$P(c|T) \sim P(U|c) \cdot P(P|c) \tag{4}$$

   Where U is the vector of unigrams of words representing the tweet T and P is the set of POS tags present in it. As was said before we assume every feature of each type conditionally independent from each other having:

   $$P(U|c) = \prod_{u \in U} P(u|c) \tag{5}$$

   $$P(P|c) = \prod_{p \in P} P(p|c) \tag{6}$$

   $$P(c|T) \sim \prod_{u \in U} P(u|c) \cdot \prod_{p \in P} p(u|c) \tag{7}$$

2. **Two-Level Nave Bayes Classifier:**
   For this attempt we used both types of features for different classification problems. From the work in [6,8] we know that objective texts are more likely to talk in third person among other characteristics, so objective vs

subjective texts could be distinguished by the POS tags present in them. Because of this we will consider for objective and subjective classification, the POS tags present in tweets as features. Therefore, subjective tweets are classified into positive or negative classes considering unigrams as features.

6 Evaluation and Results

For the evaluation we use the accuracy ratio with two distinct validation methodologies:

- We used 10-Fold Cross Validation which is a technique that is useful to evaluate a classification algorithm for a given corpora, splitting and evaluating the training set several times.
- We also manually marked 100 tweets per class in Spanish and classify them with the algorithms in order to get the same accuracy ratio, but with this technique we evaluate the accuracy of the whole system, from the data collection and feature extraction to the classification algorithms.

The accuracy ratio corresponds to:

$$accuracy = \frac{N(Correct\ classifications)}{N(All\ classifications)} \tag{8}$$

Table 4. Accuracy evaluation with 10-fold cross validation

10-Fold Cross Validation	2 Level Nave Bayes	3 Classes Nave Bayes
Without Filter	0.575	0.721
With Filter	0.614	0.787
With Filter + 4th class	0.622	0.806

Table 5. Accuracy evaluation with the hand-tagged set

Hand Tagged Set	2 Level Nave Bayes	3 Classes Nave Bayes
Without Filter	0.432	0.688
With Filter	0.484	0.707
With Filter + 4th class	0.535	0.782

In table 4 the results of the experiments with the 10-fold cross validation evaluation are shown. Independent of which classification algorithms were used, the accuracy of the system increased if we used the proposed filters. Also with the addition of the four-class rule for classifying meaningless tweets, the overall accuracy of the system showed further improvements. Regarding the algorithms, the three-class Nave Bayes Classifier had better results than the two-level algorithm for all the experiments. Also in table 5 the results of the evaluation made by the hand-tagged set of examples are shown. For this set we also obtained best results for the combined Naive Bayes classifier algorithm, with the filtered corpora and using the fourth class of meaningless tweets classification rule.

7 Case Study

We satisfactorily used the implemented system to analyze 74,066 tweets sent by Twitter users about a Chilean presidential debate held on October 29th 2013. This database was generated by querying Twitter API for all posts containing the hashtag #DebateAnatel between 8PM of October 29th and 2AM October 30th, local time. From this database we classified all tweets using the three-class Nave Bayes classification algorithm with filters and the meaningless tweet rule for the fourth class. After the classification we grouped all tweets referring to each candidate and obtained the number of positive, negative, neutral and meaningless comments. The result of this exercise is shown in table 6. We also defined a polarity score ratio to get an overall value of the sentiment polarity of tweets for each candidate.

$$polarityscore = \frac{(positive)}{(positive + negative)} \tag{9}$$

Table 6. Sentiment Frecuency and ratio for each candidate about the first presidential debate held by ANATEL

	Positive	**Negative**	**Neutral**	**Meaningless**	**Polarity Score**
Bachelet	32%	41%	9%	18%	0,44
Matthei	21%	39%	18%	22%	0,35
Parisi	34%	32%	21%	13%	0,52
MEO	48%	26%	19%	7%	0,65
Claude	37%	18%	15%	30%	0,67
Sfeir	54%	25%	17%	4%	0,68
Israel	14%	12%	45%	29%	0,54
Miranda	51%	27%	19%	3%	0,65
J-Holt	7%	48%	41%	4%	0,13

The objective of this exercise was to derive some analysis that can be generated from the developed Sentiment Classification System. The sentiment ratio explained above does not consider the number of tweets (popularity) of the candidate and it is only a measure of how many of the subjective tweets are positive. Some other ratios that can weight this value with the total number of posts in order to consider popularity may be introduced in future work.

8 Conclusion

In this study, we presented an automated approach for generating corpora for Twitter sentiment polarity detection, which is an improvement of the corpora composition by applying different filters in recognition of the problem that this methodology entails. We also found that many posts in the initial collected training set for positive and negative classes were meaningless or lacking in content, effectively causing noise within these corpora, preventing the best training results and also decreasing the system classification accuracy when trying to classify this type of tweets into a class. To solve this issue we proposed to filter those tweets that had fewer than four words excluding @mentions, #hashtags, RT tags and URLS from the collected corpora, and in the classification using this same rule to assign them to the meaningless or unclassified class. We also used our proposals to successfully implement the system and create an online application that can classify a continuous flow of tweets in real time. The results obtained validate the proposed improvements of an automated corpora collection in order to increase the number of training examples for a Twitter Sentiment Analysis algorithm. For future work we propose to train an algorithm that can generate a model to classify these meaningless tweets, and also improve the accuracy of the system by implementing bigrams and other feature-extraction algorithms.

Acknowledgments. This work was supported partially by the Corfo Innova project 13/DL2-23170, entitled *OpinionZoomt* and the Millennium Institute on Complex Engineering Systems (ICM: P-05-004-F, CONICYT: FBO16).

References

1. Duan, W., Gu, B., Whinston, A.B.: Do online reviews matter? an empirical investigation of panel data. Decision Support Systems 45(4), 1007–1016 (2008)
2. Duenas-Fernández, R., Velásquez, J.D., L'Huillier, G.: Detecting trends on the web: A multidisciplinary approach. Information Fusion 20, 129–135 (2014)
3. Go, A., Huang, L., Bhayani, R.: Twitter sentiment analysis. Entropy, 17 (2009)
4. Koppel, M., Schler, J.: The importance of neutral examples for learning sentiment. Computational Intelligence 22(2), 100–109 (2006)
5. Marrese-Taylor, E., Velásquez, J.D., Bravo-Marquez, F.: Opinionzoom, a modular tool to explore tourism opinions on the web. In: Proceedings of the The 2013 IEEE/WIC/ACM International Joint Conferences on Web Intelligence and Intelligent Agent Technology, WI-IAT 2013. IEEE Computer Society, Washington, DC (2013)
6. Pak, A., Paroubek, P.: Twitter as a corpus for sentiment analysis and opinion mining. In: LREC (2010)
7. Read, J.: Using emoticons to reduce dependency in machine learning techniques for sentiment classification. In: Proceedings of the ACL Student Research Workshop, pp. 43–48. Association for Computational Linguistics (2005)
8. Spencer, J., Uchyigit, G.: Sentimentor: Sentiment analysis of twitter data. In: Proceedings of European Conference on Machine Learning and Principles and Practice of Knowledge Discovery in Databases, pp. 56–66 (2012)
9. Twitter. Who is in twitter (2014), `https://business.twitter.com/whos-twitter/` (Online; accessed January 2014)
10. Velásquez, J.D., González, P.: Expanding the possibilities of deliberation: The use of data mining for strengthening democracy with an application to education reform. The Information Society 26(1), 1–16 (2010)

Empirical Study of Conversational Community Using Linguistic Expression and Profile Information

Junki Marui[1], Nozomi Nori[2], Takeshi Sakaki[1], and Junichiro Mori[1]

[1] School of Engineering, the University of Tokyo, Japan
{marui,sakaki,jmori}@ipr-ctr.t.u-tokyo.ac.jp
[2] Graduate School of Informatics, Kyoto University, Japan
nozomi@ml.ist.i.kyoto-u.ac.jp

Abstract. The popularization of social media exposes the structure of people's conversation - what kind of people speak with whom, on what topics and with what kinds of words. In this paper, we propose a new approach to mining conversational network by community analysis, which exploits users' profile information, interaction network and linguistic usage. Using our framework, we conducted empirical analysis on the complex relation among people's profile information, social network, and language network using a large dataset from Twitter, which covers more than 7M people. Our findings include (1) we can extract a community composed of people who use the same kinds of slangs by exploiting information from both the social network and word usage, (2) when we focus on similarity among communities in terms of both interaction and word usage, we can find specific patterns based on the people's profile information including their attributes and interests.

1 Introduction

The popularization of social media exposes the structure of people's conversation - what kind of people speak with whom, on what topics and with what kinds of words. Can we make sense of the complex relation among them - relation among people's profile information (what kind of people speak), social network (with whom people speak) and language expressions (on what topics and with what kind of words people speak)? For example, is it true that those people who interact much with each other have characteristic profile information? Is it true that those people who interact much with each other exhibit discriminative linguistic expression? By introducing community structure, which is derived from conversational interaction network, we investigate these questions. Moreover, we further investigate the relations among communities in terms of member's profile information and word usage. By treating community as the unit of analysis, we investigate the following question: is it true that if different communities have much interaction, then will the word usage also be similar among these communities?

D. Ślęzak et al. (Eds.): AMT 2014, LNCS 8610, pp. 286–298, 2014.

Investigating these questions has the potential to enhance Web-related applications such as targeted advertising and viral marketing on Online Social Networks (OSN) [4]. As pointed out in [4], one of the important problems in targeted advertising and viral marketing on OSN is to identify the adequate target audience, that is, users of the adequate demographics who are also highly *connected* among themselves. Detecting users with the right demographics enables the right product-audience matching. At the same time, the connection among people can facilitate word-of-mouth advertising. Thus, investigating the relation between connection and features that people exhibit such as interests and word usage can provide insights for the problem.

In this paper, we propose a new approach to mining conversational network by community analysis, which exploits rich information such as users' profile information, interaction network and linguistic usage. We perform these analysis as the first step to answer the above-mentioned questions.

1.1 Overview of the Analysis Framework

Below, we explain the overview of our framework that we used for the analysis on the large Twitter dataset. Our framework is divided into four main parts. (1) We extract communities from the conversation network by using only the topological information of the conversation, which becomes the basis of the following analysis. (2) We label each community by using members' profile information and evaluate them by user study; by doing this, we attempt to see the relation between connection and people's profile information including their interests and attributes. (3) We train a language model to learn *distributed representations for words* for each community. In this part, we exploit neural probabilistic language models (NPLM) to capture the "context" of word usage in each community. NPLM is based on the distributional hypothesis - similar words will be similar in the context, that is, the distribution of words that appeared with the word. After training NPLM, each word obtains a word vector as the distributed representation, which reflects the distributional hypothesis. Thus, the distributed representation for a word in a community is expected to reflect how the word is used in the community, providing us with ways to explore the differences in the use of language among conversation communities in the following part. (4) Using the trained distributed representations, we analyze the differences in the use of language among conversation communities. We will detail all of them in the "analysis framework" section.

1.2 Our Findings

A summary of our findings on the Twitter dataset is presented below.

We found that (1) communities extracted from conversational interaction network, which is solely based on the topological information, reflect member's profile information including their interests/attributes, providing a comprehensible label for each community and that (2) by exploiting NPLM, we can detect *ambiguous words among communities*, that is, words that are used for multiple meanings depending on the communities.

We also revealed that (3) when we take community as the unit of analysis, there are no clear correlations among community-level similarity in interaction and community-level similarity in word usage, but there exist specific patterns based on the labels created from people's profile information including their interests and attributes.

2 Related Work

Community detection is one of growing field in the area of social network analysis, especially after the popularization of social media. Before the social media, most researchers used topological structure of social network and ignored the node attributes which are often heterogenous. This is because datasets of social network analysis, such as *Zachary Karate Club* dataset and *American College Football* dataset, had not included rich attribute information of nodes [8]. However, it is not so difficult to extract attribute information of current social media users. There are some research that improved the performance of community detection using both topological information and attribute information on social network services. Tang et al. [7] proposed a joint optimization framework to integrate multiple data sources, which outperformed existing representative methods of community detection in social media. Guneman et al. [3] proposed a method for detecting homogeneous communities, that is, communities that are densely connected and have similar attributes, by integrating results of subgraph clustering and subgraph mining.

Among these kinds of research, research by Lim et al. [4] would be most related to our research. They detected highly interactive communities with common interests based on the behavioral information of users, the content of hashtags and mentions. They showed those communities are more cohesive and connected than communities detected by the current method. Our research is different from their research in the following two points. First, our research deals with much more users (7M) and we use deeper analysis on word usage by using the neural network based probabilistic language model (NPLM), while they analyzed about 18 K users and only considered word frequency. Second, we explored not only the characteristics of communities but also similarities among communities.

Moreover, ours can be considered as complementary research for these research; we propose a framework to explore the complex relation among social network, people's word usage and profile information by community analysis, while above-mentioned research aim to improve the performance of community detection by exploiting rich information of node attributes as well as the topological information. By using our framework, we expect we can enrich the understanding of the community with rich node attributes, which can be useful for developing methods to improve the performance of community detection.

3 Analysis Framework

3.1 Community Extraction from Conversation Network

As a conversation network, we target the mention network on Twitter. We construct a graph where an edge is created for two users (nodes) if there are bi-directional (mutual) mentions [1].

We use the Louvain method [2] for community extraction. The Louvain method is a simple and effective method for detecting communities in a large network. It is one of the most popular community detection methods for large networks. It conducts greedy optimization of modularity in several steps. We used publicly available code[2].

Applying the Louvain method, we obtain N communities. Then we select communities including more than M users so that the selected communities account for most of the whole users in the dataset, say, 95% of whole users in the dataset.

3.2 Labeling Community from User Profile Information

For each community, we extract users' profile texts, which users can set by the profile function on Twitter. Then we calculate the TF-IDF (term frequency, inverse document frequency) score [6], which is one of the most general numerical statistics for keyword extraction, to calculate the importance of words in each community and extract keywords for each community. We extract the top-L words for each community and use them as they represent the characteristics of a community. Then we label each community to characterize them by using these keywords as clues for labeling. To validate the labels, we conduct user study and evaluate the results, which will be detailed in the experiment section.

We extract the keywords of each community and label the communities as follows:

- We crawl the biography of each user on Twitter profiles.
- For each community, we aggregate the biography texts of users who belong to the community, creating one document that corresponds to the community.
- We calculate the TF-IDF score of all words in each community and extract the top-L keywords as salient keywords for each community.
- We label each community manually using phrases associated with common features of the top-L keywords.

3.3 Learning Distributed Representations for Words in Each Community

Here, in order to grasp how words are used in each community, we learn distributed representations for words in each community. As mentioned above, the distributed

[1] It is possible to weight the edge based on the number of mentions, but in this research, we consider unweighted graph for simplicity.

[2] `https://sites.google.com/site/findcommunities/`

representation expressed as a word vector reflects the context in which the word appears, and this feature can be used to analyze how the word is used differently among communities.

First, we sample $c/N_i\%$ tweets (N_i:users in the ith community, c:constant) from community members' tweets in order to create balanced corpus for each community. Then, we train a NPLM (neural probabilistic language model) for each community, and we obtain distributed representations for each word. As a implementation, we use "word2vec" [3], the fast implementation of the NPLM of Mikolov et al. [5].

Below, we briefly describe NPLM which we used to learn distributed representations for words. NPLM was first introduced in [1] by Bengio et al., aiming to overcome the curse of dimensionality caused by n-gram models. Bengio et al. avoided this difficulty by introducing a distributed representation for words, which can be expressed as word vectors. Their model predicts the next word given the last several words, called "context". They input context vectors into the neural network and obtain unnormalized log-probabilities for each word as output. The problem of this model is the training and test time. It has a non-linear middle layer and they have to calculate all the word vectors for the entire vocabulary to normalize the probability. To overcome this problem, Mikolov et al. proposed the Skip-gram model, which simplifies the neural network and adopts surrounding words as a context instead of several preceding words for the target word, as shown in Fig. 1. The posterior probability of the Skip-gram model is shown as below.

$$p(w_{t+j}|w_t) = \frac{\exp(v_{w_{t+j}}^{\mathrm{T}} \cdot v_{w_t})}{\sum_k \exp(v_{w_k}^{\mathrm{T}} \cdot v_{w_t})} \tag{1}$$

where v_{w_k} represents a word vector for word w_k, and w_t is a target word while w_{t+j} is a context word ($j \neq 0$). The posterior is expressed as a softmax function, and the number of its gradient terms is linear in vocabulary size, therefore, the training time becomes considerably long. Mikolov et al. proposed the hierarchical softmax and negative sampling to reduce calculation time. We adopt negative sampling, which is reported to perform better than the others.

Trained word vectors have the following characteristics: words having similar contexts are projected to similar vectors in the vector space. Still we cannot completely solve the problem of polysemies and homonyms in adopting NPLM; however, we focus on the characteristics that word vectors represent the context and that similar word vectors indicate similar word meanings.

3.4 Analysis on the Differences in the Use of Language among Conversation Community

Detecting Ambiguous Words among Communities. First, we address the following question: how the same word is used in different senses among

[3] Publicly available at `https://code.google.com/p/word2vec/`

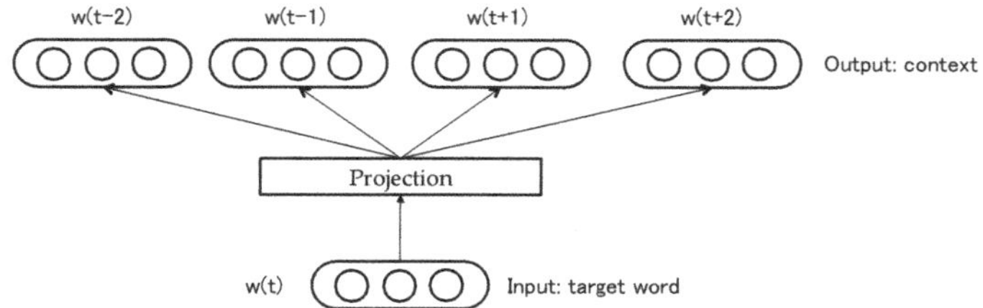

Fig. 1. The Skip-gram model. Given words $w_1, w_2, .., w_t, ., w_N$, the model learns word vectors to maximize the posterior.

communities? In this paper we use NPLM that can capture contexts for each word, which enables us to grasp the set of words that potentially co-occur with the word. Thus, by seeing how the contexts for the word are different among communities, we can expect to detect ambiguous words among communities, that is, words that are used in different meanings depending on communities.

To obtain prospective words that have multiple meanings, we calculate how similar a word is between a pair of communities. More specifically, for each word and pair of communities, we calculate cosine similarity between the word vector in one community and the word vector in the other. Then we try to detect ambiguous words from highly dissimilar words by using the context as clues, that is, the set of words that potentially co-occur with the word.

Relationship between social-network-based-similarity and word-usage-based-similarity among communities. In this part, as a unit of analysis, we adopt the individual community, which is extracted from the conversation network. Thus the word "community" means the community extracted from the conversation network, which does not change throughout the analysis.

In order to see whether communities which are close on the conversational network have similar word usage, we first define two kinds of similarities - one based on the conversation network and the other based on the word usage. We will describe them in the following.

Definition of similarity between communities based on social network: We define the social-network-based similarity between community i and community j, $Sim_{social}(i, j)$, simply as follows.

$$Sim_{social}(i, j) = \frac{|E_{i,j}|}{|V_i||V_j|} \tag{2}$$

where $|E_{i,j}|$ is the number of edges between the community i and j, and $|V_i|$ is the number of nodes (users) in the community i.

Definition of similarity between communities based on word usage: We define word-usage-based similarity between community i and community j,

$Sim_{word}(i,j)$, as follows.

$$Sim_{word}(i,j) = \frac{1}{N}\sum_{a=1}^{N}\frac{|S_{ia} \wedge S_{ja}|}{|S_{ia} \vee S_{ja}|} \tag{3}$$

where a is an index of word, w_a is a word, v_{ia} is a distributed representation for a word w_a in the community i, "·" means inner product, and $S_{ia} = \{w_b | \text{top } 30 \text{ of } v_{ia} \cdot v_{ib}\}$. N is the number of "target words" for community i and j. Target words are taken from words in the whole tweets that appear at least 100 times both in tweets in communities i and j.

Since distributed representations are trained differently in communities, w_a has different representations according to communities. S_{ia} means a set of similar words to w_a in the tweets in community i, and $Sim_{word}(i,j)$ is the average of Jaccard similarity between the sets of similar words to a target word in communities i and j.

Then, we create a correlation chart in order to obtain an overview of the relationship between these two similarities. In addition, we examine them by viewing them in terms of their labels - characteristics of the community.

4 Experiments and Results

4.1 Datasets

We collected tweets and profiles with time stamps ranging from January 1st 2012 to December 31st 2012, from 7.4M users, who were detected tweeting in Japanese by the Twitter API [4]. From this data, which contains 4.9G tweets, we extracted mentioning tweets to collect mention connections. In this period, there are 404M links of which 125M links are mutual links, i.e., both users mentioned to each other at least once.

4.2 Community Extraction from Conversation Network

Applying the Louvain method, we obtained 34,835 communities. We selected communities including more than $10K$ users, which account for 97.7% of whole users in the dataset. As a result, the total number of communities that we targeted for the analysis became 38. To balance the dataset of each community, for each community we sampled $10^6/N_i\%$(N_i:users in the ith community) from the members' tweets when $N_i > 10^4$, and when $N_i \leq 10^4$ we used all the tweets.

4.3 Characteristics in Profiles

Labeling of Communities. Here we labeled each community to characterize them by using top-20 keywords as clues for labeling. As a result, surprisingly, in all the communities, we were able to make sense of the representative words and

[4] `https://dev.twitter.com/docs/api`

label the communities. In Table 6 and Table 7 [5], we show top-3 words and label for each community. Due to space limitation, we only show top-3 words instead of top-20 words. We were able to group most of the communities into three major types (namely all except for one community): (a) same/neighboring high school, (b) same/neighboring universities and (c) interest-based communities. We also showed the type for each community in Table 6 and Table 7.

Table 1. Fraction of workers who chose the corresponding rating for each community

Com-	Rate			Com-	Rate		
munity	*adequate*	*inadequate*	*unable to judge*	munity	*adequate*	*inadequate*	*unable to judge*
1	80%	10%	10%	20	95%	5%	0%
2	100%	0%	0%	21	95%	0%	5%
3	80%	5%	15%	22	80%	20%	0%
4	95%	0%	5%	23	60%	30%	10%
5	100%	0%	0%	24	95%	5%	0%
6	90%	5%	5%	25	95%	0%	5%
7	45%	25%	30%	26	80%	5%	15%
8	95%	5%	0%	27	90%	5%	5%
9	100%	0%	0%	28	90%	5%	5%
10	70%	10%	20%	29	85%	10%	5%
11	90%	0%	10%	30	95%	5%	0%
12	95%	0%	5%	31	85%	5%	10%
13	85%	0%	15%	32	85%	5%	10%
14	90%	0%	10%	33	85%	5%	10%
15	100%	0%	0%	34	100%	0%	0%
16	100%	0%	0%	35	85%	5%	10%
17	80%	5%	15%	36	75%	5%	20%
18	75%	0%	25%	37	75%	10%	15%
19	100%	0%	0%	38	95%	5%	0%

Evaluating Labels. To validate the labels of the communities, we conducted user study using a crowdsourcing service in Japan, "Lancers" [6]. We provided the label we created and top-20 keywords for each community. Then we asked 20 highly acclaimed workers to judge the label for each community by choosing from the following options: (1) *adequate*, (2) *inadequate*, (3) *unable to judge*. For each worker, we paid USD 2.4 for completing the set of task, that is, judging the labels for all 38 communities. In case when the worker choose *inadequate* for a community, we asked the worker to provide more adequate label. The result is summarized in Table 1. For each community, we show the fraction of workers who chose the corresponding rating. We can see that for most of the communities, the majority of workers agree that the label is adequate. More specifically, for 32 out of 38 communities (84% of all the communities), more than 80% of all

[5] Due to the page limitation, we run tables 6 and 7 in the author's website (`http://junki.me/amt2014/suppl.html`)

[6] `http://www.lancers.jp`

the workers agree that the label is adequate. Although we asked workers to use search engine when they can not make sense of the word, some words are polysemous that are used differently in different communities and it might be difficult for some workers to make sense of them, producing some fraction of *unable to judge* rating. In the case where workers chose *inadequate* rating, they were required to provide an alternative label. In most of the cases, the provided labels were not different substantially from the original label; for example, they added auxiliary information to the original label.

4.4 Detecting Ambiguous Words among Communities

Here we show some sensible examples where a word is used for multiple meanings. From highly-dissimilar words, we show one example, "miito", which is the fourth dissimilar word in the rank list. We observed that many words were used differently depending on communities, but we have omitted them due to space limitation.

It may be difficult to explain this word. The pronunciation of "miito" corresponds to both the English word "meat" and "meet" for Japanese people. They have many foreign (Western) loan words whose pronunciations are similar to the original words. However, Japanese people usually do not use the word "miito", because the word is not rooted in Japan, though they can obtain the meaning of the word because they know the English word "meat"/"meet". Thus, it can be inferred that the word is a kind of slang that is used by people who share contexts for the word. With these in mind, we then conducted the analysis.

In Table 2 and Table 3, we show the top-10 context words for each community, with corresponding scores.

In community-13 (Online game fans), most of the potentially co-occurring words are words related to foods. Since the word "miito" can mean meat in English as noted above, we can infer that people in this community use "miito" to mean meat.

In contrast, in community-17 (Disneyland fans), at a glance it is not clear what the word "miito" stands for. However, it is easy to see that these words reflect the characteristics of the community, which is mainly composed of girl-like users who are enthusiastic fans of Disneyland; (1) the face mark (> _ <) and ??♡ are typically used by some girls, (2) it is also typical word usage by some girls to repeat the same word such as "do do" and "please please", (3) the word "Inn" is a slang that means to enter somewhere in Disneyland used by enthusiastic fans of Disneyland. From usage in the tweets in this community, we found that the word "miito" stands for meeting in Tokyo Disneyland. It seems that they use this loanword for expressing meet in a particular situation.

Although we used a simple measure (cosine) for the word similarity between communities and a naive method to compute them, it is possible to develop more adequate measure of similarity and an efficient computation method in order to automatically extract such *ambiguou words across communities,* which is one of the interesting topics for future work. Especially, we think such analysis could enhance cultural studies, by comparing slangs across communities.

Table 2. Context words for the word ("miito") in community-13 (Online game fans)

Word	Score
(speed-eating)	0.821099
(frozen)	0.802880
(Dos hercules; a character name)	0.800420
(crunchy)	0.792865
(lard)	0.791672
(hot drink)	0.782958
(an abbreviation of "Asai meat"; a company name)	0.778490
(white rice)	0.778290
(marbled tomato; an item name)	0.775939
(leg)	0.775758

Table 3. Context words for the word ("miito") in community-17 (Disneyland fans)

Word	Score
(in)	0.844532
(timing suits)	0.803420
(when next)	0.801293
(please please)	0.801245
(> _ <)!!!	0.792765
(do do)	0.790120
("Manamin"; nickname for a girl's name)	0.789295
??♡	0.787898
(if it is still OK)	0.785390
("Kinkan"; a fruit name)	0.782860

4.5 Similarity among Communities

Next, we extended the unit of analysis from individual user to community. First, (1) we examined what kind of communities are similar in the word usage/social network. Then, (2) we examined the relationship between the structure of conversation and the word usage among communities.

Similar community pairs in the use of language/social network. Table 4 shows the Top-10 similar community pairs based on the word usage and Table 5 shows the Top-10 similar community pairs based on the social network. From Table 4, we can see that people sharing a generation have similar word usage. More specifically, high school students have similar word usage with respect to one another, though they are not "physically" close. This held true for university students. In addition, from Table 5, we can also see that it is not always true that they have much interaction on Twitter.

Relationship between social-network-based-similarity and word-usage-based-similarity. In the previous section, we only focused on highly similar

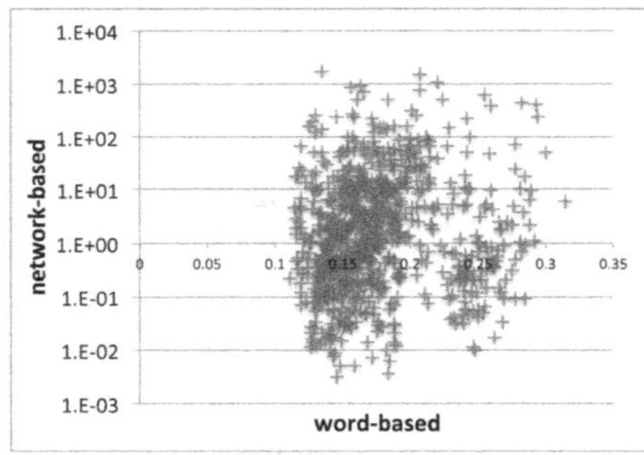

(a) Correlation between social-network-based similarity and word-usage-based similarity in entire network

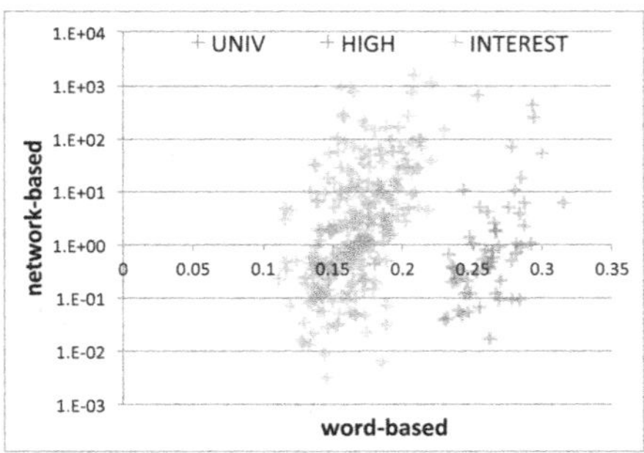

(b) Correlation between social-network-based similarity and word-usage-based similarity for high-school communities, university communities and interest-based communities

Fig. 2. Correlation between social-network-based similarity and word-usage-based similarity

communities in terms of social network and word usage. Here we would like to obtain a broad view of the relationship between these two similarities. First, we created a correlation chart that depicts the relationship between the two similarities for all community pairs. The chart is shown in Figure 2 (a). The horizontal axis corresponds to the word-usage-based similarity, Sim_{word}, and the vertical axis corresponds to the social-network-based similarity, Sim_{social}. We can see that there seem to be no correlations. But we know that there are several types of communities in the target network - high school community, university community and interest-based community. Thus we can categorize each point on the map based on the type of the community pair. For example, similarity between school communities might have a different pattern from the similarity between interest-based communities. Thus, we examined where each type of pair locates on the map.

As a result, we found clear patterns. Figure 2 (b) shows a correlation map where points are colored based on the type of community pair - blue for community pairs in the university category, red for community pairs in the high-school category and green for community pairs in the interest-based category.

We can examine the map by viewing it in terms of four types: (a) group that has similar word usage and a large amount of interaction, (b) group that has similar word usage but a small amount of interaction, (c) group that has dissimilar word usage but a large amount of interaction and (d) group that has dissimilar word usage and a small amount of interaction. Now we see that university communities can be categorized into type (a), high-school communities can be categorized into type (b) and interest-based communities belong to type (c) and (d). This observation is consistent with the tendency we saw in the previous section by using top-10 similar pairs.

Table 4. Top-10 similar community pairs based on the word usage

High schools in the Metropolitan area	High schools in east of metropolitan area
Univ in central region	Univ in southern central region
Univ in southern central region	Univ in Tokyo
Univ in central region	Univ in Tokyo
Univ in central region	High schools in northern region
Univ in southern central region	Univ in southwestern region
Univ in central region	Univ in southwestern region
High schools in east of metropolitan area	High schools in north of metropolitan area
High schools in northern region	Univ in Tokyo
High schools in the Metropolitan area	High schools in north of metropolitan area

Table 5. Top-10 similar community pairs based on the social network

Anime and music game fans	Univ in Tokyo
Anime and music game fans	"Fujoshi" [10]
Anime and music game fans	Fans of vocaloid and singers
Anime and music game fans	No Nukes, Reformists, Housewives
Anime and music game fans	Fans of "Arashi" [11] and local girl idol
Anime and music game fans	Online game fans
Anime and music game fans	"Visual kei" [9]
Univ in Tokyo	High schools in Iwate
"Visual kei" [9]	Univ in Tokyo
Fans of independent label rock'n roll band	Univ in Tokyo

From this observation, we can see that (1) communities based on the member's attributes - high-school or university - tend to have similar word usage, regardless of the interactions they have with respect to one another, while (2) communities based on the member's interests tend to have different word usage, regardless of the interactions they have with respect to one another. It is also suggested that even when people have a common attribute, students, the patterns derived from the social network and word usage can differ depending on another attribute, generation. These observations suggest the complex relations among social interaction, word usage and other factors such as their interests and attributes.

5 Conclusion

In this paper, we explored the relationship among social network, language and people's profile information, using a large dataset extracted from Twitter. Though we used a Twitter dataset mainly composed of Japanese people, we believe that the framework we proposed here can be applied to any other kind of dataset. We hope our research enriches the understanding of the complex relations among social network, language and people's profile information.

[9] A kind of Japanese rock'n roll band style.

[10] Woman who likes comics depicting male homosexual love.

[11] A Japanese boy idol.

References

1. Bengio, Y., Ducharme, R., Vincent, P., Jauvin, C.: A neural probabilistic language model. Journal of Machine Learning Research 3, 1137–1155 (2003)
2. Blondel, V., Guillaume, J., Lambiotte, R., Mech, E.: Fast unfolding of communities in large networks. Journal of Statistical Mechanics: Theory and Experiment, 10008–10019 (2008)
3. Gunnemann, S., Farber, I., Boden, B., Seidl, T.: Subspace clustering meets dense subgraph mining: A synthesis of two paradigms. In: Proceeeindgs of the 10th IEEE International Conference on Data Mining, pp. 845–850. IEEE (2010)
4. Lim, K.H., Datta, A.: Tweets beget propinquity: Detecting highly interactive communities on twitter using tweeting links. In: Proceedings of the The 2012 IEEE/WIC/ACM International Joint Conferences on Web Intelligence and Intelligent Agent Technology, WI-IAT 2012, vol. 01, pp. 214–221. IEEE Computer Society (2012)
5. Mikolov, T., Chen, K., Corrado, G., Dean, J.: Distributed representations of words and phrases and their compositionality. NIPS 2013 (2013)
6. Salton, G., McGill, M.J.: Introduction to Modern Information Retrieval. McGraw-Hill, Inc., New York (1986)
7. Tang, J., Wang, X., Liu, H.: Integrating social media data for community detection. In: Atzmueller, M., Chin, A., Helic, D., Hotho, A. (eds.) MSM/MUSE 2011. LNCS (LNAI), vol. 7472, pp. 1–20. Springer, Heidelberg (2012)
8. Zhou, Y., Cheng, H., Yu, J.X.: Graph clustering based on structural/attribute similarities. Proceedings of the VLDB Endowment 2(1), 718–729 (2009)

Multi Agent System Based Interface for Natural Disaster

Zahra Sharmeen[1], Ana Maria Martinez-Enriquez[2], Muhammad Aslam[1], Afraz Zahra Syed[1], and Talha Waheed[1]

[1] Department of Computer Science & Engineering, UET, Lahore, Pakistan
sharmeenzahra@hotmail.com, {maslam,twaheed,afrazsyed}@uet.edu.pk
[2] Department of Computer Science, CINVESTAV-IPN, D.F. Mexico
ammartin@cinvesta.mx

Abstract. Natural disasters cause devastation in the society due to unpredictable nature. Whether damage is minor or severe, emergency support should be provided within time. Multi-agent systems have been proposed to efficiently cope with emergency situations. Lot of work has been done on maturing core functionality of these systems but little attention has been given to their user interface. The world is moving towards an era where humans and machines work together to complete complex tasks. Management of such emergent situations is improved by combining superior human intelligence with efficiency of multi-agent systems. Our goal is to design and develop agents based interface that facilitates humans not only in operating the system but also in resource mobilization like ambulances, fire brigade, etc. to reduce life and property loss. This enhancement improves system adaptability and speeds up the relief operation by saving time of human-agent consumed in dealing with complex computer interface.

Keywords: Disaster management, multi agent system, system user interface, emergency situation, human machine interface.

1 Introduction

Natural disaster is an unavoidable situation [1]. It may take the form of a flood, a tsunami, a hurricane, an earthquake, a wildfire, and the most recent form, suicide attack. In 2012, a total of 6, 771 terrorist attacks occurred worldwide [2]. In Syria, 133 terrorist attacks claimed 657 lives and 1787 people were injured. There were 1, 404 terrorist attacks in Pakistan the same year, which claimed 1, 848 lives and injured over 3, 643 people. Efforts have been made by humans to somehow avoid such disaster situations for instance, use of metal detectors on entrances, installation of CCTV cameras, and other security measures but there has been no success so far. Disaster deprives people of their homes, their assets, and even their loved ones. It leaves the affected people with nothing but a hope of getting help from the society and welfare organizations.

Disaster management is a long activity spanning over three phases [3]. The first one is the prediction phase. Despite many efforts, avoiding these disasters by predicting them beforehand has not been a great success. We can only take few

D. Ślęzak et al. (Eds.): AMT 2014, LNCS 8610, pp. 299–310, 2014.

precautionary measures like warning, announcement for relocation of people, etc. to avoid loss but when these natural catastrophes hit a region, a society gets affected. Loss of lives, homes, and everyday needs leave people stressed. Preventing the loss completely is still an ideology. Emergency support, the second phase, requires instantaneous action from different institutions. The process of gathering up all units and measures is very time consuming. If proper actions are not taken at the hour of need, the life and state loss increases, resulting in causalities that could have been prevented. The third step is rehabilitation of the sufferers, which is a slow and gradual process. Rebuilding the infrastructure requires time and money. It usually proves to be a great challenge for any government.

Avoiding the disaster management stress by automating this process is required for proper handling of emergency situation [4]. This is where multi agent systems (MAS) step in. MAS have a number of intelligent agents performing different tasks and interacting with each other to handle the disaster situation efficiently. Very comprehensive architectures like Disasters2.0 [5], FMSIND [6], and ALADDIN [7], exist that address the core functionality of MAS [3]. However, little attention has been given to the user interface. Emergency relief operation is a race against time. A good system designed for managing the disaster situation facilitates the human agent to interact with system with minimum effort.

Humans are naturally more inclined towards a system that is easy to use [8]. In a disaster situation everyone is busy. There is no room for waiting or dealing with a poorly designed system interface. It is crucial for disaster management system to provide suitable interface. As an example, if MAS provides only typing facility but a situation arises where there is smoke in the environment, then user will have difficulty in operating the system in that locality. The user thus needs to change his location or wait for the smoke to clear. Such an interface degrades performance of the user.

Human agents play an important role by cooperating and communicating with disaster management institutions [9]. In countries, where bomb blasts and suicide attacks are happening every other day, the disaster situation is difficult to manage. A lot of time of the emergency institutions is wasted in getting to know about the incident as well as in identifying the spot of incident. Number of causalities due to such attacks is growing day by day. In MAS, human input proves extremely beneficial for managing disaster situation. They can inform the institutions about the disaster situation nearby. Today, the world is moving towards an era of human-machine cooperation [8]. We want to get benefit from both efficiency of automated systems and human intelligence.

Being part of a society, it is our moral responsibility to help the rescue institutions in dealing with such disasters. The disaster situation becomes more manageable when common people cooperate with the organizations. People living near the place of incident can inform MAS about the disaster via telephone call, text message, pictures, maps or any other input mode that is feasible for a common man. The system then notifies the rescue teams with the details and disaster management plan to take instant action. These timely alerts minimize the extent of loss, by saving the time consumed in identifying the location of incident.

Our motivation for conducting this research is to eradicate the mismanagement, created due to difficulty faced by rescue teams while operating the system in hurry.

MAS interface accommodates different possible scenarios and adapt the situation accordingly. Easier the user interface, more will it be adopted by the users. We target the second phase of disaster management process that is emergency support. We aim to improve the MAS architecture by proposing a user interface behind of which seamlessly multi-agents are in-charge to perform several tasks in an easy and human friendly way. This enhancement improves human to system interaction, resource mobilization, and response time of MAS. It accommodates human input not only in decision making but also in managing the situation effectively.

Section 2 reports some of the existing disaster management solutions. Section 3 presents our proposed improvement to MAS interface. Section 4 validates the results of our solution with the help of a case study. In Section 5, we conclude our work and give our future directions.

2 Related Work

Disasters2.0 [5] presents a novel approach for human and agent collaboration during disaster management. It focuses social aspect by integrating user generated information about disaster. Web client of the system provides real time summary of an activity by displaying important buildings. Users can view agents moving in the environment.

Multi-Agent Situation Management for Supporting Large-Scale Disaster Relief Operations [10] present solution to problems arising in complex situations of a disaster. It focuses on distributed solution driven management and scalability of the system. The architecture extends belief-desire-intention (BDI) system with situation awareness capability. A peer to peer overlay combined with semantic discovery mechanisms is used for large scale relief operation management. The disaster situation management constructs a real time, constantly refreshed, situational model, which is used for operation planning and updating. The decision support system is used for scheduling. Communication between different BDI-SM agents is established using foundation for intelligent agents.

A Framework of Multi-Agent Systems Interaction during Natural Disaster [6] (FMSIND) focuses on interaction among different system agents. Java application development framework is used for agent communication. Sensors are used for detecting a disaster situation. Decision support system uses learning agents, neural networks, and data warehousing components to decide the number of resources required by a site. This helps in devising the plan which is sent to the agent as well as disaster management repository to keep information up to date.

The Autonomous Learning Agents for Distributed and Decentralized Information Networks, (ALADDIN) [7] project focuses on multiple intelligent agent design, decentralized system architecture, and applications. Sensors are used for situation awareness and autonomous data collection. Research includes developing tools and techniques for intelligent agents, making progress in problematic areas such as delayed data, missing data, spurious data, and decision making. The goal of this project is to achieve robustness in multi agent architectures for disaster management.

Autonomous Notification and Situation Reporting for Flood Disaster Management [1] facilitates autonomous notification and generates situation reports for disasters using intelligent agents. A web based tool for flood management is proposed that uses multi agent concept. Water level is continuously monitored to detect any disaster situation. System signals the notification agent to send alerts.

The existing MAS for disaster management focus on system design and do not discuss the user interface of the system. Therefore, we design and develop an agent based user interface for disaster management MAS.

3 Agent Based User Interface for Disaster Management (AID)

A good user interface includes consideration of different possible situations arising due to uncertain nature of the disaster environment [8]. In this section, we present an agent based user interface for improving the usability of MAS. The more the system is human like, the better it will be at understanding and interacting with the humans.

3.1 Abstract Architecture

The focus of AID is on building a good team [9]. The effective team management is a core desirable feature since a well-managed team achieves better results than an unorganized one. Fig. 1 shows an abstract architecture of AID.

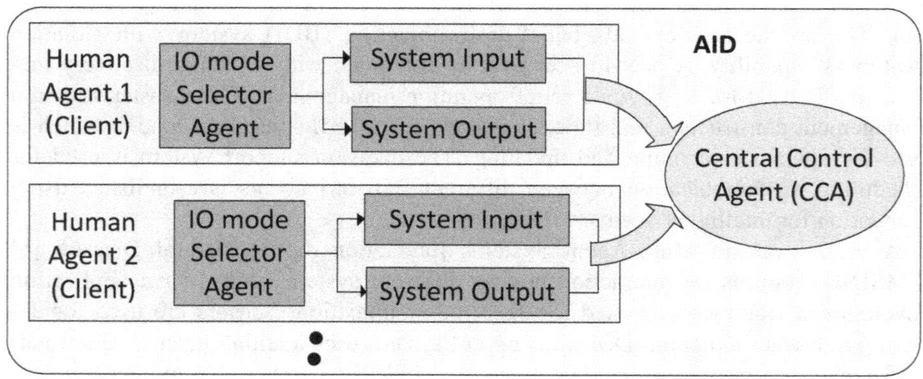

Fig. 1. Abstract architecture of AID

There is a central control agent (CCA), like a manager, administrator, or coordinator in an organization. CCA is an intelligent software agent that works in parallel with human entities for better decision making. CCA directly communicates with all agents whether intelligent software entities or humans which are performing real time activities to stay aware of their current status. A good management system ensures effective communication among the team. AID facilitates both peer to peer communication as well as centralized communication through CCA. The inter-agent communication helps the system and humans to understand each other which results in better decision making. AID provides different alternatives for input and output [11]. The system considers environment of human agent and continuously switches to the mode that best fits agent needs. It also accommodates manual requests for a particular user interface from human agent on higher priority.

3.2 System Input

AID supports different forms of input. Providing only one form of input is not enough as the disaster environment varies instantaneously. There is always a possibility of different problems arising in operating the system due to uncertainty of the environment. Suicide attacks, bomb blasts, or extremist attacks normally happen in sensitive areas where a number of foreigners visit. In such a community, people feel difficult to communicate in formal way of texting or calling due to problems like language barriers, strict protocols, unavailability of contact numbers, etc. AID is designed keeping in mind the user needs and all such environmental factors.

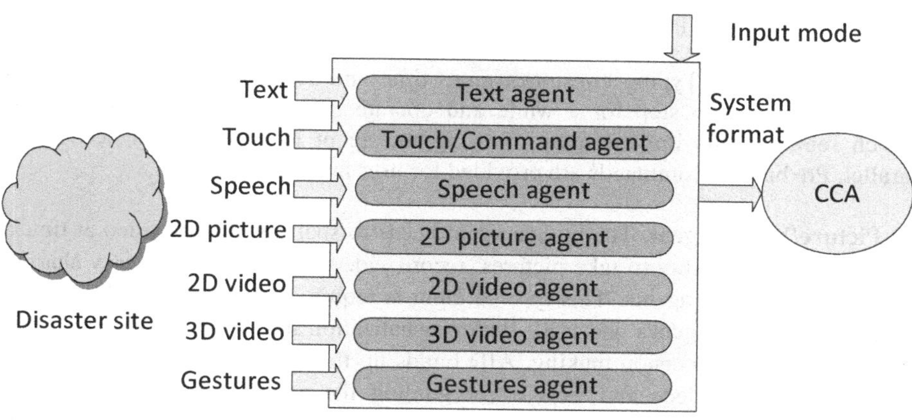

Fig. 2. AID input user interface architecture

Fig. 2 shows the architecture of input user interface where an input data is read based on the input mode of the system. The role of different input agents is to read data and convert it to a standard system format. The input mode is used for activating or deactivating the input agents. This data is passed to CCA for further processing.

Text Agent. Many a time, situation arises where vocal, visual, or touch input is not feasible to use. As an example, when entering record of agent in computer the speech may not be accurate due to noise, visual input, or touch interface doesn't seem appropriate at all on account of flashing lights. The suitable input mode here is text, which may be simple information, a message, an alert, a request, a command, etc. Text agent reads data and parses instructions if required, and take action accordingly. For saving typing effort, built-in messages are provided like suicide attack, bomb blast, building collapsed, etc. in which the user inserts the location at runtime.

Touch Agent. In a disaster situation where everyone is in a hurry, the user has no time to stop and type. AID provides an easy touch interface for giving commands with minimum effort. Most frequently used commands are more visible and more accessible. The touch interface uses universal icons which are well understood and self-descriptive, such as directions icon for map, flash icon for light, etc., so that less thinking is involved.

Speech Agent. There may be smoke due to a sudden fire which limits the visualization thus disabling the use of typing, touching, and input snapshots or video. Normally people ran in anarchy, building crash, etc. In such a situation the suitable mode of input is speech. Speech agent gets input using voice recognition application in the form of speech signals. Noise filtering and other issues are handled by the application integrated with AID for voice recognition. The speech input is validated by confirming the interpreted data from user which is later on shared over the network, parsed to text or commands for taking appropriate action. The speech is also used for one to one communication between human agents like a telephone call. In many situations, this proves fruitful for team members in describing their problem effectively. AID also supports vocal message sharing among different agents for better coordination. Typing consumes more time and effort than speaking [12]. It requires the agent to stop for a while and operate the system physically whereas speech removes this limitation by allowing the agent to continue his activity in parallel. Pre built-in commands are provided for user facility.

2D Picture/Video Agent. The human agent needs to share picture or video at times. AID provides the ability to take pictures, record videos as well as their easy sharing with CCA and other agents. This type of input is required for better visualization of the environment at agent's end [13]. Better visualization means better understanding that leads to better decision making. AID builds up the map of agent's environment from pictures or videos and uses it for guiding the agent in taking appropriate measures. As an example, AID suggests different routes to the agent for exit as well as help in forming better strategies in the rescue operation. This mode of input gives the system insight into the environment. This is particularly useful in giving environment specific instructions instead of general instructions to the agent. It saves the time of human agent that is consumed in conveying environment details.

3D Video Agent. 3D is a better approach than 2D for getting precise environment information [14]. 3D view of environment has more detail. This agent supports rotating cameras in the agent environment that are capable of changing the view like a person rotating his neck for shifting his view. Left and right images are sent to CCA which renders them into a 3D image using stereoscopy and then performs analysis of the environment on its own, in short time. CCA builds a knowledge base and gets an expert opinion on behalf of agent for sorting out an efficient rescue plan.

Gestures Agent. Body language is a crucial part of communication [15]. AID uses live streaming for getting human gestures and body language as input. Many times the communication is misinterpreted just because of lack of gestures. AID improves the human agent communication by providing a chance of having direct face to face talk.

3.3 System Output

AID continuously updates the human agent about the situation and strategies to cope with the environmental changes. The system has an effective output mechanism that gives the human agent a sense of team coordination. Similar to input, the system has different output modes for user ease [12]. AID provides both visual display and audio output, the ability to speak up.

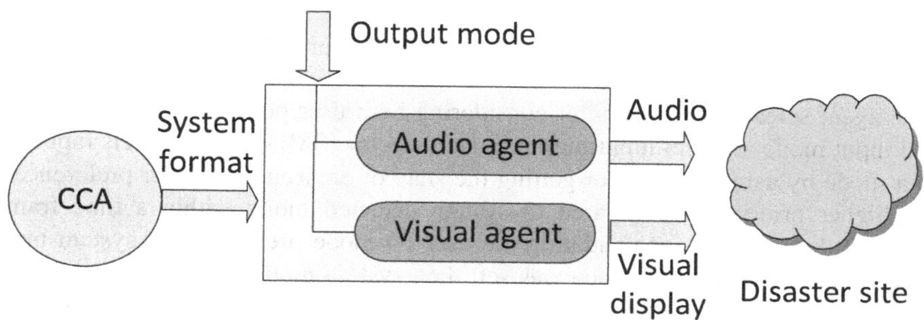

Fig. 3. AID output user interface architecture

Fig. 3 shows architecture of output user interface, data is sent to the user device in standard format. Audio and visual agents are activated or deactivated using output mode. These agents receive the content from CCA in the system format. The agent parses the content to proper output format and turns on the appropriate interface for it.

Audio Agent. Research shows that hearing a message requires less time and processing of humans as compared to reading a message visually. In many situations visual display is not a good option. The user may need to hear the instructions while working. Audio agent therefore, helps the user to continue his task without stopping to read new updates.

Visual Agent. AID supports different visual formats and adjusts its interface according to the content. It is a well-known fact that visualization develops better understanding then simply hearing anything. The human agent need not to waste time on trying to create an image of the described situation in the mind, the user can simply view it. The display is natural, well organized, and easy to understand. The better the display, more will it be pleasing for its user.

3.4 Input and Output Mode Selection

The combined architecture of mode selector agents is shown in Fig. 4.

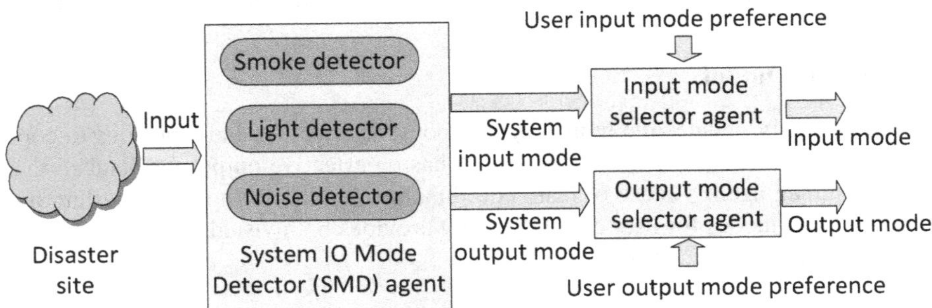

Fig. 4. AID Input/output mode selector agent architecture

The agent selects the mode after considering both user preference and system detected input mode. System input/output mode detector (SMD) agent detects input and output mode by using sensors for getting the state of environment. User preference is given higher priority as compared to system detected mode within a time frame. A time window is used for validating the user IO mode preference. If system mode was detected after user preference was set, then system mode is used instead of user preference. An alert is sent to user about change of IO mode to avoid confusion.

3.5 Adding or Removing an Agent

AID provides option for adding a new agent. All team members are informed about the new agent and its role, to keep the team updated. If the new agent has skills that meet the needs of an existing agent then it sends a help request to the former. This creates a real world team like environment where each one plays his/its role as well as mentors their peers in achieving designated tasks. The system user interface is therefore team oriented. AID also accepts requests from common people for participation in the rescue operation. A human agent can send join request to AID via its centralized service. AID validates the request and agent profile, creates a new agent, locates any existing team working nearby, adds new agent to the team, alerts the team and guides, the agent for joining his team.

AID also provides the ability to safely remove any human agent as well as any malfunctioning system agent. Unavailability of any agent does not halt the system completely. If any agent is inactive for a longer period of time, system removes that agent from the team. Fault tolerance is a highly desirable feature for any real time system, as it determines its flexibility and scalability. The system takes necessary measures to restore its performance to the minimum satisfaction level. The system alerts CCA about the decrease in performance, its cause, and the possible way to restore it back. This is useful in sending timely requests for resources and support for restoring the system to a healthy state.

3.6 Physical Proximity

In the real world, the rescue team needs to work in different distributed physical locations [4]. AID provides a sense of physical proximity to the human agents by providing features for exploring and connecting with agents working nearby. AID keeps all the agents up-to-date about the status of other agents that is any new agent joining or leaving the team on that spot. Agents explore and view other agents, communicate with them as well as help them if required. This boosts the confidence of human agents and gives them a feeling of togetherness with a physical team in a real time distributed environment. The agents share their knowledge and experience by communicating directly with each other. This induces the true essence of a team.

3.7 Centralized Services

AID provides centralized services like emergency service center, police center, intelligence agencies, etc. and their service numbers which are broadcasted to common people via mobile phone and other multimedia services. GPRS system is used by AID for validating the service user and the information they convey. GPRS is also used for agent identification. This validation is required for filtering out invalid requests and avoiding misuse of the system.

4 Case Study

We validate our solution with a scenario of a terrorist attack. Despite many measures from security agencies like installation of security cameras, security checking of individuals and vehicles in sensitive areas, etc., the situation is not controlled. One of the reasons is negligence due to the unexpected nature of these events. We create a simulated environment to perform various experiments for demonstration and get successful result in saving lives and other property losses.

In the disaster situation, 2 buildings collapsed, 50 people killed, and over 140 people were injured due to a suicide attack. There were 35 Android devices, 78 Symbian devices, 26 Apple devices and 6 Blackberry devices at the disaster site. It usually takes a lot of time in acknowledging the incident and locating the disaster spot. Using AID, the response time is reduced due to participation of common people in rescue operation. A female living near that disaster site spontaneously informs AID about the attack and its location via telephone call. AID locates the people present at the disaster site by tracking their mobile devices. It starts gathering their information such as profile and medical history in order to coordinate with rescue agents and common people onsite for initial measures via text message. AID determines the status of people and the extent of damage using services like GPRS, maps and satellite view. AID alerts the rescue organization by sending notifications. It also provides the map of disaster location to the rescue agents on their registered mobile clients. AID estimates the required number of resources such as fire brigade, ambulances, etc. and dispatches the available resources immediately. Initially, a rescue team of 15 persons was dispatched for the emergency support. There were 2 senior medical officers whereas other members had basic level training of providing emergency aid. This team of 15 persons was divided into 2 sub teams in order to work at spots A and B simultaneously. The lack of coordination between the teams often leads to unorganized activities, raising the death toll. With AID, they continuously stay in touch. Work plan and status of the human agents is monitored by CCA. If any agent faces difficulty in performing the assigned task, he requests CCA for a mentor. CCA notifies all team members about this requirement and connects the agent with an experienced agent via video chat or voice call for instructions. The system allows the agents to directly communicate with each other while carefully conducting their assigned tasks. The medical officer can view human agent performing the action in order to provide better guidance. Following the instructions received directly from the medical officer is very convenient for an agent as both parties are continuously in sync and monitoring the impact of their commands. The agent in this case does not need to waste time on asking for help, AID takes care of resolving such dependencies for the team and thus ensures a smooth rescue operation.

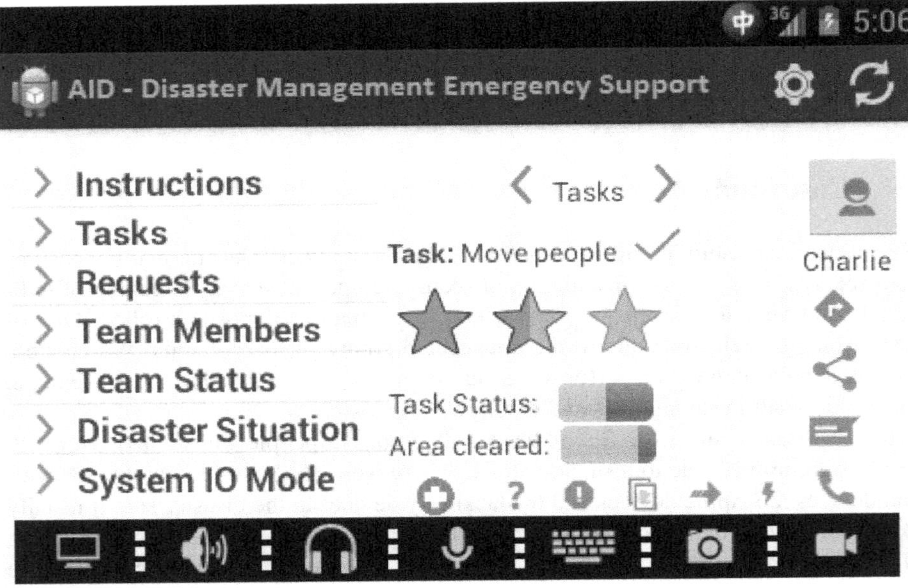

Fig. 5. AID Mobile client for Android

An android mobile interface in Fig. 5 shows a bottom bar with touch icons. Interface switches to visual mode by selecting the screen (or display) icon, which activates the visual agent. Similarly, the speaker icon turns the audio mode on. Touching keyboard changes input mode to text whereas mike icon enables voice input. Similarly, a picture or video mode is turned on by touching camera or video icon respectively. Some other icons allow the user to communicate with AID by sending commands like get directions, update task status, share environment situation, request medical aid, report blocker, request help, etc. Directions icon requests AID for map of the disaster location and best route with respect to agent's current location. The location coordinates are received in response from CCA. These coordinates are opened in a map application for better visualization. AID interface provides features for staying connected with the team via chat, live streaming, telephone call, viewing their status, responding to their help requests, sharing information with them and so on. The mobile agent continuously refreshes its state in order to stay in sync with the CCA, which shares the rescue plan with the team and updates it as per requirement. CCA not only assigns each agent task but also prioritizes their tasks after considering the environmental factors. Table.1 shows a comparison of AID with other MAS.

Table 1. A comparison of AID with other multi agent systems (MAS)

Interface Features	AID	Disaster2.0	ALLADIN	FMSIND
Multiple input/output (IO) modes	Yes	No	No	No
Team oriented	Yes	To some extent	No	No
Resource mobilization	Yes	No	No	Yes
Web/Mobile clients	Yes	Yes	No	No
Services for common people	Yes	To some extent	No	To some extent

AID is team oriented for improving the disaster management process. It considers the interaction of team members in the real world which makes AID user interface more human friendly. AID not only improves the resource mobilization process but also facilitates the human agent by providing different input and output modes.

5 Conclusion and Future Work

Lot of work is being done for designing robust MAS for disaster management but existing solutions lack a powerful user interface. The usability of a system is highly dependent on its user interface. Therefore, we have improved the MAS architecture by designing a human friendly interface. With an interface like AID, even non-technical and less educated people take part in the rescue operation to minimize the extent of loss. We further plan to extend this work by incorporating latest technologies in the system architecture and designing AID specific system components as well as the web and mobile clients for different platforms in order to facilitate the users.

References

1. Ku-Mahamud, K.R., Norwawi, N.M., Katuk, N., Deris, S.: Autonomous Notification and Situation Reporting for Flood Disaster Management. Computer and Information Science 1(3), 20 (2008)
2. National Consortium for the Study of Terrorism and Responses to Terrorism, A Department of Homeland Security Science and Technology Center of Excellence: Based at the University of Maryland. Annex of Statistical Information Country Reports on Terrorism 2012 (May 2013)

3. Basak, S., Modanwal, N., Mazumdar, B.D.: Multi-Agent Based Disaster Management System: A Review. International Journal of Computer Science & Technology 2(2), 343–348 (2011)
4. U.S. Department of Health and Human Services. A Guide to Managing Stress in Crisis Response Professions. DHHS Pub. No. SMA 4113. Rockville, MD: Center for Mental Health Services, Substance Abuse and Mental Health Services Administration (2005)
5. Puras, J.C., Iglesias, C.A.: Disasters2. 0. Application of Web2. 0 technologies in emergency situations. Proceeding of ISCRAM 9 (2009)
6. Aslam, M., Pervez, M.T., Muhammad, S.S., Mushtaq, S., Enriquez, A.M.M.: FMSIND: A Framework of Multi-Agent Systems Interaction during Natural Disaster. Journal of American Science 6(5), 217–224 (2010) (ISSN 1545-1003)
7. Adams, M.F., Gelenbe, E., Hand, D.J.: The Aladdin Project: Intelligent Agents for Disaster Management. In: IARP/EURON Workshop on Robotics for Risky Interventions and Environmental Surveillance (2008)
8. Flemisch, F., Kelsch, J., Löper, C., Schindler, A.S.J.: Automation spectrum, inner / outer compatibility and other potentially useful human factors concepts for assistance and automation. Human Factors for Assistance and Automation, 1–16 (2008)
9. Carver, L., Turoff, M.: Human-computer interaction: The human and computer as a team in emergency management information systems. Communications of the ACM 50(3), 33–38 (2007)
10. Buford, J.F., Jakobson, G., Lewis, L.: Multi-agent situation management for supporting large-scale disaster relief operations. International Journal of Intelligent Control and Systems 11(4), 284–295 (2006)
11. Jaimes, A., Sebe, N.: Multimodal human–computer interaction: A survey. Computer Vision and Image Understanding 108(1), 116–134 (2007)
12. Benoit, C., et al.: Audio-visual and multimodal speech systems. In: Handbook of Standards and Resources for Spoken Language Systems-Supplement, vol. 500 (2000)
13. Wang, Y., Liu, Z., Huang, J.-C.: Multimedia content analysis-using both audio and visual clues. IEEE Signal Processing Magazine 17(6), 12–36 (2000)
14. Yasakethu, S.L.P., et al.: Quality analysis for 3D video using 2D video quality models. IEEE Transactions on Consumer Electronics 54(4), 1969–1976 (2008)
15. Pavlovic, V.I., Sharma, R., Huang, T.S.: Visual interpretation of hand gestures for human-computer interaction: A review. IEEE Transactions on Pattern Analysis and Machine Intelligence 19(7), 677–695 (1997)

An Unsupervised Approach to Identify Location Based on the Content of User's Tweet History

Satya Katragadda, Miao Jin, and Vijay Raghavan

The Center for Advanced Computer Studies, University of Louisiana at Lafayette, Lafayette, USA
{satya,raghavan}@louisiana.edu, mjin@cacs.louisiana.edu

Abstract. We propose and evaluate an unsupervised approach to identify the location of a user purely based on tweet history of that user. We combine the location references from tweets of a user with gazetteers like DBPedia to identify the geolocation of that user at a city level. This can be used for location based personalization services like targeted advertisements, recommendations and services on a finer level. In this paper, we use convex hull and k-center clustering, to identify the location of a user at a city level. The main contributions of this paper are: (i) reliability on just the contents of a tweet, without the need for manual intervention or training data; (ii) a novel approach to handle ambiguous location entries; and (iii) a computational geometric solution to narrow down the location of the user from a set of points corresponding to location references. Experimental results show that the system is able to identify a location for each user with high accuracy within a tolerance range. We also study the effect of tolerance on accuracy and average error distance.

Keywords: Social Media, Location Analysis, Computational Geometry, Convex Hull, K-Center Clustering.

1 Introduction

Social Media usage has increased exponentially in the recent years. Increase in the usage of Twitter, a micro-blogging service, is a good example. These microblogs enable people to post smaller, frequent updates on events happening around them, as well as opinions and details of their personal life. This popularity has led to massive amounts of user-generated data, which provides variety of opportunities and challenges for research. They range from election prediction [32] to event detection [27] and stock market prediction [4]. Mining user centric data like location would help improve the accuracy of the above applications. Location also plays an important role in identifying information appropriate to the user, or events near him.

Traditionally, most of the applications identify the locations of their users by their IP addresses. The reliability of these techniques varies depending on the country and the continent[1] . Yuval et al. estimates that the accuracy of IP

[1] http://www.maxmind.com/en/city_accuracy

D. Ślęzak et al. (Eds.): AMT 2014, LNCS 8610, pp. 311–323, 2014.

prediction in United States is 44% within 40km radius and 80% within 100km radius [29]. Same methods perform better across in European countries ranging from 78% over 40km and 90% within 100km radius. However, the use of VPN networks and dynamic allocation of IP addresses by ISP mask the original location of the user. Moreover, this information is not available to the research community from either Twitter or ISP's due to privacy concerns. Thus, we require other methods to identify location of the user.

There are two ways that Twitter shares the location of the user: geotagging the tweet with location of the user when it is tweeted for those users who have their location services enabled on mobile devices and the location field of the profile. However, less than 1% of the publicly available tweets are geo-tagged due to privacy concerns [14] and about 34% of the users provide incomplete or inaccurate information in their location fields [17]. This number is further reduced when we are looking at the city level location information and the number of people using street level data is extremely small. Another important factor is that people do not update their information as they move from place to place.

We try to solve this location identification problem by identifying the location of a user using references to locations present in his/her tweets. The location here refers to the area a user resides in or a place the user is associated with or interested in at a city level. By city level we refer to n miles around a geolocation (latitude and longitude), the value of n is 30, 60 and 90 miles for this paper. We use the content of the tweets to identify location information associated from the tweets. We try to extract these location cues to predict the location of user.

One of the main challenge with this approach is the nature of social media. Tweets are usually noisy, containing a variety of information. We need to separate location information from this noise. The second problem is the ambiguity of location names. For example a location name can refer to multiple locations e.g., *Lafayette* can refer to 20 towns around the world, not to mention *Lafayette Park* or *Mount Lafayette*. This is one of the major problems with using content information from tweets, as the 140 characters for a tweet does not provide the context to identify the location. The third problem we encounter is that the interests of a user are fleeting; he may refer to multiple events around the world. A user might be talking about *St. Patrick's Day parade* in *New York* and then refer to *Crimea* in the next Tweet. We need to narrow down the location of the user. Our intuition is that the user mostly refers to things happening around him or those he is interested in, instead of the events happening across the world. Finally, the user may have multiple locations e.g., during a vacation or travel. It might be difficult to narrow down the location of the user.

Most of the previous works in this area as introduced in Section 2 concentrate on building language models or complex graph based analysis techniques to identify location information. These techniques require a lot of training data and time to process the input before identifying the location. In this paper, we solve the problem of location identification by using traditional clustering and computational geometric approaches to identify the location of a user at a city level with high accuracy without any training data. We also handle the ambiguous location references in a Tweet by considering all the relevant locations rather than just one location suggester by a gazetteer.

The rest of the paper is organized as follows: An overview of previous work is given in Section 2. Section 3 contains the methodology of our approach. Section 4 explains the data that we used in our experiments. In Section 5, we evaluate our techniques and compare them to baseline and alternative techniques. Section 6 presents some extensions and future work to our approach along with conclusions and future work.

2 Previous Work

The task of identifying geolocation from IP addresses webpages has been studied in detail in the past [5] along with location information from the content of webpages [1, 30, 34] and search query logs [2, 33]. Location was also predicted from various user-generated content like blog posts [11], Wikipedia edit logs [20], Flickr tags [8,16,24] and Facebook locations based on graph framework [3] based on the assumption that users talk more about locations they are familiar to them.

Location prediction on Twitter user based on models built using content of tweets was first studied by Cheng et al. [7]. They built a probabilistic framework to identify the location of user at a city level by identifying words specific to a particular location. This approach requires manual identification of words local to a particular location to build a statistical predictive model. They used about 4,124,660 tweets from 130,689 users to train their model and predict the list of cities in decreasing of probability of the city being the user's home city. They report an accuracy of 51% within 100 miles for 5,190 users with about 5 million tweets with about 1000 tweets per user.

Kinsella et al. [18] created language models of locations using the coordinates extracted from geotagged tweets. They predict the location of an individual tweet using the language models. The location of the user is then arrived at by aggregating all the locations of his/her tweets. They used 7.3 million tweets from Twitter Firehose that were reverse geo-coded using Yahoo Placemaker service. They used 80% of data for training, 10% for tuning and 10% for testing their model. They report a prediction accuracy of 53% for country level, 31% for state level, 29% for city level and 13% for zip code level for tweet locations. For location prediction on users, they report a accuracy of 76%, 34%, 28% and 14% for country, state, city and zip code level predictions respectively.

Hetch et al. [17] built a Multinomial Naïve Bayes model to identify words with a local focus and their respective locations. They use locations from user profile in their model, resolved using Wikipedia-based geocoder and further refined using data from ESRI and US census. They used about 99,296 users from 4 countries who had more than 10 tweets for training the model. Their model identifies the country for 2500 users with 73% accuracy and 30% accuracy for 500 users on state level. Eisenstein et al. [10] built topic models for linguistically consistent regions and states to predict location of a user. They used 9,500 users and 380,000 tweets to predict the location of users with an accuracy of 58% over a region and 24% to predict on a state level.

Mahmud et al. [21] designed an ensemble of location classifiers using tweet content, tweeting behaviours and timezone to identify location of a user at a city level. They used USGS gazetteer to reverse geocode the locations. They report an accuracy of 67% at a city level for 9500 users. Schulz et al. [28] designed a multi indicator model incorporating multiple spatial indicators, like locations from tweets, location from profile information, time zone etc. to predict the location of the tweet. They represent all these indicators in the form of a polygon to identify location of the tweet. They report that they were able to geotag 92% of tweets with a median error of 30km and predict 79% of users location in a 100 mile radius.

Unlike previous work on this topic, we use just the names of locations and landmarks mentioned in the tweet to predict the location of the user. The approaches mentioned above [7, 10, 17, 18, 21], rely on supervised and semi-supervised techniques to identify words related to a location, and then predict the location of the user. In addition, our approach also handles the disambiguation of location, unlike [21, 28] which relies on a gazetteer to identify the best match for a location reference in the tweet.

There has been some work done on identifying ambiguous location references in tweets. Gelernter et al. [12] used named-entity recognition software to extract location names from tweets and found that most of them do not perform location identification from tweets out of the box. Sultanik et al. [31] use named-entity recognizer on n-grams extracted from tweets. Their method then matches these location entries to RapidGeo gazetteer. Paradesi [23] designed a model to predict location of tweet using a POS tagger to identify noun phrases and then compared them to USGS location database to identify all the location names in a tweet. They identify the location of the tweet by calculating the distance between the location references from USGS database to the one in user profile to pick the nearest location. The author reports an accuracy of 15% from 2000 tweets in his/her dataset.

There has also been research done in identifying location of a user based on relationship and location of his friends. Nowell et al. [19] studied the influence of distance in social ties. Backstrom et al. [3] used social ties between friends on Facebook to identify location of a user. Rout et al. [25] and Sadilek [26] built a classifier to identify the location of a user based on social ties of their friends and their location. All these approaches use location fields of users to identify the location of their friends.

3 Methodology

The location identification system is made up of three main components: Extracting location-based entries from tweets, resolving ambiguity of location names and finally predicting the location of the user.

3.1 Extracting Location References

In order to identify location-based entries from tweets, we use gazetteers to assign a unique identifier to each location-based entity. DBPedia[2], Geonames[3] and US census[4] are some of the large-scale gazetteers with millions of entries for states, cities, streets and landmarks. We use DBPedia as the main resource for identifying location-based entries because it helps us in ambiguity resolution which we will explain in the next step. Each record in DBPedia is assigned a Unique Reference Identifier (URI). The problem of location identification is to match the reference to location from the tweet to a DBPedia URI.

Although DBPedia based entity recognition algorithms like DBPedia spotlight [9] are available, their performance is limited by the size of the tweet [22] and their disambiguation resolution in absence of context is error prone.

We use ark-nlp [13] a POS tagger specially designed for Twitter, to identify parts of speech in tweets with the aim of identifying noun phrases. We eliminate all adjectives, verbs and prepositions from a tweet and extract n-grams from it. The maximum number of n-grams are limited to 4 to reduce the computation and look up time. These n-grams are compared to DBPedia entries with geographic coordinates, if a match is found it is identified as a location reference. If there are multiple locations are identified for subset of an n-gram, for example *District of Columbia* vs *Columbia*, the country, we retain the largest n-gram in this case *District of Columbia*. DBPedia disambiguation pages and page redirects are also used to identify as many location-based entries as possible and retain information from the largest n-gram. When we identify the URI, we extract the geolocation (latitude and longitude) for that URI along with other information like region, state and country information associated with that URI.

3.2 Ambiguity Resolution

There are various difficulties in identifying geographic coordinates for text. Different granularity levels for a location are considered namely street - neighborhood - city - state - country. All geolocations are represented on a city level as much as possible to retain the accuracy of the model, even though we still retain geolocation from finer levels of granularity like street or neighborhood coordinates. We now divide these entries into ambiguous entries and unambiguous entries based on the following criteria.

- If we encounter multiple locations entries in the same tweet we retain all of those entries. However, if they follow the hierarchy, we consider the lowest granularity possible given they are in an order like *Seattle* and *WA* or *London, UK*. If those entries don't follow an order e.g. *Chile* and *Brazil*, we retain all those entries. These entries are categorized as unambiguous entries.

---

[2] `http://dbpedia.org/About`

[3] `http://www.geonames.org`

[4] `https://www.census.gov/geo/maps-data/data/gazetteer.html`

- In case of multiple entries for same text on different levels of granularity like *Washington D.C.* vs *Washington* state, we look at descriptive words of corresponding granularity levels like "city", "state" etc. and try to identify their granularity level and get their location. They are categorized as unambiguous entries.
- In case of ambiguous location reference like *Lafayette*, we extract all the locations with that name. All these locations are categorized as ambiguous.
- All the location entries under disambiguation section of DBPedia are categorized as ambiguous.

After all location-based entries are categorized, the granularity of the location reference is drilled down to the city level by converting state and country level entries to the city level. This drilling down is done to reduce the possibility of error in predicting the location of the user as much as possible. An extra location entry is added to all cities that have the state or country entry that match the current state or country entry. For example, for a state level entry *Texas*, all the city level location entries with state information as *Texas* for the current user will be incremented by one.

3.3 Location Prediction

Once all the location references of the user are extracted, the home location of the user needs to be identified. Geolocations for all these location references are extracted from DBPedia. We plot all these geolocation coordinates in euclidean space and try to identify the location of the user from the coordinates. The area encompassing these coordinates is considered to be the active area of the user, which he/she referred to in the recent past. The task of location identification is to predict the home location of user from the active area. To achieve this task computational geometric and clustering approaches are used to predict the location of the user.

Convex Hull Approach with Onion Peeling. The convex hull of a finite set of points Q in a plane is the smallest convex polygon P that encloses Q. convex hull is widely used in various fields like pattern recognition, GIS and image processing. Convex hull is generated for all the geolocations extracted from DBPedia for the user. The home location of the user is within the convex hull, and the most obvious solution is the centroid of the polygon. This would be the most optimal solution if area of the polygon is small. However, if the polygon is really large, as it is in the case of location references in tweets where the user talks about events half way across the world, we need to find a solution where these outliers does not affect our location identification. The number of outliers may vary for different users, so we apply onion peeling algorithm [6] to remove those outliers and predict the location of a user.

Let Q be a finite set of points in a plane. The set of convex layers of Q, denoted by $C(Q)$ is the set of convex polygons defined iteratively as follows:

1. Compute the convex hull of Q.
2. remove the geolocations on convex hull from Q if there are still geolocations in Q.
3. repeat until no more convex hulls can be formed.

The three possibilities of the remainder of points are either a polygon or a line or a single geolocation. The location of the user would be the center of polygon or the center of the line or the last remaining geolocation.

K-Center Clustering. We model the location identification as a clustering problem, the main aim is to cluster all near by geolocations together so that the largest cluster with most location references is considered to have the home location of the user, at the same time, we also need to control the size of the largest cluster as small as possible in order to reduce the average error distance. The clustering algorithm should not be sensitive to outliers, while detecting locations that are near each other. Considering all the requirements, we need a clustering algorithm that clusters all the geolocations in a dataset with a large number of outliers into a minimum number of clusters. We use k-center clustering, which is used to solve resource allocation problems.

In k-center clustering problem of a group of points P, we are required to find K points from P and the smallest radius R such that if disks with radius R are placed on those centers then every point in P is covered [15]. We calculate the location of a user from geolocation in the following steps

1. Randomly pick a geolocation x and assign it as the center of a cluster.
2. Assign all geolocations within d distance of x to the cluster.
3. repeat steps 1 and 2 with unclustered geolocation farthest from centers of all the clusters, until there are no unclustered geolocations left.
4. Predict the center of the cluster with the maximum number of geolocations as the home location of the user.

4 Data

Twitter REST API is used to collect the most recent tweets of a user (less if the user has none within the past year). We started with three profiles UL Lafayette[5], LSU[6] and Huffington Post[7]. We extract 200 most recent tweets from all public profiles of the followers of these three profiles in a depth first format. Our final data set had 2578567 tweets generated from 12,893 users. Of this 2,528 users had their location fields empty or had incomplete or nonsensical data that cannot be geotagged e.g. solar system. Table 1 shows the granularity of location information from user profiles on different levels. This location field is reverse geotagged using DBPedia to extract their geolocation (latitude and longitude

[5] https://twitter.com/ULLafayette

[6] https://twitter.com/lsu

[7] https://twitter.com/HuffingtonPost

Table 1. Granularity of location field in user profile

Granularity Level	Number of Users	Percentage of Users
Street Level	698	5.43%
City Level	8998	69.95%
State Level	213	1.66%
Country Level	456	3.55%

Table 2. Granularity of location references in tweets

Granularity Level	Number of Tweets	Percentage of Tweets
Street Level	96538	10.29%
City Level	439554	46.85%
State Level	125938	13.42%
Country Level	276294	29.45%

coordinates). In case of an ambiguity, we manually identified the location of the user based on the profile text of the user. All users whose information cannot reliably be geotagged are ignored. We finally ended up with 8740 users and 1,734,937 tweets. This geolocation for each user is considered the ground truth data and compared against the predicted location of the user. We had 516,335 (29.8%) of tweets, which had 938,314 reference to locations in them. Table 2 shows the number of location references we encountered for each granularity level.

5 Discussion of Results

5.1 Evaluation Methods

For the evaluation, we measure the accuracy of an identified location as follows
Average Error Distance (AED). The average distance between the location of the user's profile to the predicted geolocation of the user.
Accuracy (ACC). The percentage of correctly predicted locations over all the users at a city level. The tolerance value in this case is 0.
Accuracy within N miles (ACC@N). The percentage of predicted locations that are within N miles of the location in user's profile. For example, ACC@30 measures the percentage of predicted locations that are within 30 miles of location from the user profile. We use ACC@30, ACC@60, and ACC@90 to calculate accuracy within 30, 60 and 90 miles respectively. N is the tolerance value.

5.2 Prediction Models

We analyze the results for the following methods
Random Method (RDM). For each user, we randomly select a geolocation from all the location references in his history based on the probability of the locations.

Table 3. Average Distance Error

Method	Average Distance Error(Miles)
RDM	765.6
TM	632.3
COP	124.6
KCC@30	115.4
KCC@60	123.2
KCC@90	138.7

Trivial Method (TM). For each user, we simply select the geolocation that appears most frequently in user's history.
Convex Hull with Onion Peeling (COP). For each user, we select the geolocation predicted by convex hull onion peeling algorithm.
K-Center Clustering (KCC). For each user, we select the center of circle with maximum location references within the circle. Radius of the circle is the tolerance value and is same as N in Accuracy@N.

We use random and trivial methods as a baseline against convex hull and k-center clustering methods. We report the Accuracy and Accuracy@N for all the above models for all the location references in user history along with just the unambiguous entries. We also study the effect of the number of tweets in user history and their effect on accuracy of location prediction. Accuracy for precious approaches were presented in Section 2, we compare our results with their accuracy.

5.3 Results

In this section we present a detailed analysis of location identification of a user from his/her tweet history. The goal of these experiments is to understand: (i) if the accuracy of location identifier improves from usage of convex hull and k-center clustering approaches; (ii) the effect of distance on accuracy of the models; (iii) the number of tweets in the user's history affects the accuracy of the model.

Location Identification. Table 3 shows the average error distance for all the approaches with the average error decreasing considerably for convex hull and k-center approaches. This is due to the way convex hull and k-center clustering smoothes disambiguates in location references, compared to trivial method which predicts the location as the most frequent location used by the user or random method which randomly picks a location from the user's location references. For example, when a user refers to *London*, it can be the *city of London, UK* or *London, California*, a mistake in identifying correct location can offset an error of 5300 miles. It should be noted that as radius of the circle in k-center approach increases, the average error distance also increases. The average distance error for KCC@30 is 115 miles compared to 139 miles for KCC@90. This is due to the fact that as radius of cluster increases, the centre moves further away from original home location of the user.

Table 4. Accuracy of different models for various amounts of tweets

Method	Unambiguous Location References				All Location References			
	ACC	ACC@30	ACC@60	ACC@90	ACC	ACC@30	ACC@60	ACC@90
100 Tweets								
RDM	0.492	0.507	0.519	0.543	0.407	0.448	0.473	0.491
TM	**0.521**	0.528	0.537	0.557	**0.623**	0.613	0.628	0.632
COP	0.459	0.535	0.543	0.584	0.614	0.708	**0.726**	**0.728**
KCC	**0.521**	**0.559**	**0.613**	**0.626**	**0.623**	**0.713**	0.721	0.726
150 Tweets								
RDM	0.514	0.521	0.546	0.57	0.427	0.48	0.517	0.533
TM	**0.537**	0.541	0.572	0.577	**0.671**	0.684	0.694	0.712
COP	0.54	0.584	0.595	0.608	0.652	0.78	0.785	0.798
KCC	**0.537**	**0.632**	**0.647**	**0.673**	**0.671**	**0.793**	**0.805**	**0.813**
175 Tweets								
RDM	0.522	0.538	0.564	0.583	0.482	0.517	0.539	0.547
TM	0.563	0.596	0.614	0.628	**0.685**	0.702	0.711	0.729
COP	**0.571**	0.592	0.616	0.631	0.671	0.80	0.807	0.812
KCC	0.563	**0.656**	**0.668**	**0.68**	**0.685**	**0.803**	**0.814**	**0.828**
200 Tweets								
RDM	0.532	0.548	0.572	0.594	0.488	0.529	0.543	0.556
TM	**0.578**	0.617	0.631	0.639	**0.691**	0.715	0.724	0.734
COP	0.575	0.623	0.641	0.649	0.69	**0.817**	0.824	0.827
KCC	**0.578**	**0.661**	**0.673**	**0.685**	**0.691**	0.815	**0.829**	**0.831**

Table 4 shows the accuracy(ACC) and accuracy@N($ACC@N$) for all prediction models on both unambiguous location references and all the location references for all the users in the dataset. It should be noted that there is not a huge increase in accuracy of COP and KCC over baseline for unambiguous location references. This seems to be inline with the results from Mahmud et al. [21] where a pure location based classifier gave an accuracy of 64% compared to 69% for our models. The COP and KCC gives a borderline better results by smoothing the latitude and longitude coordinates over a radius of N for unambiguous location entries. However, when considering all the location references including the ambiguous references, all the methods except the random method show an increase in accuracy. This can be explained that the increased number of locations increases the factor of randomness. The convex hull and k-center algorithms out perform the baseline methods in these cases with a maximum accuracy of 83% within a 90-mile radius, compared to 79% accuray by Schulz et al. [28]. K-center approach performs better than the convex hull, but there is not a huge difference between the methods. Even though, the $ACC@N$ of method increases as the value of N increases , the average distance error also increases. Depending on the application, we have to choose optimal value of N that gives good accuracy and average error distance.

Effect of Number of Tweets. We also looked into the effect of the number of tweets on the accuracy of the methods. The accuracy of location prediction increases with the number of tweets, but does seem to taper off as the number of tweets increases beyond 175. Unlike Cheng et al. [7], we do not require 1000 tweets to predict the location of the user. There is no huge difference in execution time for generation of coordinates for both methods. The main problem is the availability of data and the freshness of the tweets. There is a huge increase jump in accuracy for *ACC@90* from 72.6% to 83.1% for k-center and 72.8% to 82.7% for convex hull method.

6 Conclusion and Future Work

In this paper, we provide a computational geometric approach to identify the location of a user based on his tweet history at a city level. Our method performs better than earlier work in this area, by retaining all ambiguous locations and predicting home location of the user, instead of identifying best possible match to the location from gazetteer. It is also important to note that our method does not require any training or manual intervention to build the model. We are able to predict the location of the user with 71% accuracy within 30 miles of home location of the user using 100 tweets and increases accuracy to 83% within 90 miles of home location using 200 tweets. We also studied the effect of increase in average error distance for k-center clustering as the value of N increases. The average distance error increases from 114 miles for N=30 to 139 miles for N=90.

In future we would like to identify the location references that are being talked about by public at large, compared to location references unique to the user. We are also interested in improving convex hull and k-center approaches by assigning weights to important location references. Instead of predicting a single location for a user, we can also include a temporal factor in identifying vacation or a visiting a new place for a user. We also hope to identify the location of the user on a finer level of granularity i.e. zip code and neighborhood level of granularity.

Acknowledgments. This work is supported in part by the awards NSF/IIP - 1160958 from the Computer and Information Science and Engineering (CISE) Directorate of the National Science Foundation and NSF CCF-1054996.

References

1. Amitay, E., Har'El, N., Sivan, R., Soffer, A.: Web-a-where: Geotagging web content. In: Proceedings of the 27th Annual International ACM SIGIR Conference on Research and Development in Information Retrieval, pp. 273–280. ACM (2004)
2. Backstrom, L., Kleinberg, J., Kumar, R., Novak, J.: Spatial variation in search engine queries. In: Proceedings of the 17th International Conference on World Wide Web, pp. 357–366. ACM (2008)
3. Backstrom, L., Sun, E., Marlow, C.: Find me if you can: Improving geographical prediction with social and spatial proximity. In: Proceedings of the 19th International Conference on World Wide Web, pp. 61–70. ACM (2010)

4. Bollen, J., Mao, H., Zeng, X.: Twitter mood predicts the stock market. Journal of Computational Science (2011)
5. Buyukkokten, O., Cho, J., Garcia-molina, H., Gravano, L., SHivakumar, N.: Exploiting geographical location information of web pages. In: Proceedings of the ACM SIGMOD Workshop on the Web and Databases (WebDB 1999), pp. 91–96 (1999)
6. Chazelle, B.: On the convex layers of a planar set. IEEE Transactions on Information Theory 31(4), 509–517 (1985)
7. Cheng, Z., Caverlee, J., Lee, K.: You are where you tweet: A content-based approach to geo-locating twitter users. In: Proceedings of the 19th ACM International Conference on Information and Knowledge Management, pp. 759–768. ACM (2010)
8. Crandall, D.J., Backstrom, L., Huttenlocher, D., Kleinberg, J.: Mapping the world's photos. In: Proceedings of the 18th International Conference on World Wide Web, pp. 761–770. ACM (2009)
9. Daiber, J., Jakob, M., Hokamp, C., Mendes, P.N.: Improving efficiency and accuracy in multilingual entity extraction. In: Proceedings of the 9th International Conference on Semantic Systems, I-Semantics (2013)
10. Eisenstein, J., O'Connor, B., Smith, N.A., Xing, E.P.: A latent variable model for geographic lexical variation. In: Proceedings of the 2010 Conference on Empirical Methods in Natural Language Processing, pp. 1277–1287. Association for Computational Linguistics (2010)
11. Fink, C., Piatko, C.D., Mayfield, J., Finin, T., Martineau, J.: Geolocating blogs from their textual content. In: AAAI Spring Symposium: Social Semantic Web: Where Web 2.0 Meets Web 3.0, pp. 25–26 (2009)
12. Gelernter, J., Mushegian, N.: Geo-parsing messages from microtext. Transactions in GIS 15(6), 753–773 (2011)
13. Gimpel, K., Schneider, N., O'Connor, B., Das, D., Mills, D., Eisenstein, J., Heilman, M., Yogatama, D., Flanigan, J., Smith, N.A.: Part-of-speech tagging for twitter: Annotation, features, and experiments. In: Proceedings of the 49th Annual Meeting of the Association for Computational Linguistics: Human Language Technologies: Short papers, vol. 2, pp. 42–47. Association for Computational Linguistics (2011)
14. Graham, M., Hale, S.A., Gaffney, D.: Where in the world are you? Geolocation and Language Identification in Twitter. CoRR abs/1308.0683 (2013)
15. Guha, S.: Tight results for clustering and summarizing data streams. In: Proceedings of the 12th International Conference on Database Theory, pp. 268–275. ACM (2009)
16. Hauff, C., Houben, G.-J.: Geo-location estimation of flickr images: Social web based enrichment. In: Baeza-Yates, R., de Vries, A.P., Zaragoza, H., Cambazoglu, B.B., Murdock, V., Lempel, R., Silvestri, F. (eds.) ECIR 2012. LNCS, vol. 7224, pp. 85–96. Springer, Heidelberg (2012)
17. Hecht, B., Hong, L., Suh, B., Chi, E.H.: Tweets from justin bieber's heart: the dynamics of the location field in user profiles. In: Tan, D.S., Amershi, S., Begole, B., Kellogg, W.A., Tungare, M. (eds.) CHI, pp. 237–246. ACM (2011)
18. Kinsella, S., Murdock, V., O'Hare, N.: I'm eating a sandwich in glasgow: Modeling locations with tweets. In: Proceedings of the 3rd International Workshop on Search and Mining User-Generated Contents, pp. 61–68. ACM (2011)
19. Liben-Nowell, D., Novak, J., Kumar, R., Raghavan, P., Tomkins, A.: Geographic routing in social networks. Proceedings of the National Academy of Sciences of the United States of America 102(33), 11623–11628 (2005)

20. Lieberman, M.D., Lin, J.: You are where you edit: Locating wikipedia contributors through edit histories. In: ICWSM (2009)
21. Mahmud, J., Nichols, J., Drews, C.: Where is this tweet from? inferring home locations of twitter users. In: ICWSM (2012)
22. Meij, E., Weerkamp, W., de Rijke, M.: Adding semantics to microblog posts. In: Proceedings of the Fifth ACM International Conference on Web Search and Data Mining, pp. 563–572. ACM (2012)
23. Paradesi, S.M.: Geotagging tweets using their content. In: FLAIRS Conference (2011)
24. Popescu, A., Grefenstette, G., et al.: Mining user home location and gender from flickr tags. In: ICWSM (2010)
25. Rout, D., Bontcheva, K., Preoţiuc-Pietro, D., Cohn, T.: Where's@ wally?: A classification approach to geolocating users based on their social ties. In: Proceedings of the 24th ACM Conference on Hypertext and Social Media, pp. 11–20. ACM (2013)
26. Sadilek, A., Kautz, H., Bigham, J.P.: Finding your friends and following them to where you are. In: Proceedings of the Fifth ACM International Conference on Web Search and Data Mining, pp. 723–732. ACM (2012)
27. Sakaki, T., Okazaki, M., Matsuo, Y.: Earthquake shakes twitter users: Real-time event detection by social sensors. In: Proceedings of the 19th International Conference on World Wide Web, pp. 851–860. ACM (2010)
28. Schulz, A., Hadjakos, A., Paulheim, H., Nachtwey, J., Mühlhäuser, M.: A multi-indicator approach for geolocalization of tweets. In: Seventh International AAAI Conference on Weblogs and Social Media (2013)
29. Shavitt, Y., Zilberman, N.: A study of geolocation databases. CoRR abs/1005.5674 (2010)
30. Silva, M.J., Martins, B., Chaves, M., Afonso, A.P., Cardoso, N.: Adding geographic scopes to web resources. Computers, Environment and Urban Systems 30(4), 378–399 (2006)
31. Sultanik, E.A., Fink, C.: Rapid geotagging and disambiguation of social media text via an indexed gazetteer. In: Proceedings of ISCRAM 2012, pp. 1–10 (2012)
32. Tumasjan, A., Sprenger, T., Sandner, P., Welpe, I.: Predicting elections with twitter: What 140 characters reveal about political sentiment. In: Proceedings of the Fourth International AAAI Conference on Weblogs and Social Media, pp. 178–185 (2010)
33. Wang, L., Wang, C., Xie, X., Forman, J., Lu, Y., Ma, W.Y., Li, Y.: Detecting dominant locations from search queries. In: Proceedings of the 28th Annual International ACM SIGIR Conference on Research and Development in Information Retrieval, pp. 424–431. ACM (2005)
34. Zong, W., Wu, D., Sun, A., Lim, E.P., Goh, D.H.L.: On assigning place names to geography related web pages. In: Proceedings of the 5th ACM/IEEE-CS Joint Conference on Digital Libraries, pp. 354–362. ACM (2005)

Sensing *Subjective Well-Being* from Social Media

Bibo Hao[1], Lin Li[2], Rui Gao[1], Ang Li[1], and Tingshao Zhu[1,*]

[1] {Institute of Psychology, University of Chinese Academy of Sciences}, CAS
[2] School of Humanities and Social Sciences, Nanyang Technological University
haobibo@gmail.com, lilindeqinchun@sina.com,
{gaorui11,liang08}@mails.ucas.ac.cn,
tszhu@psych.ac.cn

Abstract. Subjective Well-being(SWB), which refers to how people experience the quality of their lives, is of great use to public policy-makers as well as economic, sociological research, etc. Traditionally, the measurement of SWB relies on time-consuming and costly self-report questionnaires. Nowadays, people are motivated to share their experiences and feelings on social media, so we propose to sense SWB from the vast user generated data on social media. By utilizing 1785 users' social media data with SWB labels, we train machine learning models that are able to "sense" individual SWB. Our model, which attains the state-of-the-art prediction accuracy, can then be applied to identify large amount of social media users' SWB in time with low cost.

Keywords: Subjective Well-being, Social Media, Machine Learning.

The last decade has witnessed the explosion of social media, on which users generate huge volume of content every day. Because of its richness and availability, a lot of innovative research has been conducted on large scale social media data to discover patterns in sociology, economics, psychology etc., which provides a brand new way for conventional social science research. Studies have shown that people's personal traits and psychological features, such as gender, sexual orientation, personality, Intelligence Quotient and so on, can be automatically predicted through clues on social media, such as behavioral [1,2] and linguistic [3] patterns.

People pursue "good life" from ancient time to now, and the Quality of Life (QoL) is influenced by objective factors like income, jobs, health, environment, which can be measured directly with objective indicators like GDP or PM2.5 [1]. However, these objective factors cannot determine one's QoL. The key indicator of QoL is Subjective Well Being (**SWB**), encompassing emotional well-being and positive functioning [4], which refers to how people experience the quality of their lives and includes both emotional reactions and cognitive judgments.

* Corresponding author.

[1] Atmospheric Particulate Matter with diameter of 2.5 micrometers or less, which is an indicator of air pollution.

D. Ślęzak et al. (Eds.): AMT 2014, LNCS 8610, pp. 324–335, 2014.

Reliable and timely information of SWB, provides important intellectual opportunities to research scientists and policy-makers. By analyzing large population data, it will be possible to identify the trend of SWB within different groups, and figure out why some people are happy and others are not. Many governments and organizations, such as U.S.A, France, OECD (Organization for Economic Co-operation and Development) etc., have been funding surveys and research to collect people's SWB data regularly in order to support efficient decision making [5–7] and furthermore improve people's well-being.

Self-report survey is the conventional method which has been widely used to assess SWB. Surveys comprise questions such as: How happy are you with your life? Respondents answer a numerical scores (e.g., from very satisfied to very dissatisfied) in response to survey. However, questionnaire surveys, no matter in the form of paper-and-pencil, on-line etc., are costly and time consuming. What's more, due to stereotype and social desirability, participants may not provide accurate, honest answers since survey is conducted in an intrusive manner – asking questions to subjects. Besides, it is a big challenge to conduct questionnaire based surveys in large scale or carry out longitudinal study.

Recent studies focus on the prediction of psychological variables, and the predicting models are established by analyzing the features and patterns of social media users' profiles, posts, likes, friends etc. Such methods have been applied to the prediction of personality, depression, etc. SWB prediction also attracts researches' attention [8], while current works on SWB prediction are limited to prediction of groups other than individuals. Some of these work even require costly census data like "income median". Therefore, our goal in this work is to establish efficient SWB prediction model based on social media data, which is applicable for individuals.

1 Related Work

In this section, we will review foundations and studies related to our work in fields of psychology and computer science.

SWB and Its Assessment

Different from mere sentiment or simply happiness – spontaneous reflections of immediate experience, SWB is a measurement of individual's cognitive and affective evaluations of one's own life experience. The structure of SWB we used in this paper, as listed in Table 1, is composed of **emotional well-being** and **positive functioning**. Emotional well-being represents a long-term assessment towards one's life, which consists of two dimensions. Positive functioning includes multidimensional structure of psychological and social well-being, and psychological well-being encompasses six dimensions focusing on individual level.

Watson, Ryff et al., developed positive and negative affective scale (**PANAS**) [9] and psychological well-being scale (**PWBS**) [10], which are correspondent to emotional well-being and positive functioning respectively. The reliability and validity of PANAS and PWBS have been validated in long-term practices by

Table 1. Dimensions of Subjective Well-being and their description

Dimension		Description
Emotional well-being	Positive Affect	**P.A.** Experience symptoms that suggest enthusiasm, joy, and happiness for life.
	Negative Affect	**N.A.** Experience symptoms that suggest that life is undesirable and unpleasant.
Positive functioning	Self Acceptance	**S.A.** Possess positive attitude toward the self; acknowledge and accept multiple aspects of self; feel positive about past life.
	Purpose in Life	**P.L.** Have goals and a sense of direction in life; past life is meaningful; hold beliefs that give purpose to life.
	Environmental Mastery	**E.M.** Feel competent and able to manage a complex environment; choose or create personally-suitable community.
	Positive Relations with others	**P.R.** Have warm, satisfying, trusting relationships; are concerned about others welfare; capable of strong empathy, affection, and intimacy; understand give-and-take of human relationships.
	Personal Growth	**P.G.** Have feelings of continued development and potential and are open to new experience; feel increasingly knowledgeable and effective.
	Autonomy Items	**A.I.** Are self-determining, independent, and regulate internally; resist social pressures to think and act in certain ways; evaluate self by personal standards.

numerous psychological studies. In this paper, we use these two scales for SWB assessment.

Affect and Life Satisfaction Metric on Social Media
Affect and Life Satisfaction (**LS**) reflect "happiness", there are also recent studies investigated large scale social media data to metric people's affect or LS.

Quite a lot of studies use LIWC (Linguistic Inquiry and Word Count) [11], fruit carefully constructed over two decades of human research, or other similar psychological language analysis tool, to quantify psychological expression on social media. Representative works, like hedonometer (happiness indicator) through Twitter by Dodds et al. [12], twitter sentiment modeling and prediction of stock market by Bollen et al. [13], identify the sentiment (moods, emotions) in real time. Furthermore, as "face validation", the quantified metric is highly correlated to social events or economic indicators, and it can even be predictable to economic trends. By modeling people's sentiment through statuses and posts on SNS, these works demonstrate that it is applicable to sense sentiment from social media.

It is noticeable that recent works have introduced psychological as an assessment instrument for "convergent validation". Convergent validity represents to what degree a metric yielded by the model is similar to a psychological questionnaire based assessment. Kramer proposed a model of "Gross National Happiness" [14] to predict satisfactory, and used Diener and colleagues' SWL scale [15] for convergent validation. Please note that happiness defined in this work, is actually satisfaction to one's own life.

Burke and collages explored the relationship between particular activities on SNS and feelings of social capital [16]. They used Facebook Intensity Scale and UCLA loneliness scale to assess one's social capital feeling, which could be used as an evaluation of one's cognitive feeling towards getting along with others. However, on-line social well-being cannot cover the conception of SWB.

Predict Personal Traits and Mental Status via Social Media
Kosinskia [1], Schwartz [3] et al. analyzed the correlation between users' personal traits and behaviors or language usage on Facebook. in which users' personality traits are measured by using Big-Five Personality Inventory. They also build models to predict users' traits like personality through social media, which is also another evidence of "convergent validation" approach. Li et al. [2] use 839 behavioral features on microblog to predict personality, which proved the feasibility of predicting psychological variables through behavioral features. Similar works [17] are also conducted on social media like Twitter. Studies have also cast interests to the prediction of mental health status via social media. Hao [18], Choudhury [19] et al. generalize this method to prediction depression, anxiety, etc. analyzed users' both behavioral and linguistic features on microblog, and employ machine learning methods to predict depression, anxiety and other mental health status of individuals.

These pioneering works provide an innovative insight – to predict (or in another word – "sense") on-line users' psychological traits or mental health status through his/her social media records. Personality keeps relatively stable in one's life, so accumulated on-line records can be enhanced evidence to predict one's personality. While similar to mental health status, SWB is a psychological variable that varies over time. Hence, predicting one's SWB should be based on one's behaviors in a specific period of time.

Prediction of Group SWB through Social Media
Most recent work of Schwartz, Eichstaedt et al. generalize their method to LS prediction [8]. Their work used LS as a single indicator of SWB, and established model to predict the LS of each counties in the U.S.A through Twitter data. In their work, county is the unit to predict the LS, rather than individual. Their method, mainly analyze linguistic features on social media. Furthermore, their model introduced variables like "median age", "median household income" and "educational attainment", which can only be obtained via costly census.

In this work, our goal is to establish model which can predict multi-dimension SWB of individuals, considering both linguistic and behavioral features on social media. Model based on individual will provide better generalization ability to different groups, like groups with different ages, jobs and so on.

2 Method

In our study, we use both behavioral patterns and linguistic usage on social media, to identify their correlation with SWB. In order to establish models, we conduct a user study to collect user's social media data and SWB assessment. Then, we treat the modeling problem as a typical machine learning problem: to learn prediction model from social media data in which SWB is the label.

2.1 Data Collection

We ran our experiment on Sina Weibo (`http://weibo.com`), a Chinese leading social media platform with over 300 million users where more than 100 million microblogs are posted or reposted (retweeted) every day.

In the October of 2012, we randomly sent inviting messages to about twenty thousand Weibo users who fulfill our requirements of "active". Active users are defined as users who have posted more than 500 microblogs before recruiting. Such active users have a relatively long term usage of social media, and their Weibo statuses provide adequate information for analysis.

Users who were willing to participate our experiment are guided to a web APP (`http://ccpl.psych.ac.cn:10002`). Participants were then guided to agree an informed consent and fill psychological questionnaire. Finally, 1785 adult volunteers (female:1136) filled the PANAS and PWBS survey to assess their Emotional Well-being and Positive Functioning as SWB, and their social media data were all downloaded through Sina Weibo API one month after the user study. Figure 1 illustrates the distribution of participants' age and SWB distribution. Our dataset contains users' social media data and SWB score.

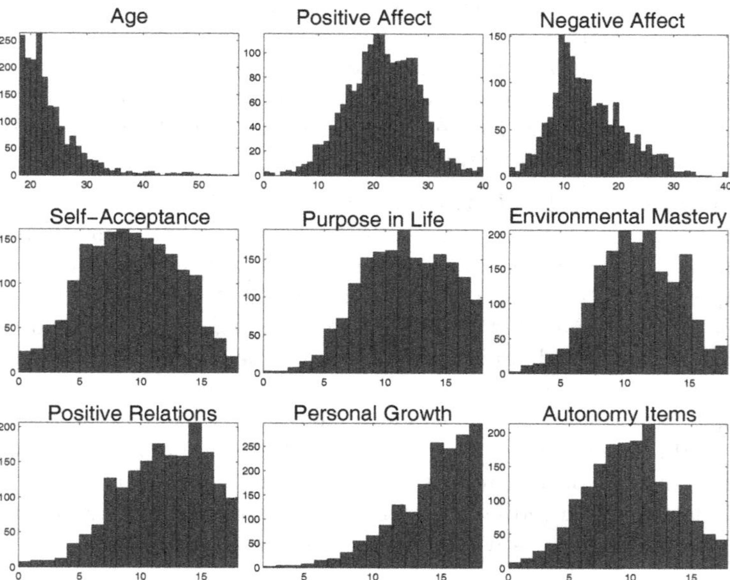

Fig. 1. Distribution of Age and SWB Dimensions, where X-axis represents age or the score of each SWB dimension and Y-axis represents the number of participants

2.2 Experiment Design

In previous studies, researchers proposed different indicators (indexes) based on social media data, and use psychological assessment as verification. These indexes are defined subjectively, and the procedure is always guided by researches' intuition of what factors in social media data may be correlated to target variable. Actually, it is possible to build a prediction model by applying machine learning methods. In this paper, we take the task of predicting users' psychological variable based on their social media behaviors as a typical machine learning problem. To do so, we extract features from users' social media data, and train a machine learning model to predict the target variable (i.e., users' psychological variable).

Golden Standard. Core part of SWB sensing system is the procedure of "learning" patterns of SWB from social media behavioral features. To evaluate reliability of established model, we use Pearson's correlation coefficient. In social psychology, Pearson's correlation coefficient is a well-recognized measurement for convergent validity, which is used to compare the relevance between two assessing instruments or methods [20,21]. Specifically, we calculate the correlation coefficient between psychological scales assessed SWB $\mathbf{Y}$, and SWB sensed by a predicting model $\hat{\mathbf{Y}}$.

$$\gamma = \rho(\hat{\mathbf{Y}}, \mathbf{Y}) = \frac{\mathbf{Cov}(\hat{\mathbf{Y}}, \mathbf{Y})}{\sqrt{\mathbf{Var}(\hat{\mathbf{Y}})\mathbf{Var}(\mathbf{Y})}}.$$

The higher a model's γ is, the closer the model can reach original scale. In the work of Schwartz, Eichstaedt et al. [8], they also adopted the same standard.

When measuring a psychological variable with different assessment instruments or methods, correlation coefficient between different instruments or methods, is typically around 0.39 to 0.68 [22], i.e.: $\gamma \in [0.39, 0.68]$. As a comparison, random guess (uniform distribution) yields $\gamma \in [-0.05, 0.05]$.

Feature Extraction. Our assumption in this study is that one's SWB has impacts on one's behavioral or linguistic patterns on social media. To predict SWB, we adopted demographic, behavioral and linguistic features to build the predicting model.

Demographic Features (D, 3 Features). In our case, we use ***gender***, ***age***, and ***category of living place***[2] categorized by population density as demographic features. Although other demographic information, like "educational attainment", can be quite useful for SWB prediction, they are actually unavailable on Weibo or many other social media platform. The three features we extracted from social media profile are available in users' profile on most social media platform.

[2] We categorize living place in mainland China to a) First-tier Cities: provincial capital and municipality cites, sample size $N_3 = 1009$; b) Other cities, $N_2 = 650$; c) Rural areas, $N_1 = 126$. When using this features for regression, we simply let: $(LivingPlace = 3, 2, 1)$ respectively, which can be seen as an indicator of population density. Similarly, gender are set to 1 (male) and 0 (female).

Notably, in our dataset, when applying Student's t-test, we find that:

- Except for `Negative Affect`, people live in first-tier cities, score significantly higher that people live in other areas in 7 dimensions of SWB ($p < 0.005$);
- For `Positive Affect` and `Autonomy Items`, male users score significantly higher than female users ($p < 0.005$);
- For `Negative Affect`, male users score significantly lower ($p < 0.01$);
- Slight correlation occurs between users' age and SWB dimensions: `Environmental Mastery` 0.15, `Autonomy Items` 0.15 and `Negative Affect` -0.11.

These founding have also been reported by previous research [23]. As found in our dataset, in China, living in first-tier city seems to offer a "happier" life, although it means to be in a more competitive environment. This might be caused by a comprehensive effect of income, education etc.

Behavioral Features (B, 26 Features). We extract behavioral features from user profile and microblogs, including:
- Interaction with other users, like following, friends and bi-following count;
- Express patterns, like microblog count, repost ratio within all statuses;
- Privacy protection, like whether enable geographical information, whether allowing "strangers" to comment;
- Personalization to social media access, like the length of nickname (on Weibo, users can change their nickname at any time).

These features generally describe users' implicit behavioral patterns on social media, and they are available in user's detailed profile and microblog posts.

Linguistic Features (L, 88 Features). SWB comprises abstract dimension like `Autonomy Items`, `Purpose in Life`, we believe such patterns might be implied in users' linguistic expression in microblogs. Like many previous studies, we use an improved version of LIWC, SCLIWC – Simplified Chinese version LIWC optimized for microblog [24], to acquire users' linguistic patterns. SCLIWC's dictionary categorizes words by psychological attributes, like Social Process, Percept, Personal Concern, etc.

Since SWB may vary over time, we extract linguistic features according to particular time period – one week before and one week after the survey (denoted by $\pm 1Week$). This is because our preliminary trial on different time point, like 2 weeks before ($-2Week$), 2 weeks after ($+2Week$) filling questionnaire, with simple linear regression algorithm reveals $\pm 1Week$ performs best. We also compare the relevance of predicted SWB using 6 different combinations of feature set $\{D, B, L\}$, and we take the regression model using only $\{D\}$ as baseline model.

Feature Analysis. Among all the 117 features, we chose some features which are correlated with 8 SWB dimensions in relatively high level and listed them in Table 2. The table shows the correlation coefficient between features and SWB dimensions, from which we can see some behavioral and linguistic features on social media, are positively or negatively correlated with SWB dimensions. For example, users using more first pronoun word "I" in language tend to have

Table 2. Correlation coefficient between 8 SWB dimensions and some features

	P.A.	N.A.	P.G.	P.L.	E.M.	P.R.	A.I.	S.A.
L: Usage of "I"	-.24	-.13	-.35	-.25	-.25	-.24	-.22	-.17
B: N(BiFollowers) / N(Friends)	-.22	-.17	-.34	-.25	-.22	-.22	-.21	-.16
L: Usage of Pronoun	-.24	-.14	-.34	-.23	-.25	-.24	-.22	-.17
B: N(BiFollowers)/N(Followers)	-.24	-.17	-.34	-.25	-.23	-.24	-.22	-.18
B: Domain Name Contains Digits (bool)	-.24	-.17	-.32	-.23	-.22	-.23	-.19	-.17
B: User Description Contains "I" (bool)	-.24	-.15	-.31	-.24	-.24	-.23	-.19	-.17
B: Using Personalized Avatar (bool)	-.22	-.15	-.30	-.22	-.21	-.22	-.19	-.16
B: Usage of Past Sense Words †	-.22	-.13	-.29	-.21	-.21	-.22	-.19	-.17
B: Usage of "We"	-.18	-.13	-.28	-.17	-.19	-.18	-.18	-.13
B: Allowing strangers' comments (bool)	-.19	-.13	-.26	-.19	-.19	-.18	-.17	-.14
L: Usage of Question Mark	-.19	-.12	-.26	-.19	-.19	-.19	-.17	-.14
L: Usage of Semicolon	-.19	-.13	-.26	-.19	-.19	-.19	-.16	-.15
B: N(Statuses)	+.23	+.08	+.25	+.20	+.23	+.21	+.21	+.17
B: Usage of Present Sense Words †	-.16	-.11	-.22	-.17	-.16	-.17	-.13	-.13
L: Usage of Anxiety Words	-.12	-.07	-.21	-.18	-.16	-.15	-.15	-.10
L: Usage of Friend Words	-.13	-.09	-.20	-.15	-.13	-.13	-.10	-.09
L: Usage of Discrepancy Words	-.18	-.10	-.20	-.15	-.15	-.17	-.15	-.11
B: Allow Strangers to send Message (bool)	-.12	-.11	-.20	-.16	-.14	-.14	-.12	-.11
L: Usage of Negative Words	-.15	-.09	-.20	-.15	-.14	-.16	-.10	-.11
D: Category of Living Place	+.20	-.01	+.19	+.17	+.23	+.19	+.20	+.18
D: Age	+.12	-.05	+.13	+.11	+.23	+.18	+.21	+.15
D: Gender (1 for male and 0 for female)	+.11	-.03	+.01	+.05	+.04	+.05	+.15	+.04

† Tense marking words are only available in Chinese.

lower `Personal Growth`; users who posted more statuses on social media tend to have higher `Environmental Mastery`. Such conclusions are in accordance with people's intuition, which can also be seen as a face validation of our method.

Learning Algorithm. SWB assessed by questionnaire survey comprises 8 dimensions, whose values are integers. To build the SWB prediction model, our goal is, for each SWB dimension, learn a function to maximize γ.

Since we didn't find algorithms targeting on maximizing Pearson's Correlation Coefficient, we treated this problem as regression and tried following algorithms:

- **Stepwise Regression**: Choose predictive variables Using F-test.
- **LASSO** (Least Absolute Shrinkage and Selection Operator): Using L1 norm to prevent overfitting, good at reducing feature space.
- **MARS** (Multivariate Adaptive Regression Splines) non-parametric regression technique.
- **SVR** (Support Vector Regression) We used LibSVM implementation.

As SWB is widely used in social science and psychology, although non-linear models may performs better, it will be less interpretable that which factors impact SWB and how they contribute to SWB. Whereas in linear models, it is much easier to figure out which factors impact SWB, and their contributions.

Since the range of different features values are quite different, data are normalized to keep features range in [0, 1]: $X'_{i,j} = (X_{i,j} - minValue_j)/(maxValue_j - minValue_j)$. Besides, we apply **5-fold cross validation** to take most advantage of data and avoid potential overfitting on each algorithm.

3 Results

As shown in Table 3, we compare the performance of models trained by 4 algorithm on 6 combination of feature set. In the left part, column **Feature Set** refers to feature combination, for example, $B + D$ means to use Behavioral Features and Linguistic Features for training and testing. Column **Algorithm** describes which algorithm is used to train learning model. In the right part, each cell shows the γ value. Darker cell background color means better performance.

At bottom of the table, there are **"Feature Set Baseline"** – models trained with only 3 demographic features. Feature set baseline model perform poorly on each dimension, the γ value is around 0.2, which is a very weak correlation. Additionally, performance in the case of random guess is basically $\gamma \in [-0.05, +0.05]$.

Best performance of learning model is listed in the row of "**Best Sensing Result**". It can be seen that our model performs fairly well in 7 dimension of SWB (except for `Negative Affect`). As mentioned before, in social psychology research, to particular psychological variable, when a new developed assessing instrument or method achieves the standard of $\gamma \in [0.39, 0.68]$ with an existing reliable assessing method, it is fair to say the new developed method has equivalent utility with the existing one. As a comparison, work [8] predicts SWB of groups at the level of $\gamma \in [0.264, 0.535]$ using social media data, and work [5] predict SWB of groups at level of $\gamma = 0.598$ using objective data. Hence, our SWB prediction model has attained the state-of-the-art standard.

4 Discussion

In our experiment, we tried 4 algorithms on 6 feature set combinations to establish models. Experiment results show that feature set combination is significant for model performance.

Comparison of Features Set Combinations. Models using only B, L or $B + L$ actually perform no better than baseline. While adding demographic features into training feature set will improve the model performance to different extent. Especially when we use all feature set ($D+B+L$), sensing model achieve best performance. Although demographic feature set only contains age, gender and category of living place, adopting these factors into feature set to train model will improve the model performance significantly. This phenomenon also echoes to the work of [8], after they added control factors like age, sex, monocytes, income and educational attainment into model, prediction accuracy accrues from 0.307 to 0.535.

Table 3. γ Values: Pearson's Correlation Coefficient between SWB "sensed" by our model and assessed by PANAS/PWBS questionnaire scales

Feature Set	Algorithm	Emotional Well-being		Positive Functioning					
		P.A.	N.A.	S.A.	P.L.	E.M.	P.R.	P.G.	A.I.
B	StepWise	.24	.16	.21	.13	.22	.18	.21	.21
	LASSO	.16	.11	.15	.00	.19	.14	.00	.16
	MARS	.16	.08	.14	.04	.05	.14	.07	.13
	SVR	.19	.13	.12	.06	.17	.13	.10	.15
L [±1Week]	StepWise	.22	.16	.15	.23	.20	.16	.23	.25
	LASSO	.17	.10	.15	.16	.14	.00	.22	.16
	MARS	.19	.10	.12	.14	.15	.14	.21	.12
	SVR	.11	.09	.08	.21	.20	.12	.17	.18
D+B	StepWise	.27	.22	.23	.16	.30	.21	.20	.30
	LASSO	.20	.16	.19	.11	.24	.18	.12	.25
	MARS	.19	.06	.17	.02	.20	.12	.06	.23
	SVR	.13	.13	.11	.21	.13	.17	.24	.07
D+L [±1Week]	StepWise	.24	.19	.22	.26	.25	.20	.22	.28
	LASSO	.20	.13	.19	.23	.22	.20	.23	.25
	MARS	.16	.07	.04	.09	.06	.17	.14	.18
	SVR	.24	.11	.16	.16	.30	.22	.19	.24
B+L [±1Week]	StepWise	.23	.21	.18	.22	.18	.19	.21	.26
	LASSO	.24	.20	.11	.20	.17	.18	.24	.20
	MARS	.10	.03	.09	.07	.04	.16	.00	.10
	SVR	.18	.13	.14	.11	.22	.19	.10	.22
D+B+L [±1Week]	StepWise	.45	.26	.35	.45	.41	.45	.51	.40
	LASSO	.38	.26	.29	.34	.35	.34	.42	.35
	MARS	.40	.24	.30	.43	.45	.38	.60	.40
	SVR	.41	.27	.30	.35	.38	.39	.49	.34
Best Sensing Result		.45	.27	.35	.45	.45	.45	.60	.40
Feature Set Baseline (Only D)	StepWise	.23	.17	.31	.27	.23	.30	.25	.20
	LASSO	.14	.10	.21	.14	.09	.21	.15	.13
	MARS	.16	.03	.27	.19	.07	.21	.12	.13
	SVR	.19	.14	.28	.19	.08	.22	.23	.14

Comparison of Algorithms. We adopted algorithms of linear and non-liner, parametric and non-parametric, while in most cases, linear algorithm already perform fairly well. In the same feature set of $D + B + L$, StepWise Regression performs better than other algorithms on 4 dimensions. And MARS achieved $\gamma = 0.6$ on dimension of Personal Growth, which is the best performance in all combinations. Models trained using algorithm of LASSO contain relatively less features, for example, in the combination of $D + B + L$, 44 features enter the final model to predict dimension of `Personal Growth`.

Limitation of This Work. Like many other studies on social media, our model also requires adequate user data for analysis. In our experiment, we set the standard of "active user" as posting more than 500 microblog posts. This limitation can be overcame along with users posting accumulating posts.

Weibo users accounts for more than a quarter of Chinese population, which means, there are Weibo users in, if not every village, nearly every county. SWB tendency of different groups can provide practical opportunities to policy-makers. But social media users, surely cannot cover all population.

5 Conclusions

In this paper, we established models to sense social media users' individual SWB, without survey or costly census data. The established models, which attain the state-of-the-art prediction standard, have equivalent utility with well-designed psychological scales. This approach of psychological assessment, can predict one's SWB by automatically by analyzing his/her social media data in a non-invasive manner, and makes it feasible to assess users' psychological features, in large scale and timely.

Core of the paradigm in this study, is to "learn" sensing (prediction) model from social media data and label data of psychological assessment. Patterns and the interaction structures of explicit or implicit variables in social media, can be automatically learned with algorithms (if they can be represented in the feature space). Such paradigm avoids subjective bias of "designing an index" from numerous features, in which case significant patterns may be hard to be discovered and adopted in the final model. Besides, model is self-verified in the machine learning procedure using techniques like cross validation.

It is our will that the methods in this study can inspire subsequent research in the area of conventional psychology or social sciences. More empirical analysis on real data, leads to more reliable conclusion, and such conclusion can be used to improve the public welfare.

Acknowledgements. The authors gratefully acknowledges the generous support from National High Technology Research and Development Program of China (2013AA01A606), and Strategic Priority Research Program (XDA06030800).

References

[1] Kosinskia, M., Stillwella, D., Graepelb, T.: Private traits and attributes are predictable from digital records of human behavior. PNAS 110(15), 5802–5850 (2013)

[2] Li, L., Li, A., Hao, B., Guan, Z., Zhu, T.: Predicting active users' personality based on micro-blogging behaviors. PloS One 9(1), e84997 (2014)

[3] Schwartz, H.A., Eichstaedt, J.C., et al.: Personality, gender, and age in the language of social media: The open-vocabulary approach. PLoS ONE 8(9), e73791 (2013)

[4] Keyes, C.L.M., Magyar-Moe, J.L.: The measurement and utility of adult subjective well-being. In: Positive Psychological Assessment. American Psychological Association, pp. 411–425 (2003)

[5] Oswald, A.J., Wu, S.: Objective confirmation of subjective measures of human well-being: Evidence from the usa. Science 327(5965), 576–579 (2010)

[6] Stiglitz, J.E., Sen, A., Fitoussi, J.P., et al.: Report by the commission on the measurement of economic performance and social progress. Paris: Commission on the Measurement of Economic Performance and Social Progress (2010)
[7] OECD: OECD Guidelines on Measuring SubjectiveWell-being. OECD Publishing (2013), `http://dx.doi.org/10.1787/9789264191655-en`
[8] Schwartz, H.A., Eichstaedt, J.C., Kern, M.L., Dziurzynski, L., Agrawal, M., Park, G.J., Lakshmikanth, S.K., Jha, S., Seligman, M.E., Ungar, L., et al.: Characterizing geographic variation in well-being using tweets. In: Seventh International AAAI Conference on Weblogs and Social Media, ICWSM 2013 (2013)
[9] Watson, D., Clark, L.A.: Development and validation of brief measures of positive and negative affect: The panas scales. Journal of Personality and Social Psychology 54(6), 719–727 (1998)
[10] Ryff, C.D., Keyes, C.L.M.: The structure of psychological well-being revisited. Journal of Personality and Social Psychology 69(4), 719–727 (1995)
[11] Pennebaker, J.W., Stone, L.D.: Words of wisdom: Language use over the life span. Journal of Personality and Social Psychology 85(2), 291–301 (2003)
[12] Dodds, P.S., Harris, K.D., Kloumann, I.M., Bliss, C.A., Danforth, C.M.: Temporal patterns of happiness and information in a global social network: Hedonometrics and twitter. PLoS ONE 6(12), e26752 (2011)
[13] Bollen, J., Mao, H.: Twitter mood as a stock market predictor. IEEE Computer 44(10), 91–94 (2011)
[14] Kramer, A.D.I.: An unobtrusive behavioral model of "gross national happiness". In: CHI, pp. 287–290 (2010)
[15] Diener, E., Emmons, R.A., Larsen, R.J., Griffin, S.: The satisfaction with life scale. Journal of Personality Assessment 49, 71–75 (1985)
[16] Burke, M., Marlow, C., Lento, T.: Social network activity and social well-being. In: Proceedings of the SIGCHI Conference on Human Factors in Computing Systems, pp. 1909–1912. ACM (2010)
[17] Quercia, D., Lambiotte, R., Stillwell, D., Kosinski, M., Crowcroft, J.: The personality of popular facebook users. In: Proceedings of the ACM 2012 Conference on Computer Supported Cooperative Work, pp. 955–964. ACM (2012)
[18] Hao, B., Li, L., Li, A., Zhu, T.: Predicting mental health status on social media. In: Rau, P.L.P. (ed.) CCD/HCII 2013, Part II. LNCS, vol. 8024, pp. 101–110. Springer, Heidelberg (2013)
[19] De Choudhury, M., Gamon, M., Counts, S., Horvitz, E.: Predicting depression via social media. In: AAAI Conference on Weblogs and Social Media (2013)
[20] Brackett, M.A., Mayer, J.D.: Convergent, discriminant, and incremental validity of competing measures of emotional intelligence. Personality and Social Psychology Bulletin 29(9), 1147–1158 (2003)
[21] Duckworth, A.L., Kern, M.L.: A meta-analysis of the convergent validity of self-control measures. Journal of Research in Personality 45(3), 259–268 (2011)
[22] Graham, J.R.: Assessing personality and psychopathology with interviews. In: Handbook of Psychology: Assessment Psychology, vol. 10, p. 487 (2003)
[23] Diener, E., Suh, E.M., Lucas, R.E., Smith, H.L.: Subjective well-being: Three decades of progress. Psychological Bulletin 125(2), 276 (1999)
[24] Gao, R., Hao, B., Li, H., Gao, Y., Zhu, T.: Developing simplified chinese psychological linguistic analysis dictionary for microblog. In: Imamura, K., Usui, S., Shirao, T., Kasamatsu, T., Schwabe, L., Zhong, N. (eds.) BHI 2013. LNCS, vol. 8211, pp. 359–368. Springer, Heidelberg (2013)

Detecting Stay Areas from a User's Mobile Phone Data for Urban Computing

Hui Wang[1], Ning Zhong[1,2], Zhisheng Huang[1,3], Jiajin Huang[1], Erzhong Zhou[1], and Runqiang Du[1]

[1] International WIC Institute, Beijing University of Technology, Beijing, China
[2] Dept. of Life Science and Informatics, Maebashi Institute of Technology Maebashi, Japan
[3] Dept. of Computer Science, Vrije University of Amsterdam Amsterdam, The Netherlands
hui.wang.bjut@gmail.com, zhong@maebashi-it.ac.jp, hjj@emails.bjut.edu.cn

Abstract. Nowadays, mobile phones are often used as an attractive option for large-scale sensing of human behavior, providing a source of real and reliable data for urban computing. As it is known to all, a user's behavior often happened at some places where the user stayed over a certain time interval for a trip. For understanding a user's behavior effectively, we need to detect the places where the user stayed over a certain time interval and we call these places stay areas. In this paper, we propose a method for detecting the stay areas from a user's mobile phone data. The proposed method can tackle the complicated situations that the general method cannot deal with effectively. Through experimental evaluation, the proposed method is shown to deliver excellent performance.

Keywords: Stay Area, Mobile Phone Data, Urban Computing.

1 Introduction

Urban computing is emerging as a concept where every sensor, device, person, vehicle, building, and street in the urban areas can be used as a component to probe city dynamics to further enable city-wide computing, which aims to enhance both human life and urban environment smartly [1].

In recent years, some studies have been performed for urban computing based on large numbers of GPS trajectories, which can be used to sense human behaviors and activities [2–9].

In addition, mobile phones are often used as another attractive option for large-scale sensing of human behaviors and activities, due to the huge amount data that may be collected at the individual level, and to the possibility to obtain high levels of accuracy in time and space. These features make mobile phone data ideal candidates for a large range of applications, such as urban computing. We will take mobile phone data as our main data source to probe city dynamics for urban computing.

D. Ślęzak et al. (Eds.): AMT 2014, LNCS 8610, pp. 336–346, 2014.

In order to probe city dynamics for urban computing by using mobile phone data, we need to understand the behavior of a mobile phone user first. As it is known to all, a user's behavior often happened at some places where the user stayed over a certain time interval for a trip. For understanding a user's behavior effectively, we need to detect the places where the user stayed over a certain time interval and we call these places stay areas. Obviously, it is important to detect the stay areas of a user for a trip exactly.

So far, many studies have been conducted on mobile phone data for urban computing. Caceres and Calabrese exploited mobile phone data to acquire high-quality origin-destination information for traffic planning and management respectively [10, 11]. Calabrese developed a real-time urban monitoring system to sense city dynamics which range from traffic conditions to the movements of pedestrians throughout the city, using mobile phones [12]. Ying mined the similarity of users by using their mobile phone data [13]. Ying and Lu predicted the next location of the user with mobile phone data respectively [14, 15]. Liu studied the annotation of mobile phone data with activity purposes, which can be used to understand the travel behavior of users [16]. However, the detection of the stay areas is not discussed in detail or it is ignored completely in these studies above. In this paper we aim to detect the stay areas exactly, which is totally different from these studies above.

The remainder of this paper is organized as follows: We give the preliminary to our work in Section 2. Then the problem is stated in Section 3. In Section 4, we describe the proposed method, in detail. The empirical evaluation for performance study is presented in Section 5. Finally, we conclude our work in Section 6.

2 Foundation

In this section, we give the foundation about mobile phone data obtained from the GSM (Global Systems for Mobile Communications) network.

To begin with, we clarify the concepts of location area and cell briefly. In a GSM network, the service coverage area is divided into smaller areas of hexagonal shape, referred to as cells (as shown in Figure 1). In each cell, a base station (namely, an antenna) is installed. And within each cell, mobile phones can communicate with a certain base station. In other words, a cell is served by a base station. A location area consists of a set of cells that are grouped together to optimize signaling, which is identified distinctively by a location area identifier (LAI) in the GSM network. A cell is also identified uniquely by a cell identity (CI) in a location area. That is to say, a cell in a GSM network is identified by a LAI and a CI. For convenience, we use CID to indicate a cell uniquely in a GSM network instead of a LAI and a CI. In urban areas, cells are close to each other and small in area whose diameter can be down to one hundred meters, while in rural areas the diameter of a cell can reach kilometers.

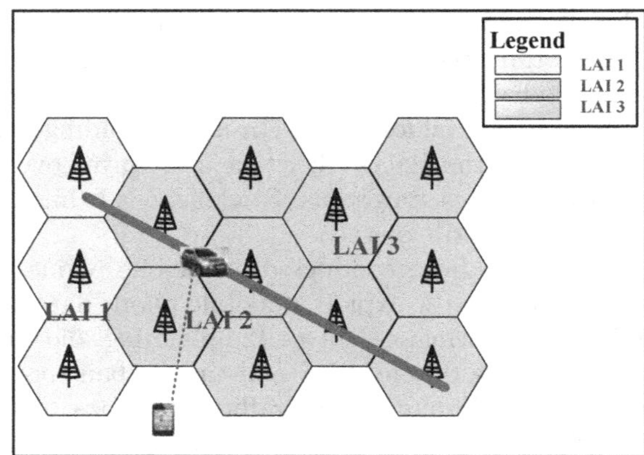

Fig. 1. The concepts of location area and cell

When a mobile phone corresponds with the GSM network, the signal sent by the mobile phone contains the location information (in the form of a CID) of the mobile phone. In order to provide service for mobile phones effectively, the location information will be stored by the GSM network. That the mobile phone sends a signal to the GSM network is triggered by one of the following events:

- The mobile phone is switched on or switched off.
- The mobile phone receives or sends a short message.
- The mobile phone places or receives a call (both at the beginning and end of the call).
- The mobile phone connects the Internet (for example, browsing the web).
- The mobile phone moves into a cell belonging to a new location area, which is called Normal Location Updating.
- The mobile phone during a call is entering into a new cell, which is called Handover.
- The timer set by the network comes to an end when there is no any event mentioned above that happened to the mobile phone, which is called Periodical Location Updating.

3 Problem Statement

In this section, we define some terms used in this paper and give the representation of our problem.

Definition 1 (Mobile Phone Record Set and Mobile Phone Trajectory). *A mobile phone record set is a sequence of records R:* $r_1 \rightarrow r_2$, ..., $\rightarrow r_n$*, each of which* r_i $(1 \leq i \leq n)$ *contains a LAI(*r_i*.LAI), a CI(*r_i*.CI), a timestamp(*r_i*.T), an event(*r_i*.E), and an antenna(*r_i*.A) with the constraint*

($r_i.T \leq r_{i+1}.T$ $\forall 1 \leq i \leq n-1$), representing that an event $r_i.E$ happens in cell $r_i.CI$ belonging to location area $r_i.LAI$ at time $r_i.T$ and the antenna $r_i.A$ (whose latitude and longitude can be obtained) is located in cell $r_i.CI$.

As shown in the left part of Figure 2, $r_1, r_2, \ldots, r_{12}$ constitute a mobile phone record set. We can connect these cells within r_i ($1 \leq i \leq 12$) into a mobile phone trajectory according to their time serials, as shown in the right part of Figure 2.

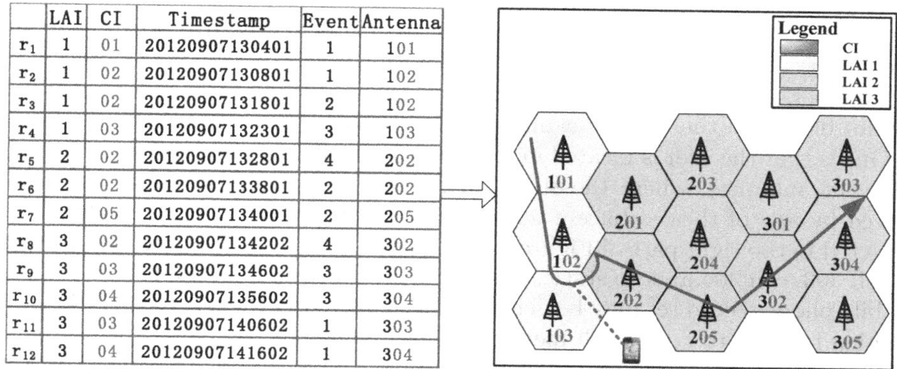

	LAI	CI	Timestamp	Event	Antenna
r_1	1	01	20120907130401	1	101
r_2	1	02	20120907130801	1	102
r_3	1	02	20120907131801	2	102
r_4	1	03	20120907132301	3	103
r_5	2	02	20120907132801	4	202
r_6	2	02	20120907133801	2	202
r_7	2	05	20120907134001	2	205
r_8	3	02	20120907134202	4	302
r_9	3	03	20120907134602	3	303
r_{10}	3	04	20120907135602	3	304
r_{11}	3	03	20120907140602	1	303
r_{12}	3	04	20120907141602	1	304

Fig. 2. The example of a mobile phone record and a mobile phone trajectory

Definition 2 (Stay Area). *A stay area stands for a geographic region where a user stayed over a certain time interval. A stay area carries a particular semantic meaning, such as a place we work/live in, a business district we walk around for shopping, or a spot we wander in for sightseeing. Owing to the fact that the user cannot be located accurately, a stay area can be regarded as a cell or a group of consecutive cells. The detection of a stay area depends on two scale parameters, a time threshold θ_t representing the minimum stay time interval, and a distance threshold θ_d representing the maximum distance between any two antennas in the stay area. Given a mobile phone record set R: $r_1 \rightarrow r_2, \ldots, \rightarrow r_n$, if there exists a sub-sequence R': $r_{i'} \rightarrow r_{i'+1}, \ldots, \rightarrow r_{j'}$, where $r_{j'}.T - r_{i'}.T \geq \theta_t$ and $\forall i' \leq i \leq j \leq j'$, Distance($r_i.CID, r_j.CID$) $\leq \theta_d$, a stay area can be defined as s=\{$r_{i'}.CID$, $r_{i'+1}.CID$, ..., $r_{j'}.CID$\}. The function Distance($r_i.CID$, $r_j.CID$) indicates the Euclidean distance between the antenna $r_i.A$ and the antenna $r_j.A$.*

Problem Representation. With the definitions above, the main problem we are addressing in this paper is formulated as follow: Given a mobile phone set R: $r_1 \rightarrow r_2, \ldots, \rightarrow r_n$ of a user, if there exists a sub-sequence R': $r_{i'} \rightarrow r_{i'+1}, \ldots, \rightarrow r_{j'}$, where R' can be regarded as a stay area, and our task is to detect all these sub-sequences.

4 Proposed Method

As mentioned earlier, the stay area plays an important role in understanding a user's behavior. In order to understand a user's behavior, we need to detect the stay areas first.

The general method for detecting the stay areas is made up of two steps:

1) Deriving the stay time interval in a cell by calculating the difference between the times of the first event and the last event in this cell.

2) Setting a time threshold and judging a cell as a stay area when the stay time interval in this cell is greater than the time threshold.

However, it is complicated in the real physical world and the general method cannot deal with the two following situations effectively.

One is that the user is located in the overlapped signal coverage area of several adjacent antennas, where the signal drift may exist (namely the user can be served by any of these adjacent antennas according to their signal quality). As depicted in the right part of Figure 2, the user may be located in the intersection of cell 303, cell 304 and be static. Although the user is static, the corresponding mobile phone record set may be generated as depicted in the left part of Figure 2 ranging from r_9 to r_{12}, which means the user is moving.

The other situation is that the user may be wandering in a business district or a spot, where several antennas may be situated. As shown in the right part of Figure 2, cell 102, cell 103, and cell 202 may be situated in a business district and the user is randomly wandering among them for shopping. The mobile phone record set may be generated as shown in the left part of Figure 2 ranging from r_2 to r_6.

Assuming that we take the general method to deal with these two situations, the stay interval time in cell 102, cell 103, cell 202, cell 303, and cell 304 will be 10 minutes, 0, 10 minutes, 0, and 0 respectively. Then we set the time threshold as 30 minutes and no cell will be judged as a stay area. In fact, the user is staying 30 minutes in the intersection of cell 303 and cell 304 or the user is wandering 30 minutes among cell 102, cell 103, and cell 202.

Algorithm 1 shows the procedure of detecting the stay areas for the general and particular situations.

In Algorithm 1, the input data are a mobile phone record set of a user, a time interval, and a distance threshold. The output data are a set of stay areas of the user. The variables i' and j' in line 1 indicate that there exists a sub-sequence R': $r_{i'} \rightarrow r_{i'+1}, \ldots, \rightarrow r_{j'}$ satisfying the constraint $\forall i' \leq i \leq j \leq j'$, Distance($r_i.CID, r_j.CID$) $\leq \theta_d$. Lines 4-9 checks whether the distance constraint between cell $r_k.CID$ and cell $r_{j'+1}.CID$ is also satisfied. If the distance constraint is not be satisfied and the time interval between $r_{j'}.T$ and $r_{i'}.T$ is greater than the time threshold, these cells contained in the sub-sequence r' are judged as a stay area s and are inserted into the set S (lines 13-16). The main idea of this algorithm is to try and detect the stay area continuously and the time complexity of this algorithm is $O(n^2)$.

Algorithm 1. StayAreaDetection(R, θ_t, θ_d)

Require:
A mobile phone record set R: $r_1 \rightarrow r_2, ..., \rightarrow r_n$, a time interval threshold θ_t and a distance threshold θ_d

Ensure:
A set of stay areas $S=\{s\}$

```
i'=1, j'=i';
while (i' < n − 1) and (j' < n) do
   k=i';
   while k ≤ j' do
      if Distance(r_k.CID, r_{j'+1}.CID) > θ_d then
         break;
      end if
      k++;
   end while
   if k > j' then
      j'++;
   else
      Δt=r_{j'}.T-r_{i'}.T;
      if Δt > θ_t then
         s ← {r_{i'}.CID, r_{i'+1}.CID, ..., r_{j'}.CID};
         S.Insert(s);
         i'=j'+1;
      else
         i'=k+1;
      end if
      j'=i';
   end if
end while
return S;
```

5 Experiments

In this section, we conduct a series of experiments to evaluate the performance for the proposed stay areas detection method, using the simulated mobile phone data. First, we present the simulation model for generating the mobile phone data. Secondly, we introduce the evaluation methodology for the proposed method. Finally, we give the experimental results and some discussions on them.

5.1 Simulation Model

To evaluate the performance of the proposed method, we simulate the conditions of the real world to generate the required mobile phone data. The simulation model is referred by Du [17].

The simulation process is made up of three main steps:

1) A subject depicts some points on the map (as shown in Figure 3) according to his/her real trip in a city and gives a starting time for this trip. And we can connect these points sequentially and form a trajectory for the subject.

2) The subject annotates his/her trajectory with his/her stay areas. Meanwhile, the subject also needs to give an estimated stay time interval for each stay area.

3) The simulation model generates the corresponding mobile phone data for the subject according to the principle of communication introduced in Section 2, based on the base station data in this city. In order to simulate the real world exactly, we also consider some random factors in the simulation process to reflect the various situations in the real world.

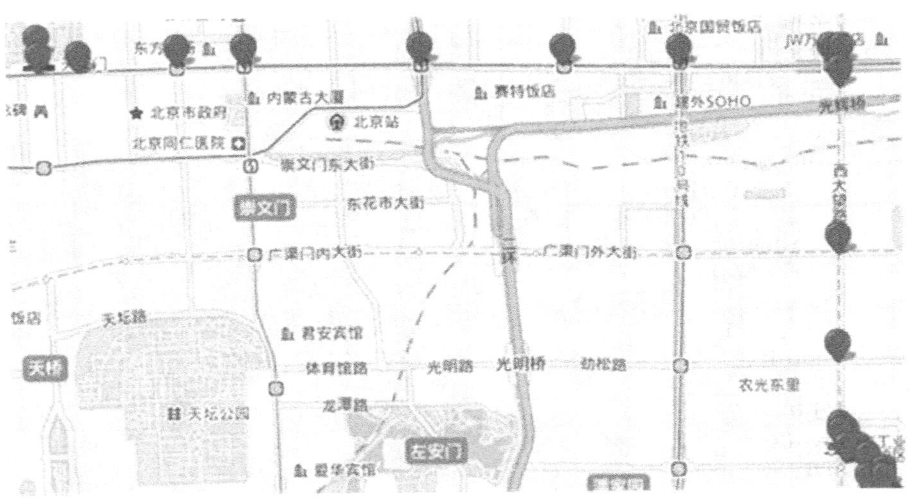

Fig. 3. An interface of the simulation system for generating the mobile phone data

We invited 9 subjects to take part in our experiments and used the real base station data in Beijing as the parameters of the simulation model. These subjects depicted their trajectories on the map and annotated the mobile phone data with their stay areas (as shown in Figure 4). We employed these simulated mobile phone data as our ground truth to evaluate the performance of our proposed stay area detection method.

5.2 Evaluation Methodology

We use Precision, Recall and F-measure to evaluate the performance of the proposed method (called *M1* for short) and the general method (introduced in Section 4, called *M0* for short) as our baseline.

	LAI	CI	Timestamp	Event	Antenna	StayArea
r_1	1	01	20120907130401	1	101	No
r_2	1	02	20120907130801	1	102	Yes
r_3	1	02	20120907131801	2	102	Yes
r_4	1	03	20120907132301	3	103	Yes
r_5	2	02	20120907132801	4	202	Yes
r_6	2	02	20120907133801	2	202	Yes
r_7	2	05	20120907134001	2	205	No
r_8	3	02	20120907134202	4	302	No
r_9	3	03	20120907134602	3	303	Yes
r_{10}	3	04	20120907135602	3	304	Yes
r_{11}	3	03	20120907140602	1	303	Yes
r_{12}	3	04	20120907141602	1	304	Yes

Fig. 4. An example of the mobile phone data with annotations generated by the simulation model

The *Precision*, *Recall* and *F-measure* are defined as Equations (1), (2) and (3), where p^+ and p^- indicate the number of correct predictions and incorrect predictions about the stay areas respectively, and $|R|$ indicates the total number of stay areas.

$$Precision = \frac{p^+}{p^+ + p^-} \tag{1}$$

$$Recall = \frac{p^+}{|R|} \tag{2}$$

$$F - measure = \frac{2 \times Precison \times Recall}{Precision + Recall} \tag{3}$$

5.3 Experimental Results and Discussions

Figures 5, 6 and 7 present *Precision*, *Recall* and *F-measure* of the general method and the proposed method changing over the time threshold θ_t defined in Algorithm 1 respectively. In Figures 5, 6 and 7, let the distance threshold θ_d in Algorithm 1 be 700 meters.

In Figure 5, we observe that the *Precison* performance of these two methods is improved as the parameters θ_t increases respectively.

In Figure 6, we also observe that the *Recall* performance of these two methods declines as the parameters θ_t increases respectively.

In Figures 5, 6 and 7, the *Precision*, *Recall* and *F-measure* performances of that of the proposed method are much better than the general method. Through experimental evaluation, the proposed stay area detection method is shown to deliver excellent performance.

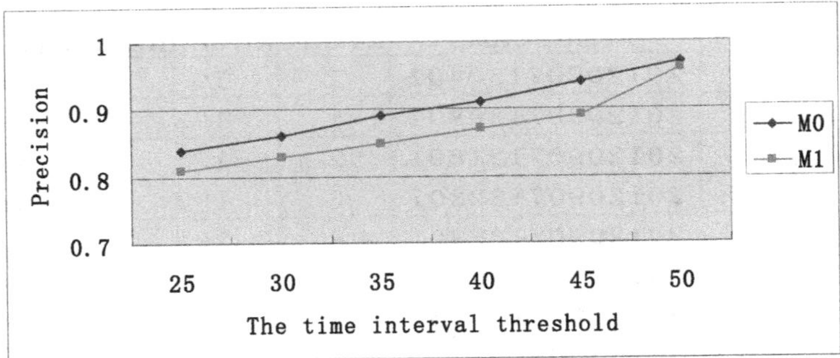

Fig. 5. Precision changing over the time interval threshold

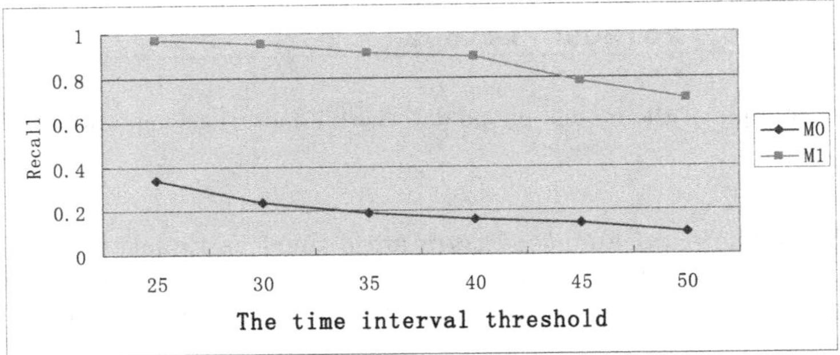

Fig. 6. Recall changing over the time interval threshold

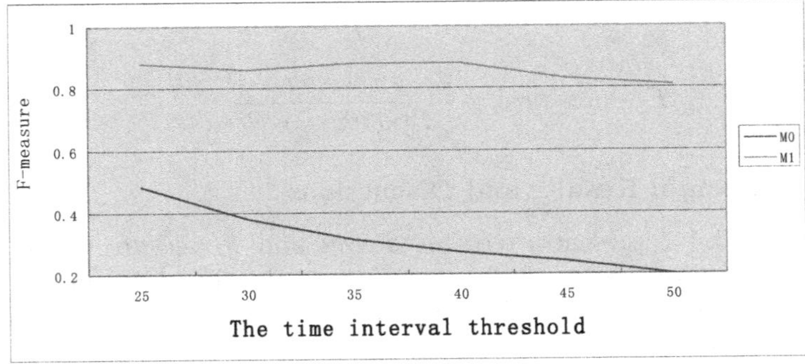

Fig. 7. F-measure changing over the time interval threshold

Figure 8 presents Precision, Recall and F-measure of the proposed method changing over the distance threshold θ_d defined in Algorithm 1 respectively. In Figure 8, let the time interval threshold θ_t in Algorithm 1 be 40 minutes.

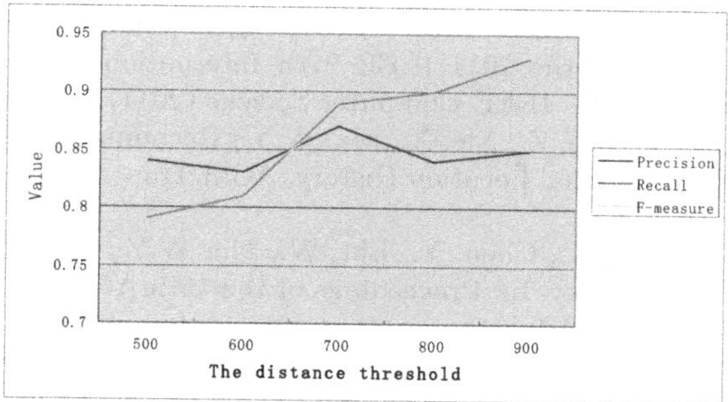

Fig. 8. Precision, Recall and F-measure changing over the distance threshold

Through these experimental results, we can observe that the detection of the stay areas depends on the parameters θ_t and θ_d. However, the parameters θ_t and θ_d are related to the distribution of the antennas in a city. In order to detect the stay areas exactly, we need to try different values for the parameters θ_t and θ_d repeatedly according to the mobile reality dataset in a city.

6 Conclusion

Stay areas play an important role in understanding the behavior of a user. In this paper, we proposed a stay area detection method which is used to detect the places where a user stayed over a certain time interval from a user's mobile phone data. Experimental results show that the performance of the proposed method is better than that of the general method. In future, we will detect a user's stay areas by taking into account the related geographic information, such as POIs.

Acknowledgements. This work is supported by the Beijing Natural Science Foundation (No. 4132023), the National Natural Science Foundation of China (No. 61105118, No. 61272345), the International Science & Technology Cooperation Program of China (No. 2013DFA32180) and the Beijing Nova Program (No. Z121110002500000, No. Z131 107000413120).

References

1. Zheng, Y., Zhou, X.: Computing with Spatial Trajectories. Springer-Verlag New York Inc. (2011)
2. Yuan, J., Zheng, Y., Xie, X., Sun, G.Z.: T-Drive: Enhancing Driving Directions with Taxi Drivers' Intelligence. IEEE Transactions on Knowledge and Data Engineering 25(1), 220–232 (2011)
3. Wei, L.-Y., Zheng, Y., Peng, W.-C.: Constructing Popular Routes from Uncertain Trajectories. In: Proceedings of the 18th ACM SIGKDD International Conference on Knowledge Discovery and Data Mining, pp. 195–203. ACM (2012)

4. Chen, Z.B., Shen, H.T., Zhou, X.F.: Discovering Popular Routes from Trajectories. In: Proceedings of the 2011 IEEE 27th International Conference on Data Engineering, pp. 900–911. IEEE Computer Society (2011)
5. Zheng, Y., Zhang, L., Ma, Z., Xie, X., Ma, W.Y.: Recommending Friends and Locations Based on Individual Location History. ACM Transactions on the Web 5(1), 1–44 (2011)
6. Li, Q., Zheng, Y., Xie, X., Chen, Y., Liu, W., Ma, W.Y.: Mining User Similarity Based on Location History. In: Proceedings of the 16th ACM SIGSPATIAL International Conference on Advances in Geographic Information Systems, pp. 1–10. ACM (2008)
7. Pang, L.X., Chawla, S., Liu, W., Zheng, Y.: On Detection of Emerging Anomalous Traffic Patterns Using GPS Data. Data & Knowledge Engineering 87, 357–373 (2013)
8. Tang, L.A., Zheng, Y., Yuan, J., Han, J., Leung, A., Peng, W.C., Porta, T.F.L.: A Framework of Traveling Companion Discovery on Trajectory Data Streams. ACM Transactions on Intelligent Systems and Technology 5(1), 3 (2013)
9. Xiao, X., Zheng, Y., Luo, Q., Xie, X.: Inferring social ties between users with human location history. Journal of Ambient Intelligence and Humanized Computing 5(1), 3–19 (2014)
10. Caceres, N., Wideberg, J.P., Benitez, F.G.: Deriving Origin Destination Data from A Mobile Phone Network. IET Intelligent Transport Systems 1(1), 15–26 (2007)
11. Calabrese, F., Lorenzo, G.D., Liu, L., Ratti, C.: Estimating Origin-Destination Flows Using Mobile Phone Location Data. IEEE Pervasive Computing 10(4), 36–44 (2011)
12. Calabrese, F., Colonna, M., Lovisolo, P., Parata, D., Ratti, C.: Real-Time Urban Monitoring Using Cell Phones: A Case Study in Rome. IEEE Transactions on Intelligent Transportation Systems 12(1), 141–151 (2011)
13. Ying, J.J.-C., Lu, E.H.-C., Lee, W.-C.: Mining User Similarity from Semantic Trajectories. In: Proceedings of the 2nd ACM SIGSPATIAL International Workshop on Location Based Social Networks, pp. 19–26. ACM (2010)
14. Ying, J.J.-C., Lee, W.-C., Weng, T.-C.: Semantic Trajectory Mining for Location Prediction. In: Proceedings of the 19th ACM SIGSPATIAL International Conference on Advances in Geographic Information Systems, pp. 34–43. ACM (2011)
15. Lu, E.H.-C., Tseng, V.S., Yu, P.S.: Mining Cluster-Based Temporal Mobile Sequential Patterns in Location-Based Service Environments. IEEE Transactions on Knowledge and Data Engineering 23(6), 914–927 (2011)
16. Liu, F., Janssens, D., Wets, G., Cools, M.: Annotating Mobile Phone Location Data with Activity Purposes Using Machine Learning Algorithms. Expert Systems with Applications 40(8), 3299–3311 (2013)
17. Du, R., Huang, J., Huang, Z., Wang, H., Zhong, N.: A System to Generate Mobile Data Based on Real User Behavior. In: Huang, Z., Liu, C., He, J., Huang, G. (eds.) WISE Workshops 2013. LNCS, vol. 8182, pp. 48–61. Springer, Heidelberg (2014)

Towards Robust Framework for On-line Human Activity Reporting Using Accelerometer Readings

Michał Meina[1,3], Bartosz Celmer[2], and Krzysztof Rykaczewski[1,3]

[1] Faculty of Mathematics, Informatics and Mechanics
University of Warsaw, Warsaw, Poland
[2] Section of Computer Science
The Main School of Fire Service, Warsaw, Poland
[3] Faculty of Mathematics and Computer Science
Nicolaus Copernicus University, Toruń, Poland

Abstract. This paper investigates subsequent matching approach and feature-based classification for activity recognition using accelerometer readings. Recognition is done by similarity measure based on Dynamic Time Warping (DTW) on each acceleration axis. Ensemble method is proposed and comparative study is executed showing better and more stable results. Our scenario assumes that activity is recognized with very small latency. Results shows that hybrid approach is promising for activity reporting, i.e. different walking patterns, using of tools. The proposed solution is designed to be a part of decision support in fire and rescue actions at the fire ground.

1 Introduction

Activity recognition of a person in motion is an important task in many fields such as ubiquitous computing, medical diagnosis [6,25], or inertial-based dead reckoning. One of the possible approaches concern using accelerometers mounted on different parts of the human body that reports its acceleration in three axes. Readings from such devices can be employed in activity estimation using, for example, machine learning. Recent advances in mobile technology make those devices cheaper and more precise, which makes them more accessible for large-scale application and broad scientific research. In this paper, we investigate subsequent matching approach of time series readings for the problem of activity recognition.

Before going into the details let us clarify the notation. By *series* we mean a sequence $\sigma = [a_1, \ldots, a_N]$, where $N \in \mathbb{N}$ is the length of the time series and $a_i := a(t_i) = \big(x(t_i), y(t_i), z(t_i)\big)$, $i = 1, \ldots, N$, are accelerations in local inertial system. Each acceleration in the time series we call a *signal.*

The problem of on-line activity recognition can be stated as finding of a function c which assigns to fragments of time series (from different parts of the body) a tag, i.e.

$$c_m(h) = c(a^\sigma_{h-m}, \ldots, a^\sigma_h, a^\tau_{h-m}, \ldots, a^\tau_h, \ldots) \rightsquigarrow \mathfrak{T}, \tag{1}$$

D. Ślęzak et al. (Eds.): AMT 2014, LNCS 8610, pp. 347–358, 2014.

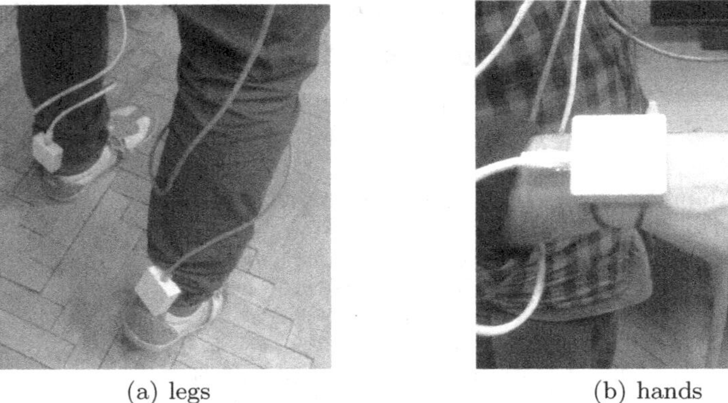

(a) legs (b) hands

Fig. 1. Mounting points of accelerometers

where $\mathfrak{T}$ is a set of predefined activity descriptions (tags, see Section 3), $h = m+1, \ldots, N$, $m \geq 1$ is an activity recognition *window size*, and σ, τ, $\ldots$ denotes that signal is collected from hand, leg, pelvis etc. Moreover, *lead time* (i.e. the time after the action which is spend on calculations) minimizing issue is crucial for interactive and real-time reporting applications.

In the context of the activity recognition systems few aspects should be considered. Firstly, ability of the classification scenario to generalize between different individuals (*subjects*). It is noticeable that different subjects have different walking/gesture patterns to such extend that it can be even used for user identification [7,8]. Secondly, specificity of whole system is important. For instance, model-based approach assumes existence of very specific events in time series that are not easy to discover for general application. We try to overcome this problem by building a framework that combines approaches (1) and (2) described above.

We evaluated our method experimentally using custom-build recording device with sensors mounted on different part of human body. Results that concers different walking patterns shows that described method can fullfil on-line processing requirements outputing the same or better results than state-of-the-art methods.

The contributions of this paper are as follows: (1) study of distance-based approach for activity recognition, (2) comparison study of different approaches, and (3) preliminary proposition of a hybrid method.

2 Related Work

In the context of this paper, we can consider three general approaches for time series classification: (1) feature-based, (2) distance-based, and (3) model-based. First group combines feature extraction and windowing techniques with machine learning (for detailed description see survey [19]). In the second case predefined set of patterns is matched to incoming series using some distance measurements.

This approach is proven to be efficient in, for example, hands gesture recognition [12]. Model-based approach assumes that acceleration changes denote some events in movement (e.g. detachment of the heel from the ground [3] etc.). Such events can be identified by various methods and its order of appearance can be described using statistical models.

The first papers concerning estimation of behaviour using accelerometers were related to daily activities [5,18]. Those approaches were based on feature extraction from time series. Moreover, with the spread of low-cost accelerometers there were attempts to use higher number of devices, e.g. for all segments of the body [21].

The common solution in the most recent works is to combine feature extracting methods with machine learning algorithms [17]. For example, using this idea in [4] authors defined characteristics such as: mean, standard deviation, skewness, kurtosis and eccentricity. Afterwards, these parameters are processed by a simple neural network, which gives the possibility to recognize states such as: standing, sitting, lying-back, lying-on, walking, running, running upstairs, and downstairs. In the above-mentioned paper the authors rate this method to be around 85%–90% effective. Unfortunately, continued work and re-tests carried out after a period of approximately 1.5 years showed that the quality of the classifier was significantly weaker.

Signal is usually processed in relatively small parts of the data called moving windows. The sub-problem, therefore, is to answer the question how to apply the window. The simplest solution is to use fixed size of moving window. More sophisticated methods are based on event detection [9]. Another usage of the event detection may be the development of movement language [15,20,22]. Exhaustive survey [16] gives a state-of-the-art, which shows that nowadays methods have about 80% up to 95% accuracy.

Various authors propose to put sensors in different places of the human body. The most natural are: top of the foot, ankle, elbow, wrist, head, waist. For example, in some papers data obtained from a sensor worn on the waist allow to detect step, or even estimates its length [2,23].

Another way to recognize a human behaviour is to perform a subsequence matching against predefined time series patterns. This can be done with, for instance, simple Euclidean distance measurement. Taking into consideration the ability of the signal to shift or scale a better way to do this is to use Dynamic Time Warping [13] or Longest Common Subsequence [10]. Those solutions are very promising [1].

3 Data Acquisition

Accelerometer readings for our experiments are collected by a prototype recorder of our design. It consists of five ADXL345 accelerometers (mounted on hands, legs and torso), Arduino board, and SD card. Sensors was able to determine linear acceleration in range [-8g, 8g] at 100hz with 13 bits of resolution.

Data were collected from five subjects. Each recording session consisted of two phases: training data collection and test movement. Firstly, subjects were asked to perform some action (such as walking, running, etc.) over the straight line. *Training data* recorded in this way can be considered as "ideal". In the *test data* collection phase, however, subjects were able to move freely over the predefined path. Speed, time and form of movement was completely up to them with exception that all of the previously recorded activities should be performed. By that means the test time series was distorted, introducing unknown deviation from training data such as turnings and *transition states.*

When considering normal human activity it is very difficult to specify clear boundaries between different activities and, therefore, it is challenging to perform good quality tagging of test data. This is due following reasons: (1) the activities are often mixed together in short periods, and (2) the existence of transient states between activities. Furthermore, evaluation on "ideal" data can be misleading and one can easily achieve almost one hundred percent accuracy. Therefore, proposed above data acquisition method is closer to the real-life applications than cross-validated testing on training sample (i.e. training on data with exclusion of the tested person).

In this paper, we are using the following self-describing tags: `walk`, `run`, `stairs_up`, `stairs_down`, `stairs_up_run`, `stairs_down_run`.

4 Pattern Extraction and Matching

The whole process of constructing patterns using the training set is illustrated in Figure 2. Following intuition from [12, Section 2], we assume that activity can be either composed of consecutive events or one event can denote whole activity. Examples of such activities are: walking (which is composed of individual steps), or opening doors (which is an indivisible event). Those different training sets are illustrated in Figure 2 as *Training Set A* and *Training Set B*, respectively.

Readings from the three axes of relative coordinate system are transformed so that we end up with only one time series, i.e. multivariate series is transformed into univariate one. For this reason, we use the following mapping: $\widehat{a}(t) := \sqrt{x(t)^2 + y(t)^2 + z(t)^2}$, $t = t_1, \ldots, t_N$. Such transformation is widely used in similar applications. Roughly speaking, it describes spectral energy of the system [19, Section 4.3].

Subsequent matching approach in our case means that patterns from training data are discovered at first. They represent events that unambiguously identify an activity. Patterns within one tag are then divided into classes, due to length of the sample. More formally, it can be said that with training data $\{p^{i,j,k} := [b_1^{i,j,k}, \ldots, b_{l_{j,k}}^{i,j,k}]\}_{i,j,k} \subset \mathcal{S}_{l_{j,k}}$ we associate pattern $\hat{p}^{j,k} := [\hat{b}_1^{j,k}, \ldots, \hat{b}_{l_{j,k}}^{j,k}] \in \mathcal{S}_{l_{j,k}}$, where $\mathcal{S}_N$ is the space of all series of length N, index $i \in \{1, \ldots, i_{j,k}\}$ enumerates examples in training set, $k \in \{1, \ldots, k_j\}$ indicates classes, and $j \in \mathfrak{T}$. This pattern is a representation of the considered training data set; see Figure 4.

4.1 Pattern Extraction

In the case of multi-event activities, training series at first need to be divided into cyclic events (example of such segmentation is depicted in Figure 3). Event can

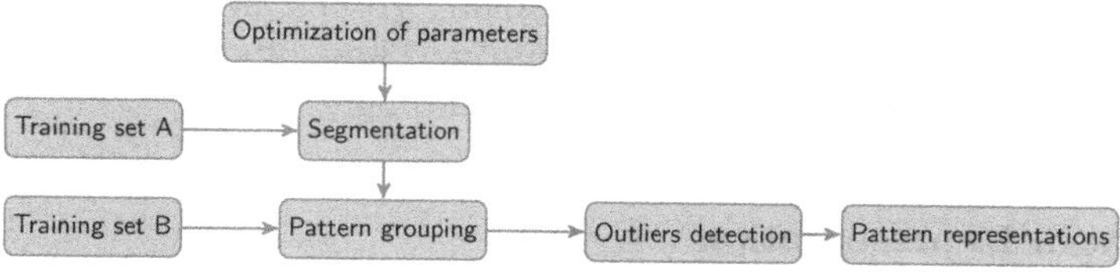

Fig. 2. Pattern Extraction pipeline

be considered as shortest subsequence with highest autocorrelation coefficient. Formally, we can express the length of the event in training series $\widehat{a}(t)$ as:

$$\text{event\_length}\big(\widehat{a}(t)\big) := \underset{i}{\operatorname{argmax}} \operatorname{corr}\big(\widehat{a}(t), \widehat{a}(t)^i\big), \tag{2}$$

where $\widehat{a}(t)^i$ is "series shift" by i (i.e. $\widehat{a}^i(t_j) := a(t_{i-j})$, $j = i+1, \ldots, N$). It is a matter of choice which autocorrelation number one can choose (first peak or the maximum peak). For example, considering torso readings first peak denotes steps, while position of maximum peek stands for the length of two steps for all tested subjects (due to common asymmetric gait).

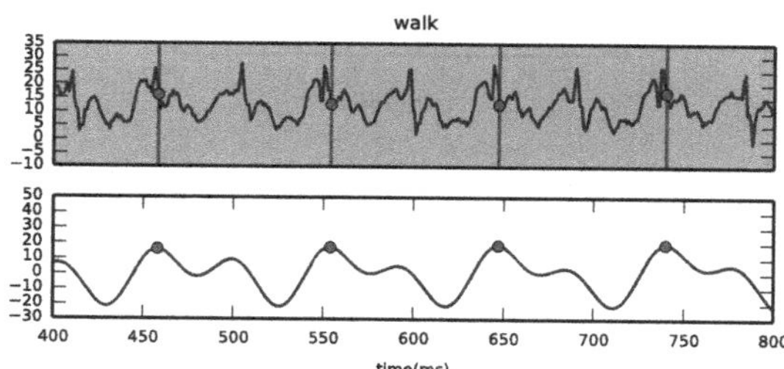

Fig. 3. Segmentation of the training signal. Upper row is the mapped signal $\widehat{a}(t)$ and discovered segments. Lower row contains wavelet filter.

Segmentation in the training data is created using filtered signal $\widehat{a}(t)$. Wavelet filter is used with the base function being the *Mexican hat* of length equal to $m := \text{event\_length}\big(\widehat{a}(t)\big)$. Afterwards, in the filtered signal we are seeking for a local maxima or minima. We assume that different patterns have their end and start markers exhibiting some acceleration peaks. However, if the action has several such peaks it may be difficult to find the corresponding one. Thus, we solved this issue by searching the extremes which distance from the current state is not greater that m.

Clustering of the patterns is obtained by hierarchical-agglomeration clustering with dendrogram cutting factor $t = 0.5 \times \max(\mathit{distances})$. By *outliers* we mean such observations that are contained in "small" groups, i.e. those that number of elements is less than 10% of the all examples. These often come from "edges" of the signal or are mistakes in tagging or recording.

4.2 Pattern Representation

There are two main goals for choosing the representation of patterns: (1) performance of classification, and (2) lower computational cost. Unfortunately, these objectives are mutually exclusive. In the first case the function that is optimized determines the distances of individual groups of tags, while in the second case we minimize the amount of compared patterns.

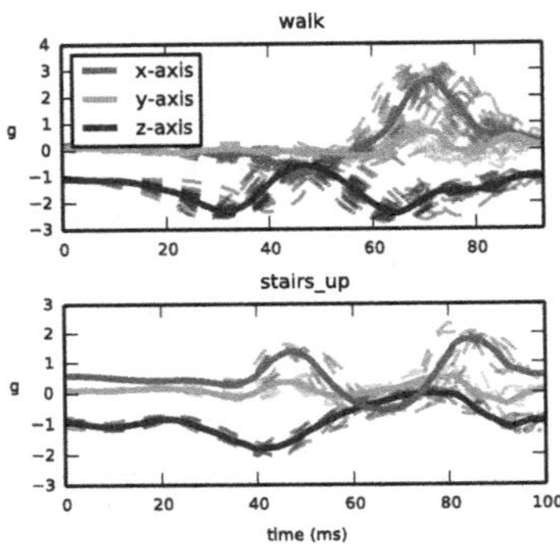

Fig. 4. Example of two patterns from one cluster; dotted line is a original example and thick solid line is a pattern representation (in this case it is *centroid*)

In this paper, we do not compare all of the patterns but only representatives of each clusters. There are a lot of possibilities of choosing the right representative and for the means of this paper we are using centroids of the clusters. Example is depicted in Figure 4.

4.3 Distance Measurements

This section concerns comparing of time series. One of the most popular methods for comparing the patterns is *Dynamic Time Warping* (DTW). DTW is a method that finds an optimal alignment between two given time series. Intuitively, time series are warped in a non-linear manner to fit each other.

Following this idea as *similarity measure* for patterns p_1 and p_2 we take measure ρ defined as

$$\rho(p_1, p_2) := \alpha_x \rho_1(x_1, x_2) + \alpha_y \rho_2(y_1, y_2) + \alpha_z \rho_3(z_1, z_2), \tag{3}$$

where ρ_l are DTWs and constants $\alpha_x, \alpha_y, \alpha_z \in (0,1)$, $\alpha_x + \alpha_y + \alpha_z = 1$, are chosen by simultaneous maximisation of the expressions

$$\Delta_i := \frac{1}{|\mathfrak{T}|} \sum_{\substack{i \neq j \\ j \in \mathfrak{T}}} \sum_{k \in \{1,\ldots,k_j\}} \rho\left(\hat{p}^{i,k}, \hat{p}^{j,k}\right), \quad i \in \mathfrak{T}. \tag{4}$$

By brute force optimisation method we end up with constants $(0.3, 0.1, 0.6)$, which, roughly speaking, favours match z and x axes rather than y axis.

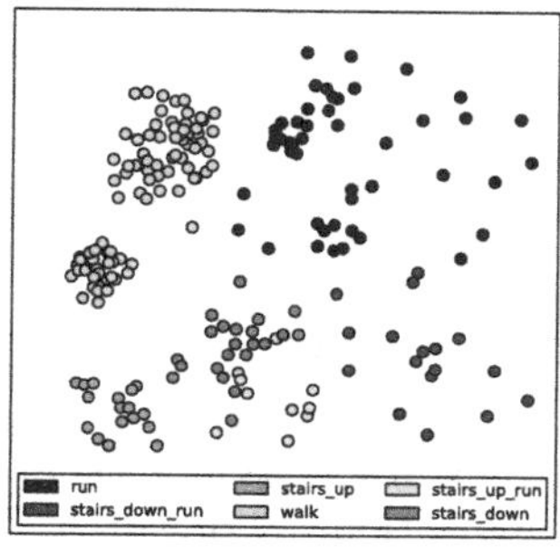

(a) Patterns distances.

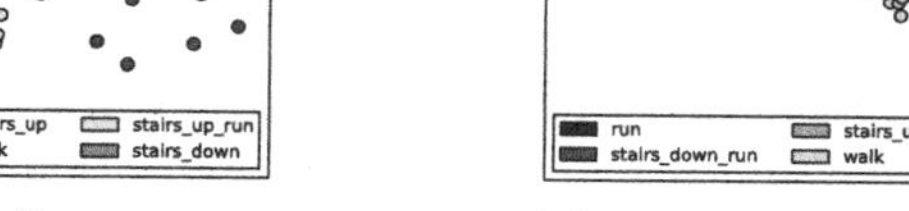

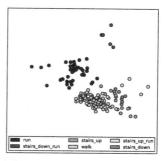

(b) Feature distances using peek-to-peek detection.

Fig. 5. Separation of examples in both methods

4.4 Subsequent Matching Approach

In the subsequent matching approach as a classifier we choose

$$c_m(h) := \underset{j \in \mathfrak{T}}{\operatorname{argmin}} \left\{ \min_{k \in \{1,\ldots,k_j\}} \frac{\rho\left(\hat{p}^{j,k}, [\widehat{a}_{h-m+1}, \ldots, \widehat{a}_h]\right)}{m} \right\}. \tag{5}$$

It is easy to see that we classify each time point by means of its moving window. Later, such classifier we call `1nn-centr`. Other variants of it will appear in Section 6, where we discuss the issue of combining of this approach with feature-based approach.

Remark. When you look closely on the last expression you will notice that it is just 1-NN classifier but over the set of representations.

5 Feature-Based Approach

This section describes a feature-based approach threaten in this work as a baseline for activity recognition due to its popularity in related papers. This approach consists of steps leading to a list of features calculated using a sliding window.

Basically, *moving window* is a subset of time series composed by consecutive elements (i.e. $[\widehat{a}_q, \widehat{a}_{q+1}, \ldots, \widehat{a}_r] \subset \sigma$). Essentially, there are three possibilities how it can be obtained: (1) we can assume that the window can have fixed size, (2) window can have starting and ending trigger, and (3) there can be events (e.g. peaks that can be interpreted as touching the ground by heel) indicating beginnings and endings [19, Section 4]. Second and third approaches are different in such a way that the first needs two markers, and the latter just one. Since the first variant is most straightforward we focused on this one.

Having moving window of length $m \in \mathbb{N}$ *sliding window* $\mathcal{W}(f, \cdot, m)\colon \mathcal{S}_N \to \mathcal{S}_{N-m}$ with feature $f\colon \mathbb{R}^m \to \mathbb{R}$ is defined in the following way

$$\mathcal{W}(f, \sigma, m) := \left[f(\widehat{a}_1, \ldots, \widehat{a}_m), f(\widehat{a}_2, \ldots, \widehat{a}_{m+1}), \ldots, f(\widehat{a}_{N-m+1}, \ldots, \widehat{a}_N)\right]. \quad (6)$$

Since time series is noisy, it have to be smoothed first. For this reason, we utilize Kalman filter, because it has large signal-to-noise ratio and correlation coefficient [24]. It is worth to note that other filter (such as, e.g., wavelet filter) can be chosen as well. However, in case of wavelet filter one has to choose specific component of the wavelet decomposition and concentrates on it.

Relying on the results from [11], we use the list of features as follows: *(a)* total variation, *(b)* fractional dimension, *(c)* standard deviation, *(d)* mean deviation, and *(e)* mean. Therefore, in the training set each tag has five features. Training dataset can now be classified using machine learning techniques.

In order to determine the level of similarity of different activities we projected dataset onto the plane using Multidimensional Scaling (MDS). In Figure 5 the reader can see MDS on the set of similarity measures between: *(a)* patterns, and *(b)* features. Clearly, the case (a) is more easily separable than (b), even in the case of many subjects.

6 Experimental Results

We tested classification performance over the collected dataset using reading from single accelerometer as it is described in Section 3. In order to perform comparative evaluation of feature-based and distance-based approach we constructed simplified classifier in the following way

$$c_m(h) = c(a^{\sigma}_{h-m}, \ldots, a^{\sigma}_h) \rightsquigarrow \mathfrak{T}, \quad h = m+1, \ldots, N. \quad (7)$$

Evaluation was performed in three settings: (1) one-to-one — accuracy and recall was measured for each person, using just his training data, (2) leave-one-out — cross-validated evaluation is performed, e.g. classifiers are trained on four subjects and tested on the excluded one, (3) all-to-one — all of the training series were used for classifiers learning. First and third scenario can be considered as personalized system, the second case, however, shows how each classifier generalize between humans.

Table 1. Confusion matrix for the Majority classifier

	run	stairs_down_run	stairs_up	walk	stairs_down	stairs_up_run
run	**.77 ± .19**	.02 ± .07	.00 ± .01	**.14 ± .18**	.00 ± .00	.07 ± .08
stairs_down_run	**.41 ± .34**	**.57 ± .36**	.00 ± .00	.01 ± .05	.00 ± .00	.01 ± .05
stairs_up	.05 ± .15	.00 ± .00	**.76 ± .31**	.04 ± .06	.00 ± .00	**.15 ± .26**
walk	.05 ± .06	.00 ± .00	.02 ± .02	**.90 ± .07**	.00 ± .00	.03 ± .03
stairs_down	.00 ± .00	.00 ± .00	.00 ± .00	.00 ± .00	**1. ± .00**	.00 ± .00
stairs_up	**.47 ± .30**	.00 ± .00	.06 ± .14	.09 ± .10	.00 ± .00	**.38 ± .28**

6.1 Hybrid Approach

In order to assess our approach we created two more classifiers. Both are versions of the classifier described in Section 4.4. The first one is called 1nn-patterns and is given by the formula

$$c_m(h) := \underset{j \in \mathfrak{T}}{\operatorname{argmin}} \left\{ \min_{\substack{i \in \{1,\ldots,i_{j,k}\} \\ k \in \{1,\ldots,k_j\}}} \frac{\rho(p^{i,j,k}, [a_{h-m+1}, \ldots, a_h])}{m} \right\}. \tag{8}$$

The approach which combines features and patterns is achieved by modifying the similarity measurements by the following formula

$$c_m(h) := \underset{j \in \mathfrak{T}}{\operatorname{argmin}} \left\{ \min_{k \in \{1,\ldots,k_j\}} \frac{pr_{\mathsf{lsvm}}(j; \sigma_h) \cdot \rho(\hat{p}^{j,k}, \sigma_h)}{m} \right\}, \tag{9}$$

where $\sigma_h := [a_{h-m+1}, \ldots, a_h]$ and $pr_{\mathsf{lsvm}}(j; \sigma_h)$ denotes classification probability of tag $j \in \mathfrak{T}$ for window σ_h using Linear SVM classifier, which we choose, because it was fast and has good classification score. Such classifier we called 1nn-centr*.

Remark. Proposed recipe for a hybrid approach is very simple and promising method for activity recognition.

This method can easily be modified to work on-line, since the most computationally demanding operation is DTW (with complexity of $\mathcal{O}(m^2)$). In the worst case one can apply any method of time series indexation.

6.2 Classification of Walk Patterns

Tables 1 and 2 contain confusion matrices for all subjects in all settings for Majority classifier over features and 1nn-pattern classification, respectively. Bold values indicates those scores which are greater than 0.1. It is worth to note that both classifiers fail in different way. For instance, 1nn-patterns generates the largest number of false positives in the case of `stairs_down`, while Majority gives very good results when considering this tag. This suggests that those classifiers are independent to some extent and ensemble classification is promising approach.

Figure 6 depicts classification accuracy and recall. Results indicates that for each setting 1nn-patterns achieves better accuracy by 3.25%, 4.58% and 6.78% on each setting, respectively. Surprisingly, even for leave-one-out settings is better, which indicates that it has good generalization ability. This classifier, however, is

Table 2. Confusion matrix for 1nn-patterns classifier

	run	stairs_down_run	stairs_up	walk	stairs_down	stairs_up_run
run	**.79 ± .19**	.01 ± .04	.01 ± .02	.07 ± .13	**.11 ± .07**	.01 ± .04
stairs_down_run	.03 ± .07	**.67 ± .30**	.00 ± .00	.02 ± .06	**.18 ± .17**	**.10 ± .21**
stairs_up	.00 ± .00	.02 ± .08	**.80 ± .24**	.03 ± .08	**.12 ± .10**	.04 ± .10
walk	.01 ± .02	.01 ± .02	.00 ± .01	**.88 ± .08**	.09 ± .04	.01 ± .04
stairs_down	.00 ± .00	.00 ± .00	.00 ± .00	.00 ± .00	**1. ± .00**	.00 ± .00
stairs_up	.03 ± .11	**.15 ± .24**	.01 ± .03	.06 ± .05	**.14 ± .07**	**.61 ± .30**

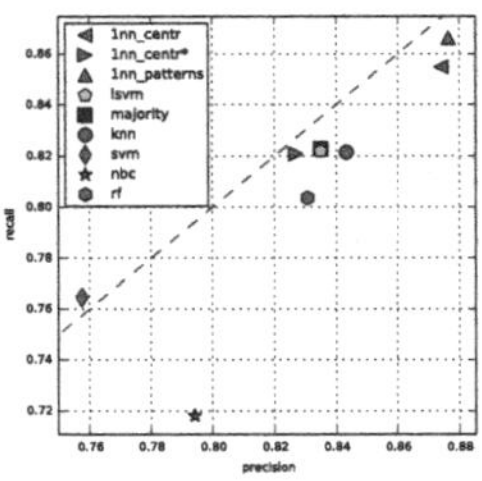

(a) one-to-one 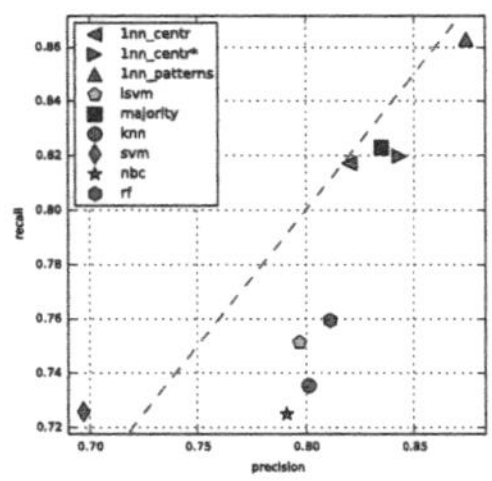 (b) leave-one-out 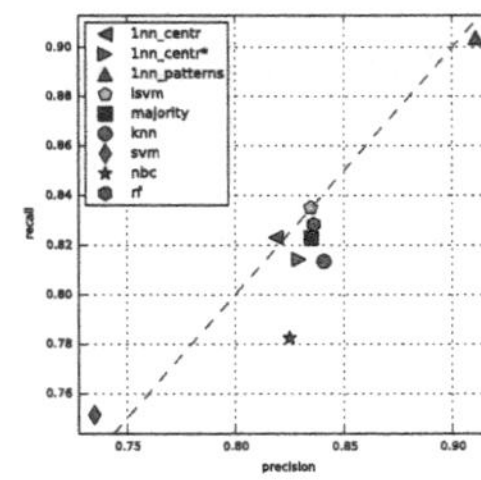 (c) all-to-one

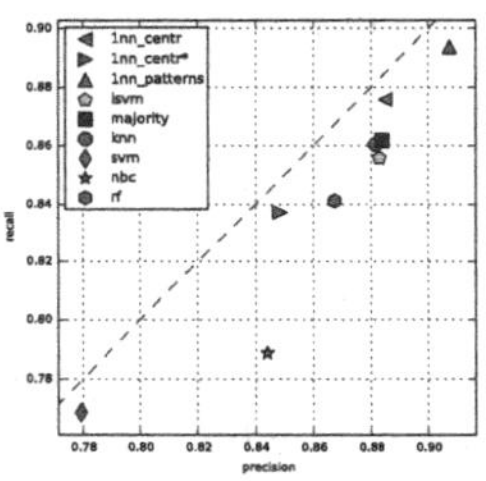

(d) one-to-one 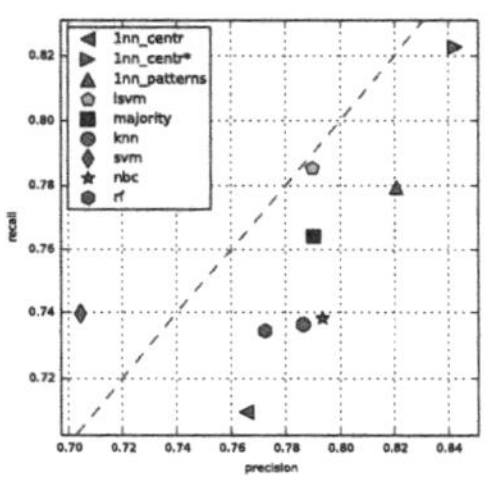 (e) leave-one-out 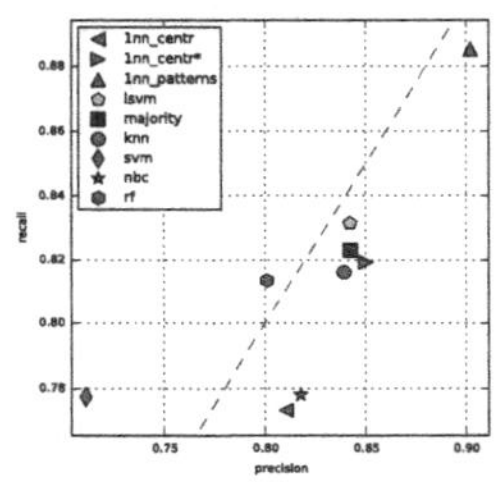 (f) all-to-one

Fig. 6. Precision vs recall of classification based upon readings from right leg (upper row, subfigures a–c) and left leg (lower row, subfigures d–e). Non-triangles are feature-based classification over combined signal $\widehat{\sigma}$ with moving window of the size of 101 (approximately one second), smoothed by the Kalman filter. Triangles depicts pattern-based classification (Section 6), whole pattern set (1nn-patterns), representations (1nn-centr), and combined lsvm classifier with 1nn-centr according to formula (9).

not applicable in on-line processing due to computation cost as opposed to 1nn-centr, which is quite fast, however, it does not give the expected results (standard deviation of the precision is high). 1nn-centr* gives more stable results which are better or, at least, not worse than features-based classifiers. An interesting case is Figure 6 (e), where 1nn-centr* is the best.

7 Conclusion & Future Work

In this paper, we proposed a framework oriented toward on-line activity recognition. We compared two approaches for this task using readings from accelerometers mounted on different parts of the body. Our experimental results shows that hybrid approach (using feature-based and subsequent matching approach) is promising for activity reporting.

This work is a part of larger project called *ICRA*[1], in which it is a submodule responsible for fire fighters activity reporting. One of the goal of the ICRA system is to assess risks at the emergency scene. In the most cases the risk is related to the activities performed by fire fighters. Therefore, the activity recognition plays a pivotal rôle in risk assessment [14].

As part of the further work we would like to prepare a set of data, which will be held during simulated operations with the encounter of fire. Subsequently, we will also test the way in which our framework performs for single-event activities, like, for example, operating with tools and environment.

Acknowledgement. The research was supported by Polish National Science Centre (NCN) grants DEC-2011/01/B/ST6/03867 and DEC-2012/05/B/ST6/03215, and by the Polish National Centre for Research and Development (NCBiR) — Grant No. O ROB/0010/03/001 in the frame of Defence and Security Programmes and Projects: "Modern engineering tools for decision support for commanders of the State Fire Service of Poland during Fire&Rescue operations in the buildings".

References

1. Akl, A., Valaee, S.: Accelerometer-based gesture recognition via dynamic-time warping, affinity propagation, & compressive sensing. In: IEEE ICASSP, pp. 2270–2273. IEEE (2010)
2. Alvarez, D., González, R.C., López, A., Alvarez, J.C.: Comparison of step length estimators from wearable accelerometer devices. In: 28th Annual International Conference of the IEEE, EMBS, pp. 5964–5967. IEEE (2006)
3. Auvinet, B., Berrut, G., Touzard, C., Moutel, L., Collet, N., Chaleil, D., Barrey, E.: Reference data for normal subjects obtained with an accelerometric device. Gait & Posture 16(2), 124–134 (2002), `http://www.sciencedirect.com/science/article/pii/S096663620100203X`
4. Baek, J., Lee, G., Park, W., Yun, B.-J.: Accelerometer signal processing for user activity detection. In: Negoita, M.G., Howlett, R.J., Jain, L.C. (eds.) KES 2004. LNCS (LNAI), vol. 3215, pp. 610–617. Springer, Heidelberg (2004)
5. Bouten, C.V., Koekkoek, K.T., Verduin, M., Kodde, R., Janssen, J.D.: A triaxial accelerometer and portable data processing unit for the assessment of daily physical activity. IEEE Transactions on Biomedical Engineering 44(3), 136–147 (1997)
6. Foerster, F., Smeja, M., Fahrenberg, J.: Detection of posture and motion by accelerometry: A validation study in ambulatory monitoring. Computers in Human Behavior 15(5), 571–583 (1999)
7. Gafurov, D., Bours, P., Snekkenes, E.: User authentication based on foot motion. Signal, Image and Video Processing 5(4), 457–467 (2011)
8. Gafurov, D., Helkala, K., Søndrol, T.: Biometric gait authentication using accelerometer sensor. Journal of Computers 1(7), 51–59 (2006)

[1] `www.icra-project.org`

9. Guralnik, V., Srivastava, J.: Event detection from time series data. In: Proceedings of the Fifth ACM SIGKDD International Conference on Knowledge Discovery and Data Mining, pp. 33–42. ACM (1999)
10. Han, T.S., Ko, S.-K., Kang, J.: Efficient subsequence matching using the longest common subsequence with a dual match index. In: Perner, P. (ed.) MLDM 2007. LNCS (LNAI), vol. 4571, pp. 585–600. Springer, Heidelberg (2007)
11. Jahankhani, P., Kodogiannis, V., Revett, K.: EEG signal classification using wavelet feature extraction and neural networks. In: IEEE JVA, pp. 120–124 (2006)
12. Junker, H., Amft, O., Lukowicz, P., Tröster, G.: Gesture spotting with body-worn inertial sensors to detect user activities. Patt. Recog. 41(6), 2010–2024 (2008)
13. Keogh, E., Ratanamahatana, C.A.: Exact indexing of dynamic time warping. Knowledge and Information Systems 7(3), 358–386 (2005)
14. Krasuski, A., Jankowski, A., Skowron, A., Slezak, D.: From sensory data to decision making: A perspective on supporting a fire commander. In: IEEE/WIC/ACM International Joint Conferences on Web Intelligence (WI) and Intelligent Agent Technologies (IAT), vol. 3, pp. 229–236. IEEE (2013)
15. Lee, Y.-S., Cho, S.-B.: Activity recognition using hierarchical Hidden Markov Models on a smartphone with 3D accelerometer. In: Corchado, E., Kurzyński, M., Woźniak, M. (eds.) HAIS 2011, Part I. LNCS, vol. 6678, pp. 460–467. Springer, Heidelberg (2011)
16. Mannini, A., Sabatini, A.M.: Machine learning methods for classifying human physical activity from on-body accelerometers. Sensors 10(2), 1154–1175 (2010)
17. Mayagoitia, R.E., Nene, A.V., Veltink, P.H.: Accelerometer and rate gyroscope measurement of kinematics: An inexpensive alternative to optical motion analysis systems. Journal of Biomechanics 35(4), 537–542 (2002)
18. Meijer, G.A.L., Westerterp, K.R., Verhoeven, F.M.H., Koper, H.B.M., ten Hoor, F.: Methods to assess physical activity with special reference to motion sensors and accelerometers. IEEE Trans. on Biomedical Engineering 38(3), 221–229 (1991)
19. Preece, S.J., Goulermas, J.Y., Kenney, L.P.J., Howard, D., Meijer, K., Crompton, R.: Activity identification using body-mounted sensors' review of classification techniques. Physiological Measurement 30(4), R1 (2009)
20. Pylvänäinen, T.: Accelerometer based gesture recognition using continuous HMMs. In: Marques, J.S., Pérez de la Blanca, N., Pina, P. (eds.) IbPRIA 2005. LNCS, vol. 3522, pp. 639–646. Springer, Heidelberg (2005)
21. Ravi, N., Dandekar, N., Mysore, P., Littman, M.L.: Activity recognition from accelerometer data. In: AAAI, vol. 5, pp. 1541–1546 (2005)
22. Sant'Anna, A., Wickstrom, N.: Developing a motion language: Gait analysis from accelerometer sensor systems. In: 3rd International Conference on Pervasive Computing Technologies for Healthcare, pp. 1–8. IEEE (2009)
23. Shin, S.H., Park, C.G., Kim, J.W., Hong, H.S., Lee, J.M.: Adaptive step length estimation algorithm using low-cost MEMS inertial sensors. In: IEEE SAS, pp. 1–5. IEEE (2007)
24. Wang, W., Guo, Y., Huang, B., Zhao, G., Liu, B., Wang, L.: Analysis of filtering methods for 3D acceleration signals in body sensor network. In: ISBB, pp. 263–266 (November 2011)
25. Weiss, A., Herman, T., Plotnik, M., Brozgol, M., Maidan, I., Giladi, N., Gurevich, T., Hausdorff, J.M.: Can an accelerometer enhance the utility of the Timed Up & Go Test when evaluating patients with Parkinson's disease? Medical Engineering & Physics 32(2), 119–125 (2010)

Privacy-Preserving Emotion Detection for Crowd Management

Zeki Erkin[1], Jie Li[2], Arnold P.O.S. Vermeeren[2], and Huib de Ridder[2]

[1] Cyber Security Group, Department of Intelligent Systems, Delft University of Technology, 2628 CD, Delft, The Netherlands
[2] Persuasive Experience Research, Industrial Design Engineering, Delft University of Technology, 2628 CD, Delft, The Netherlands
{z.erkin,j.li-2,a.p.o.s.vermeeren,h.deridder}@tudelft.nl

Abstract. Emotion detection plays a vital role in crowd management as it enables social event organizers to detect the actions of masses and react accordingly. There are several approaches to detect emotions in a crowd, including surveillance cameras, human observers and sensors. One other approach to gather emotion data is self-reporting. A recent study showed that self-reporting is feasible, reliable and efficient. However, there is a strong privacy concern among people that risks the use of such self-reporting mechanisms in wide use. In this work, we address the privacy aspect of self-reporting mechanism and propose a cryptographic approach that hides the sensitive data from the organizers but permits to compute statistical data for crowd management. The feasibility of using cryptography in real life for privacy protection is also investigated in terms of complexity.

Keywords: Emotion detection, crowd management, privacy, homomorphic encryption.

1 Introduction

Crowd management has become an important aspect of today's social life. Particularly in events with massive attendance, safety, security and guidance of the attendants are very important aspects. There are three phases of crowd management, before the event, during the event and after the event. Different parties like the event organizers, medical health support, fire department and the security team plan each step of the event carefully to mitigate possible problems that may arise.

Among many aspects of this procedure, measuring crowd emotion during the event has many valuable insights for the management of the crowd as well as assessing the success of the event afterwards. Emotions influence and serve as a predictor for human behavior [1]. As noted by Arnold in [2], emotions can be essentially characterized as "felt action tendencies", which could be understood as impulses that motivate people to move towards the stimulus appraised as beneficial and avoid the one appraised as harmful. That is to say, through

D. Ślęzak et al. (Eds.): AMT 2014, LNCS 8610, pp. 359–370, 2014.

understanding crowd emotions, crowd managers can judge crowd members intentions and predict their behaviors so as to act accordingly.

A recent study in this field presented a self-reporting mechanism to collect data from the attendants of an event [3]. In that work, the authors designed a non-intrusive software application for mobile phones to gather data that consist of a unique identifier for the device, location of the device, the emotion of the attendant and the perceived emotion of the crowd in proximity of the attendant, and drew interesting conclusions in terms of real-time crowd emotion maps based on self-report emotions from different areas of the event. The collected data reflected the real situation as we observed. In general, the amount of emotion reports in an area reflected the crowdedness of that area. Attendants' movement and emotional changes were consistent with the activities at the event. For example, when the performance of a stage stopped, it is observed that the amount of emotion reports declined in that area. More negative emotions popped up in an area on the emotion map when we received some spontaneous complaints from the attendants about the unsatisfactory performance at that area. Real-time crowd emotion maps do not only provide information about crowd size and density, but also rich emotion information for crowd managers to predict crowd behavior and prepare for the possible incidents. In addition, attendants' emotion reaction to the activities in the event can be used as an indicator for evaluating the event.

Among many aspects of crowd emotion detection, the authors point out issues related to trust and privacy as a serious concern. It is reported that many attendants were concerned about a number of possible misuse cases. Two most relevant concerns for this work are as follows:

- *Tracking.* Individuals can be tracked during and after the event using the software.
- *Identity linking.* Individuals and their emotion feedback can be linked and used against them.

Particularly, the identity linking problem observed to be a serious concern that demotivates the use of the software. This is a valid concern, considering that the emotion detection can be performed for any kind of events like political congresses, protests, uprisings, and riots as seen in many countries recently. This serious concern on privacy also clearly indicates that without proper security and privacy mechanisms employed, emotion detection for crowd management using software and even sensor technology can be perceived as a privacy invasion and not be accepted by the individuals. Obviously, for the better management of the crowd and for reducing the security and safety risks, a proper privacy protection mechanism should be provided to establish trust among people.

In this paper, we address the identity linking problem of the emotion detection and propose a scientific solution to protect the privacy of individuals. We define two entities: 1) a server, which collects data or the crowd management, and 2) users, whose data are collected using the software in [3]. Our goal is to protect the privacy of the users by hiding the sensitive emotional data from the server. While the server cannot observe the emotions of any user, it can still draw conclusions based on statistical data such as the histogram of the emotional data for a specific location. We achieve our goal by deploying techniques known from cryptography. More precisely, the privacy sensitive data of the users are only given to the server in the encrypted form. Without having the decryption key, the server can still process the encrypted data for crowd management.

Our contributions are as follows:

1. We present a self-reporting mechanism for emotion detection that is privacy-preserving in a server-client model.
2. We propose a cryptographic protocol based on existing tools and optimized in terms of bandwidth using data packing [4].
3. We provide a complexity analysis to show that the proposed cryptographic protocol is feasible to be deployed in real life.

The rest of the paper is organized as follows: Section 2 presents the related work in this field. Section 3 provides background information for emotion detection software from [3], as well as the cryptographic tools that are used in this paper. Section 4 describes the privacy-preserving emotion detection protocol. Section 5 presents security and complexity analyzes. Section 6 provides a number of open questions for further research. Finally, Section 7 draws conclusions.

2 Related Work

To the best of our knowledge, privacy for emotion detection has not been addressed before in literature. Protecting privacy sensitive data, on the other hand, is a well-known topic and studied in different perspectives. Anonymization and perturbation techniques have been used in data mining [5,6]. Unfortunately, anonymization is not a possible solution for our case since the mobile devices can easily be identified by the server. Data perturbation techniques are not suitable either as they provide privacy at the cost of adding noise to the original data, which is not desirable in our scenario. Another approach in literature provides a way to perform the desired service based on privacy-sensitive data using cryptography [7]. This approach is based on cryptographic tools like homomorphic encryption [8] and multi-party computation techniques [9].

The main idea in cryptographic approach is to hide the sensitive data using encryption. The encrypted data is given to the server, which does not have the decryption key, and yet it can process the encrypted data using the homomorphic properties of the encryption scheme. This line of research has been applied effectively in many different domains, including but not limited to e-voting [10], biometric-medical data processing [11,12], recommender systems [13,14] and data clustering [15,16].

3 Preliminaries

In this section, we discuss the emotion detection in [3], briefly introduce the cryptographic tools we use, and present our security and privacy requirements.

3.1 Emotion Detection

We assume that user i sends the tuple T to the server using the software in [3]:

$$T = < ID, t, \ell, e_i, c_i > , \tag{1}$$

where ID is the unique mobile phone identifier, t is the time of the report, ℓ is the code for the location, e_i is the user emotion, and c_i is the user's observation on the crowd. The mobile software works as follows. After installing the application, the software determines the location of the user. However, the measurement is not precise due to technological challenges and thus, the user is asked to state their location –one of the six stages– as a precaution. The software, then, asks every half an hour about the emotional state of the user and the perceived emotional state of the crowd around that user. The reporting mechanism is designed carefully with an circular emotion detector, corresponding colors for emotions and cartoon characters. To motivate self-reporting, a game component is also added to the application where a number of reports are rewarded later. The collected data is then processed for analysis; an example is given in Figure 1. We refer readers to the original work [3] on the design of user interface and the details about the data collection.

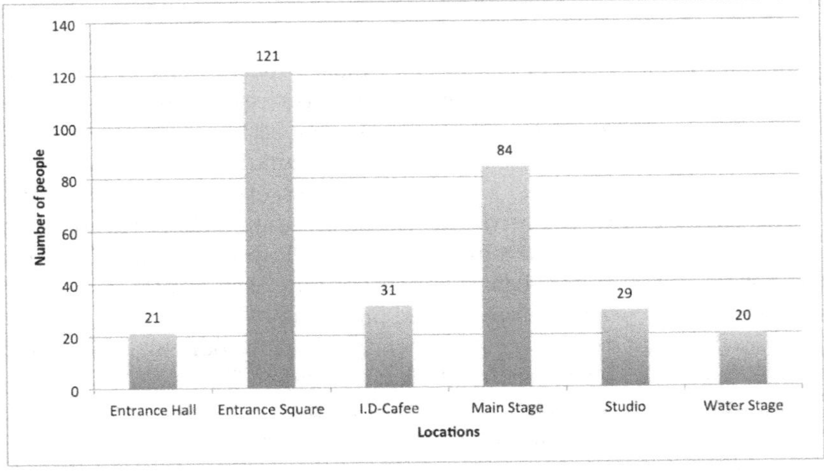

Fig. 1. One of the analysis from [3]: Amount of reports in six areas

3.2 Homomorphic Encryption

Encryption schemes are designed to make messages unreadable for anyone who does not have the decryption key. The output of an encryption function, called cipher text, looks completely random and thus, it is not possible to deduce meaningful information from it. Public key encryption schemes also output cipher text that requires the decryption key of the recipient to obtain the plain text message. Even though it is not feasible to decipher the cipher text without the decrytion key, public key encryption schemes do preserve some structure after encryption that we can exploit. For example, Paillier encryption scheme [17] is additively homomorphic, meaning that if two cipher text are given, it is possible to obtain an encryption of the sum of original messages by simply multiplying their cipher texts. Given two messages, m_1 and m_2,

$$\mathcal{D}_{sk}(\mathcal{E}_{pk}(m_1) \cdot \mathcal{E}_{pk}(m_2) = m_1 + m_2) , \tag{2}$$

where $\mathcal{E}_{pk}(\cdot)$ and $\mathcal{D}_{sk}(\cdot)$ are encryption and decryption functions with public key and decryption key, respectively. Consequently, any plain text m can also be scaled with a public constant c due to the homomorphic property:

$$\mathcal{D}_{sk}(\mathcal{E}_{pk}(m)^c) = mc . \tag{3}$$

By using the additively homomorphic property, it is possible to realize linear functions using only encrypted inputs. In addition to homomorphism, the Paillier encryption scheme is also semantically secure. That is there is a random factor in the encryption function, which causes different cipher texts even the same message is encrypted. We refer interested user to [17] for more information on the encryption scheme.

In this paper, we use the threshold version of the Paillier encryption scheme [18]. The threshold version of the scheme is similar to the original one however, in order to decrypt a cipher text, a pre-defined number of users should participate. This means that each user in the system has a share of the decryption key and only with the contribution of any of the pre-defined number of users, the cipher text can be decrypted.

3.3 Privacy Requirements and Security Assumptions

In this work, we assume that the software transmits a unique identifier for each mobile device, along with the location and time information, user's emotion data, and the crowd emotion. Our aim is to hide the last three from the server, while it is still possible for the server to compute a number of functions on the encrypted versions of the data:

1. Total number of users for a specific location,
2. Emotion distribution of users in each location.

We assume that both the server and the users are honest-but-curious, a notion used in cryptography, meaning that they perform tasks according to the protocol description but they are curious at the same time and they might try to extract more information than they are entitled to from the communication.

4 Private Emotion Detection

In this section we describe the cryptographic protocol for privacy-preserving emotion detection. We assume that every mobile device has a unique identifier, assigned by the software, and a Paillier public key pk and a corresponding decryption key share sk_i. We design our system for an event with N participants, where there are L different locations and E number emotions defined.

In our proposal, the server is not trusted with the raw data of the users. Furthermore, the mobile devices are considered to be limited, and thus the number of operations in terms of encryption and decryption should be kept minimum. In the following, we describe data formatting and collection, both of which are handled locally by the mobile devices, and data analysis, which is done by the server.

4.1 Data Formatting

The software formats the data before sending it to the server as follows: the location data is represented as a binary value which is constructed with L compartments of size $\log_2(N)$-bits:

$$\ell_i = (\ell_{i,1}||\ell_{i,2}||\ldots||\ell_{i,L}) \ , \tag{4}$$

where $||$ is the concatenation operator. The software sets the value to 1 for the location of the user, and all others to 0. In a similar fashion, it also creates the following values:

$$\begin{aligned} e_i &= (e_{i,1}||e_{i,2}||\ldots||e_{i,E}) \ , \\ c_i &= (c_{i,1}||c_{i,2}||\ldots||c_{i,E}) \ , \end{aligned} \tag{5}$$

where e_i is the emotion indicator for user i and c_i is the emotion indicator of the crowd around user i. Only the value $e_{i,j}$ is set to 1 if the user's emotion code is j. This is also true for c_i. These two values are consisting of E compartments of size $\log_2(E)$ bits.

In addition to above values, the software also computes the following:

$$\begin{aligned} \tilde{e}_i &= \ell_i \times e_i \ , \\ \tilde{c}_i &= \ell_i \times c_i \ . \end{aligned} \tag{6}$$

Notice that in addition to the information sent in the original work, we now have two new values, $\tilde{e}_i$ and $\tilde{c}_i$. The former indicates the emotion of the user i at his/her current location, the later indicates the crowd emotion at that location as perceived by user i. These two values are also consist of L compartments; $L-1$ of them having $\log_2(N)$ zeros.

4.2 Data Collection

The software sends the tuple T' to the server at time slot t:

$$T' =< ID_i, t, \mathcal{E}_{pk}(\ell_i), \mathcal{E}_{pk}(e_i), \mathcal{E}_{pk}(c_i), \mathcal{E}_{pk}(\tilde{e}), \mathcal{E}_{pk}(\tilde{c_i}) > \ . \tag{7}$$

While the first 5 elements of the tuple are same as it is defined in the original work, we send the encrypted versions. We also send two additional piece of information for the ease of computation on the server side: $\mathcal{E}_{pk}(\tilde{e}), \mathcal{E}_{pk}(\tilde{c_i})$. These terms actually define the user and crowd emotion per location, respectively. Note that each encryption is indistinguishable than the others as the encryption scheme is semantically secure as explained in Section 3.2.

4.3 Data Analysis

Upon receiving the tuple, the server can perform the following functions on the encrypted data. We assume that $M < N$ users sent their data.

1. The total number of people in each location:

$$\mathcal{E}_{pk}(\ell) := \mathcal{E}_{pk}(\sum_{i=1}^{M} \ell_i) = \prod_{i=1}^{M} \mathcal{E}_{pk}(\ell_i) . \tag{8}$$

The resulting ℓ is in the form of integer values for each location in a concatenated form:

$$\ell = (\sum_{i=1}^{M} \ell_{i,1} || \sum_{i=1}^{M} \ell_{i,2} || \dots || \sum_{i=1}^{M} \ell_{i,L}) . \tag{9}$$

Recall that the server has ℓ in the encrypted form, thus he cannot access its content.

2. The total number of users per each emotion:

$$\mathcal{E}_{pk}(e) := \mathcal{E}_{pk}(\sum_{i=1}^{M} e_i) = \prod_{i=1}^{M} \mathcal{E}_{pk}(e_i) , \tag{10}$$

where e is

$$e = (\sum_{i=1}^{M} e_{i,1} || \sum_{i=1}^{M} e_{i,2} || \dots || \sum_{i=1}^{M} e_{i,E}) .$$

3. The emotion distribution per location for the users:

$$\mathcal{E}_{pk}(\tilde{e}) := \mathcal{E}_{pk}(\sum_{i=1}^{M} \tilde{e}_i) = \prod_{i=1}^{M} \mathcal{E}_{pk}(\tilde{e}_i) . \tag{11}$$

4. The emotion distribution per location for the crowd:

$$\mathcal{E}_{pk}(\tilde{c}) := \mathcal{E}_{pk}(\sum_{i=1}^{M} \tilde{c}_i) = \prod_{i=1}^{M} \mathcal{E}_{pk}(\tilde{c}_i) . \tag{12}$$

Figure 2 illustrates the main idea of computing $\tilde{e}$ with an example, ignoring the encryption. In this example, we assume that there are 2 locations and 3 emotions. Clearly, ℓ gives the number of people in each location, e gives the total number of people for each emotion. However, $\tilde{e}$, provides more detailed

$$\begin{array}{l}
\ell_1 = (1||0) \quad \ell_2 = (1||0) \quad \ell_3 = (0||1) \\
e_1 = (1||0||0) \; e_2 = (0||0||1) \; e_3 = (0||0||1) \\
\hline
\ell = \ell_1 + \ell_2 + \ell_3 = (2||1) \\
e = e_1 + e_2 + e_3 = (1||0||2) \\
\hline
\tilde{e}_1 = \ell_1 \times e_1 = (1|0|0||0|0|0) \\
\tilde{e}_2 = \ell_2 \times e_2 = (0|0|1||0|0|0) \\
\tilde{e}_3 = \ell_3 \times e_3 = (0|0|0||0|0|1) \\
\hline
\tilde{e} = \tilde{e}_1 + \tilde{e}_2 + \tilde{e}_3 = (\underbrace{1|0|1}_{location\ 1} \; || \; \underbrace{0|0|1}_{location\ 2})
\end{array}$$

Fig. 2. Illustration of the output of data analysis

information on emotions per location: there are 2 people in the first location with emotions 1 and 3, and only one person in location 2 with emotion 2. $\tilde{c}$ can be visualized in the same manner.

Recall that none of the above values can be obtained in plain text by the server as it does not have the decryption key. In order to decrypt the cipher texts, the users run a joint decryption protocol. This means that users also receive messages from the server. At the end of the decryption protocol, the server obtains all the statistical data for that specific time interval.

5 Analysis

In this section, we provide an informal discussion on our proposed method for emotion detection and give complexity analysis, which is important for the deployment of the protocol for real life.

5.1 Security

In this section, we provide an informal discussion on the security of our proposal. Recall that we assume all involved parties, the server and the users, are honest-but-curious. Our goal is to hide the individual emotion data from the server. It is clear that as long as the underlying cryptographic primitive, that is the threshold encryption scheme, is secure, the server cannot learn the content of the encrypted messages from the users. As the encryption scheme is also semantically secure, the server cannot distinguish cipher text, even if the same value is encrypted by different users. The users, on the other hand, do not obtain additional information, for example about other users' emotions, other than the aggregated data that may be broadcasted by the server after the joint decryption protocol.

Even though, our protocol is conceptually simple, and provably secure, there are a number of possibilities for the server to extract more information. These actions are not supposed to be performed under our security assumption, nevertheless we would like to address them here.

In the first type of attack, the server can present a specific encrypted message of a user as the aggregated data and ask users to decrypt it. These users can avoid this attack by performing only one decryption per time slot. However, remember that the joint decryption protocol is performed by a subset of users and thus, the server can ask users from the remaining users to decrypt the encrypted message. This can be avoided by setting the number of users required in the threshold decryption to $N/2 + 1$ users. By this way, there will be overlapping users for the decryption so that these users can check whether there are more than one decryption per time slot or not.

If such malicious acts are expected from the server, a better approach is to use zero-knowledge proofs (ZKP) for the verification of the performed actions of the server. However, using ZKPs are costly in terms of computation and communication, and can be overwhelming for the mobile device owners.

5.2 Complexity

Our proposal is designed to have as minimum number of operations as possible on the user mobile device. We assume that the operations on the plain text data is negligible compared to the operations on the encrypted data. Therefore, in Table 1, we only provide the complexity of operations in the encrypted domain and the amount of encrypted data transmission. Note that multiplication and exponentiation are over modulo n^2, where n is a very large number.

Table 1. Complexity analysis per operation

	Server	**User**
Encryption	-	$\mathcal{O}(1)$
Decryption	-	$\mathcal{O}(1)$
Multiplication	$\mathcal{O}(M)$	-
Exponentiation	-	-
Data	$\mathcal{O}(1)$	$\mathcal{O}(1)$

As seen in Table 1, the complexity is reasonable low. Each user encrypt 5 messages if s/he reports emotion, and participates in 1 decryption per time slot. The server, on the other hand, computes the aggregated data by performing $5M$ multiplications in each time slot. Note that $M = N$ in the worst-case scenario, since there will be less people reporting their emotions per time slot. As for the data transmission, each user sends 5 encryption to the server and server broadcasts 5 cipher texts for decryption. In summary, the overall protocol is quite efficient in terms of computation and communication.

6 Discussion

In this section, we address a number of open issues in terms of privacy preserving emotion detection.

6.1 Application Scenario

The protocol described previously protects the private data of the participants, while it is still feasible for the server to analyze the data for crowd management. However, there is a strong dependency on the involvement of people in the protocol, particularly for the decryption of the encrypted values. This fact introduces two challenges:

1. **On-line processing only.** It is not reasonable to expect people to use the software after the event. Therefore, all measurements should be performed in real time. This means that the duration of collecting data has to be determined carefully. Short intervals can be overwhelming for the users in terms of processing, and long intervals cannot provide useful information.
2. **Time constraint.** Consequently, it is also essential to associate each encryption with the corresponding time-slot so that the server cannot combine encrypted messages from different time slots and ask users to decrypt for deducing more information than it should have.

In the following, we address these two challenges and also provide a direction for off-line processing.

On-line Processing with Time Constraint. Our proposal requires 2 round of communication between the server and the users: 1) the users sends their data to the server, and 2) the server runs the joint decryption protocol to obtain the aggregated data. As noted before, a user can send emotion data at any time during the event. However, it is essential to have the state of the crowd for a given time interval. Therefore, we assume that the event is divided into certain time slots and for each time interval a different generator for the encryption is used. Using different generators for each time slot guarantees that only the data provided for that time slot are aggregated by the server.

Off-line Processing. In a server-client model, it is not possible to process the collected data without the help of the users. Unfortunately, in our application scenario, users go off-line after the event. To be able to process data after the event, a third semi-trusted entity (STE) is required. Assuming that such an entity exists, we have two options to be able to process data off-line:

1. The users send their partial keys to the STE at the end of the event so that for any computations can be performed with its help afterwards.
2. The protocol can be changed in such a way that users submit their data encrypted using the public key of the STE. In that approach, the users are not required to participate in any computations. The server and the STE, on the other hand, run a similar protocol to obtain the aggregated emotional data.

In either case above, there is a strong assumption that the server and the STE are not colluding, that is they act according to the protocol and they do not co-operate to reveal the privacy-sensitive data of the users.

6.2 Location Privacy

Even though our protocol is secure and privacy-preserving, it only allows the server to deduce statistical data. However, an essential component in crowd management is emotion maps, where users are monitored in space and time. This is very important especially for emergency procedures and security countermeasures. Unfortunately, determining the exact location of every single individual along with their emotion state for crowd management creates a trade-off in terms of privacy protection. Although there are ways to hide the location of users in a crowd, for example using Mixnets [19], it becomes impossible to create emotion maps, which require the exact location of the users. Therefore, we envision that it is not possible at the moment to hide both the emotion state and the location of the users for crowd management.

7 Conclusions

Emotion detection for crowd management presents itself as a powerful tool to understand the state of the people. Based on the information gathered in real time, authorities can have a clear idea about the event and react fast to sudden changes in the crowd. In an ideal case, emotion detection should be transparent to the people. However, due to technological challenges, it is not feasible to collect reliable emotion data without user's participation. Therefore, self-reporting tools have been developed for creating emotion maps with the help of people participating in an event. Unfortunately, without any privacy protection mechanism, it is not desirable for people to use such self-reporting tools. In this paper, we present a way to protect the privacy-sensitive data, in this case emotions, from the server, which would like to process for crowd management. Our proposal is to use cryptography to hide the private data and enable the server only obtain aggregated data. We achieve this goal by using cryptographic tools such as homomorphic encryption and increase the efficiency of the system by employing data packing technique. The resulting protocol is quite efficient to be used in real systems as it is shown in the complexity analysis.

References

1. Levenson, R.W.: The intrapersonal functions of emotion. Cognition & Emotion 13(5), 481–504 (1999)
2. Arnold, M.B.: Emotion and Personality: Psychological Aspects, vol. 1. Colombia University Press, New York (1960)
3. Li, J., Erkin, Z., de Ridder, H., Vermeeren, A.: A field study on real-time self-reported emotions in crowds. In: Proceedings of ICT OPEN 2013, Eindhoven, The Netherlands (2013)
4. Bianchi, T., Piva, A., Barni, M.: Composite signal representation for fast and storage-efficient processing of encrypted signals. IEEE Transactions on Signal Processing (2009)

5. Agrawal, R., Srikant, R.: Privacy-preserving data mining. In: SIGMOD 2000: Proceedings of the 2000 ACM SIGMOD International Conference on Management of Data, vol. 29(2), pp. 439–450. ACM Press, New York (2000)
6. Lindell, Y., Pinkas, B.: Privacy preserving data mining. Journal of Cryptology, pp. 36–54 (2000)
7. Lagendijk, R.L., Erkin, Z., Barni, M.: Encrypted signal processing for privacy protection: Conveying the utility of homomorphic encryption and multiparty computation. IEEE Signal Process. Mag. 30(1), 82–105 (2013)
8. Melchor, C.A., Fau, S., Fontaine, C., Gogniat, G., Sirdey, R.: Recent advances in homomorphic encryption: A possible future for signal processing in the encrypted domain. IEEE Signal Process. Mag. 30(1), 108–117 (2013)
9. Goldreich, O.: Foundations of Cryptography II. Cambridge University Press (2004)
10. Hirt, M.: Receipt-free k-out-of-l voting based on elgamal encryption. In: Chaum, D., Jakobsson, M., Rivest, R.L., Ryan, P.Y.A., Benaloh, J., Kutylowski, M., Adida, B. (eds.) Towards Trustworthy Elections. LNCS, vol. 6000, pp. 64–82. Springer, Heidelberg (2010)
11. Barni, M., Failla, P., Lazzeretti, R., Sadeghi, A.R., Schneider, T.: Privacy-preserving ecg classification with branching programs and neural networks. IEEE Transactions on Information Forensics and Security 6(2), 452–468 (2011)
12. Erkin, Z., Franz, M., Guajardo, J., Katzenbeisser, S., Lagendijk, I., Toft, T.: Privacy-preserving face recognition. In: Goldberg, I., Atallah, M.J. (eds.) PETS 2009. LNCS, vol. 5672, pp. 235–253. Springer, Heidelberg (2009)
13. Erkin, Z., Veugen, T., Toft, T., Lagendijk, R.L.: Generating private recommendations efficiently using homomorphic encryption and data packing. IEEE Transactions on Information Forensics and Security 7(3), 1053–1066 (2012)
14. Kononchuk, D., Erkin, Z., van der Lubbe, J.C.A., Lagendijk, R.L.: Privacy-preserving user data oriented services for groups with dynamic participation. In: Crampton, J., Jajodia, S., Mayes, K. (eds.) ESORICS 2013. LNCS, vol. 8134, pp. 418–442. Springer, Heidelberg (2013)
15. Jagannathan, G., Wright, R.N.: Privacy-preserving distributed k-means clustering over arbitrarily partitioned data. In: KDD, pp. 593–599 (2005)
16. Beye, M., Erkin, Z., Lagendijk, R.L.: Efficient privacy preserving k-means clustering in a three-party setting. In: IEEE Workshop on Information Forensics and Security, pp. 1–6 (2011)
17. Paillier, P.: Public-Key Cryptosystems Based on Composite Degree Residuosity Classes. In: Stern, J. (ed.) EUROCRYPT 1999. LNCS, vol. 1592, pp. 223–238. Springer, Heidelberg (1999)
18. Damgård, I.B., Nielsen, J.B.: Universally composable efficient multiparty computation from threshold homomorphic encryption. In: Boneh, D. (ed.) CRYPTO 2003. LNCS, vol. 2729, pp. 247–264. Springer, Heidelberg (2003)
19. Chaum, D.: Untraceable electronic mail, return addresses, and digital pseudonyms. Commun. ACM 24(2), 84–88 (1981)

Cellular Automaton Evacuation Model Coupled with a Spatial Game

Anton von Schantz and Harri Ehtamo

Systems Analysis Laboratory, Aalto University School of Science
P.O. Box 11100, FI-00076 Aalto, Finland
{anton.von.schantz,harri.ehtamo}@aalto.fi

Abstract. For web-based real-time safety analyses, we need computationally light simulation models. In this study, we develop an evacuation model, where the agents are equipped with simple decision-making abilities. As a starting point, a well-known cellular automaton (CA) evacuation model is used. In a CA, the agents move in a discrete square grid according to some transition probabilities. A recently introduced spatial game model is coupled to this CA. In the resulting model, the strategy choice of the agent determines his physical behavior in the CA. Thus, our model offers a game-theoretical interpretation to the agents' movement in the CA.

Keywords: Real-time; evacuation simulation; cellular automaton; spatial game.

1 Introduction

To avoid losses, e.g., in evacuation situations, the rescuing authorities should make timely and accurate decisions. A successful operation requires real-time safety analysis to forecast various disasters and accidents that may take place in events involving human crowds. Thus, safety simulations should be computationally light enough to run in real-time, e.g., in the internet. Recent research sites aiming at these goals are [17, 18].

Our ultimate goal is to create a computationally light evacuation simulation model suited for web-based real-time analyses. Our focus in this paper is on two computational evacuation models: the cellular automaton (CA) model [7–9] and the social-force model [10]. FDS+Evac is a validated evacuation simulation software based on the social-force model [6]. In FDS+Evac, the agents' exit selection is modeled using optimization and game theory [2].

Computationally very light CA model is especially suitable to simulate moving agents in traffic jams and evacuation situations. Hence, it could be used to develop web-based tools to simulate these matters as well. Although, agent movement in the CA model is rather realistic resembling granular flow, it lacks agents' explicit decision-making abilities. In CA the agents move according to some transition probabilities defined by the so called static and dynamic floor fields. The influence of the floor fields on the transition probabilities depend on

D. Ślęzak et al. (Eds.): AMT 2014, LNCS 8610, pp. 371–382, 2014.

two parameters, or coupling constants, resulting in different behaviors of the crowd.

So far, in the CA literature [11–15], game theory has been used to solve a conflict situation, i.e., a situation where several agents try to move simultaneously to the same cell.

In this paper, we couple the spatial game defined in [5] to CA. In our approach, each agent plays the Hawk-Dove game in his neighborhood leading to two types of strategies for each agent described by two possible values of coupling constants. In our model, the agent does not just choose his strategy when in conflict, but optimizes it constantly to minimize his evacuation time.

2 Cellular Automaton Model

The agents' movement is simulated with a CA introduced by Schadschneider et al. [9]. Next, we give brief overview of the CA model. In the model, the agents are located in a room divided into cells, so that a single agent occupies a single cell. At each time step of the simulation, the agent can move to one of the unoccupied cells orthogonally next to him, i.e., in the *Moore neighborhood*, where the transition probabilities associated with the diagonal cells are set to zero.[1]

2.1 Movement in the CA

The transition probabilities depend on the values of the static and dynamic floor field in the cells. The *static floor field* S is based on the geometry of the room. The values associated with the cells of S increase as we move closer to the exit, and decrease as we move closer to the walls. On the other hand, the *dynamic floor field* D represents *virtual paths* left by the agents. An agent leaving a cell, causes the value of D in that cell to increase by one unit. Over time, the virtual path decays and diffuses to surrounding cells. The values of the fields D and S are weighted with two *coupling constants* $k_D \in [0, \infty)$ and $k_S \in [0, \infty)$.

Now, for each agent, the transition probabilities p_{ij}, for a move to a neighbor cell (i, j) are calculated as follows

$$p_{ij} = N e^{k_D D_{ij}} e^{k_S S_{ij}} (1 - \xi_{ij}), \tag{1}$$

where

$$\xi_{ij} = \begin{cases} 1 & \text{for forbidden cells (walls and occupied cells)} \\ 0 & \text{else} \end{cases}$$

and the normalization

$$N = \left[\sum_{(i,j)} e^{k_D D_{ij}} e^{k_S S_{ij}} (1 - \xi_{ij}) \right]^{-1}.$$

[1] Also called *von Neumann neighborhood.*

The agents' desired movement directions are updated with a *parallel update scheme*, i.e., the directions are updated simultaneously for all agents. In a *conflict situation*, i.e., a situation where several agents try to occupy the same cell, all the agents are assigned equal probabilities to move, and with probability $1 - \mu$ one of the agents is allowed to move to the desired cell. Here, $\mu \in [0, 1]$ is a friction parameter, illustrating the internal pressure caused by conflicts. The impact of the friction parameter is depicted in Figure 1.

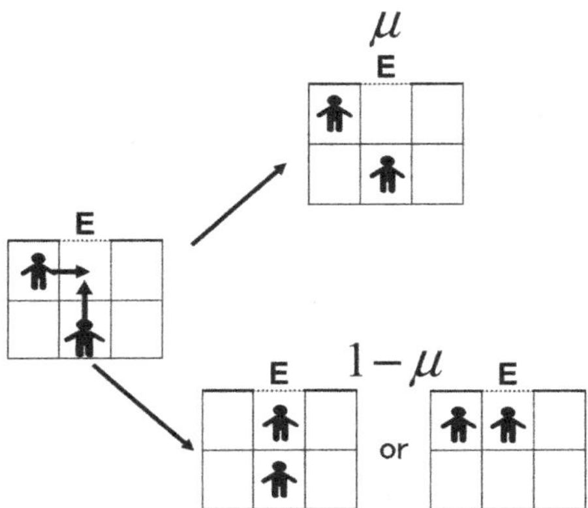

Fig. 1. The impact of friction parameter on the agents movement. With probability μ neither of the agents get to move, and with probability $1 - \mu$ the other agent moves. Here, E refers to the exit cell.

A cell is assumed to be 40 cm × 40 cm. The maximal possible moving velocity for an agent, who does not end up in conflict situations, is one cell per time step, i.e., 40 cm per time step. Empirically the average velocity of a pedestrian is about 1.3 m/s. Thus, a time step in the model corresponds to 0.3 s.

2.2 Different Crowd Behaviors

In [8], Schadschneider showed that by altering the coupling constants k_S and k_D different crowd behaviors can be observed. He named the different crowd behaviors ordered, disordered and cooperative. In Figure 2, the coupling constant combinations responsible for different regimes are plotted in a schematic phase diagram.

In the *ordered regime*, the agents move towards the exit using the shortest path. The regime is called ordered, because the movement of the agents is in a sense deterministic. In the disordered regime, the agents just blindly follow other agents' paths, whether the path they are following is leading to the exit or not. In this study, we are only focusing on ordered and cooperative behavior, as disordered behavior is thought to occur mainly in smoky conditions. Between

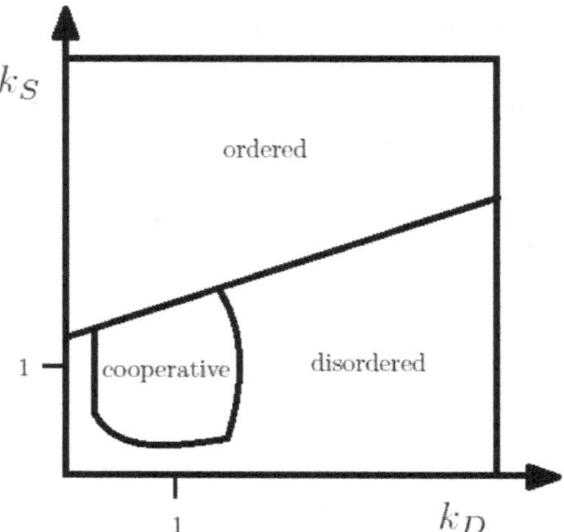

Fig. 2. Altering the coupling constants k_S and k_D, in the CA model, produces different crowd behaviors

the ordered and disordered regime is the *cooperative regime* around the values $k_D = k_S = 1$. There, the agents move towards the exit using paths of higher flow, i.e., paths where the amount of conflict situations is small.

Consequently, for a freely moving agent, ordered behavior makes the agent evacuate fastest. However, a sufficiently large μ causes a faster-is-slower phenomenon, where a crowd of ordered agents will evacuate slowest. The reason is that ordered agents cross paths often, which causes conflicts that slow down the evacuation. In the cooperative regime, even though the whole crowd moves to the paths of higher flow, there will not be as much conflicts as in the ordered regime. If too many agents get into conflicts in a path of higher flow, the path ceases to be a path of higher flow and the agents change path.

3 Spatial Evacuation Game

Next, we present the spatial game defined by Heliövaara et al. in [5]. It should be noted that the spatial game and CA are two separate models. In the game, n_a agents, indexed by i, $i \in I = \{1, ..., n_a\}$, are in an evacuation situation, and located in a discrete square grid. Each agent has an *estimated evacuation time* T_i, which depends on the number λ_i of agents between him and the exit, and on the capacity of exit β. T_i is defined as

$$T_i = \frac{\lambda_i}{\beta}. \tag{2}$$

Each agent has a *cost function* that describes the risk of not being able to evacuate before the conditions become intolerable. The cost function $u(T_i)$ is a function of T_i. The shape of the cost function depends on the parameter T_{ASET}, *available safe egress time*, which describes the time, in which the conditions in the building become intolerable. Additionally, a parameter T_0 describes the time difference between T_{ASET} and when the agents start to play the game.

The agents interact with other agents in their Moore neighborhood. Each agent can choose to play either *Patient* or *Impatient*. Let us denote the average evacuation time of agent i and j, $T_{ij} = (T_i + T_j)/2$. In an impatient vs. patient agent contest, an impatient agent i can overtake his patient neighbor j. This reduces agent i's evacuation time by $\triangle T$ and increases j's evacuation time by the same amount. The cost of i is reduced by $\triangle u(T_{ij})$ and increased for j by the same amount. Here

$$\triangle u(T_{ij}) = u(T_{ij}) - u(T_{ij} - \triangle T) \simeq u'(T_{ij})\triangle T. \tag{3}$$

In a patient vs. patient agent contest, the patient agents do not compete with each other, they keep their positions and their costs do not change. In an impatient vs. impatient agent contest, neither agent can overtake the other, but they will face a conflict and have an equal chance of getting injured. The risk of injury is described by a cost $C > 0$, which affects both agents. The constant C is called the *cost of conflict*. We assume that $u'(T_{ASET}) = C$. Also, we assume that $u'(T_{ij}) > 0$. Thus, based on Equation 3, we have $\triangle u(T_{ij}) > 0$. Now, an illustration of a quadratic cost function can be drawn (see Figure 3).

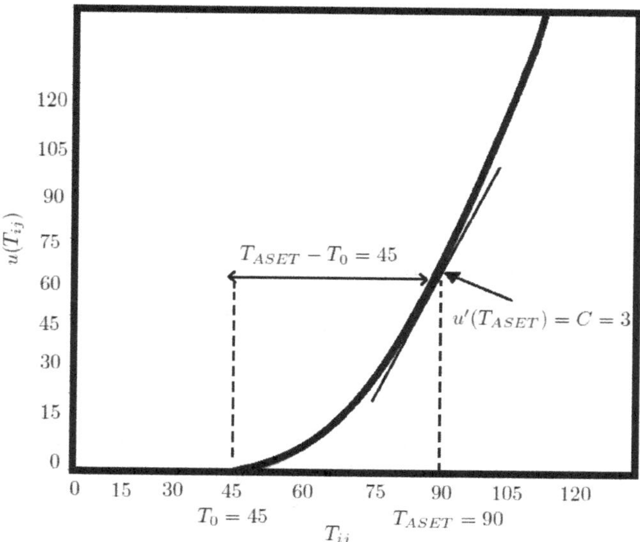

Fig. 3. Illustration of the parameters of the cost function. The function in the figure has the parameter values: $T_{ASET} = 90, T_0 = 45, C = 3$.

From the aforementioned assumptions, a 2×2 game matrix can be constructed:

		Agent 2	
		Impatient	Patient
Agent 1	Impatient	$C/\triangle u(T_{ij}), C/\triangle u(T_{ij})$	$-1, 1$
	Patient	$1, -1$	$0, 0$

Here, all the elements of the more intuitive form of the game matrix have been divided by $\triangle u(T_{ij})$. When a particular pair of strategies is chosen, the costs for the two agents are given in the appropriate cell of the matrix. The cost to agent 1 is the first cost in a cell, followed by the cost to agent 2.

Because this is a cost matrix, the agents want to minimize their outcome in the game. Depending on the number $C/\triangle u(T_{ij})$, the matrix game, considered as a one-shot game, is a Prisoner's Dilemma game or a Hawk-Dove game. In addition to pure Nash equilibria (NE) the latter has mixed strategy NE. These equilibria are analyzed in detail in [5].

3.1 Update of Strategies

During a simulation round, all n_a agents update their strategies once, so that a simulation round consists of n_a iteration periods. Hence, on an iteration period t, there is only one agent updating its strategy once. The strategies are updated with a *shuffle update scheme*, i.e., the order in which the strategies are updated is randomized. At this point, we do not assume the agents to move. In the next section, it is explained how the game is coupled to the CA model presented in the previous section. Thus, do not confuse a simulation round or iteration period of the game with a time step in the CA.

The total cost for an agent is the sum of the costs against all of his neighbors, and the agent's *best-response strategy* is a strategy that minimizes his total cost. The agents are *myopic* in the sense that they choose their strategies based on the previous iteration period of the game, not considering the play of future iteration periods. The best-response strategy $s_i^{(t)}$ of agent i on iteration period t is given by his best-response function BR_i, defined by

$$s_i^{(t)} = BR_i(s_{-i}^{(t-1)}; T_i, T_{-i}) = \arg\min_{s_i' \in S} \sum_{j \in N_i} v_i(s_i', s_j^{(t-1)}; T_{ij}). \tag{4}$$

Here, N_i is the set of agents in agent i's Moore neighborhood. Note that when we couple the game model to the CA, the N_i will change as agent i moves in the square grid. The function $v_i(s_i', s_j^{(t-1)}; T_{ij})$ gives the loss defined by the evacuation game to agent i, when he plays strategy s_i', and agent j has played strategy $s_j^{(t-1)}$ on iteration period $(t-1)$. That is, $v_i(s_i', s_j^{(t-1)}; T_{ij})$ is equal to the corresponding matrix element. Here, $s_{-i}^{(t-1)}$ is used to denote the strategies of all other agents than agent i on iteration period $t-1$, and T_{-i} includes the estimated evacuation times of these agents.

Simulations in [5] have been done with an experimental (undocumented) version of FDS+Evac software [6]. There, playing the game actually changes the physical behavior of the agents. Impatient agents do not avoid contacts with other agents as much; they accelerate faster to their target velocity, and move more nervously. Whereas, patient agents avoid contact with other agents.

4 Cellular Automaton Evacuation Model Coupled with a Spatial Game

There are similarities between the presented spatial game and CA model. As noted above, impatient agents end up in conflicts by competing with other agents, whereas patient agents avoid conflicts. The description of impatient agents resembles the movement of agents in the ordered regime of CA; recall Section 2.2. Agents in the ordered regime are set to move towards the exit using the shortest path, and thereby have a tendency to get into conflicts. On the other hand, the description of patient agents resembles the movement of agents in the cooperative regime. Agents in the cooperative regime move towards the exit using paths of higher flow, i.e., paths where the amount of conflict situations is small, and thereby have a tendency to avoid conflicts.

From the aforementioned observations, we propose a model, where we couple the CA model with the spatial evacuation game. In our model, we let the strategy choice of playing Impatient result in ordered behavior, i.e., the agent to move towards the exit using the shortest path, and playing Patient in cooperative behavior, i.e., the agent to move towards the exit using paths of higher flow. For an agent playing Impatient, the coupling constants are set to $k_S = 10$, $k_D = 1$, and for an agent playing Patient $k_S = 1$, $k_D = 1$. The coupling constant values chosen to represent ordered and cooperative behavior are chosen to be such that they are clearly inside the appropriate regimes in Figure 2. The effect of strategy choice on the agent's behavior is depicted in Figure 4.

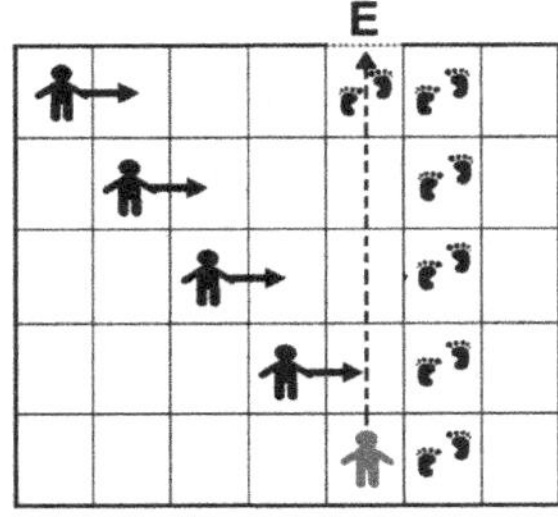

(a) If the agent plays Impatient, he moves towards the exit using the shortest path, regardless of the awaiting conflict situation.

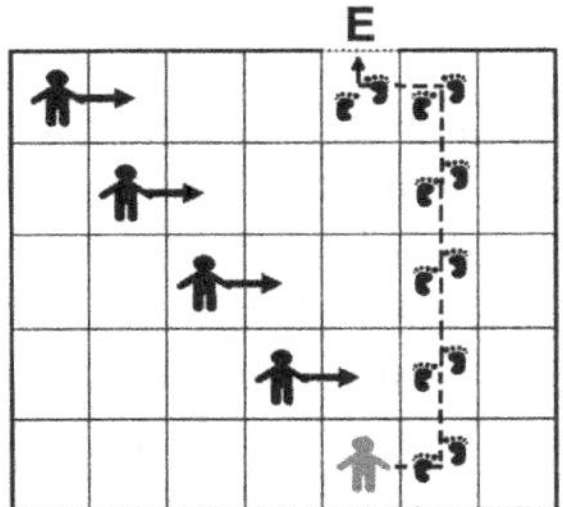

(b) If the agent plays Patient, he moves towards the exit using the path of higher flow, avoiding the awaiting conflict situation.

Fig. 4. Effect of strategy choice on the agent's behavior

It should be noted, that the strategy choice the agent makes, does not reflect an optimal path to the exit, i.e., it is not an optimal strategy for the whole evacuation over time. Rather, the strategy choice is optimal in a snapshot of the evacuation against his immediate neighbors (actually the whole crowd is in an NE in a snapshot [5]).

4.1 Model Description

Next, a step-by-step description of our model is given. In the beginning of the simulation, the agents are located randomly in the room. None of the agents play the game, and all agents are considered patient.

Step 1. At the beginning of each time step, T_i is calculated for $i = 1, ..., n_a$. If $T_i > T_{ASET} - T_0$, the agent i plays the game.

Step 2. The agents' strategies are updated with the shuffle update scheme. The agents observe the strategies of the other agents in their Moore neighborhood, and choose a best-response strategy according to Equation 4.

Step 3. The agents' behavior is updated in the CA model, to correspond to their strategy choice. This is done by altering the agents' coupling constants as follows:

(a) Playing Impatient results in ordered behavior. The agents coupling constants are set to $k_D = 1.0$ and $k_S = 10.0$.

(b) Playing Patient results in cooperative behavior. The agents coupling constants are set to $k_D = 1.0$ and $k_S = 1.0$.

Step 4. The agents move in the CA.

Step 5. Go to Step 1. This procedure is repeated until all agents have evacuated the room.

Remark 1 : Here, a time step refers to a time step in the CA, i.e., the agents are able to move once.

Remark 2 : In Step 2, the shuffle update scheme is repeated multiple times, to ensure that the agents are in an equilibrium configuration all the time. Figure 5 illustrates a snapshot of the evacuation in such a configuration. Note that because the estimated evacuation times of the agents increase farther from the exit, the proportion of impatient agents do so; this is explicitly shown in [5]. More such simulations, with different patient and impatient agent densities, can be found in [1], [5]. The convergence of the best-response dynamics in the spatial Hawk-Dove game has previously been studied in [16].

5 Evacuation Simulations

We have presented an evacuation model, where the agents' coupling constants appear as a result from the game the agents play. In the following, we illustrate how the agents behave in a typical evacuation simulation. Additionally, we show

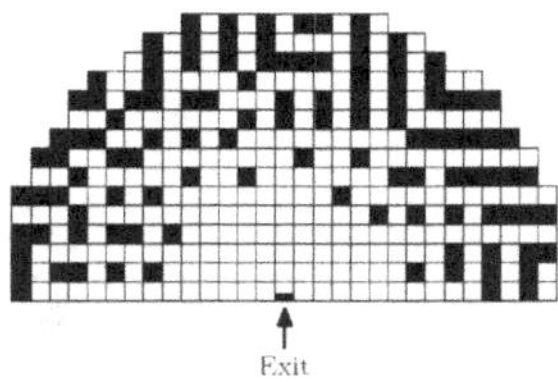

Fig. 5. An equilibrium configuration for 378 agents with parameter values $T_{ASET} = 450$ and $T_0 = 400$. Black cells represent impatient agents and white patient.

that the faster-is-slower effect, already found in the original formulation [9], now appears as a result of the game the agents play. The result is compared to a similar analysis made by Heliövaara et al. with an experimental (undocumented) version of FDS+Evac [5].

5.1 Evacuation of a Large Room

Here, we simulate a typical evacuation situation, i.e., the evacuation of a large room. In Figure 6 there are three snapshots from different stages of this evacuation simulation. The black squares represent impatient agents and the white patient.

As can be seen, the agents form a half-circle rather quickly in front of the exit. Notice, that the agents play their equilibrium strategies at each snapshot of the simulation. At these snapshots, the impatient agents move towards the exit using the shortest path, whereas the patient agents use a path of higher flow.

5.2 Faster-is-Slower Effect

Some people experience the evacuation situation more threatening than others, and thus start to behave more impatiently in relation to the other people. It is striking that our model describes this feature of human beings. It is clearly seen in Figure 5; see also the explanation in Remark 2.

In [5] the dependence of the proportion of impatient agents on egress flow was studied with an experimental (undocumented) version of FDS+Evac. The agents were set in a half-circle in front of the exit, and they updated their strategies until equilibrium was reached. Afterwards, the agents' strategies were fixed, the exit was opened and the agents start to evacuate. The same simulations were run with our model. Here, we want to demonstrate that both models describe qualitatively the faster-is-slower effect. The results of the simulations with these two models can be seen in Figure 7.

It is clearly seen, from both Figures 7 (a) and (b), that the more agents behave impatiently, the smaller the egress flow is. Since the effective velocity of an impatient agent is larger than that of a patient, a faster-is-slower effect can be distinguished. In the experimental version of FDS+Evac, this is caused by impatient agents pushing harder towards the exit, which results in jams and reduced

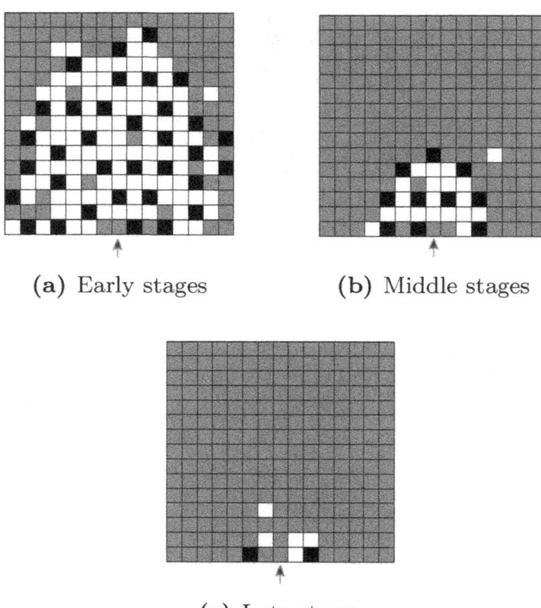

(a) Early stages (b) Middle stages

(c) Late stages

Fig. 6. Snapshots of the simulation in different stages of the evacuation process. The black squares represent impatient agents and the white patient.

flows [5]. In our model, it is caused by impatient agents moving straight towards the exit, resulting in more conflict situations and slowing down the evacuation. The quantitative differences can be explained by the different geometries of both the agents and the exits. Also, the velocities of the agents are different in the two models.

6 Discussion and Conclusions

We introduced a CA evacuation model, where the agents are equipped with simple decision-making abilities. For the simulation of the agents' movement, we used the simulation platform by Schadschneider et al. [9]. In it, ordered and cooperative crowd behaviors can be obtained by altering the coupling constants k_D and k_S. To provide decision-making abilities, we coupled it with a spatial game introduced by Heliövaara et al. [5].

In our model, the choice of strategy actually changes the physical behavior of the agent in the CA. Patient agents move towards the exit using paths of higher flow, i.e., have a tendency to get avoid conflicts, whereas impatient agents move towards the exit using the shortest path, i.e., have a tendency to get into conflicts.

In the original model by Schadschneider et al., the values of the coupling constants should be fixed before simulation starts. In our formulation, the agents' coupling constants depend on their strategy choice in the spatial game. Moreover,

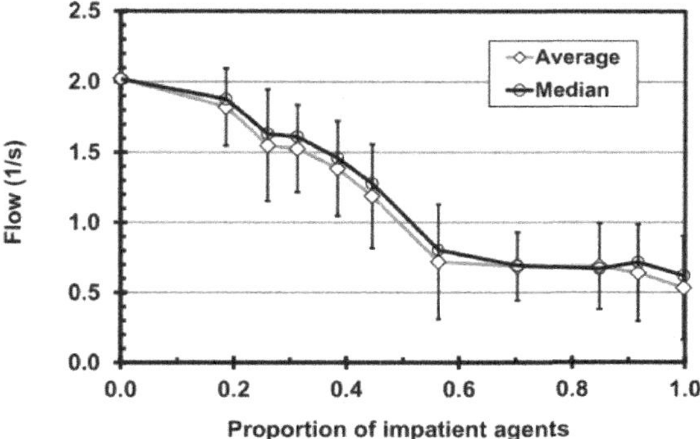

(a) Simulations with the experimental version of FDS+Evac [5] (a 0.8 m wide exit).

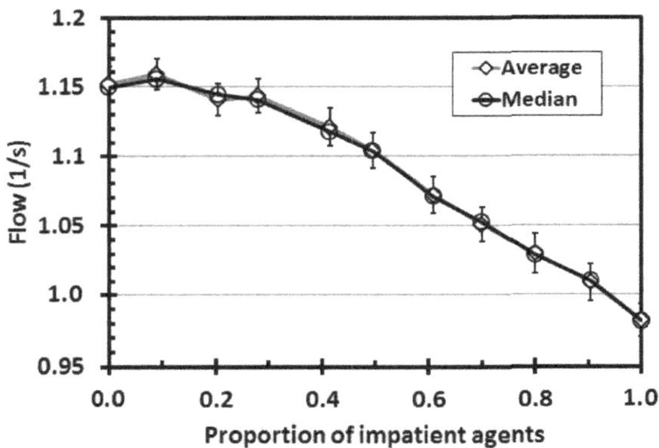

(b) Simulations with our model (a 0.4 m wide exit).

Fig. 7. Average egress flow for 200 agents with different proportion of impatient agents in the population. In the simulations, 11 different values of T_{ASET} were used. Note that the vertical scales in the figures differ.

the agents' parameters change dynamically according to their perception of the surrounding conditions, i.e., the risk of not being able to evacuate in time, and the behavior of neighboring agents.

In the end of the numerical section, we noticed that our model in some aspects give qualitatively similar results as in [5]. To map the full potential of our model, further comparisons with evacuation simulation software should be done. Since our model is computationally light, it could be used for web-based real-time safety analyses.

References

1. von Schantz, A.: Modeling Egress Congestion Using a Cellular Automaton Approach. Master's Thesis (2014), http://sal.aalto.fi/fi/julkaisut/opinnaytetyot/mas
2. Ehtamo, H., Heliövaara, S., Korhonen, T., Hostikka, S.: Game Theoretic Best-Response Dynamics for Evacuees' Exit Selection. Advances in Complex Systems 13, 113–134 (2010)
3. Heliövaara, S., Korhonen, T., Hostikka, S., Ehtamo, H.: Counterflow Model for Agent-Based Simulation of Crowd Dynamics. Building and Environment 48, 89–100 (2012)
4. Heliövaara, S., Kuusinen, J.-M., Rinne, T., Korhonen, T., Ehtamo, H.: Pedestrian Behavior and Exit Selection in Evacuation of a Corridor - an Experimental Study. Safety Science 50, 221–227 (2012)
5. Heliövaara, S., Ehtamo, H., Helbing, D., Korhonen, T.: Patient and Impatient Pedestrians in a Spatial Game for Egress Congestion. Physical Review E 87, 012802 (2013)
6. Korhonen, T., Hostikka, S.: Fire Dynamics Simulator with Evacuation: Fds+Evac. Tech. rep., VTT Technical Research Centre of Finland (2009)
7. Burstedde, C., Klauck, K., Schadschneider, A., Zittartz, J.: Simulation of Pedestrian Dynamics Using a Two-Dimensional Cellular Automaton. Physica A: Statistical Mechanics and its Applications 295, 507–525 (2001)
8. Kirchner, A., Schadschneider, A.: Simulation of Evacuation Processes Using a Bionics-Inspired Cellular Automaton Model for Pedestrian Dynamics. Physica A: Statistical Mechanics and its Applications 312, 260–276 (2002)
9. Kirchner, A., Nishinari, K., Schadschneider, A.: Friction Effects and Clogging in a Cellular Automaton Model for Pedestrian Dynamics. Physical Review E 67, 056122 (2003)
10. Helbing, D., Farkas, I., Vicsek, T.: Simulating Dynamical Features of Escape Panic. Nature 407, 487–490 (2000)
11. Zheng, X., Cheng, Y.: Conflict Game in Evacuation Process: A Study Combining Cellular Automata Model. Physica A: Statistical Mechanics and its Applications 390, 1024–1050 (2011)
12. Zheng, X., Cheng, Y.: Modeling Cooperative and Competitive Behaviors in Emergency Evacuation: A Game-Theoretical Approach. Computers & Mathematics with Applications 62, 4627–4634 (2011)
13. Hao, Q.Y., Jiang, R., Hu, M.B., Jia, B., Wu, Q.S.: Pedestrian Flow Dynamics in a Lattice Gas Model Coupled with an Evolutionary Game. Physical Review E 84, 036107 (2011)
14. Shi, D.M., Wang, B.H.: Evacuation of Pedestrians from a Single Room by Using Snowdrift Game Theories. Physical Review E 87, 022802 (2013)
15. Bouzat, S., Kuperman, M.N.: Game Theory in Models of Pedestrian Room Evacuation. Physical Review E 89, 032806 (2014)
16. Sysi-Aho, M., Saramäki, J., Kertész, J., Kaski, K.: Spatial Snowdrift Game with Myopic Agents. The European Physical Journal B 44, 129–135 (2005)
17. http://www.emergencymgmt.com/training/Online-Evacuation-Tool-Plan.html (accessed April 24, 2014)
18. http://2013.nycbigapps.com/project/113/mesa-the-massive-evacuation-simulation-application (accessed April 24, 2014)

Shape from Motion Revisited

Władysław Skarbek

Warsaw University of Technology, Faculty of Electronics and Information Technology,
Nowowiejska 15/19, 00-665 Warsaw, Poland
`W.Skarbek@ire.pw.edu.pl`

Abstract. This brief tutorial paper on Shape from Motion (SfM), the profound 3D object modeling method, focuses on mathematical background for the batch scenario. Error bounds for pixels with respect to depth change are derived to analyze the applicability of orthographic projection versus perspective projection. Key geometric properties, used for SfM algorithms design and analysis, are stated and proved. Moreover, the case of measurement matrix with rank two, for non planar shapes is fully characterized and its role in shape ambiguity explored. Other sources of SfM ambiguity are presented what justifies the definition of ambiguous error function used for nonlinear optimization of rotation coefficients. Experiments refer to head pose identification and 3D face animation and show the good visual accuracy of SfM approach for digital 3D face projects.

Keywords: shape from motion, structure from motion, SfM ambiguity, head pose identification, 3D face animation.

1 Introduction

Shape from Motion (a.k.a. Structure from Motion `SfM`) is the profound 3D object modeling method which uses only representative image points in sequence of the single camera frames. It offers also object's pose recovery.

The seminal paper introducing the approach belongs to Tomasi and Kanade [1] and it was published in 1992. In its basic form, `SfM` admits only modeling of rigid objects with option of uniform scaling, i.e. the object can change its size but evenly in each direction wrt its geometric center, considered only for representative points. It also assumes orthographic projection for camera imaging process. Moreover, it needs a batch of frames to start the procedure of modeling.

Few researchers attempted to generalize the original approach in various ways, for instance: considering nonrigid objects modeling [2,3,4], replacing orthographic by projective camera [5], incomplete representative point sets [9], and incremental model building [9].

For more than twenty years the method was of moderate interest of researchers because the concept of representative points was not clear neither from theoretical nor practical point of view. For instance, how can we define the representative points for faces in order to get the reasonable approximation of 3D surface of human face in real time?

D. Ślęzak et al. (Eds.): AMT 2014, LNCS 8610, pp. 383–394, 2014.

Why real time is important? If time is not of interest then we have other more robust and accurate methods, for instance stereo vision approach. The rough shape together with its poses for each frame, identified in one consistent algorithm, comprised the unique advantage and benefit of `SfM`.

In subsection 1.1 the orthographic projection versus perspective projection is discussed. There are given error bounds for pixels, as a function of relative depth change for which the applicability of orthographic projection is still valid. The mathematical notation is introduced in subsection 1.2, and `SfM` problem is strictly defined as an optimization problem, with a set of orthogonal constraints.

The section 2 is the core section of the paper and it includes the presentation of key geometric properties, used for SfM algorithms design and analysis. Beside the holistic equation, the rank three case, the property of rank two measurement matrix for non planar shapes is fully characterized and its role in shape ambiguity explored. Other sources of SfM ambiguity are shown what justifies the definition of ambiguous error function used for nonlinear optimization of rotation coefficients. Having all necessary formulas derived and proved, in subsection 2.7 all steps of `SfM` algorithm are given.

Experiments, described in section 3, refer to head pose identification and 3D face animation.

1.1 Orthographic versus Perspective Projection

The classical `SfM` assumes orthographic model for the camera which is based on orthographic projection. There is a projection plane and the perpendicular to this plane projection direction.

The orthographic projection can be described by the simple formulas if only the projection plane coordinates XY are compatible with the world coordinates XYZ where moving object to be modeled exists.

Namely, the *orthographic projection* of the point $P = P(X, Y, Z)$ onto the projective plane $Z = f$ is the point $p = p(x, y, z)$ which is the nearest point from the projective plane to P, i.e. $x = X, y = Y, z = f$.

The compatibility of the image and world coordinates is achieved by mapping pixel coordinate system as follows:

1. setting the world X axis parallel to x_{pix} axis which is parallel to the pixel rows,
2. setting the world Y axis parallel to y_{pix} axis which is parallel to the pixel columns,
3. setting the world Z axis perpendicular to the imaging plane of pixels,
4. anchoring the world coordinates origin point at the origin of pixel coordinate system[1] with respect to XY and at the focus distance f from the projective plane with respect to Z coordinate[2].

---

[1] For instance left upper corner of pixel array or its geometric center.

[2] It means that $Z = f$ for the projective plane. It is irrelevant for the orthographic projection, but it gets meaning at comparison with the central projection onto the plane $Z = f$.

However, in computer vision, the cameras are usually modeled by pinhole camera which works according the central (perspective) projection. It means there is a projection plane and a point of central projection beyond it. The pinhole model is more accurate in description of world to image geometric relationships. Does it mean that the classical `SfM` theory and the related algorithms are not applicable?

The answer is `NO`. There are spatial conditions and application scenarios when `SfM` fully satisfies application requirements. Like Newtonian mechanics is widely applicable for small velocities of masses, `SfM` is quite useful for small Z relative movements of 3D objects moving in the image near its central part. There are two very essential reasons for this not obvious result:

- *image plane discretization* – the small central pixels shifts may correspond to big depth changes in space,
- *scale ambiguity* – `SfM` modeling is always scale ambiguous, i.e. the obtained model is the scaled version of the original object with the unknown true scale.

In order to justify the above claims, let us define a simple pinhole camera model which is also simple for the above special choice of the coordinate systems.

Namely, *the central (perspective) projection* of the point $P = P(X, Y, Z)$ onto the projective plane $Z = f$ is the point $p = p(x, y, z)$ which is the intersection of a ray, i.e. straight line interval, which in turn begins at P and ends at the origin $O = O(0, 0, 0)$ of the coordinate system. Therefore: $x = fX/Z, y = fY/Z, z = f$.

Having x, y coordinates of the projected point, its *pixel central coordinates* are obtained by scaling them with pixel width s_x and pixel height s_y, respectively: $x_{pix} = x/s_x = (fX/s_x) \cdot (1/Z)$, $y_{pix} = (fY/s_y) \cdot (1/Z)$.

Now, suppose our point $P(X, Y, Z)$ moves to the point $P'(X, Y, Z')$, i.e. it changes only its depth Z to Z'. What can we say in this case of the pixel change in the image plane? The answer is quite easy:

$$\begin{aligned} x_{pix} - x'_{pix} &= \tfrac{fX}{s_x}\left(\tfrac{1}{Z} - \tfrac{1}{Z'}\right) = x_{pix}\left(1 - \tfrac{Z}{Z'}\right) \\ y_{pix} - y'_{pix} &= \tfrac{fY}{s_y}\left(\tfrac{1}{Z} - \tfrac{1}{Z'}\right) = y_{pix}\left(1 - \tfrac{Z}{Z'}\right) \end{aligned}$$

Hence the change of $x_{pix} \neq 0$ in the image plane which is below δ pixels (e.g. $\delta = 1$) corresponds to the relative depth change which is less than δ/x_{pix}. The similar conclusion we get for the change of y_{pix}. Jointly the conclusion is as follows: *In the central (perspective) projection the relative change $1 - Z'/Z$ of point depth Z to Z' modifies the image not more than δ pixels in each direction if and only if it is bounded by $\delta/max(|x_{pix}, |y_{pix}|)$:*

$$|x_{pix} - x'_{pix}| < \delta,\ |y_{pix} - y'_{pix}| < \delta \iff \left|\frac{Z}{Z'} - 1\right| < \frac{\delta}{\max(|x_{pix}|, |y_{pix}|)}$$

From the above bounds we see that the maximal relative change of image is the upper bound for the relative change of depth. Therefore the 5% relative shift of pixel gives a freedom for 5% change of depth.

For instance the human face observed by the camera in the window 100×100 pixels from a distance of about $100cm$, can change its distance by at most $5cm$ if we accept the reprojection error of about two pixels.

Therefore, we can use the orthographic projection instead of the central projection as such depth changes results in acceptable pixel errors. However, then we implicitly change the scale on average by about $(s_x + s_y)(Z_{min} + Z_{max})/(4f)$ what in turn is also acceptable due to `SfM` scale ambiguity.

1.2 SfM Problem Definition

`SfM` (Shape from Motion) is a theory and practice of 3D object modeling using camera. The object is moving in front of the camera. We register N_F consecutive frames. Our goal is now to identify not only the shape represented by the fixed number of points N_P but the object's movements represented by affine transformations, as well. Namely, for the discrete time f we want to compute the unknown scaling factor c_f, the unknown rotation $\bar{R}_f \in \mathbb{R}^{3\times 3}$, and the unknown displacement $\bar{d}_f \in \mathbb{R}^3$, $f \in [1, N_F]$.

In order to define the problem rigorously, let us introduce the mathematical notation for matrices representing the discrete 3D shape, the affine movements, and the discrete shape images.

1. The shape matrix $\mathcal{P} \in \mathbb{R}^{3\times N_P}$ of the modeled object stores in its N_P columns all 3D points of its discrete representation:
$$\mathcal{P} \doteq \begin{bmatrix} X_1 \ , \ldots , \ X_{N_P} \\ Y_1 \ , \ldots , \ Y_{N_P} \\ Z_1 \ , \ldots , \ Z_{N_P} \end{bmatrix} .$$
2. For the moment f, the observed characteristic points of the object which are found in the camera frame f, are stored in the measurement matrix $W_f = [W_f(1), \ldots, W_f(N_P)]$. They are related to the shape matrix by the components of the affine transform:
$$W_f = c_f R_f \mathcal{P} + d_f \mathbb{1}_{N_P}^t$$
where for $f = 1, \ldots, N_F$, R_f, d_f are the parts of $\bar{R}_f$, and $\bar{d}_f$, respectively
$$\begin{array}{l} \bar{R}_f \doteq \begin{bmatrix} r_1(f) \\ r_2(f) \\ r_3(f) \end{bmatrix}, \ r_3(f) = r_1(f) \times r_2(f), \ R_f \doteq \begin{bmatrix} r_1(f) \\ r_2(f) \end{bmatrix} \in \mathbb{R}^{2\times 3} \\ \bar{d}_f \doteq \begin{bmatrix} d_1(f) \\ d_2(f) \\ d_3(f) \end{bmatrix}, \ d_f \doteq \begin{bmatrix} d_1(f) \\ d_2(f) \end{bmatrix} \end{array} \quad (1)$$
3. Definitions for centering of measurements and shape:
$$\left. \begin{array}{l} w_f^c \doteq \frac{1}{N_P} \sum_{p=1}^{N_P} W_f(p) = W_f \mathbb{1}_{N_P}/N_P \\ [\bar{X}, \bar{Y}, \bar{Z}] \doteq \frac{1}{N_P} \sum_{p=1}^{N_P} [X_p, Y_p, Z_p] = \mathcal{P}\mathbb{1}_{N_P}/N_P \end{array} \right\} \longrightarrow \left\{ \begin{array}{l} W_f^c \doteq W_f - w_f^c \mathbb{1}_{N_P}^t \\ \mathcal{P}^c \doteq \mathcal{P} - [\bar{X}, \bar{Y}, \bar{Z}]^t \mathbb{1}_{N_P}^t \end{array} \right.$$

4. The point centering operation leads [3] from the affine to the linear relationships for each frame $f = 1, \ldots, N_F$:

$$W_f^c = c_f R_f \mathcal{P}^c \tag{2}$$

In terms of the above definitions, the `SfM` problem can be defined formally:
Given: $W_f^c \in \mathbb{R}^{2\times N_P}$ *for* $f \in [1, N_F]$.
Find: $c_f \in \mathbb{R}^+$, $R_f \in \mathbb{R}^{2\times 3}$, *for* $f \in [1, N_F]$, *and* $\mathcal{P}^c \in \mathbb{R}^{3\times N_P}$ *satisfying the equation* (2) *constrained by the orthogonality condition:*

$$R_f R_f^t = I_2 \doteq \begin{bmatrix} 1 & 0 \\ 0 & 1 \end{bmatrix} \tag{3}$$

2 Shape from Motion – Batch Mode

`SfM` in batch mode takes all frames $f = 1, \ldots, N_F$ and aggregates all individual equations (2) into the single *holistic measurement equation* of the form:

$$\mathcal{W}^c = \mathcal{R}\mathcal{P}^c \tag{4}$$

where $\mathcal{W}^c \doteq \begin{bmatrix} W_1^c \\ \vdots \\ W_{N_F}^c \end{bmatrix} \in \mathbb{R}^{2N_F \times P}$, $\mathcal{R} \doteq \begin{bmatrix} c_1 R_1 \\ \vdots \\ c_{N_F} R_{N_F} \end{bmatrix} \in \mathbb{R}^{2N_F \times 3}$

2.1 Theorem on Equality of Linear Spans: span $[\mathcal{W}^c]$ = span $[\mathcal{R}]$

The first key theorem claims that for the shape $\mathcal{P}$ which is not flat, i.e. is not a subset of a plane, the linear spans of matrices $\mathcal{W}^c$ and $\mathcal{R}$ are equal. In other words, the linear subspace span $[\mathcal{W}^c]$ which is spanned by the columns of the matrix $\mathcal{W}^c$ is equal to the linear subspace span $[\mathcal{R}]$ which is spanned by the columns of the matrix $\mathcal{R}$.

Let us notice that the inclusion span $[\mathcal{W}^c] \subseteq$ span $[\mathcal{R}]$ directly follows from the holistic equation $\mathcal{W}^c = \mathcal{R}\mathcal{P}^c$. Namely, each column in $\mathcal{W}^c$ is the linear combination of the columns in $\mathcal{R}$. Therefore each linear combination of the columns in $\mathcal{W}^c$ is the linear combination of the columns in $\mathcal{R}$.

The opposite inclusion is also true but then we will use non-planarity of the shape. If the shape is not planar then after its centering the set of points stored in columns of the matrix $\mathcal{P}$ is not included in any linear subspace of dimension two. It means that the matrix $\mathcal{P}^c$ has rank three. Therefore there exists three columns i, j, k in $\mathcal{P}^c$ such that the square matrix $\mathcal{P}_{ijk}^c \doteq [s_i^c, s_j^c, s_k^c]$ exhibits the following properties: (a) $\mathcal{P}_{ijk}^c$ is invertible, (b) $\mathcal{W}_{ijk}^c \doteq \mathcal{R}\mathcal{P}_{ijk}^c$, (c) $\mathcal{R} = \mathcal{W}_{ijk}^c (\mathcal{P}_{ijk}^c)^{-1}$. From the last equation we get directly span $[\mathcal{R}] \subseteq$ span $[\mathcal{W}^c]$.

[3] It follows from the equations $W_f = c_f R_f \mathcal{P} + d_f \mathbb{1}_{N_P}^t$ and $w_f^c = W_f \mathbb{1}_{N_P}/N_P = (c_f R_f \mathcal{P} + d_f \mathbb{1}_{N_P}^t)\mathbb{1}_{N_P}/P = c_f R_f [\bar{X}, \bar{Y}, \bar{Z}]^t + d_f$.

Having proved the equality of linear spans of matrices $\mathcal{R} \in \mathbb{R}^{2N_F \times 3}$, $\mathcal{W}^c \in \mathbb{R}^{2N_F \times N_P}$

$$\mathtt{span}\left[\mathcal{W}^c\right] = \mathtt{span}\left[\mathcal{R}\right] \tag{5}$$

we immediately conclude on the rank of our measurement matrix:

$$2 \leq \mathtt{rank}\left[\mathcal{W}^c\right] = \mathtt{rank}\left[\mathcal{R}\right] \leq 3 \tag{6}$$

While the upper bound follows from the number of columns in $\mathcal{R}$, the lower bound we get from orthogonality of two rows in the matrix, for instance, of R_1 which creates the first two rows in $\mathcal{R}$.

2.2 Rank Two Analysis for $\mathcal{R}$

Is the rank 2, the lower bound for $\mathtt{rank}\left[\mathcal{R}\right]$ ever attained if $N_F > 1$? The answer is: `YES`. Namely, we can extend any initial rotation $R_0 = [a, b]^t$ by arbitrary number of different rotations which being collected into $\mathcal{R}$ create always the rank two matrix.

The construction is the following. For any $\alpha \in [0, 2\pi]$ we define: $R(\alpha) \doteq \left[\cos(\alpha)a + \sin(\alpha)b, -\sin(\alpha)a + \cos(\alpha)b\right]^t$.

It is easy to show that:

1. the orthogonality of $R(\alpha)$ is valid: $R(\alpha)R(\alpha)^t = I_2$,
2. the third row of the full rotation matrix $\bar{R}(\alpha)$ equals to $(a \times b)^t$,
3. the rows of $R(\alpha)$ belong to the plane $(a \times b)^t x = 0$,

From the defined family of the rotation matrices $\{R(\alpha)\}_{\alpha \in [0, 2\pi)}$ we select $N_F > 1$ elements: $R_f = R((\pi(f-1)/N_F), f \in [1, N_F]$. Obviously $R_1 = R(0) = R_0$. Therefore the sequence begins from the initial matrix R_0. Let now $\mathcal{R} \doteq (R_1, \ldots, R_{N_F})$. Then we have

1. $\mathtt{span}\left[\mathcal{R}^t\right] = \mathtt{span}\left[a, b\right]$, $\mathtt{rank}\left[\mathcal{R}\right] = 2$,
2. The rotation $\bar{R}^t_{f+1}\bar{R}_f$ rotates around the axis $a \times b$, for any $f \in [1, N_F)$.

We conclude that the relative rotations having common rotation axis, lead to the case of rank two for the measurement matrix – the case ignored in the `SfM` algorithms presented in the literature.

The above construction of the matrix $\mathcal{R}$ of rank two was based on the family of rotations sampled uniformly in α domain. It is easy to show that any rank two matrix $\mathcal{R}$ can be obtained by, generally nonuniform sampling of the family $R(\alpha)$.

It seems that the case of rank two is important from the practical point of view. The following shape ambiguity result shows that in our `SfM` applications planar relative rotations should be avoided. Though the rank two has solutions of the holistic equation (4), we cannot guarantee consistent shape. Namely any recovered point can be shifted by the plane normal $a \times b$ and still the equation is valid. It follows from the following reasoning.

1. If there are two shapes $\mathcal{P}^c_1, \mathcal{P}^c_2$ satisfying (4) then $\mathcal{R}(\mathcal{P}^c_1 - \mathcal{P}^c_2) = \mathbb{0}$.

2. If $\texttt{rank}[\mathcal{R}] = 2$ and $\mathcal{R}X = \mathbb{0}$, $X \in \mathbb{R}^{3\times N_P}$ then $X = (a \times b)c^t$ where for the vector $c \in \mathbb{R}^{N_P}$ we can assume any value.

The ambiguity of the shape $\mathcal{P}^c$ in the direction of common axis $a \times b$ for the relative rotations cannot be removed without additional knowledge. [4]

2.3 Solution of SfM Holistic Equation

For the reasons explained in the previous section, we exclude the rank two case from our measurement matrix $\mathcal{W}^c$, i.e. $\texttt{rank}[\mathcal{W}^c] = 3$.

Then we know that there is an orthogonal base, i.e. a set of unit length, mutually orthogonal vectors $U_3 = [u_1, u_2, u_3]$ in the subspace $\texttt{span}[\mathcal{W}^c]$ if $\texttt{rank}[\mathcal{W}^c] = 3$.

1. We find the coefficients $C_W \in \mathbb{R}^{3\times N_P}$ of linear combinations for $\mathcal{W}^c$ such that $\mathcal{W}^c = U_3 C_W$
2. Both above tasks, i.e. finding the orthogonal base $U \in \mathbb{R}^{m\times 3}$ for a matrix $X \in \mathbb{R}^{m\times n}$ and the coefficients matrix $C_X \in \mathbb{R}^{3\times n}$ wrt to U is easily solved using restricted SVD decomposition $X \stackrel{svd}{=} U_r \Sigma_r V_r^t$, where $r = \texttt{rank}[X]$: $U \doteq U_r, C_X \doteq \Sigma_r V_r^t$.
3. Since by (5) the columns of the matrix $\mathcal{R}$ belong to the subspace $\texttt{span}[\mathcal{W}^c]$, there exists the coefficient matrix C_R wrt to the base U_3 :

$$\mathcal{R} = U_3 C_R \tag{7}$$

4. Having $\mathcal{R}$, the shape $\mathcal{P}^c$ is defined uniquely: $\mathcal{R}\mathcal{P}_1^c = \mathcal{R}\mathcal{P}_2^c \longrightarrow \mathcal{R}(\mathcal{P}_1^c - \mathcal{P}_2^c) = \mathbb{0}_{2N_F\times N_P} \longrightarrow$ the matrix $(\mathcal{P}_1^c - \mathcal{P}_2^c)$ consists of any kernel vectors for $\mathcal{R}$. However for rank three the kernel of $\mathcal{R}$ is trivial, i.e. consists of $\mathbb{0}$ only.
5. Having (from somewhere) the coefficient matrix C_R, by (5) we get: $\mathcal{W}^c = U_r C_W = (U_r C_R)(C_R^{-1} C_W) = \mathcal{R}\underbrace{(C_R^{-1} C_W)}_{\mathcal{P}^c}$.

   Therefore from the uniqueness of $\mathcal{P}^c$, we get the solution for shape:

$$\mathcal{P}^c = C_R^{-1} C_W \tag{8}$$

In order to complete solutions (7)(8) we have to find the matrix of coefficients C_R. It is the topic of the next section.

2.4 Finding of Coefficient Matrix C_R

We reduced solving SfM to establishing coefficients for the linear combinations of the vectors in the orthogonal base U_3 which define the columns of the matrix $\mathcal{R}$. There are three vectors to describe and there are three elements in the orthogonal base U_3. Therefore $C_R \in \mathbb{R}^{3\times 3}$.

[4] However, we can prove that we can avoid this kind of ambiguity if the centering operation is not performed.

We know U_3 but we do not know $\mathcal{R}$. What constraints can be imposed on C_R to get this coefficient matrix? The answer is: use the orthogonality conditions for R_f, $f \in [1, N_F]$. To this goal we decompose the matrix U_3 into stack of blocks of size 2×3 :

1. Let $U_3 = (A_1, \ldots, A_{N_F})$, $A_f \in \mathbb{R}^{2\times 3}$, $f \in [1, N_F]$. [5]
2. Then the equation $\mathcal{R} = U_3 C_R$ is decomposed into N_F equations: $c_f R_f = A_f C_R$, $f \in [1, N_F]$.
3. Hence the orthogonality constraints for R_f impose the constraints for C_R in the form of N_F quadratic equations wrt elements of C_R :
$$A_f C_R C_R^t A_f^t = c_f^2 I_2,\ f \in [1, N_F].$$
4. We solve the above equations using Levenberg Marquard Method (LMM) - the nonlinear optimization technique. Namely, if we get an approximation $\widetilde{C}_R$ then its error $E(\widetilde{C}_R)$ is given by the formula:
$$E(\widetilde{C}_R) \doteq \sum_{f=1}^{N_F} \left[\left(\frac{\alpha_f - \beta_f}{\alpha_f + \beta_f} \right)^2 + \left(\frac{\gamma_f}{\alpha_f + \beta_f} \right)^2 \right] \tag{9}$$
where
$$A_f \widetilde{C}_R \widetilde{C}_R^t A_f^t \doteq \begin{bmatrix} \alpha_f & \gamma_f \\ \gamma_f & \beta_f \end{bmatrix}$$
5. The initial approximation $\widetilde{C}_R$ for LMM is set to a neutral state: $\widetilde{C}_R = \mathbb{1}_{3\times 3}$.

2.5 Finding Scaling Factor c_f and Rotation Matrix R_f

Observe that $\|A_f C_R\|^2 = \|c_f R_f\|_F^2 = 2c_f^2$.

Hence, the scaling coefficients are given by the formula:
$$c_f = \frac{\|A_f C_R\|_{N_F}}{\sqrt{2}},\ f \in [1, N_F] .$$
Then
$$R_f = \frac{A_f C_R}{c_f},\ f \in [1, N_F] .$$
and the approximation of the full rotation matrix can be obtained as follows: [6]
$$\bar{R}_f = \begin{bmatrix} R_f \\ \text{cross}(R_f) \end{bmatrix},\ f \in [1, N_F] .$$

2.6 Sources for Ambiguity of SfM Solution

We already know that fixing the rotations and scalings by freezing the matrix $\mathcal{R}$ makes the centered shape unique, provided the measurement matrix $\mathcal{W}^c$ is of rank three.

Therefore, the possible ambiguity can arise only at changing of the matrix $\mathcal{R}$. It may refer to scaling or rotation matrices. Namely, the actual sources for SfM solution ambiguity follows from:

[5] The matrix block notation $(X_1, \ldots, X_k)$ denotes stacking the block X_{i-1} over the block X_i, $i = 2, \ldots, k$.

[6] The $cross(R)$ returns the cross vector in the transposed form for the rows of $R \in \mathbb{R}^{2\times 3}$.

1. Change of sign for $\mathcal{R}$ implies the central symmetry of the shape:
$$\mathcal{W}^c = \mathcal{R}\mathcal{P}^c = (-\mathcal{R})(-\mathcal{P}^c).$$
2. Change of scale by the factor k implies the scale of shape by the factor $1/k$:
$$W_f^c = (kc_f)R_f\left(\frac{\mathcal{P}^c}{k}\right),\ f = 1,\ldots,N_F.$$
3. Change of the rotations by pre-composing with common rotation $Q^t \in \mathbb{R}^{3\times 3}$ implies the inverse rotation Q of the shape: [7]
$$\mathcal{W}^c = \mathcal{R}Q^t(Q\mathcal{P}^c) \longrightarrow R'_f = R_f Q^t,\ f \in [1, N_F].$$
4. Decentering of the centered shape $\mathcal{P}^c$ can be made uniquely only in XY plane. The depth value $d_3(f)$ can be arbitrary. Actually we can move the shape to arbitrary centroid $[\bar{X}, \bar{Y}, \bar{Z}]^t$:
$$d_f \doteq w_f^c + c_f R_f[\bar{X}, \bar{Y}, \bar{Z}]^t$$

2.7 Summary of SfM Algorithm

The summary of all steps for the described above steps could be presented as follows:

1. Input the centered measurement matrix $\mathcal{W}^c$.
2. If $\texttt{rank}\,[\mathcal{W}^c] < 3$ then return none.
3. Find the orthogonal base $[U_3, \Sigma_3, V_3] \xleftarrow{svd} \mathcal{W}^c$.
4. Compute the coefficients $C_W \leftarrow \Sigma_3 V_3^t$.
5. Let $(A_1, \ldots, A_{N_F}) \leftarrow U_3$.
   Then $C_R \leftarrow LMM(E(), \mathbb{1}_{3\times 3})$,
   where the error function $E()$ is given by (9).
6. For $f \in [1, F]$ compute:
   (a) the temporary matrix: $A'_f \leftarrow A_f C_R$,
   (b) the scaling factor: $c_f \leftarrow \|A'_f\|_F/\sqrt{2}$,
   (c) the rotation matrix: $R_f \leftarrow A'_f/c_f$,
7. Let $\mathcal{R} \leftarrow (c_1R_1, \ldots, c_{N_F}R_{N_F})$.
   Then the centered shape: $\mathcal{P}^c \leftarrow C_R^{-1}C_W$.
8. Return $\mathcal{P}^c$; $(c_1, R_1), \ldots, (c_{N_F}, R_{N_F})$.

3 Application: Digital 3D Face

Experiments with Shape from Motion refer to head pose identification, 3D geometry modeling, and 3D face animation coupled with on-line texturing using camera frames.

The characteristic points for the face were defined using Active Orientation Modeling (AOM [10,11]), the method which is an effective modification of the celebrated Active Appearance Modeling (AAM [12]).

[7] This property could be used to obtain the rotations relatively to the shape orientation in the selected frame. For instance in the first frame $f = 1 : Q = R_1 \longrightarrow R'_f = R_f R_1^t,\ f \in [1, N_F]$. Then, the reference shape is also modified to $(\mathcal{P}')^c \doteq R_1\mathcal{P}^c$.

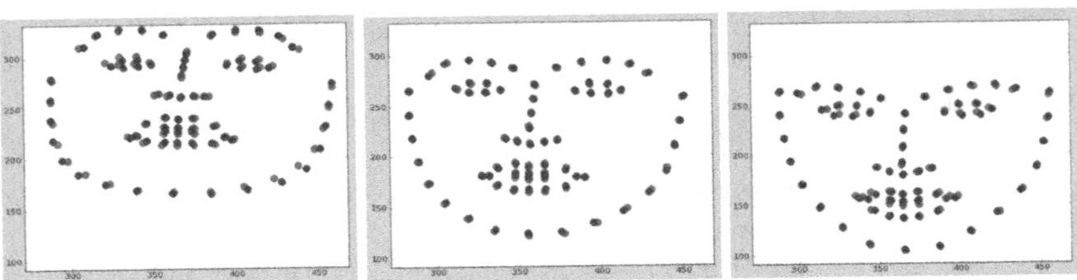

Fig. 1. Screen window shots for the facial characteristic points reprojected from the recovered 3D face model (blue) shown along the original marks generated by the marking procedure (red)

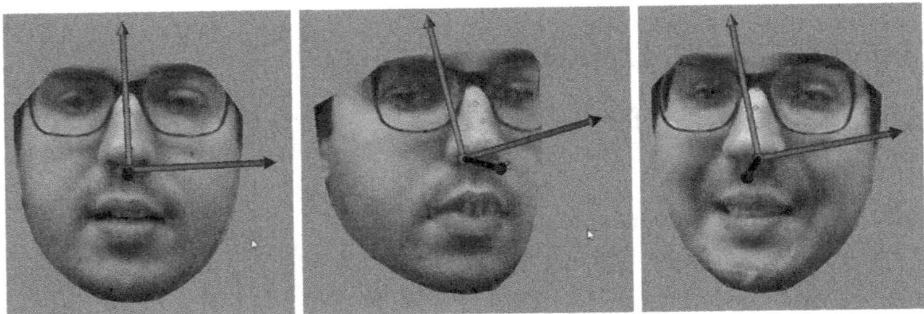

Fig. 2. Screen window shots of different views of 3D model textured by the original facial images together with the orientation axes located in front of the animated 3D face model in the fixed 3D spatial position

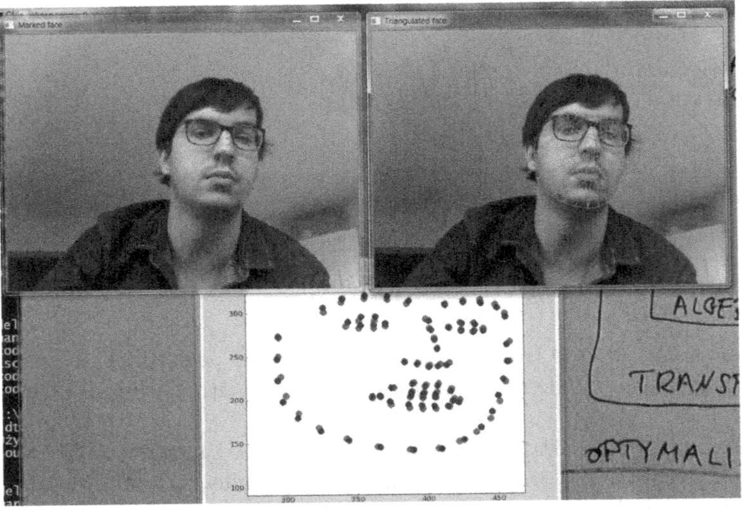

Fig. 3. Screen shot of testbed application. It shows the automatically marked facial image (left), the facial characteristic points reprojected from the recovered 3D face model (bottom), and triangles to be used for 3D visualization (right).

In our experiments we observed the good visual accuracy of SfM approach. Therefore it could be recommended for digital 3D face core tasks: head pose identification, 3D geometry modeling and 3D animation with realistic visualization (cf. Fig.2).

A sample screen shot of our testbed application is given in Fig.3.

4 Conclusions

There are several factors which lead to the renewal of interest in `SfM`. One of them is the recent progress in computer technologies as the fast multi-core CPU and GPU extensions. Also the arising of new fast algorithms for representative points detection and tracking [6,7,8]. Finally, the new `SfM` options like incremental model building from incomplete data [9].

This paper is a brief tutorial which could be used also as a supporting material for a lecture on `SfM`. Its contents mainly refer to mathematical aspects of `SfM` for rigid models. For readers interested in nonrigid case the papers [15-22] could be recommended, too. However, this area is less convincing in theory and practice.

Experiments show that `SfM` could be recommended for core tasks of 3D face processing: head pose identification, 3D geometry modeling and 3D animation with realistic visualization.

References

1. Tomasi, C., Kanade, T.: Shape and Motion from Image Streams under Orthography. International Journal of Computer Vision (IJCV) 9, 137–154 (1992)
2. Akhter, I., Sheikh, Y., Khan, S., Kanade, T.: Trajectory Space: A Dual Representation for Nonrigid Structure from Motion. IEEE Transactions on Pattern Analysis and Machine Intelligence 33(7), 1442–1456 (2011)
3. Bregler, C., Hertzmann, A., Biermann, H.: Recovering non-rigid 3D shape from image streams. In: IEEE Conf. on Computer Vision and Pattern Recognition, Hilton Head, North Carolina (CVPR), vol. 2, pp. 690–696 (2000)
4. Xiao, J., Chai, J., Kanade, T.: A closed form solution to non-rigid shape and motion recovery. International Journal of Computer Vision (IJCV) 67, 233–246 (2006)
5. Hartley, R.I., Vidal, R.: Perspective Nonrigid Shape and Motion Recovery. In: Forsyth, D., Torr, P., Zisserman, A. (eds.) ECCV 2008, Part I. LNCS, vol. 5302, pp. 276–289. Springer, Heidelberg (2008)
6. Xiao, J., Moriyama, T., Kanade, T., Cohn, J.F.: Robust Full-Motion Recovery of Head by Dynamic Templates and Re-registration Techniques. International Journal of Imaging Systems and Technology 13(1), 85–94 (2003)
7. Jang, J.S., Kanade, T.: Robust 3D Head Tracking by Online Feature Registration. In: The IEEE International Conference on Automatic Face and Gesture Recognition (2008)
8. Yan, J., Lei, Z., Yi, D., Li, S.Z.: Learn to Combine Multiple Hypotheses for Accurate Face Alignment. In: IEEE International Conference on Computer Vision Workshops (ICCVW), pp. 392–396 (2013)
9. Ryan, K., Balzano, L., Wright, S.J., Taylor, C.J.: Online Algorithms for Factorization-Based Structure from Motion. In: IEEE Winter Conference on Applications of Computer Vision (WACV) (2014)

10. Tzimiropoulos, G., Alabort-i-Medina, J., Zafeiriou, S., Pantic, M.: Generic Active Appearance Models Revisited. In: Lee, K.M., Matsushita, Y., Rehg, J.M., Hu, Z. (eds.) ACCV 2012, Part III. LNCS, vol. 7726, pp. 650–663. Springer, Heidelberg (2013)
11. Kowalski, M., Naruniec, J.: Evaluation of active appearance models in varying background conditions (Conference Proceedings). In: Photonics Applications in Astronomy, Communications, Industry, and High-Energy Physics Experiments (2013)
12. Cootes, T., Edwards, G., Taylor, C.: Active Appearance Models. IEEE Transactions on Pattern Analysis and Machine Intelligence (1998)
13. Hartley, R., Zisserman, A.: Multiple View Geometry in Computer Vision, 2nd edn. Cambridge (2004)
14. Ma, Y., Soatto, S., Kosecka, J., Sastry, S.: An Invitation to 3D Vision: From Images to Geometric Models. Springer (2003)
15. Mahamud, S., Hebert, M., Omori, Y., Ponce, J.: Provably-convergent iterative methods for projective structure from motion. In: Conference on Computer Vision and Pattern Recognition, vol. I, pp. 1018–1025 (2001)
16. Aanaes, H., Kahl, F.: Estimation of deformable structure and motion. In: ECCV Workshop on Vision and Modelling of Dynamic Scenes (2002)
17. Torresani, L., Bregler, C.: Space-time tracking. In: Heyden, A., Sparr, G., Nielsen, M., Johansen, P. (eds.) ECCV 2002, Part I. LNCS, vol. 2350, pp. 801–812. Springer, Heidelberg (2002)
18. Xiao, J., Kanade, T.: Non-rigid shape and motion recovery: Degenerate deformations. In: Conference on Computer Vision and Pattern Recognition, pp. 668–675 (2004)
19. Hartley, R.I., Schaffalitzky, F.: Reconstruction from projections using Grassmann tensors. In: Pajdla, T., Matas, J(G.) (eds.) ECCV 2004. LNCS, vol. 3021, pp. 363–375. Springer, Heidelberg (2004)
20. Xiao, J., Kanade, T.: Uncalibrated perspective reconstruction of deformable structures. In: IEEE International Conference on Computer Vision, pp. 1075–1082 (2005)
21. Vidal, R., Abretske, D.: Nonrigid shape and motion from multiple perspective views. In: Leonardis, A., Bischof, H., Pinz, A. (eds.) ECCV 2006. LNCS, vol. 3952, pp. 205–218. Springer, Heidelberg (2006)
22. Oliensis, J., Hartley, R.: Iterative extensions of the Sturm/Triggs algorithm: convergence and nonconvergence. IEEE Transactions on Pattern Analysis and Machine Intelligence 29 (2007)

Using Kinect for Facial Expression Recognition under Varying Poses and Illumination

Filip Malawski[1], Bogdan Kwolek[1], and Shinji Sako[2]

[1] AGH University of Science and Technology, 30-059 Krakow, Poland
{fmal,bkw}@agh.edu.pl
[2] Nagoya Institute of Technology, Japan
sako@mmsp.nitech.ac.jp

Abstract. Emotions analysis and recognition by the smartphones with front cameras is a relatively new concept. In this paper we present an algorithm that uses a low resolution 3D sensor for facial expression recognition. The 3D head pose as well as 3D location of the fiducial points are determined using Face Tracking SDK. Tens of the features are automatically selected from a pool determined by all possible line segments between such facial landmarks. We compared correctly classified ratios using features selected by AdaBoost, Lasso and histogram-based algorithms. We compared the classification accuracies obtained both on 3D maps and RGB images. Our results justify the feasibility of low accuracy 3D sensing devices for facial emotion recognition.

Keywords: Facial Image Analysis; Depth Maps Analysis.

1 Introduction

The use of emotion recognition in the age of smartphones with front cameras is a new concept. Through the specialized software the camera can record and then transmit the user facial expressions to a remotely located analysis center. The analysis center equipped with emotion recognition software will be able to perform analysis and recognition of user emotions. What it means is that as people see a news item, or watch a TV show or see an advertisement, it will be possible to know how they are feeling or seeing the current media coverage. Through visual analysis of the face articulations the future technology will be able to decipher the user's facial expressions and to tell whether he/she is happy, sad, angry, tense, relaxed, or depressed. In consequence, it will be possible to infer about the relevance of message or information selection for a specific user.

Emotion recognition offers a new direction to media analysis as well as human machine communication. It will allow finding out what the customers are truly feeling about the delivered messages, and not what they say about their feelings. This is because many research studies have shown that people may hide their true emotions during surveys or filling the questionnaires. Emotion recognition technology should not depend upon what the customers say they are feeling, but it should capture their facial expressions to find out their true emotions.

D. Ślęzak et al. (Eds.): AMT 2014, LNCS 8610, pp. 395–406, 2014.

As media analysis progresses further, in the near future it might be possible to analyze the true emotions that the targeted news and movies are generating.

The potential applications of facial expression recognition (FER) concern not only media technology but also include service robotics, virtual reality, games etc. Most of the previous work focused primarily on 2D domain, see survey [1]. 2D domain-based facial expression classification algorithms have demonstrated remarkable performance in controlled conditions, particularly in constant lighting conditions and with small head pose variations. On the other hand, the 3D data based approaches are invariant to changes mentioned above and therefore current research focuses on 3D modalities [2]. One of the most popular methods for 3D FER is based on the distances between certain facial landmarks and their changes that occur during facial articulations. The focus on such approaches is because the facial geometry is invariant to illumination and imaging conditions.

As demonstrated in another survey [3], a considerable attention has been drawn on 3D FER after publication of the BU-3DFE dataset. The discussed dataset is in fact a testbed for benchmarking the 3D approaches to FER. It was captured using 3DMD setup, and consists of 3D models with 20000 to 30000 polygons depending on the face size. As mentioned in [2], all publicly available databases for 3D-based FER were recorded in controlled conditions with the use of similar setups, i.e. using devices that allow data recordings with high precision and accuracy, and with very low noise level. In this context it is worth mentioning that most databases were recorded using 3D acquisition systems, which are based on structured light technologies, such as the Minolta Vivid 900/910 series. For instance, [4] presents a FER system, which is capable of operating at several frames per second using data acquired from a precise 3D scanner.

Automatic 3D FER recognition from image sequences of low resolution or video quality that is offered by current consumer cameras, smartphone cameras or service robot cameras is very challenging research problem, with many potential applications in media technology. However, little work has been done in the area of 3D face analysis using 3D consumer cameras. One exception is work by Li et al. [5], who recently demonstrated that on the basis of RGBD images acquired by the Kinect sensor it is possible to achieve high face recognition rates. To the best of our knowledge, no significant work has been done in the area of facial articulations analysis using RGBD images delivered by currently available low cost 3D sensors, which are now utilized or will be utilized in modern game consoles, smart TV, service robots, etc., or even in future smartphones.

Affective computing is the study of systems and devices that can recognize, interpret, process, and simulate human affects. A motivation for such research is desire of simulating and utilizing empathy. In general, the machine assisting humans in daily activities should interpret their emotional state and adapt its behavior to them, giving an appropriate response considering context and emotions. An example of utilization of emotions in context of robotics is robot Kismet, which has been developed in MIT [6]. The robot Nao is a recently developed humanoid robot, which is equipped with two cameras. Among others, the robot is capable of connecting with the Internet and searching the requested content, for instance latest news. Our aim is to equip this service robot with ability of analysis of facial expressions during presentation to the user the messages or news, which have been found in Internet in response to he/she requests.

2 Face Detection and Tracking

Active Shape Models (ASMs) are statistical models of the shape of objects, which iteratively deform to fit to an example of the object in a new image [7]. The shape of an object is represented by a set of points. The ASM model is trained using contours (surfaces in 3D) in training images. It finds the main variations in the training data using Principal Component Analysis, which enables to automatically decide if a contour is a good object contour. ASMs are frequently used to analyze 2D and 3D images of faces. In face tracking application the user should collect training images and then represent all shapes with a set of landmarks in order to form a Point Distribution Model (PDM). The ASM works by alternating two stages, where in the first one it generates a suggested shape by looking in the image around each point for a better position for the point, whereas in the second one it conforms the suggested shape to the PDM. Figure 1a depicts locations of 116 facial points, which were determined by ASM.

Depth is very useful cue to attain reliable face detection and tracking since face may not have consistent color and texture but has to occupy an integrated region in space. Kinect sensor provides both color and dense depth images. It combines structured light with depth from focus and depth from stereo. The sensor consists of infrared laser-based IR emitter, an infrared camera and a RGB camera. The IR camera and the IR projector compose a stereo pair with a baseline of approximately 75 mm. A known pattern of dots is projected from the IR laser emitter. Since there is a distance between laser and sensor, the images correspond to different camera positions, and this in turn allows us to use stereo triangulation to calculate each spec depth. It captures the depth and color images simultaneously at a frame rate of about 30 fps. The RGB stream has size 640×480 and 8-bit for each channel, whereas the depth stream is 640×480 resolution and with 11-bit depth. The field of view is $57°$ horizontally and $43°$ vertically, the minimum measurement range is about 0.6 m, whereas the maximum range is somewhere between 4-5 m.

Together with the sensor it is delivered Kinect for Windows SDK and the Face Tracking SDK, which enable developing applications capable of tracking human faces in real time. The face tracking engine determines 3D positions of semantic facial feature points as well as 3D head pose. It tracks the 3D location of 121 points, which are depicted on Fig. 1b. Additionally, the Face Tracking SDK fits a 3D mask to the face. The 3D model is based on the Candide 3 model [8], which is a parameterized 3D face mesh specifically developed for model-based coding of human faces. This 3D model is widely used in head pose tracking [9].

As already indicated, Kinect sensor allows low cost sensing with high capture speed. However, the 3D maps provided by Kinect are very noisy and have relatively low resolution in comparison to typical devices utilized in facial expression recognition. In consequence, many important fiducial points such as

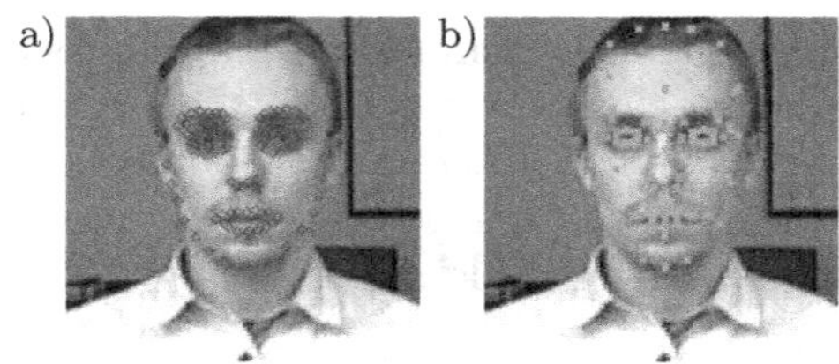

Fig. 1. Locating facial features, a) on gray images using ASM, b) on depth maps using Kinect Face SDK

eye and mouth corners are not too precisely locatable. Even more, some fiducial markers undergo occlusion, particularly the points that are located close to the nose. To the best of our knowledge, there exists only one work [5] that was published recently, in which noisy images acquired by Kinect have been used for face analysis.

3 Automatic Feature Selection

The most popular method in static image-based FER consists in using characteristics distances between certain facial landmarks as well as their changes that occur due to face articulations. For instance, BU-3DFE dataset provides the location of 83 facial points, which together with their distances are widely used in static facial analysis [2]. In [10,11] the classification was done using the distances among all pairs of the available features. An average expression recognition rate was equal to 93.7% and 83.5%, respectively. In a method discussed in [12] six characteristic distances extracted from the distribution of 11 facial feature points from the available points resulted in an average recognition rate of 91.3%. Tang and Huang [13] proposed an automatic feature selection method that is based on maximizing the average relative entropy of the marginalized class-conditional distributions of the features. Tens of the features are automatically selected from a pool determined by all possible line segments between the 83 landmarks. In another work [14] they selected a pool of 96 discriminative features including not only the normalized distances, but also additionally the slopes of line segments connecting a subset of 83 landmarks. In general, the discussed methods rely on features extracted from the locations of facial points provided by 3D databases. In such approaches, typically, a small subset of the selected subset features gives relatively good classification accuracy.

The Kinect sensor provides both depth and color images, and in order to perform a classification of facial expressions the fiducial features should be extracted first. In our approach the location of all facial features was determined using the methods discussed in Section 2. Because of noisy depth maps, all points were subjected to automatic feature selection. In contrast to methods relying on high quality measurements, in the classification we employ relatively large number of the features. In 3D FER we employ the slopes of line segments connecting 121 points. Thus, the total number of all features is $n_i = 7260$. The slopes were chosen since they give slightly better results in comparison to the distances between the features. Given the estimated pose, the head together with the determined face points were rotated to the canonical frontal pose. The size of each face was then scaled accordingly to its width and the distance to the camera. The faces were captured using Kinect for Windows and Kinect for Xbox 360. The Kinect for Windows supports near mode, which enables the camera to see objects as close as 60 centimeters in front of the device without losing accuracy or precision. For Kinect for Xbox 360 the minimal distance of the object to the sensor is about 1 m. Thus, the faces were captured in two sessions from the distance of about 60 cm and 1 m to the cameras. A typical size of the face on the depth map acquired in such a way was about 280×280 pixels for Kinect for Windows, and 160×160 pixels for Xbox Kinect.

3.1 Histogram-Based Feature Selection

Given a set of training depth images for each considered facial expression, we calculate for each facial point a pool of histograms. A histogram reflects a distribution of the slopes between a given facial point and one of the remaining points within a specific class. This means, that for a given line connecting two facial points the histogram bins are incremented using the training data of a given class. Each histogram consists of 20 bins and is normalized prior the feature selection. For each pair of the emotions we calculate the distances between the corresponding histograms. The distances are then sorted. The larger the distances between corresponding histograms representing a pair of facial expressions, the more discriminative is the equivalent feature. Given such sorted lists of the distances (divergences or ratios) between histograms we choose the assumed number of features together with their corresponding histograms. Such features are then utilized in the classification. It is worth mentioning that in the classification stage we considered also optionally such a pool of the features, additionally extended about their symmetric counterparts. The histograms were compared using:

- histogram intersection
- Kullback-Leibler divergence
- Earth Mover's Distance (EMD)
- Fisher's ratio

Since none of the mentioned above method achieved superior results, we decided to use Kullback-Leibler divergence in comparison of the histograms.

3.2 Feature Selection with AdaBoost

The idea of boosting is to select and then combine several classifiers, which are often referred to as weak learners into a more powerful one using a voting procedure. AdaBoost is a supervised algorithm and it learns such strong classifier by selecting only those individual features that can best discriminate among classes [15]. Although AdaBoost was developed as a method to improve the classification accuracy by combining such weak learners, it has also been utilized for feature selection in detection and classification tasks [16]. During training, incorrectly classified training samples are weighted more to redirect focus of the training on them by subsequent weak learners.

In our boosting-based algorithm for feature selection, the finite set of features is considered as the space of the weak learners. The base learner is the decision stump, a one-decision two-leaf decision tree. This means that each feature is considered as a boolean predictor. The learning of the decision stump means selecting a feature and a threshold. Thus, the training of a weak learner simply consists of selecting the one with the minimum error rate. The input of the algorithm is a set of the training examples $\{I_i, i = 1, \ldots, N\}$ and their associated labels $\{l_i, i = 1, \ldots, N\}$, where N is the number of the training examples, and $l_i \in \{0, 1\}$. Each training example is represented by the initial feature set $\{x_i^{(j)}, j = 1, \ldots, n_i\}$. At the beginning, the AdaBoost initializes the weights of the training samples w_i to $\frac{1}{2N_p}, \frac{1}{2N_n}$, where N_p and N_n stand for the number of the positive and negative examples, respectively. Afterwards, in each round it selects one feature as weak classifier, the feature that achieves the highest score with respect to the actual weight, and updates the weights of the training examples. It decreases the weights of the examples that were correctly classified by an optimal classifier for the selected feature. In consequence, in the next iteration the classifiers will focus on the examples that were misclassified. Each selected feature forms a weak classifier h_k that is parameterized by a threshold θ_k and output label u_k. The goal is to select T features, which have the best ability to discriminate the considered samples into the desired classes. The error of the classifier is calculated over the entire examples set as a weighted sum of the absolute differences between the output of the threshold-based binary classifier $h(x_i, \theta_i, u_i)$ and the class label l_i in the following manner: $\varepsilon(x, \theta, u, w) = \sum_{i=1}^{N} w_i \, |h(x_i, \theta_i, u_i) - l_i|$. Given the feature $x_i^{(j)}$ and the corresponding w_i the error of the classifier depends solely on threshold θ. The AdaBoost selects a feature with the optimal threshold θ^*, which is calculated as follows $\theta^* = \text{argmin}_\theta \, \varepsilon(x, \theta, u, w)$. The whole feature selection procedure is shown as Algorithm 1.

3.3 Feature Selection via the Lasso

The least absolute shrinkage and selection operator, which is known as Lasso, permits computationally efficient feature selection based on linear dependency between input features and output values. It is an L1 penalized regression technique introduced by Tibshirani [17]. Lasso commonly gives sparse solutions due to the L1 penalty so it is an alternative to model or subset selection. Any features that have non-zero regression coefficients can be seen as selected by the Lasso algorithm. Lasso solves the following regularized optimization problem:

$$min_\beta h(\beta) = \frac{1}{2}\|y - X\beta\|_2^2 + \lambda\|\beta\|_1, \quad \text{where } \lambda \geq 0 \tag{1}$$

Algorithm 1. AdaBoost algorithm for feature selection

1. **INPUT:**
   - Training data: $\mathcal{D} = \{(x_i^{(j)}, l_i), i = 1, \ldots, N, j = 1, \ldots, n_i\}$, $x_i^{(j)} \in \mathbf{R}$, $l_i \in \{0, 1\}$
   - The initial feature set: $\mathcal{F}$
   - Weak learner: $\mathcal{L}$ that learns binary classifier $h(x) : \mathbf{R} \mapsto \{0, 1\}$
   - The desired number of features: T
2. **OUTPUT:**
   - The sequence of selected features: $\{\hat{x}^{(1)}, \hat{x}^{(2)}, \ldots, \hat{x}^{(T)}\}$
3. **Algorithm**
4. Initialize the distribution $\mathcal{D}$: $w_i = \frac{1}{2N_p}, \frac{1}{2N_n}, i = 1, \ldots, N$
5. $\mathcal{F}' = \mathcal{F}$
6. **for** $t = 1$ to T **do**
7. Normalize the weights w_i of the examples
8. Select a pool of classifiers $H^{(t)}(x)$ from $\mathcal{F}'$
9. Train binary classifiers $H^{(t)}(x)$
10. Select classifier $h^{(t)}(x)$ having highest evaluation score with respect to weights
11. Remove feature $\hat{x}^{(t)}$ corresponding to selected classifier $h^{(t)}(x)$ from $\mathcal{F}'$: $\mathcal{F}' = \mathcal{F}' \backslash \hat{x}^{(t)}$
12. Compute the error rates $\varepsilon^{(t)} = \sum_{i=1}^{N} w_i^{(t)} |(h^{(t)}(x_i) - l_i)|$
13. Update the weight as $w^{(t)} = \begin{cases} w^{(t)} \frac{\varepsilon^{(t)}}{1-\varepsilon^{(t)}} & \text{if} \quad h^{(t)}(x_i) = l_i \\ w^{(t)} & \text{otherwise} \end{cases}$
14. **end for**

and β is a $p \times 1$ vector, y is a $n \times 1$ vector, and X is a $n \times p$ matrix. The penalty term in (1) is a L1-norm penalty or simply the sum of the absolute values of the components of β. This penalty term encourages sparsity in the components of the solution vector and thus automatically leads to feature selection. Additionally, the penalty term regularizes the solution vector β and hence prevents overfitting. If $p > n$, the Lasso selects at most n variables and the number of selected features is bounded by the number of samples.

4 Experiments

We conducted extensive experiments to evaluate the usefulness of depth map acquired by Kinect for facial expressions recognition. The data for the evaluation of the proposed algorithm were recorded in two sessions using Kinect for Xbox 360 and Kinect for Windows. Each dataset consists of 2520 images of 10 individuals with variations in pose and illumination. The images in the datasets contain three basic facial expressions: normal, smile and anger. Each dataset consists of 90 images in frontal pose and normal illumination (30 images for each expression), 30 images in frontal pose and dark face (10 images for each expression), 66 images in normal illumination and various poses, and 66 images in non-frontal poses and poor illumination. Figure 2 illustrates some example RGB images that were acquired in the considered illumination conditions using

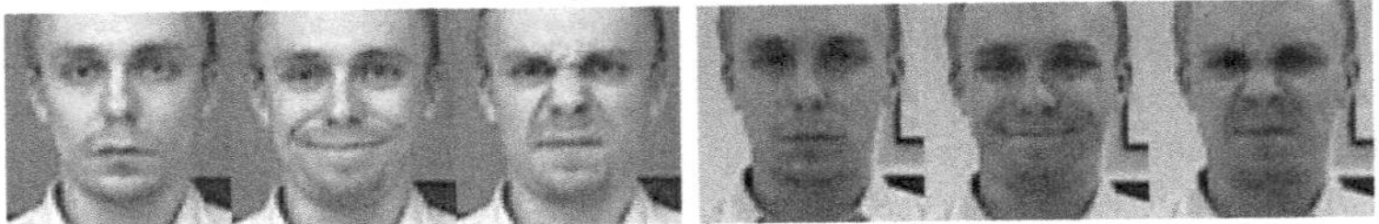

Fig. 2. Facial expressions in two different illumination conditions

Kinect for Xbox 360. The first three images contain expressions shot in normal conditions, whereas the next ones contain the expressions in poor illumination.

Figure 3 depicts facial expressions in the considered head poses. As we can observe, the change in the head poses is quite considerable. The Face Tracking SDK estimates the user's head pose and returns tree angles: pitch, roll, and yaw, which describe its orientation. Using such angles the head is rotated to the canonical pose as well as is scaled according to its distance to the camera.

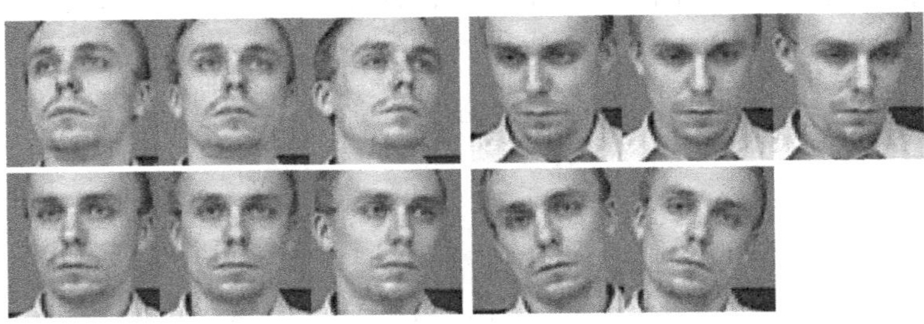

Fig. 3. Facial expressions in various poses

Table 1 shows the correctly classified ratio that was obtained in 10-fold validation on datasets acquired by Kinect for Xbox 360. The features were extracted using histograms with Kullback-Leibler (KL) divergence, AdaBoost (AB) and sparse (SP). The correctly classified ratio was determined using Naïve Bayes (NB), random forests (RF) and Support Vector Machine (SVM). In nonlinear SVM-based classification, the most important parameter is the soft-margin constant c, which controls the trade-off between complexity of decision rule and frequency of error. A smaller value of c allows to ignore points close to the boundary, and increases the margin. In nonlinear SVM, the (Gaussian) radial basis function kernel, or RBF kernel, is a popular kernel function, which usually gives good results. The parameter σ controls how quickly an increased distance causes the value of the kernel to fall toward zero. By the use of cross-validation and grid-search on c and σ parameters [18], the best prediction results were obtained by SVM with $c = 1$ and $\sigma = 0.01$.

Using the discussed classifiers we evaluated the classification accuracy of facial expressions for different numbers of the selected features. The best number of the selected features is shown in the third row of Tab. 1. The classification performance was evaluated for normal head pose and normal illumination conditions,

Table 1. Correctly classified ratio [%] using 3D data acquired by Kinect for Xbox 360

	NB			RF			SVM		
feature sel.	HI	AB	SP	HI	AB	SP	HI	AB	SP
# features	100	800	100	400	200	400	200	400	100
normal	72.8	70.4	73.0	76.4	79.2	78.6	80.6	81.6	72.8
dark	74.8	75.2	78.0	75.0	78.6	75.6	77.8	80.0	77.0
pose	62.2	66.0	71.4	68.8	66.0	66.4	66.6	69.2	69.2
pose dark	62.6	63.4	69.0	70.6	69.0	71.5	66.0	67.8	64.8
average	68.1	67.8	72.9	71.9	73.2	73.0	72.8	74.7	71.0

dark face, non-frontal pose, and non-frontal pose shot in poor illumination, see subsequent rows in Tab. 1. The average correctly classified ratio is shown in the last row of the table. As we can see, the best classification accuracy was achieved for normal pose and normal illumination conditions via Support Vector Machine classifier, which has been trained on features extracted by AdaBoost. The classification accuracy is equal to 81.6%. Slightly worse correctly classification ratio was obtained in the case of poor lighting (dark). A considerable decrease of the classification performance can be observed for non-frontal head poses. For features selected automatically without extending them by symmetric counterparts the classification accuracy is slightly worse.

Table 2 presents the correctly classified ratio that has been achieved using 3D data acquired by Kinect for Windows. As we can observe, owing to the near mode of the sensor, the results are far better. The SVM-based classification was performed using the same parameters, i.e. $c = 1$ and $\sigma = 0.01$. As we can see, for the discussed sensor the best result was obtained by SVM operating on features selected by sparse-based feature selection. As it was shown in Tab.1, the best result for Xbox Kinect has also been achieved by SVM on the features selected by AdaBoost algorithm. The discussed results demonstrate that the number of features required to achieve favorable classification accuracy is quite high.

Since the SVM gave the best result in the terms of correctly classified ratio we conducted grid-searching on c and σ parameters to achieve even better performance. We found that the best correctly classified ratio is achieved for $c = 100$ and $\sigma = 0.01$, see results shown in Tab. 3. As we can notice, for normal head pose and illumination conditions the best CCR is achieved on features selected by AdaBoost and it is equal to 87%, whereas for normal head pose and poor illumination the best CCR is achieved on features selected on the basis of histogram and KL divergence, and it is equal to 85%. The best, averaged CCR is greater than 80%. A recognition rate of 80% using only noisy depth data provided by Kinect is a good starting point for further research in this area. However, a remarkable decrease of the classification performance can be observed for non-frontal head poses. On the other hand, 3D data provided by Kinect are precise enough to cope with the non-frontal face poses and therefore our future work will focus more on this issue.

Table 2. Correctly classified ratio [%] on 3D data acquired by Kinect for Windows

	NB			RF			SVM		
feature sel.	HI	AB	SP	HI	AB	SP	HI	AB	SP
# features	400	800	800	100	200	800	400	400	100
normal	86.2	80.0	85.0	83.2	82.2	83.8	83.2	84.0	85.2
dark	85.8	83.0	86.0	85.2	84.0	81.4	84.6	82.8	87.0
pose	73.4	65.8	71.8	72.8	70.8	72.2	72.4	70.2	75.0
pose dark	74.6	69.0	74.2	73.6	70.4	70.0	73.2	69.6	73.0
average	**80.0**	74.5	79.3	**78.7**	76.9	76.9	78.4	76.7	**80.1**

Figure 4 depicts the selected features, which gave the best correctly classified ratios for NB, RF, SVM with $c = 1$ and $\sigma = 0.01$, and SVM with $c = 100$ and $\sigma = 0.01$. The figures correspond to the best average results, which were achieved by each of the considered classifier. In Tab. 2 and 3 the best results of each classifier were typeset in bold. As we can see, the selection is done accordingly with our intuition. In particular, the selected pairs of the features and their corresponding lines concern the points that during facial articulations undergo significant misalignments.

The depth-based facial expressions classification was compared with classification using RGB images acquired by Kinect. Using the angles between the fiducial points, which were determined by ASM and then selected by AdaBoost, the classification accuracy was about 10% worse for the normal face pose and normal illumination conditions. A small decrease in efficiency was observed for dark pose, whereas the decrease in efficiency for non-frontal pose was about 25%. For non-frontal poses and normal illumination quite similar classification accuracy was obtained via matching of SIFT descriptors. However, in case of illumination change the classification accuracy was worse in comparison to accuracy obtained by ASM.

The complete FER system was implemented in C++/C#. The recognition performance was evaluated using WEKA software. The system runs in real-time on an ordinary PC with Intel Core i5 2.5 GHz CPU. Computing the 3D positions of semantic facial feature points as well as 3D head pose by Face

Table 3. The best correctly classified ratio [%] on 3D data

	SVM		
feature sel.	Hist.	AdaBoost	Sparse
# features	100	100	100
normal	85.8	87.0	84.2
dark	85.0	84.8	81.4
pose	74.4	74.0	70.2
pose dark	76.6	76.6	70.4
average	80.5	**80.6**	76.6

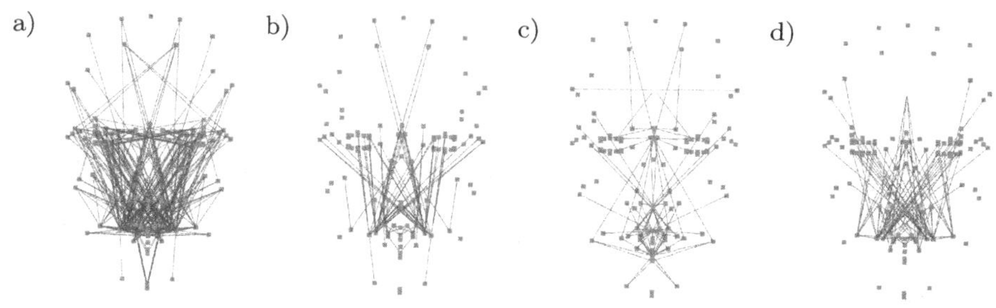

Fig. 4. The selected features, which gave the best CCR for the evaluated classifiers: a) NB, b) RF, c) SVM with $c = 1$ and $\sigma = 0.01$, d) SVM with $c = 100$ and $\sigma = 0.01$

Tracking SDK is the most computationally expensive part of our algorithm. On ordinary PC/laptop computer the processing time of single frame is 40-50 ms. The classification time is far shorter. For instance, the SVM operating on 100 features takes 0.2 ms, the RF operation on the same number of features requires 0.1 ms, whereas NB operating on 400 features takes 1.7 ms.

5 Conclusions

We have proposed an approach for facial expression recognition using only depth information provided by consumer level 3D image sensor. The best recognition accuracy is equal to 87% and it was obtained using AdaBoost-based feature selection and SVM. In particular, we demonstrated that 3D information is very useful in case of non-frontal head poses. The results suggests that using low-cost 3D sensors a promising recognition accuracy of facial expressions can be obtained even in situation of poor lighting conditions and considerable pose variations. Although depth maps provided by low cost 3D sensors like Kinect are very noisy, they still might be useful for facial expression recognition, particularly in case of non-frontal head poses.

Acknowledgments. A part of this study was supported by JSPS KAKENHI (Grant-in-Aid for Scientific Research) Grant Number 25350666, FY 2014 Researcher Exchange Program between JSPS and PAN, and by the National Science Center (NCN) within the research project N N516 483240.

References

1. Pantic, M., Rothkrantz, L.J.M.: Automatic analysis of facial expressions: The state of the art. IEEE Trans. Pattern Anal. Mach. Intell. 22(12), 1424–1445 (2000)
2. Sandbach, G., Zafeiriou, S., Pantic, M., Yin, L.: Static and dynamic 3d facial expression recognition: A comprehensive survey. Image Vision Comput. 30(10), 683–697 (2012)

3. Fang, T., Zhao, X., Ocegueda, O., Shah, S., Kakadiaris, I.A.: 3D facial expression recognition: A perspective on promises and challenges. In: FG, pp. 603–610 (2011)
4. Tsalakanidou, F., Malassiotis, S.: Real-time 2d+3d facial action and expression recognition. Pattern Recogn. 43(5), 1763–1775 (2010)
5. Li, B., Mian, A., Liu, W., Krishna, A.: Using Kinect for face recognition under varying poses, expressions, illumination and disguise. In: WACV, pp. 186–192 (2013)
6. Brooks, R.A., Breazeal, C., Marjanovic, M., Scassellati, B., Williamson, M.M.: The cog project: Building a humanoid robot. In: Nehaniv, C.L. (ed.) CMAA 1998. LNCS (LNAI), vol. 1562, pp. 52–87. Springer, Heidelberg (1999)
7. Cootes, T.F., Taylor, C.J., Cooper, D.H., Graham, J.: Active shape models - their training and application. Comput. Vis. Image Underst. 61(1), 38–59 (1995)
8. Ahlberg, J.: Candide-3 - an updated parameterised face. Technical report, Dept. of Electrical Engineering, Linkping University, Sweden (2001)
9. Kwolek, B.: Model based facial pose tracking using a particle filter. In: Int. Conf. on Geometric Modeling and Imaging: New Trends, pp. 203–208. IEEE Comp. Soc. (2006)
10. Soyel, H., Demirel, H.: Optimal feature selection for 3D facial expression recognition with geometrically localized facial features. In: Int. Conf. on Soft Computing, Comp. with Words and Perceptions in Syst. Anal., Dec. and Control, pp. 1–4 (2009)
11. Sha, T., Song, M., Bu, J., Chen, C., Tao, D.: Feature level analysis for 3D facial expression recognition. Neurocomputing 74(12-13), 2135–2141 (2011)
12. Soyel, H., Demirel, H.: Facial expression recognition using 3d facial feature distances. In: Kamel, M.S., Campilho, A. (eds.) ICIAR 2007. LNCS, vol. 4633, pp. 831–838. Springer, Heidelberg (2007)
13. Tang, H., Huang, T.: 3D facial expression recognition based on automatically selected features. In: Conf. CVPR, pp. 1–8 (2008)
14. Tang, H., Huang, T.: 3D facial expression recognition based on properties of line segments connecting facial feature points. In: Conf. FG, pp. 1–6 (2008)
15. Freund, Y., Schapire, R.E.: A decision-theoretic generalization of on-line learning and an application to boosting. In: Vitányi, P.M.B. (ed.) EuroCOLT 1995. LNCS, vol. 904, pp. 23–37. Springer, Heidelberg (1995)
16. Viola, P., Jones, M.J.: Robust real-time face detection. Int. J. Comput. Vision 57(2), 137–154 (2004)
17. Tibshirani, R.: Regression shrinkage and selection via the lasso. Journal of the Royal Statistical Society Series B 58(1), 267–288 (1996)
18. Hsu, C.W., Chang, C.C., Lin, C.J.: A practical guide to support vector classification. Technical report, Department of Computer Science, National Taiwan University, Taipei 106, Taiwan (2010)

High Accuracy Head Pose Tracking Survey

Błażej Czupryński and Adam Strupczewski

Warsaw University of Technology, Poland

Abstract. Head pose estimation is recently a more and more popular area of research. For the last three decades new approaches have constantly been developed, and steadily better accuracy was achieved. Unsurprisingly, a very broad range of methods was explored - statistical, geometrical and tracking-based to name a few. This paper presents a brief summary of the evolution of head pose estimation and a glimpse at the current state-of-the-art in this field.

Keywords: Head pose estimation, 3D tracking, Face analysis, Optical flow, Tracking, Feature matching, Model matching.

1 Introduction

Head pose estimation is becoming a more and more popular topic in recent years. This is because it has a wide variety of uses in human-computer interaction. The development of smartphones, tablets and other mobile consumer electronics presents a large area for various applications of head pose tracking algorithms. Furthermore, head pose estimation is a very important element of eye gaze tracking when no infra-red sensors are used - that is only using a simple web camera. Some approaches [1] directly link head pose tracking to eye gaze tracking as a means of providing the initial field of view. Both head pose estimation and eye gaze tracking seem to be the future of computer vision.

Computer vision researchers have been exploring the topic of head pose estimation for the last 20 years. A lot of improvement has been made over the years, but still no single method is capable of estimating the head pose both accurately and robustly in all situations. There are generally two categories of head pose estimation algorithms: coarse head pose estimation, which tend to be more robust but not necessarily accurate, and fine head pose estimation, which demonstrate high accuracy. The focus of this paper are accurate methods, with other approaches presented as background information and not covered in much detail.

2 General Classification of Approaches

Over the past two decades a lot of different techniques have been tried. A good and fairly exhaustive overview is presented in [2]. Recently, considerable advances have been made in high accuracy tracking [3,4], which are covered in more detail

D. Ślęzak et al. (Eds.): AMT 2014, LNCS 8610, pp. 407–420, 2014.

in section 6 and 7. A lot of recent focus was also on approaches using a depth camera [5], which however is not the focus of this paper.

One way to divide the existing approaches could be by the importance of statistics and pre-learned models in comparison to the importance of geometrical measurements. This could lead to five groups:

- **Solely statistical approaches.** These include template based methods that compare new images to a set of head images with known poses [6], detector based methods that have many head detectors each tuned to a different pose [7], nonlinear regression methods that map the image data to a head pose measurement [8] and manifold embedding methods that seek low dimensional manifolds to model continuous variation in head pose [9].
- **Approaches using flexible models.** These include elastic graphs [10], active shape models [11] and active appearance models [12]. All methods attempt to fit a non-rigid model to the face in the image and derive the head pose from the model's parameters.
- **Geometrical approaches.** These methods first locate certain facial features and based on their relative locations determine the head pose [13].
- **Tracking approaches.** These track the face movement between consecutive frames and from this infer the pose changes. There are many variations among these methods - the tracked elements include features [14], models [15] or dynamic templates [16].
- **Hybrid approaches.** These approaches combine two or more of the previously mentioned approaches and tend to provide best robustness and accuracy. A good example is [17], where the authors combined a template database, real-time feature tracking and the Kalman filter to achieve very high head pose estimation precision.

The above groups will be covered in more detail in the following sections. The last two groups are the main point of focus of this article, as they allow the highest accuracy.

3 Statistical Approaches

The most simple statistical approach is using appearance templates and assigning the pose of the most similar template to the query. Normalized cross-correlation [18] or mean squared error over a sliding window [19] can be used for image comparison. Appearance tamplates have a number of advantages, most importantly: simplicity, easy template set expansion and independence of image resolution. However, they only allow to estimate a discrete set of poses corresponding to the pre-annotated database. They are also not very efficient, especially with large template sets. The biggest problem of the approach is the assumption that similarity in image space corresponds to similarity in pose space. This is often not true, as identity and illumination might influence the dissimilarity more than a change in pose. To resolve this problem, more sophisticated distance metrics have been proposed. Convolutions with a Laplacian-of-Gaussian filter [20] emphasize facial contours, while convolutions with Gabor wavelets [6] emphasize directed features such as the nose (vertical) and mouth (horizontal). A recent development focused on mobile devices uses Hu moments calculated from facial pixels and their projection using the Fischer linear discriminant matrix to determine similarity with reference templates [21]. It is assumed however, that only five different poses are distinguished by the system.

A somewhat similar method to appearance templates is using arrays of detectors. One such approach using an ensemble of weak classifiers trained with Adaboost is described in [7]. Instead of matching the query to a set of templates, the query is run through a set of pre-trained detectors and the classifier with highest confidence determines the query's pose. Compared to appearance templates, detector arrays are trained to ignore the appearance variations and are sensitive to pose variations only. Furthermore, they solve the problem of face detection and localization - which remains a separate task in case of appearance templates. Big disadvantages of this scheme are the necessary scale of training, binary output of detectors and small accuracy - in practice at most 12 different detectors can be trained [2], which limits the pose estimation resolution to 12 states. Due to these constraints, detector arrays have not become very popular, and few papers propose usable systems. One usable approach is described in [7].

Another group of statistical approaches is constituted by nonlinear regression methods. These aim to find a mapping from the image space to a pose direction. If such a mapping existed given sample labeled data, a pose estimate could be found for any new data using the trained model. In practice such nonlinear regression is most often exploited using support vector regressors or neural networks. The first approach performs well if dimensionality can be reduced using PCA [22] or feature data extracted at facial feature points [23]. The second approach with neural networks can use either multi-layer perceptrons (MPL) [24], which in principle work similarly to detector arrays, or locally-linear maps (LLM), which first select a centroid for the query, and then perform linear regression for pose refinement [25]. These methods can work fast and be fairly accurate. The main disadvantages of nonlinear regression methods are the need for long, sophisticated training and the very big sensitivity to head localization. Inaccurate head localization will lead to a complete failure of these methods.

The last group of statistical approaches that should be mentioned are manifold embedding methods. The aim of these methods is to reduce the dimensionality of the input face image so that it can be placed on a low-dimensional manifold constrained only by allowable head poses. The placement then defines the head pose. Initially, PCA and LDA have been explored as dimensionality reduction techniques and pose-eigenspaces were created for projecting the input [9]. Alternatively, pose-eigenspaces could be substituted by SVMs, which have shown to perform best when combined with local Gabor binary patterns [26]. Different manifold embedding approaches perhaps perform even better. The most representative of these is probably isometric feature mapping [27]. Despite relatively good robustness, techniques using manifold embedding suffer for the same reason as appearance templates - they easily capture appearance and not only pose variations. Furthermore, good models require a lot of training and ideally a complete set of poses from all people in the database [28].

All in all, statistical approaches demonstrate good robustness in pose estimation, but most often do not provide very high accuracy. The techniques which are capable of providing more than just a few discrete pose states require complicated training and are not guaranteed to work in cases that differ greatly from the training data. Despite their multiple advantages, statistical approaches do not seem to be the best way of estimating the head pose with high accuracy.

4 Flexible Model Approaches

Statistical methods treat head pose estimation as a specific signal processing task with the 2D image as input. Flexible models aim to fit a non-rigid model to certain facial features so as to uncover the facial structure in each case. Depending on how the process of fitting is performed and what the final result is, flexible models come in several forms. Early work in this area was related to Elastic Bunch Graphs. These deformable graphs of local feature points such as eyes, nose or lips could be first matched to labeled training images to establish a reference pose set. The same graph was later matched to the query image, and based on the mutual feature locations a best fit from the training data could be found [10]. An advantage of feature-domain comparisons instead of face image comparisons is a much stronger link to pose similarity. Unfortunately, similarly as appearance templates, these approaches only allow to distinguish between a discrete set of poses and become difficult with large amounts of training data.

Potentially much higher accuracy can be achieved by using active shape models (ASMs) and active appearance models (AAMs). The first technique involves fitting a specific shape to new data, where the fitted shape is a linear combination of some pre-trained eigenspace of shapes [29]. This allows to combine greedy local fitting with the constraint of a model. Adding appearance to the shape and combining the two proved to be a better solution when fitting these models to faces [30]. Once such a model, being originally a set of 2D points, is fitted to the face image, it is possible to infer the pose based on the relative locations of these points. One possibility is to use simple linear regression on these points [31]. Another possibility is to use a combined 2D and 3D Active Appearance Model, where an inherent 3D model is used to constrain the fitting of 2D points [12]. This allows to infer the head pose directly. Finally, it is possible to use structure from motion algorithms to infer the 3D locations of model points and directly calculate the pose [32]. In fact, this technique could even be used without flexible models by matching feature points detected on the face and using them for reconstruction. With increased processing power becoming available in recent years, a real-time implementation of this approach might be a promising line of research.

A more recent development is based on a combination of tree models and a shared pool of parts [33]. A set of facial landmarks is localized in the query image, thus simultaneously providing head detection, landmark localization and pose estimation. The authors argue that their approach is capable of fitting the elastic models much better then AAMs and CLMs (Constrained Local Models). As the proposed system saturates previous benchmarks, a new face database was created for the purpose of evaluation - Annotated Faces in the Wild (AFW). The reported results are very impressive, but they all refer to coarse pose estimation - an error is assumed only when the detected pose deviates by more than 15 from the annotation.

To sum up, flexible models have a big and still largely unexplored potential to provide both good robustness and accuracy of head pose estimation. While their precision is limited by the feature fitting accuracy and so depends on the input image resolution and quality, they may be a good starting point for use with refining algorithms.

5 Geometric Approaches

Geometric approaches derive the head pose from the geometric configuration of facial features depending on how the face is positioned in relation to the silhouette of the head or how big the nose deviation is from bilateral symmetry. One group of approaches uses the outer corners of the eyes, the outer corners of the mouth and the nose position [34]. The center points between the eye corners and between the lip corners project a line. The distance of the nose from this line, as well as the angle, gives information about the pose. This is illustrated in Figure 1. A recent paper describes the implementation of this algorithm on contemporary hardware using a cascade of detectors for feature localization [35].

Another set of methods uses also the inner eye corners for measurements [36]. One of the most recent methods [13] uses the parallelism of lines between inner eye corners, outer eye corners and lip corners. A so called *vanishing point* can be calculated as the intersection of the eye line and the lip line. The head pose can

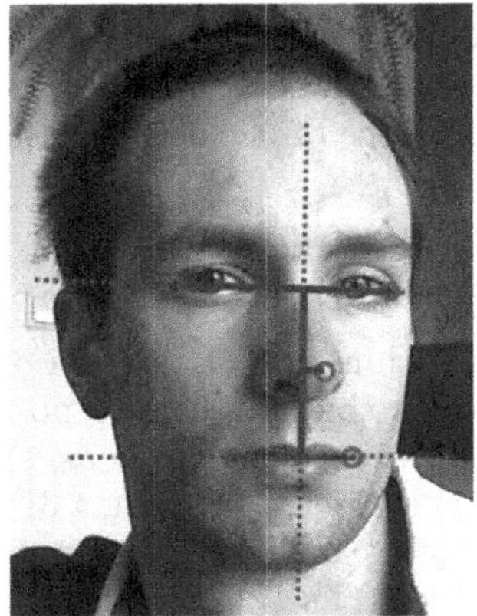

Fig. 1. Geometric approaches use relative feature positions

be estimated based on its location. Furthermore, the ratios of the lines can be used to infer the pitch of the head. Moreover, an EM algorithm with a Gaussian mixture model was proposed to account for the variation among different people.

Geometric approaches usually require few calculations and are simple. They are also capable of providing a highly accurate estimation, but their accuracy strongly depends on the accuracy of facial feature location estimation. At present, the most accurate detectors have an error of at least 1-2 pixels even with high quality images. This is a lower bound on the potential accuracy of geometric methods. In our experience the 2-pixel feature location error can cause a considerably large, 5-10 degree head pose angle estimation error for typical webcam face images. It means that when using a simple webcam, geometric methods are unable to provide an accurate head pose estimation because of limited feature detection accuracy. A further limitation of these methods is the ability to estimate the head pose only when all the facial features are visible and detectable.

6 Tracking Approaches

Tracking methods can be used to determine the head pose by accumulating estimations of the relative head movement between subsequent video frames. This approach assumes observing smooth motion of the head so it is applicable only when a continuous video stream containing the head motion is available. Tracking-based head pose estimation methods are capable of providing very accurate results, as they can calculate very small pose changes between successive frames and thus outperform other approaches [2]. The main disadvantage of these methods is the need of tracking initialization during algorithm start-up and after tracking gets lost. This means that it is necessary to use some different head pose estimation algorithm in the initialization step. To simplify this step, the tracking algorithm can assume that the head pose is frontal at initialization - which can be provided by a frontal face detection algorithm. Another drawback of tracking approaches is that they are very accurate only in the short-term. They are also sensitive to occlusions and illumination changes. Therefore, many extensions of tracking algorithms have been proposed to improve long-term stability and robustness, mainly in the presence of occlusions or large out-of-plane rotations [16,15,3].

6.1 Model-Based 3D Tracking

In ordinary 2D tracking algorithms the 3D motion of the object is modeled as a 2D transformation. To recover the full 3D head pose, which has six degrees of freedom (translations and rotations along three axes), a 3D tracking algorithm is required [37]. This can be achieved by utilizing a 3D shape model of an object. The translation and rotation of the model is estimated by analyzing the projected 2D images of the modeled object. The 2D translations can be unprojected from the image into the 3D scene to account for 3D motion. Several approaches have been proposed to solve this problem. [38,16,37,3,17].

The model used in tracking can be rigid or non-rigid (deformable). This paper focuses only on the rigid model approach, as non-rigid models are more useful for emotion recognition than high-accuracy head pose tracking. In general, the shape of the human head can be modeled either using a precise 3D model or approximated by a geometric primitive. The selection of the model type directly affects the accuracy and robustness of the tracker, but is something separate from the tracking algorithm itself. Approaches using both generic and user-specific precise head models have been proposed [39,38]. The main advantage of a precise face model is the capability to provide a very accurate pose estimation. However, this is only possible when combined with precise initialization to closely fit the model to the observed face. When the model is not well aligned to the face, tracking errors rise significantly. Moreover, the alignment usually degrades during tracking due to tracking error accumulation [17].

Algorithms based on approximating the head shape with geometrical primitives are more robust. The most simple one is a planar surface, but the recovery of out-of plane rotation in this case is inaccurate due to lack of depth in the model. From all geometric primitives, the best choice seems to be using a cylinder [16,15,3]. It possesses depth, is very robust to initialization errors and can work well despite appearance changes between different people.

6.2 3D Tracking under Perspective Projection

As mentioned in section 6.1, introducing a known 3D model into 2D tracking allows to recover the full 6-degree of freedom motion of the tracked object. A straightforward method of solving this problem was proposed in [3], based on [16]. Having a set of

- 2D image points at time $t-1$ along with their 3D coordinates in the real world given by a model,
- corresponding 2D image points at time t

it is possible to recover model pose at time t. If the model's pose at time $t-1$ is known, the updated pose at time t can be represented as the previous pose transformed by a motion vector $\mu = [t_x, t_y, t_z, \omega_x, \omega_y, \omega_z]$, which represents translation and rotation using twist representation. To describe correspondences between coordinates of the object in 3D space and their projections on the imaging plane of the camera, a simplified perspective projection camera model given only by the focal length f can be assumed. Considering a single tracked point, the locations p_{t-1}, p_t of this point are observed in the image at times $t-1$, t respectively. The 3D position P_{t-1} of this point at time $t-1$ is obtained by unprojecting p_{t-1} into 3D world coordinates using depth from the model, which has known orientation. The new 3D position estimated during tracking can be expressed as $P_t = M \cdot P_{t-1}$, where M is a transformation matrix based on vector μ:

$$M = \begin{bmatrix} 1 & -\omega_z & \omega_y & t_x \\ \omega_z & 1 & -\omega_x & t_y \\ -\omega_y & \omega_x & 1 & t_z \\ 0 & 0 & 0 & 1 \end{bmatrix} \tag{1}$$

This situation is depicted in Figure 2. Given the transformation matrix M, the projection of P_t can be expressed using the previous position P_{t-1} and motion parametrized by vector μ:

$$p'_t = \begin{bmatrix} x_t \\ y_t \end{bmatrix} \frac{f}{z_t} = \begin{bmatrix} x_{t-1} - y_{t-1}\omega_z + z_{t-1}\omega_y + t_x \\ x_{t-1}\omega_y + y_{t-1}\omega_x - z_{t-1}\omega_x + t_y \end{bmatrix} \frac{f}{-x_{t-1}\omega_y + y_{t-1}\omega_x + z_{t-1} + t_z} \tag{2}$$

Now, two forms of the projection of point P_t are available. The first one is the observed location p_t and the second is p'_t, which was obtained by estimated motion of the previous location p_{t-1} based on 3D model and motion vector μ. Assuming N such point pairs have been collected, the goal is to compute motion vector μ, which minimizes the sum of distances between the observed points $p_{i,t}$ and the estimated points $p'_{i,t}$.

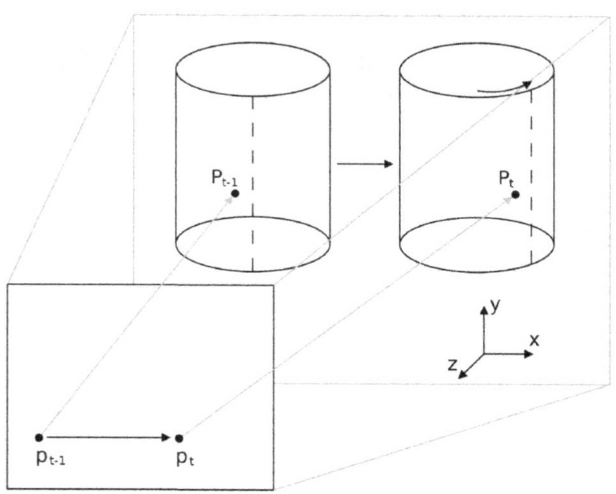

Fig. 2. Cylinder tracking under perspective projection

6.3 Finding Correspondences

For the presented algorithm to work, a set of corresponding point pairs from two images is necessary. One way to get it is by using a local feature point extraction algorithm. In [3] the authors use the SIFT algorithm, proposed in [40]. Feature points are extracted independently for each frame and then matched by finding pairs with the most similar descriptors.

Although independent feature extraction and matching for each frame is possible, it not not the fastest method. A more efficient method is tracking feature points in consecutive frames. A technique of 3D pose estimation by applying optical flow to track 2D feature points and assuming a cylindrical head model was proposed in [16]. It uses a mesh of evenly distributed points as keypoints and works directly on image luminance. The algorithm operates on the same motion vector μ as introduced in section 6.2. Let the intensity of the image at point p and time t be denoted as $I(p,t)$. Let $F(p,\mu)$ be a function which maps point p into a new location p' using vector μ, according to the motion model described in section 6.2. This mapping function was already introduced in equation 2. The region containing all considered face pixels is denoted as Ω. Computing the motion vector between two frames based on luminance can be expressed as the minimization of the sum of luminance differences between the face image from previous frames and the current face image transformed by mapping function F:

$$\min\left(\sum_{p\in\Omega}\left(I\left(F\left(p,\mu\right),t\right)-I\left(p,t-1\right)\right)^2\right) \tag{3}$$

In [16] the motion vector μ is computed using the Lucas-Kanade method:

$$\mu=\left(\sum_{\Omega}w\left(I_pF_\mu\right)^T\left(I_pF_\mu\right)\right)^{-1}\sum_{\Omega}w\left(I_t\left(I_pF_\mu\right)^T\right) \tag{4}$$

where I_t and I_p are temporal and spatial image gradients, while w is a weight assigned to each point. F_μ denotes the partial differential of F with respect to μ at $\mu=0$:

$$F_\mu=\begin{bmatrix}-xy & x^2+z^2 & -yz & z & 0 & -x\\ -\left(y^2+z^2\right) & xy & xz & 0 & z & -z\end{bmatrix}\frac{f}{z^2} \tag{5}$$

Motion is computed iteratively. After each iteration, the model is transformed using the computed motion vector μ and the weights for all points are updated.

A hybrid combination of tracking a mesh of points with feature point matching is also possible. The approach of [41] can be used for adjusting mesh points to good corner points within so called *cells*. Optical flow can be used to track these points and feature matching is not necessary, while at the same time the tracked points are not random. Although promising at first, assessing the impact of this modification on tracking accuracy is difficult, as there is no linear relationship between mesh tracking accuracy and the distinctiveness of mesh points.

6.4 Long-Term Stability

Tracking in a frame-to-frame fashion provides accurate pose estimation in the short-term, but it tends to lose accuracy over time due to drift error accumulation. To obtain the head pose in the actual frame, motion vectors have to be accumulated starting from the beginning of tracking. Motion estimation errors are accumulated together with measurement errors. Thus, the drift error increases in time. Moreover, the frame-to-frame tracking is not able to recover after large errors, which occur in presence of occlusions or when the head is temporarily out of the camera's view.

In order to achieve long-term robustness and stability, the drift error has to be compensated and the algorithm has to be able to recover after tracking loss. This is usually done by introducing an alternative tracking algorithm capable of performing reinitialization. This auxiliary method periodically performs tracking to stored reference frames (templates) instead of the previous frame. Such compensation of errors greatly increases the pose estimation robustness and general performance in real-world videos. In [16] the initial frame with a specified pose is stored as a reference keyframe, and the tracking to this keyframe is called re-registration. Tracking to reference frames should be performed from time to time, in two cases:

- when normal frame-to-frame tracking is working for a long time and drift error has accumulated,
- when there is a large error in frame-to-frame tracking caused by abrupt movement, occlusion etc. and traditional tracking has failed

7 Hybrid Approaches

Tracking to a reference frame can be performed simultaneously with frame-to-frame tracking giving a head pose estimation algorithm which is robust and accurate at the same time. This approach was presented in [3], where the Kalman filter is applied to combine motion estimation of the frame-to-frame tracker and the reference frame tracker. The authors proposed a robust tracking system, where a head feature point database stores a set of reference templates. Each template is a set of SIFT points captured at a known head pose. The templates are extracted at certain head poses during the initial stage of tracking, where a frame-to-frame tracker is used to determine the head pose. The initial template is stored for a frontal pose. Additionally, it contains generic facial feature points from a Bayesian Tangent Shape Model [42], which is used for the initial alignment of the 3D head model. Once the template database is filled, successive input frames are tracked in parallel using previous frames and template frames from the database. SIFT points from the input frame are matched with all templates independently, and the template with the largest number of matches is used to track the model. Separate motion vectors are estimated using the previous frame. In both cases tracking is done using an algorithm described in section 6.2. Finally, the two motion vectors are combined using the Kalman filter.

Depending on the tracking errors, bigger confidence is assigned either to the continuous tracking method or to the template tracking method, giving the final resulting pose vector estimated by the filter. The biggest gain in using the Kalman filter is that the estimation is smooth regardless of abrupt temporary errors. If there is an occlusion and the frame-to-frame tracker suddenly gets lost, the Kalman filter will use the estimate from the template tracker and such disturbances will be barely noticed.

A similar method is described in [17]. Here however, the authors propose to simultaneously use feature-based tracking and intensity-based tracking. The aim is to minimize the error from both methods. This is presented as a nonlinear optimization problem, with different weights assigned to the intensity tracking and feature tracking based on time and tracking conditions. The reported results are impressive and demonstrate better accuracy and robustness than each method alone, especially in demanding test sequences.

A more recent publication [43] attempts to use the Kalman filter to predict the head motion from previous motion, in order to support texture-based optical flow tracking. The authors claim that in comparison to pure texture-based tracking, a significant speed improvement can be achieved.

One of the most sophisticated hybrid tracking approaches is described in [4]. The system combines three types of tracking using linear regression:

- static person-independent pose estimation
- static user-dependent pose estimation based on reference templates
- differential motion-based estimation

As shown in section 8, the proposed hybrid approach is very competitive and achieves accuracy of head pose angle estimation close to the state-of-the-art. This is despite having a completely different architecture than most contemporary head pose estimation research.

8 Conclusion

To conclude it has to be noted that hybrid approaches provide the best accuracy and robustness. Table 1 shows a brief comparison of various tracking methods in terms of precision. For a reliable comparison, the head pose estimation methods were compared on the same dataset - Boston University dataset with uniform lighting. Most results are averaged for all 45 uniform lighting test sequences of the dataset, but some only for a subset of it - as mentioned in the remarks.

If only one method was to be chosen, then tracking based approaches are favored. The numerical comparison in Table 1 clearly shows their superiority [16,17]. The recent improvements of feature point estimation have allowed methods using them [35] to reach precision close to that of tracking methods.

It must be remembered though, that real-world situations and use cases place big importance on tracking robustness - something which is not necessarily shown with numbers. At the moment it seems that if computational resources are available, the most accurate and robust head pose estimation methods are hybrid tracking methods combining several tracking-based methods with filters [4]. For maximum robustness, combining tracking methods with a static head pose estimation method has emerged as a popular trend in research [16,3,17].

As for future improvement of tracking accuracy, a fusion of 3D model tracking and feature point based pose estimation seems to be a promising line of research.

Table 1. Head pose estimation accuracy on Boston University Dataset - as reported by authors

Method	Roll [deg]	Pitch [deg]	Yaw [deg]	Remarks
Robust Full-Motion Recovery of Head by Dynamic Templates and Re-registration Techniques [16]	1,4	3,2	3,8	Boston Uni (45 uniform) + Optotrack
Robust, Real-Time 3D Face Tracking from a Monocular View [17]	1,4	3,5	4,0	Boston Uni - uniform example, error values estimated from chart
Fasthpe: A recipe for quick head pose estimation [35]	3,03	5,27	3,91	Boston Uni (18 sequences)
Robust 3D Head Tracking by View-based Feature Point Registration [3]	3,44	4,22	2,07	Boston Uni (45 uniform)
Generalized Adaptive View-based Appearance Model: Integrated Framework for Monocular Head Pose Estimation (GAVAM) [4]	3,67	4,97	2,91	Boston Uni (45 uniform)
Fast, reliable head tracking under varying illumination: An approach based on registration of texture-mapped 3d models [15]	9,8	6,1	3,3	Boston Uni (45 uniform)

As shown in [3] matching features to templates can work very well. Considering the recent developments in real-time 3D reconstruction [44,45], it seems feasible to create and refine an online 3D head model and use it for continuous, highly accurate head pose tracking. If appropriately robust feature points can be chosen or developed for facial images, this approach could yield the most accurate head pose estimation yet.

References

1. Valenti, R., Sebe, N., Gevers, T.: Combining head pose and eye location information for gaze estimation. IEEE Transactions on Image Processing (2012)
2. Murphy-Chutorian, E., Trivedi, M.: Head pose estimation in computer vision: A survey. IEEE Transactions on Pattern Analysis and Machine Intelligence (2009)
3. Jang, J., Kanade, T.: Robust 3d head tracking by online feature registration. In: The IEEE International Conference on Automatic Face and Gesture Recognition (2008)
4. Morency, L., Whitehill, J., Movellan, J.: Generalized adaptive view-based appearance model: Integrated framework for monocular head pose estimation. In: 8th IEEE International Conference on Automatic Face Gesture Recognition (2008)
5. Fanelli, G., Gall, J., Van Gool, L.: Real time head pose estimation with random regression forests. In: 2011 IEEE Conference on Computer Vision and Pattern Recognition (2011)

6. Sherrah, J., Gong, S., Ong, E.J.: Face distributions in similarity space under varying head pose. Image and Vision Computing 19 (2001)
7. Viola, M., Jones, M., Viola, P.: Fast multi-view face detection. In: Proc. of Computer Vision and Pattern Recognition (2003)
8. Gourier, N., Hall, D., Crowley, J.: Estimating face orientation from robust detection of salient facial structures. In: FG Net Workshop on Visual Observation of Deictic Gestures (2004)
9. Srinivasan, S., Boyer, K.: Head pose estimation using view based eigenspaces. In: Proceedings of 16th International Conference on Pattern Recognition (2002)
10. Kruger, N., Potzsch, M., Malsburg, C.: Determination of face position and pose with a learned representation based on labelled graphs. Image and Vision Computing 15 (1997)
11. Lanitis, A., Taylor, C., Cootes, T., Ahmed, T.: Automatic interpretation of human faces and hand gestures using flexible models. In: International Workshop on Automatic Face- and Gesture-Recognition (1995)
12. Xiao, J., Baker, S., Matthews, I., Kanade, T.: Real-time combined 2d+3d active appearance models. In: Proceedings of the 2004 IEEE Computer Society Conference on Computer Vision and Pattern Recognition (2004)
13. Wang, J., Sung, E.: EM enhancement of 3d head pose estimated by point at infinity. Image and Vision Computing 25 (2007)
14. Yao, P., Evans, G., Calway, A.: Using affine correspondence to estimate 3-d facial pose. In: Proceedings of International Conference on Image Processing (2001)
15. La Cascia, M., Sclaroff, S., Athitsos, V.: Fast, reliable head tracking under varying illumination: an approach based on registration of texture-mapped 3d models. IEEE Transactions on Pattern Analysis and Machine Intelligence 22 (2000)
16. Xiao, J., Kanade, T., Cohn, J.: Robust full-motion recovery of head by dynamic templates and re-registration techniques. In: Proceedings of the Fifth IEEE International Conference on Automatic Face and Gesture Recognition (2002)
17. Liao, W., Fidaleo, D., Medioni, G.: Robust, real-time 3d face tracking from a monocular view. EURASIP Journal on Image and Video Processing (2010)
18. Beymer, D.: Face recognition under varying pose. In: Proceedings of IEEE Computer Society Conference on Computer Vision and Pattern Recognition (1994)
19. Niyogi, S., Freeman, W.: Example-based head tracking. In: Proceedings of the Second International Conference on Automatic Face and Gesture Recognition (1996)
20. Gonzalez, R.C., Woods, R.E.: Digital Image Processing (2001)
21. Ren, J., Rahman, M., Kehtarnavaz, N., Estevez, L.: Real-time head pose estimation on mobile platforms. Journal of Systemics, Cybernetics and Informatics 8 (2010)
22. Li, Y., Gong, S., Sherrah, J., Liddell: Support vector machine based multi-view face detection and recognition. Image and Vision Computing 22 (2004)
23. Ma, Y., Konishi, Y., Kinoshita, K., Lao, S., Kawade, M.: Sparse bayesian regression for head pose estimation. In: 18th International Conference on Pattern Recognition (2006)
24. Zhao, L., Pingali, G., Carlbom, I.: Real-time head orientation estimation using neural networks. In: Proceedings of International Conference on Image Processing (2002)
25. Zhang, M., Li, K., Liu, Y.: Head pose estimation from low-resolution image with hough forest. In: 2010 Chinese Conference on Pattern Recognition (2010)
26. Ma, B., Zhang, W., Shan, S., Chen, X., Gao, W.: Robust head pose estimation using lgbp. In: 18th International Conference on Pattern Recognition (2006)

27. Raytchev, B., Yoda, I., Sakaue, K.: Head pose estimation by nonlinear manifold learning. In: Proceedings of the 17th International Conference on Pattern Recognition (2004)
28. Yan, S., Zhang, Z., Fu, Y., Hu, Y., Tu, J., Huang, T.: Learning a person-independent representation for precise 3D pose estimation. In: Stiefelhagen, R., Bowers, R., Fiscus, J.G. (eds.) RT 2007 and CLEAR 2007. LNCS, vol. 4625, pp. 297–306. Springer, Heidelberg (2008)
29. Cootes, T., Taylor, C., Cooper, D., Graham, J.: Active shape models-their training and application. Computer Vision and Image Understanding 61 (1995)
30. Matthews, I., Baker, S.: Active appearance models revisited. Int. J. Comput. Vision 60 (2004)
31. Cootes, T., Walker, K., Taylor, C.: View-based active appearance models. In: Proceedings of Fourth IEEE International Conference on Automatic Face and Gesture Recognition (2000)
32. Gui, Z., Zhang, C.: 3d head pose estimation using non-rigid structure-from-motion and point correspondence. In: IEEE Region 10 Conference on TENCON (2006)
33. Zhu, X., Ramanan, D.: Face detection, pose estimation, and landmark localization in the wild. In: 2012 IEEE Conference on Computer Vision and Pattern Recognition (2012)
34. Gee, A., Cipolla, R.: Determining the gaze of faces in images. Image and Vision Computing 12 (1994)
35. Sapienza, M., Camilleri, K.: Fasthpe: A recipe for quick head pose estimation. In: Technical Report (2011)
36. Horprasert, T., Yacoob, Y., Davis, L.: Computing 3-d head orientation from a monocular image sequence. In: Proceedings of the 2nd International Conference on Automatic Face and Gesture Recognition (1996)
37. Lepetit, V., Fua, P.: Monocular model-based 3d tracking of rigid objects. Found. Trends. Comput. Graph. Vis (2005)
38. Malciu, M., Preteux, F.: A robust model-based approach for 3d head tracking in video sequences. In: Proceedings of 4th IEEE International Conference on Automatic Face and Gesture Recognition (2000)
39. Lu, L., Zhang, Z., Shum, H., Liu, Z., Chen, H.: Model- and exemplar-based robust head pose tracking under occlusion and varying expression. In: 2001 IEEE Conference on Computer Vision and Pattern Recognition (2001)
40. Lowe, D.: Distinctive image features from scale-invariant keypoints. Int. J. Comput. Vision 60 (2004)
41. Matas, J., Vojir, T.: Robustifying the flock of trackers. In: 16th Computer Vision Winter Workshop, Mitterberg, Austria (2011)
42. Zhou, Y., Gu, L., Zhang, H.: Bayesian tangent shape model: Estimating shape and pose parameters via bayesian inference. In: Proceedings of the 2003 IEEE Computer Society Conference on Computer Vision and Pattern Recognition (2003)
43. Wang, Y., Gang, L.: Head pose estimation based on head tracking and the kalman filter. Physics Procedia (2011), 2011 International Conference on Physics Science and Technology
44. Stühmer, J., Gumhold, S., Cremers, D.: Real-time dense geometry from a handheld camera. In: Goesele, M., Roth, S., Kuijper, A., Schiele, B., Schindler, K. (eds.) DAGM 2010. LNCS, vol. 6376, pp. 11–20. Springer, Heidelberg (2010)
45. Wu, C.: Towards linear-time incremental structure from motion. In: 2013 International Conference on 3D Vision, pp. 127–134 (2013)

Facial Expression Data Constructed with Kinect and Their Clustering Stability

Angdy Erna[1], Linli Yu[2], Kaikai Zhao[2], Wei Chen[2], and Einoshin Suzuki[1]

[1] Department of Informatics, ISEE, Kyushu University, Japan
[2] Graduate School of Systems Life Sciences, Kyushu University, Japan

Abstract. In this paper, we construct facial expression benchmark data of 100 persons using Kinect face tracking application and study the stability of the benchmark data in terms of clustering. Kinect with its Software Development Kit applications has enabled low-cost constructions of various benchmark data on humans. We devised multi-lingual instruction sheets on 25 expressions, collected data from 115 persons, and carefully inspected and labeled the outcome to construct the data. The benchmark data consist of 263,106 instances, each of which includes 6 animation units, 11 shape units, and an image file all provided by the application. Out of the 263,106 instances, we labeled 62,500 of them as 1 of the 25 expressions and investigated their clustering stabilities to the 17 features. We show that the most frequently used clustering algorithm: k-means achieves the average normal mutual information about 0.92 as an evidence of the stability of our facial expression data.

Keywords: Kinect, Face Tracking Application, *k*-means Algorithm, Instruction Sheets.

1 Introduction

One of the important ways humans display emotions is through facial expressions [1]. With the proliferation of human-computer intelligent interaction systems including those in the ambient intelligence domain [2], the problem of automatically recognizing human emotions through his/her facial expressions has been attracting much attention of the research community [3, 4, 1, 5–8]. The degree of significance of such recognitions as intelligent media technology keeps on increasing, which necessitates fundamental research on the methods as well as benchmark data.

Kinect is a low-cost, calibration-free, and powerful sensing input device initially developed for video game console Microsoft Xbox 360. Together with its applications in the Software Development Kit (SDK), Kinect has enabled constructions of powerful human monitoring systems at low cost. Benchmark data built under the same environment naturally strengthen such constructions substantially. We had studied past benchmark data on facial expressions [9, 4, 1], all of which were built without Kinect, and decided to construct multi-national data with richer expressions with Kinect.

D. Ślęzak et al. (Eds.): AMT 2014, LNCS 8610, pp. 421–431, 2014.

The rest of this paper is organized as follows. Section 2 explains the construction of our benchmark data on facial expression. Section 3 studies its clustering stability with the k-means algorithm [10]. Section 4 concludes.

2 Building Facial Expression Benchmark Data

2.1 Design Principles

From past benchmark data [9, 4, 1], 100 persons seemed to be the minimum requirement. It is well-known that young students are usually more open to be part of the benchmark data than older faculty members [1]. One of the strengths of the authors is that we mostly consist of Indonesian and Chinese students studying in Japan. We set our goal on collecting data from at least 100 persons, mainly targeting at university students from diverse countries.

There has been a debate between deliberate and authentic facial expressions in collecting facial expression benchmark data. Deliberate facial action tasks typically differ in appearance and timing from the authentic facial expressions induced through events in the normal environment of the person [1]. The latter expressions are more appropriate as base data in developing a human monitoring system. However, collecting authentic facial expressions leads to highly skewed distributions of expressions in the resulting data, e.g., genuine sadness or fear are difficult to collect [1]. As a first step to study the feasibility of building a monitoring system with Kinect, which is a relatively new sensing input device, we decided to collect deliberate facial expressions.

Most of the past facial expression benchmark data adopted action units of the facial action coding system (FACS) [11] and its revision. Kinect face tracking application [12], developed for displaying an animated face, has kept only 6 of the 64 action units [11] and call them animation units. In addition to the 6 animation units (AUs), it also measures 11 shape units (SUs), which will be explained in the Appendix. This reduction is a significant loss of information compared to the suits of action units. We have decided to keep these 17 features as they are in order to investigate their feasibility for recognizing facial expressions.

To make the investigation detailed, we decided to collect a larger number of expressions than past benchmark data[1]. Instead of starting from basic emotions such as neutral, joy, surprise, disgust [1], we adopted 25 facial expressions posed as a challenge to artists illustrating facial expressions [13]. This increase of the number of expressions can possibly degrade the quality of the benchmark data, since the at least 100 subjects vary. To reduce the degree of varieties of facial expressions among the individuals for the fundamental research of the new sensing input device, we devised instruction sheets each of which includes comics of a male and a female expressing 1 of the 25 facial expressions using illustrations from [13, 14], which will be explained in the next section.

[1] CMU-Pittsburg Data [9] have 23 expressions.

Fig. 1. Two examples of the instruction sheets

2.2 Collecting Data from 100 Persons

We collected Facial Expression Data from 115 persons from November 2013 to January 2014. As we had expected, finding collaborators who act facial expressions was the highest hurdle. The persons were mostly university students affiliated with Kyushu University, Kyushu Institute of Technology, Waseda University (Kitakyushu Campus), Kyushu Sangyo University, and Nagoya University. The first four universities are located in Fukuoka prefecture whereas the last one is located in a remote prefecture about 760km distant by train. 8 Indonesian visitors to Kyushu University also participated in the data collection.

In the data collection process, the participant sat in front of a Kinect. The Kinect was placed at height about 1.2m from the ground and about 1.1m distant from the subject. An experimenter sat near the participant and monitored the process of data collection. For every second, Kinect took about 20-30 images. The first four authors of this paper served as an experimenter.

As we explained in the previous section, each person expressed 25 kinds of facial expressions taken from the 25 Essential Expressions Challenge [13]. The 25 expressions are shown in Table 1, in which each expression is accompanied with an example image from a different person. Note the diversity of the persons included in our facial expression data.

Each of the instruction sheet, as explained in the previous section, contained comics of a male and a female taken from Karmajello and DevianArt artists [13, 14]. Each sheet also contains the corresponding facial expression in four languages: English, Japanese, Chinese, and Indonesian. Figure 1 shows two examples of the instruction sheets. We asked each person to express the instructed facial expression within 10 seconds and give a sign when they change to another expression.

2.3 Labeling Data

The data collection on 115 persons were not always successful: several persons did not follow a part of the instructions. As the result, we had to discard 15 persons from our data. The remaining 100 persons are dominated by Chinese (65%),

Table 1. 25 expression samples from our facial expression data

1. Happy	2. Sad	3. Pleased	4. Angry	5. Confused
				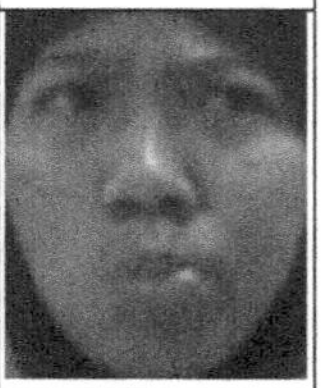
6. Tired	7. Shocked/ Surprised	8. Irritated	9. WTF?!	10. Triumph
	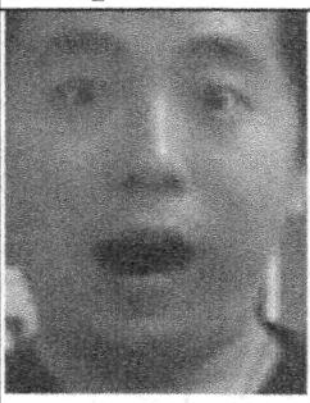		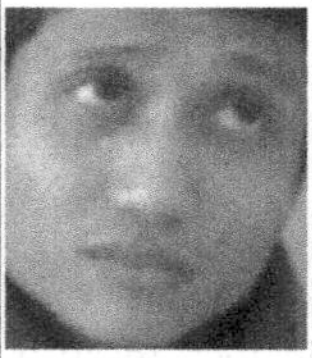	
11. Fear	12. Bereft	13. Flirty	14. Serious	15. Silly
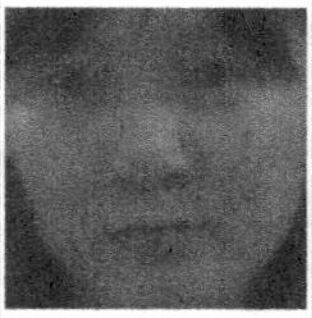				
16. Hollow	17. Incredulous	18. Confident	19. Fierce	20. Despondent/ Pouty

21. Drunk	22. Rage	23. Sarcastic	24. Disgusted	25. Ill/ Nauseous
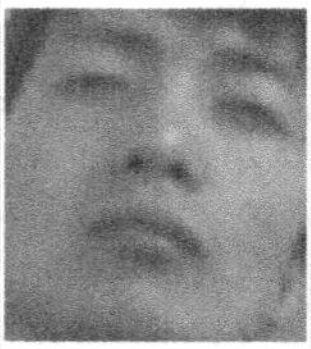	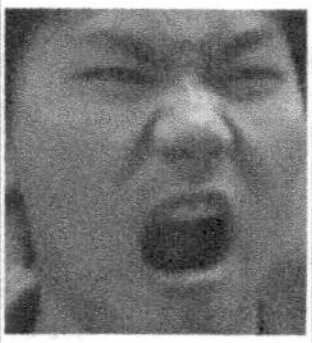			

Table 2. Example of an instance in our facial expression data

AUs		*SUs*		*OTHERS*			
AU0 ...	*AU5*	*SU0* ...	*SU10*	*ExID*	*PID*	*GEN*	*NAT*
-0.223450 ...	-0.012707	-0.410510 ...	-0.258990	1	95	0	1
0.527321 ...	-0.142106	-0.211530 ...	0.197830	4	2	0	3
-0.068765 ...	0.112506	0.171442 ...	0.123350	22	4	1	2

Japanese (18%), and Indonesians (12%). The other 5% are 2 Swedish, 1 Malaysian, 1 Singaporean, and 1 Vietnamese.

More importantly, the use of a sign for separating different expressions was a failure. The signs were simply missing in the data or not informative enough due to ambiguous facial expressions. We manually picked up 25 most expressive images for each expression of a participant and labeled the images. As the result, we have labeled 625 images for each person. A part of the remaining 200,606 images were also labeled, though their facial expressions are not as expressive as those in the 625 images.

Each image is associated with 21 features: 6 AUs, 11 SUs, expression ID, person ID, gender, and nationality. Table 2 shows 3 examples of the values of the 21 features.

3 Experiments for Investigating Clustering Stability of the Data

3.1 Experiments with 25 Clusters

One might suspect that our facial expression data are not as useful as they look due to the limitation to the 17 features: 25 facial expressions might not be distinguishable with the 17 features. Other natural questions that arise include difference of facial expressions between genders and among nationalities. We give answers to this suspicion and these questions by using clustering [15].

Given a set of instances, divisive clustering, which belongs to unsupervised learning, returns a division of the set, in which each subset is called a cluster. It is expected that each cluster contains similar instances and instances from different clusters are dissimilar. The k-means algorithm [10] is the most commonly used divisive clustering algorithm and can handle instances each of which is represented as a point in a Euclidean space. The algorithm is adequate for a problem which contains spherical clusters in the Euclidean space and each of the clusters consists of a similar number of instances.

If the k-means algorithm can separate the 25 facial expressions in our data, the fact proves that the expressions are at least distinguishable with the features and hence suggests that these features are appropriate for expressing the 25 facial expressions. We intuitively define that a data set possesses a high clustering

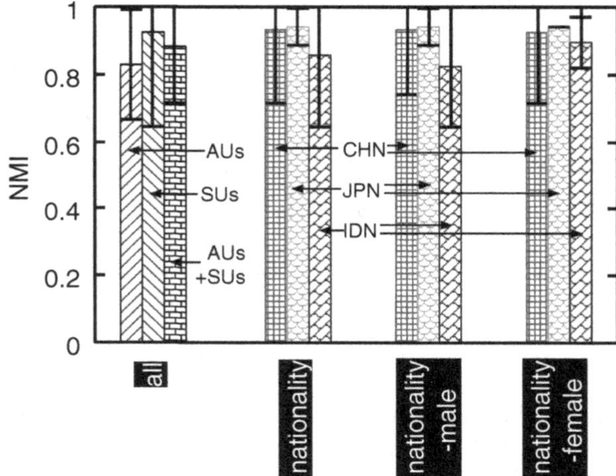

Fig. 2. NMI results

stability if a basic clustering algorithm such as the k-means achieves accurate clustering. Normalized Mutual Information (NMI), which is defined in terms of the cluster assignments and pre-existing labeling of the dataset, is the de facto standard evaluation measure for the degree of accuracy of the clustering result. In this paper, we adopt NMI to measure the degree of the clustering stability of a data[2].

We first applied the k-means algorithm to our 62,500 labeled instances with $k = 25$, i.e., the number of output clusters is 25. Three kinds of features were tested: 6 AUs, 11 SUs, and 6 AUs + 11 SUs. We followed the common protocol so the labels, i.e., the facial expressions, were not given to the k-means algorithm and were used only in calculating NMI. As a hill climbing algorithm, the k-means algorithm needs criteria for judging the convergence. We set the criteria as no move in the $k = 25$ clusters in a subsequent iteration or 100 iterations conducted. The leftmost 3 boxes of Figure 2 show the results of the experiments, where the upper and lower ends of a bar represent the maximum and minimum values, respectively. We see that using 11 SUs yields the highest average NMI (0.92) compared with the other cases (6 AUs: 0.83, 6 AUs+ 11 SUs: 0.88). We decided to use the 11 SUs as features for further experiments.

Our next target is analyses on different nationalities. We adopted Chinese, Japanese, and Indonesians and ignored other groups due to the lack of credibility arising from their small numbers of instances. The fourth, fifth, and sixth boxes from the left in Figure 2 show the results. From the Figure we see that Chinese and Japanese show NMI slightly higher than the overall NMI (0.92) while Indonesians not. Their smaller number of instances (12) compared with the other two groups (41 and 18) is possibly a part of the cause.

[2] Another kind of analysis would be to use supervised learning methods such as SVM and neural networks, which is out of the scope of this paper.

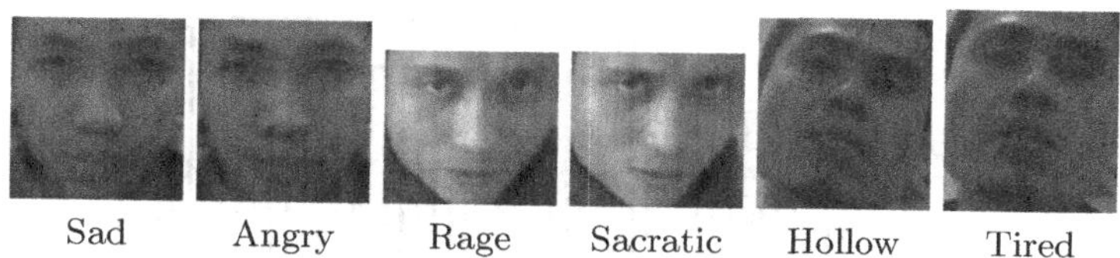

Fig. 3. Examples of ambiguous expressions among the Indonesians

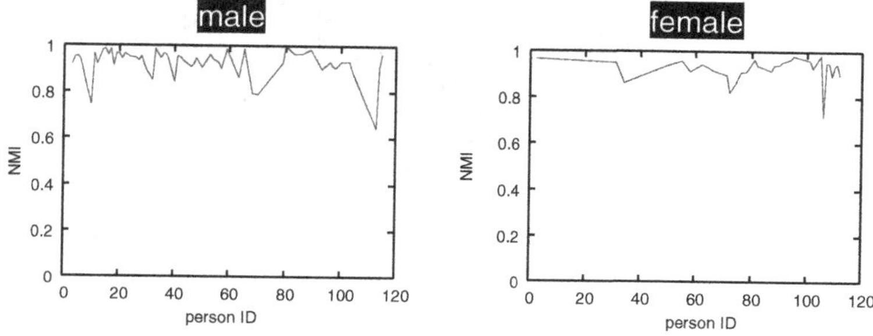

Fig. 4. NMI results on genders per person

By manual inspection on the Indonesians with low NMI, we have found that there are ambiguous expressions Sad - Angry, Rage - Sarcastic, and Hollow - Tired. Figure 3 shows several examples. The universality of facial expressions has been a subject of a long debate [16]. We admit that we need more investigation before drawing a nationality-specific conclusion.

Figure 4 shows the clustering results on different genders per person. As expected, the difference between male and female was not significant, e.g., their average NMI values are 0.93 and 0.92, respectively. We believe that even the difference of the minimum NMI in the Figure is attributed to the diversity of the individuals.

To confirm our belief, we further investigated males and females among Chinese, Japanese, and Indonesians. The rightmost 6 boxes of Figure 2 show the results of the experiments. We see that there is no significant differences between genders in terms of NMI for Chinese and Japanese[3]. The boxes show that NMIs of males and females vary among Indonesians, which needs further investigation on new data due to their small numbers.

3.2 24 Cluster Experiments

The results of the experiments in the previous section signify that most instances of each expression typically form a spherical cluster, as the output $k = 25$ clusters

[3] There is only one Japanese female so the maximum and minimum values are equivalent to the average value in the rightmost but one box.

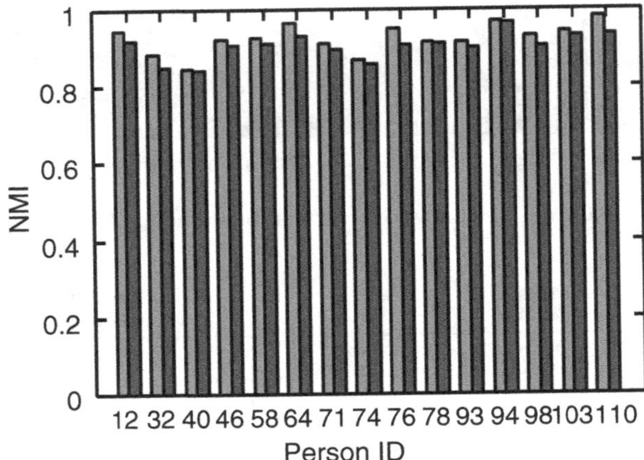

Fig. 5. Persons with increasing NMI with the combination of Bereft and Despondent, where person IDs are aligned on the x-axis; green bar (24 clusters), red bar (25 clusters)

show high NMI in terms of the 25 facial expressions. Note that these results do not mean that instances from intuitively similar expressions form similar spherical clusters, as the high NMI was obtained by handling the 25 facial expressions independently.

To test this hypothesis, we applied the k-means algorithm with $k = 24$ and measured the NMI of the output clusters in terms of 3 kinds of 24 classes, each of which were obtained by combining Bereft and Despondent, Sad and Ill, or Angry and Irritated. Preliminary manual inspection suggested promising outcome. However the corresponding NMI, 0.9168, 0.9158, and 0.9183, respectively, are slightly lower than the the overall NMI (0.9248).

Such a combination forms a double-size class and is hence harmful for the k-means algorithm. We know, however, from our experience that a smaller number of true classes typically result in the increase of NMI. The latter fact seems to be more important than the former in this case. We consider that the hypothesis is denied, probably due to the fact that the 11 shape units tend to have similar values in consecutive images and thus so for each expression. This lesson teaches us that our 25 facial expressions are a kind of "atomic" elements and a simple combination of them deteriorates accuracy[4].

Figure 5 shows results on 15 persons with increasing NMI with the combination of Bereft and Despondent. Investigating common reasons for these 15 persons corresponds to informative learning whereas contrasting these 15 persons against the remaining 85 persons corresponds to discriminative learning [17]. Our benchmark data opens new avenues for facial expression analysis especially in machine learning and data mining.

---

[4] We applied the k-means algorithm also with $k = 23$ by combining Happy, Pleased and Triumph. We obtained lower NMI (0.9065) than the cases with $k = 24$.

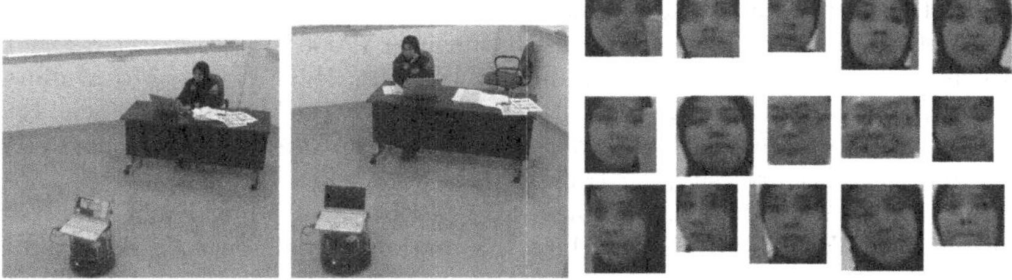

Fig. 6. Snapshots of our face monitoring robot and several examples of face images taken by the robot

4 Conclusions

We constructed facial expression benchmark data of 100 persons using Kinect face tracking application and studied its stability in terms of clustering experimentally. The experimental results were most satisfactory, revealed the "atomic" nature of each facial expression, and signified new research avenues related to machine learning and data mining. Utilizing traditional feature extraction methods in computer vision for other kinds of classes such age [18] is highly relevant.

Our future work includes the informative and discriminative learning on individuals in terms of similar expressions as well as application of the resulting outcome to our human monitoring autonomous mobile robots [19]. Figure 6 shows snapshots of our face monitoring robot and several examples of face images taken by the robot. Note that images are much smaller, e.g., 42×37 pixels, 39×35 pixels, whereas those of our benchmark data are around 90×90 pixels. Moreover, the robot can observe another person's face during the monitoring. Our facial expression benchmark data of 100 persons would be highly useful in handling such a real situation.

Acknowledgments. We appreciate the 115 collaborators who acted facial expressions. Without them this work would not have been possible. We are grateful to Vasile-Marian Scuturici for providing us an executable version of the face tracking application used for static Kinect. The face monitoring robot including the executable version of the face tracking application used for the onboard Kinect are from Yutaka Deguchi and Ryousuke Kondo. Thanks to Mustikasari, who agreed to use her pictures in this paper. This work was partially supported by the grants-in-aid for scientific research (24650070, 25280085) from the Japanese Ministry of Education, Culture, Sports, Science and Technology.

References

1. Sebe, N., Lew, M., Sun, Y., Cohen, I., Geners, T., Huang, T.: Authentic Facial Expression Analysis. In: Image and Vision Computing, vol. 25(12), pp. 1856–1863. Elsevier (2007)

2. Aarts, E., Encarnacao, J.: Into Ambient Intelligence. In: True Vision - The Emergence of Ambient Intelligence, pp. 1–16. Springer (2006)
3. Hossny, M., Filippidis, D., Abdelrahman, W., Zho, H., Fielding, M., Mullins, J., Wei, L., Creighton, D., Puri, V., Nahavandi, S.: Low Cost Multimodal Facial Recognition via Kinect Sensors. Technical report, Centre for Intelligent Systems Research (CISR), Deakin University, Australia (2013)
4. Lyons, M.J., Akamatsu, S., Kamachi, M., Gyoba, J.: Coding Facial Expressions with Gabor Wavelets. In: Proc. Third International Conference on Face and Gesture Recognition (FG), pp. 200–205 (1998)
5. Valstar, M.F., Jiang, B., Mehu, M., Pantic, M., Schrer, K.: The First Facial Expression Recognition and Analysis Challenge. In: Proc. International Conference on Face and Gesture Recognition (FG), pp. 921–926 (2011)
6. Verma, A., Sharma, L.K.: A Comprehensive Survey on Human Facial Expression Detection. Int'l J. Image Processing 7(2), 171–182 (2013)
7. Yong, C.Y., Sudirman, R., Chew, K.M.: Facial Expression Monitoring System Using PCA-Bayes Classifier. In: Future Computer Sciences and Application (ICFCSA), pp. 187–191 (2011)
8. Yu, L.: Constructing Facial Expressions Data Measured with Kinect for Human Monitoring. Master's thesis, Graduate School of Systems Life Sciences, Kyushu University, Japan (2014)
9. Kanade, T., Cohn, J., Tian, Y.: Comprehensive Database for Facial Expression Analysis. In: Proc. Fourth IEEE International Conference on Automatic Face and Gesture Recognition (FG), pp. 46–53 (2000)
10. MacQueen, J.: Some Methods for Classification and Analysis of Multivariate Observations. In: Proc. Fifth Berkeley Symp. on Math. Statist. and Prob., vol. 1, pp. 281–297 (1967)
11. Ekman, P., Friesen, W.V.: Facial Action Coding System: Investigator's Guide. Consulting Psychologists Press, Palo Alto (1978)
12. Microsoft: Face Tracking `http://msdn.microsoft.com/en-us/library/jj130970.aspx` (cited October 2013)
13. Lee, A.: 25 Essential Expressions Challenge, `http://karmajello.com/culture/art/illustrated-facial-expressions-gallery.html` (cited October 2013)
14. RikoJasmine: 25 Essential Expressions Challenge, `http://rikojasmine.deviantart.com/art/25-Essential-Expressions-124330040` (cited October 2013)
15. Han, J., Kamber, M.: 7. In: Data Mining: Concepts and Techniques, 2nd edn., pp. 383–460. Morgan Kauffman, San Fransisco (2006)
16. Jack, R.E.: Culture and Facial Expressions of Emotions. Visual Cognition 21(9-10), 1248–1286 (2013)
17. Rubinstein, Y.D., Hastie, T.: Discriminative vs Informative Learning. In: Proc. KDD, pp. 49–53 (1997)
18. Zhou, H., Miller, P., Zhang, J.: Age classification using radon transform and entropy based scaling SVM. In: Proc. 22nd British Machine Vision Conference (BMVC 2011) (2011)
19. Suzuki, E., Deguchi, Y., Takayama, D., Takano, S., Scuturici, V.-M., Petit, J.-M.: Towards Facilitating the Development of a Monitoring System with Autonomous Mobile Robots. In: Information Search, Integration and Personalization. Springer (accepted for publication)
20. Ahlberg, J.: Candide-3 An Updated Parameterised Face. Technical report, Linkoping University (January 2001)

A Animation Units and Shape Units

AUs and SUs were defined based on the Candide 3 Model [20]. The AUs are delta from the neutral shape. Each AU is expressed as a numeric weight varying between -1 and +1 [12]. The SUs estimate particular shapes of the user's head; the neutral position of their mouth, brows, eyes, and so on. A Shape Unit defines a deformation of a standard face towards a specific face. The shape parameters should be invariant over time, but specific to each individual[5]. Animation parameters naturally varies over time, but can be used for animating different faces.

Tables 3 and 4 show definitions of the SUs and the AUs in the face tracking SDK application, respectively. Note that they are slightly different from those in the Candide 3 Model.

Table 3. Definition of SUs

SU name	SU number in Candide-3
Head height	0
Eyebrows vertical position	1
Eyes vertical position	2
Eyes, width	3
Eyes, height	4
Eye separation distance	5
Nose vertical position	8
Mouth vertical position	10
Mouth width	11
Eyes vertical difference	n/a
Chin width	n/a

Table 4. Definition of AUs (See `http://msdn.microsoft.com/en-us/library/jj130970.aspx` for images)

AU name and value	AU value interpretation
Neutral Face (all AU is 0)	Neutral face
AU0 - Upper Lip Raiser (In Candide3 this is AU10)	0=neutral, covering teeth; 1=showing teeth fully; -1=maximal possible pushed down lip.
AU1 - Jaw Lowerer (In Candide3 this is AU26/27)	0=closed; 1=fully open; -1= closed, like 0.
AU2 - Lip Stretcher (In Candide3 this is AU20)	0=neutral; 1=fully stretched (joker's smile); -0.5=rounded (pout); -1=fully rounded (kissing mouth).
AU3 - Brow Lowerer (In Candide3 this is AU4)	0=neutral; -1=raised almost all the way; +1=fully lowered (to the limit of the eyes)
AU4 - Lip Corner Depressor (In Candide3 this is AU13/15)	0=neutral; -1=very happy smile; +1=very sad frown
AU5 - Outer Brow Raiser (In Candide3 this is AU2)	0=neutral; -1=fully lowered as a very sad face; +1=raised as in an expression of deep surprise

[5] We observed that SUs do change over time, but more slowly than the AUs.

Eye-Gaze Tracking-Based Telepresence System for Videoconferencing

Bartosz Kunka[1], Adam Korzeniewski[1], Bożena Kostek[2], and Andrzej Czyżewski[1]

[1] Multimedia Systems Dept.
[2] Audio Acoustics Laboratory,
Gdańsk University of Technology, Gdańsk, Poland
{kuneck,adamkorz,andcz}@sound.eti.pg.gda.pl
bokostek@audioacoustics.org

Abstract. An approach to the teleimmersive videoconferencing system enhanced by the pan-tilt-zoom (PTZ) camera, controlled by the eye-gaze tracking system, is presented in this paper. An overview of the existing telepresence systems, especially dedicated to videoconferencing is included. The presented approach is based on the CyberEye eye-gaze tracking system engineered at the Multimedia Systems Department (MSD) of Gdańsk University of Technology (GUT), as well as on a standard PTZ security camera communicating with the computer by the TCP/IP protocol. Technical aspects of the developed system prototype including two different use cases (one-way and two-way configuration of system) are described. Moreover, a discussion related to the gathered user's experience as well as to difficulties and opportunities concerning the proposed approach are included.

Keywords: Telepresence; active media applications; videoconferencing system; eye-gaze tracking; CyberEye.

1 Introduction

Nowadays, videoconferencing sessions are becoming increasingly important, especially in business relations and remote training. This paper presents the idea of the use of eye-gaze tracking system in videoconferencing system. Due to eye-gaze tracking technology it is possible to freely change a field of view of the pan-tilt-zoom (PTZ) camera located in a room of the second interlocutor by gaze. This approach provides the user with a sense of sharing a virtual space with remote participants. Therefore, applying – within this paper – terms such as teleimmersion or telepresence is justified. In this context eye-gaze tracking teleconference system may be listed as one of the active media applications. There are many commercial eye-gaze tracking systems available on the market. The employed system called CyberEye was developed at the Multimedia Systems Department (MSD). The CyberEye is a hands-free as well as low-cost video-based eye-gaze tracking system. Eye-gaze tracking-based telepresence systems for videoconferencing are likely to be popularized only for relatively inexpensive eye-gaze tracking systems. Thus, systems such as CyberEye could introduce

D. Ślęzak et al. (Eds.): AMT 2014, LNCS 8610, pp. 432–441, 2014.

new functionality and possibilities to videoconferencing systems because of its availability. It is worth mentioning that employing eye-gaze tracking interfaces in videoconferencing systems for telepresence is a novel approach. The users are immersed in the created telereality due to possibility of interactive changing of field of view during the videoconferencing session just by gazing. The presented application may use other gaze tracking systems if they are adapted to the above given assumptions.

The paper is organized as follows. In Section 2, some related studies and similar videoconferencing systems were presented. In Section 3, the architecture for the proposed videoconferencing system employing eye-gaze tracking technology, the prototype of the proposed system and other use cases utilizing two PTZ cameras were described. Further, in Section 4, some difficulties as well as opportunities and other applications for eye-gaze tracking-based telepresence system were considered.

2 Related Work

Generally, telepresence means a virtual presence in a remote environment. In terms of user experience, telepresence fulfill the definition of the active media applications. In this Section various realizations of this idea are presented. There are two main types of telepresence realizations. First, when we try to simulate someone presence in the desired localization using simplified two-dimensional or three-dimensional avatars [1]. Remote presence is then quite easily realized and could be very efficient but without interaction or the real sense of touch. The second realization is related to a virtual-real presence utilizing a visual copy of remote participant. The European project VIRTUE (VIRtual Team User Environment) aimed at implementing such a viewpoint-adaptive scheme for realizing the virtual-real presence concept and succeeded in this approach [2]. It is achieved by properly prepared, realistic, three-dimensional views of a remote participant who is presented in real time to a local participant with appropriate view perspective. It should be emphasized that a proper perception of virtual person is a very important aspect. There are three visual cues essential to three-dimensional perception: the motion parallax cue, the stereo depth cue, and the eye lens accommodation cue [3]. The motion parallax cue is the most important for proper 3-D perception and can easily be provided by an adaptive system which changes a viewpoint of recreated remote participant by aligning it in line with local participant viewpoint [3].

The eye-gaze tracking-based telepresence system for videoconferencing presented in the paper can also be used as an adaptive scheme for realization of the virtual-real presence concept. Eye-contact should be considered as the most characteristic feature of interpersonal interaction. Thus, eye-gaze tracking can be employed in order to simulate remote participant's attention to a local participant. Further presented works are dedicated to topics identifying the problem of proper interpretation of the user's gaze direction.

The exact process of visual perception related to gaze scanning has not been explained for a long time but major features used as indicators can be determined. Gibson and Pick observed in 1963 that the perception of gaze-direction should involve the perception of both head position and ocular-position [4]. There is a head-turn effect, i.e. when the target's head was rotated in one direction, participant tended to perceive gaze to be rotated in the opposite direction. It confirms that perception associated with gaze direction involves an interaction between eye and face appearance. Similar results were obtained by Cline in other experiments [5] and connected with faces on TV screen by Anstis *et al.* [6]. In more recent work, experiments with two-dimensional pictures were performed [7]. Gaze has also been studied in immersive virtual environments. Ability to discern gaze target of an avatar with tracked eyes, played back through a stereo display, showed a significant improvement from the tracking and reproduction of eye gaze over simple head gaze [8]. Simplifying the representation of a virtual human to make the eyes more apparent resulted in observers noticing eye gaze less than with a more realistic human face [9].

Gaze cues regulate conversation between humans, provide observers feedback and information about their behavior, enable to express emotions and attitudes, and facilitate goal setting [10]–[12]. Thus, a main advantage of teleconferencing systems offering a face-to-face communication is the maintenance of gaze awareness and eye contact. In the system proposed within the paper change of the camera perspective in the horizontal and in the vertical plane is performed by focusing the user's gaze on the marginal regions of the camera's field of view.

3 System Architecture

The idea of employing the eye-gaze tracking system in videoconferencing in order to enhance immersion of the user in remote environment was implemented in the prototype stand and was tested. The core of the developed videoconferencing system prototype is the CyberEye enabling to control the PTZ camera remotely. The videoconferencing system envisaged also requires the infrastructure of the TCP/IP network. However, it can be assumed that the access to the network is provided. In a further part of this Section, the CyberEye system has been presented and possible use cases of our proposed approach have been described.

3.1 CyberEye

The CyberEye is based on infrared (IR) illumination similarly to the proposed earlier gaze trackers [13], as well as most of commercial systems. IR illumination does not disturb the user's attention as well as his/her interaction with the computer. Utilization of IR light improves image processing significantly. First, contrast between the pupil and the iris in IR image is much higher compared to the standard grayscale image taken utilizing only visible light illumination.

Moreover, IR sources generate unique corneal reflections on the eye image, called glints. They support precision of gaze direction estimation.

The presented gaze-tracking system comprises of five components which have been shown in Fig. 1:

- modified webcam used to acquire image of the user – camera sensitive to infrared, with customized lenses (focal length of 12 mm) and infrared band-pass filter;
- IR LEDs ring mounted around the camera lens;
- 4 groups of IR LEDs fixed on display corners;
- IR LEDs driver – the USB controlled device allowing separate activation of all mentioned IR modules;
- software for image processing and control of the IR LEDs driver and webcam driver.

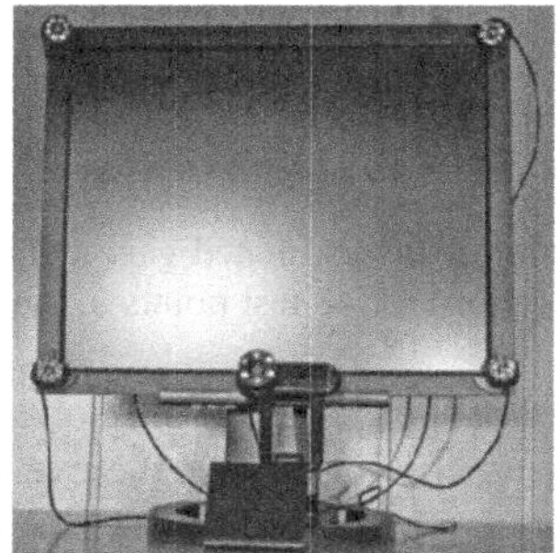

Fig. 1. Hardware setup of the CyberEye system

Time resolution of the CyberEye system is determined by parameters of the employed webcam. Information about the position of the point of visual fixation is refreshed with the same frequency (every 200 ms), thus the time resolution of the CyberEye is 5 Hz. Angular resolution of the CyberEye system has been estimated within the experiment involving ca. 120 students sitting at a distance of 60 cm in front of the computer screen. According to the obtained results, mean angular resolution of the CyberEye is about 3.32° (horizontal plane) and 3.38° (vertical plane). Figure 2 shows an example of distribution of gaze points estimated by the system for a randomly selected participant of the study. Red points represent actual gaze points – test points being focused during the experiment. It should be noted that the results of evaluation of the CyberEye's accuracy shown in Fig. 2 represent the case when the user successfully completed the calibration process. The plot includes measurement for 48 test points whereas, according to our assumptions, sufficient accuracy of the CyberEye system means distinguishing between 9 areas.

More information on hardware and software parts of the CyberEye system is to be found in some earlier publications of the authors [14]–[17]. Also some papers dedicated to utilization of eye-gaze tracking technology in PTZ cameras operating were published [18].

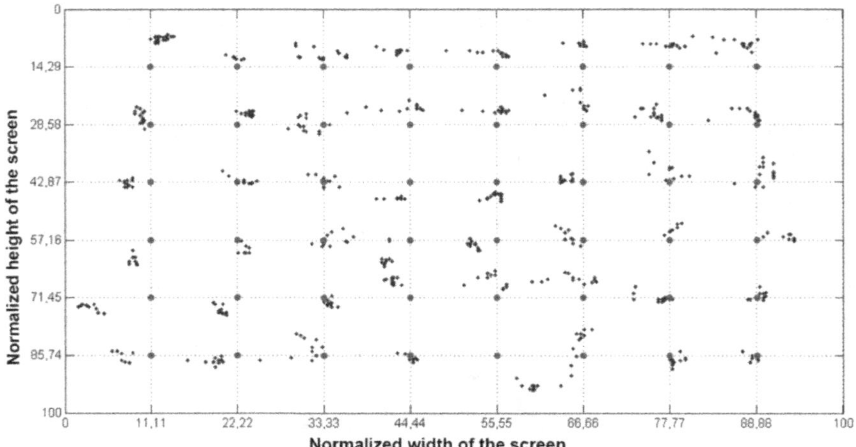

Fig. 2. CyberEye accuracy: the plot of the estimated gaze points (blue dots) and the true gaze points (red points placed at intersection) for 48 test points in the screen

It was necessary to perform proper programming changes in the PTZ controller in order to enable to control the camera directly by the eye-gaze tracking system. In our prototype system we used a camera produced by Axis. A producer provided us with a free SDK (Software Development Kit) dedicated to camera controlling, thus it was possible to perform appropriate modifications. Fig. 3 shows the scheme of communication between personal computer (PC), the eye-gaze tracking system (EGT) and the PTZ camera.

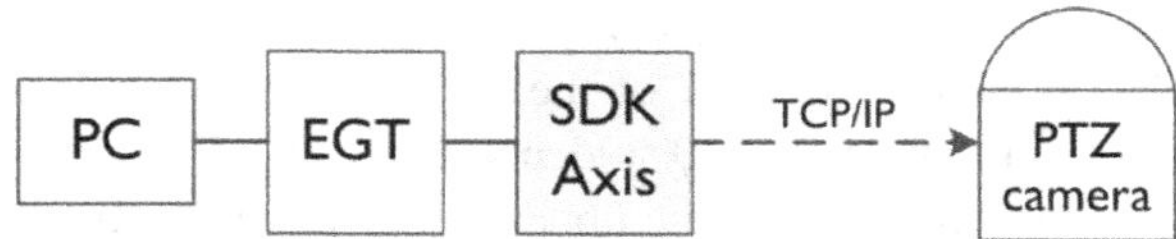

Fig. 3. A scheme of communication between components of the engineered prototype system

3.2 Use case Utilizing a One-Way Configuration

Within the development of the prototype of our eye-gaze tracking-based telepresence system for videoconferencing it was assumed that only one PTZ camera is employed. Such a configuration is more reliable in training remote employees than typical videoconferencing. Nevertheless, this simplified configuration is sufficient for testing the proposed approach. Observations related to the system working in real conditions are to be sufficiently reliable to be applied to the development of the videoconferencing system using two PTZ cameras. Fig. 4 shows the scheme of connections between specific components of the engineered system.

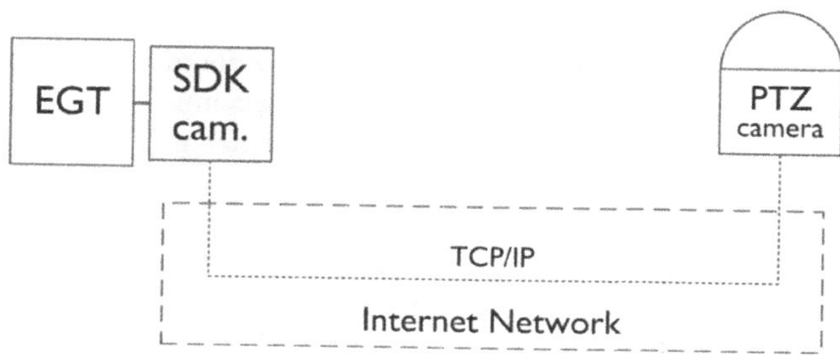

Fig. 4. Scheme of the developed system prototype with one PTZ camera used (one-way)

a)

b)

PTZ camera →

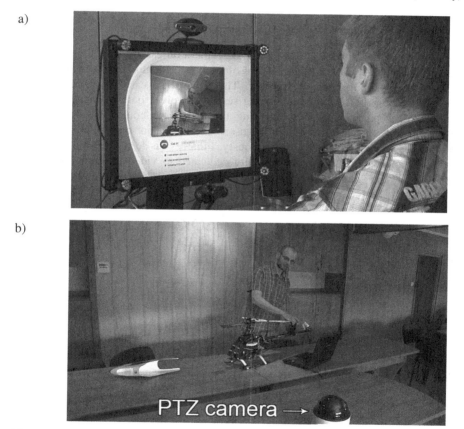

Fig. 5. Test of the developed videoconferencing system: a) the user controlling the PTZ camera by his gaze; b) remote environment with the PTZ camera

Configuration of the engineered system was validated and tested in real conditions. We prepared a special application enabling the user to connect with some IP address (in this case with the PTZ camera) and connecting eye-gaze tracking functionality with the SDK camera. The use case tested was associated with assembling the remotely-controlled helicopter. The user operating the PTZ camera by his gaze could freely change virtual field of view. It was possible to zoom in/zoom out a field of viewing and looking at for example mounting small elements. The option of zooming in/zooming out is activated when the user focuses his/her gaze on the center of the camera preview. Then two buttons appear – "+" for zoom in and "-" for zoom out. This functionality impacts the greatest the user's sensation of immersion in a remote environment. Fig. 5a presents the user participating in the remote training and controlling the PTZ camera by the CyberEye system (location No. 1). Fig. 5b shows location No. 2, the so-called remote environment, with the PTZ camera placed in the bottom part of the image.

3.3 Use Case Utilizing Two-Way Configuration

The proposed approach enables to use the PTZ camera on each side of the videoconferencing session. It means that immersion of each interlocutor into a remote environment of the second person is possible. The typical use case, also presented in Fig. 6, assumes that the videoconferencing session involves only two interlocutors. Obviously, other participants not operating the CyberEye could also be present. Moreover, in case of the PTZ camera the user can control not only one camera but several cameras, apparently not simultaneously. Finally, we do not have to consider the proposed approach as a system typically dedicated to videoconferencing. There are many use cases to employ the eye-gaze tracking in telepresence-based system.

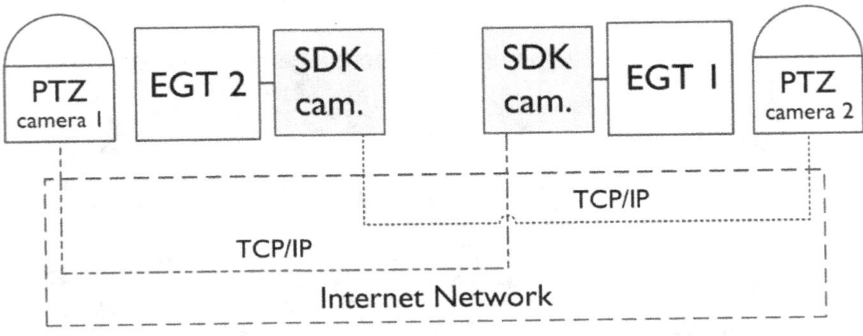

Fig. 6. Configuration of the proposed system for videoconferencing (two-way), where: *PTZ camera 1* – pan-tilt-zoom camera of interlocutor No. 1; *EGT 2* – eye-gaze tracking system operated by the interlocutor No. 2; *SDK cam.* – software development kit modified for controlling the PTZ camera by the EGT system

4 Difficulties and Advantages

The presented approach to the eye-gaze tracking-based telepresence videoconferencing brings some limitations that should also be considered. A key difficulty is the fact that the user operating the camera remotely should have a fixed position relatively close to the eye-gaze tracking system. In situation when this system is stationary (not worn on head) it's necessary to operate it with an almost fixed head position. Furthermore, the user should be focused on it for most of time. In some special situations this disadvantage can be turned into advantage and such exception is also described. Another strong drawback of the presented eye-gaze tracking system is its sufficient accuracy while maintaining a low cost of the system [17]–[21]. Contrarily, the relatively low cost of the CyberEye enables to use it on both sides of the teleconference connection. The designed eye-gaze tracking-based system is inexpensive and easy to use. Of course, locating the PTZ camera of a high quality in a remote environment can make the system more expensive.

It is also necessary to go through the calibration process. If the PTZ camera is fixed, i.e. has a permanent location, calibration of the teleconferencing system can be performed only once. However, each time the teleconference connection is established, the system based on eye-gaze tracking technology requires a short and uncomplicated calibration process. However, these additional operations related to activating the presented system are easy to perform.

Probably, most of additional requirements will reveal themselves while using it in various test scenarios depending on two types of configuration of the presented system (one-way and two-way). While using the one-way direction system version, i.e. when only one of the participants controls the remote camera using eye-gaze tracking, the most relevant scenario associates one person with a group of people with whom interaction is made.

Such a system can be used nowadays in distance learning (e-learning), as well. The PTZ camera can be placed in a large classroom filled in with students and the teacher can be presented on the screen. In a situation when the room is very large and a typical wide-angle fixed-focus camera lens is able to capture only the overall picture without enough details, it's almost impossible to distinguish between individual students. The eye-gaze tracking-based telepresence system can be used by the teacher to focus his/her gaze on e.g. a student who wants to ask a question. A hand lifted up is a typical behavior in such a situation and it can be seen in a wide-angle view. Then, a teacher can focus his/her attention on the appropriate part of the screen containing the lifted hand resulting in zooming out a field of view and moving the PTZ camera to the exact location of a student. This would enable the individual to ask the person inquiring leading to seamless identification which may be important for the teacher.

Another use case of the one-way eye-gaze tracking-based telepresence system is having a remote interactive conversation between a person with a limited mobility (permanently or temporarily) and a person without any restriction of movement. Apparently, typical videoconferencing system can be used in such a situation but sometimes a feedback from those persons with the limited mobility is needed. Focusing a field of view of the camera on a movable interlocutor keeps tracking of the conversation or simply shows an interest in the topic. This application can easily be extended to situations of therapy in which a patient is physically disabled, and on the other side of the system there is a therapist who cannot be present in the patient's room. Employing the one-way eye-gaze tracking-based telepresence system gives an opportunity to create a series of tests on the side of professionals therapists (doctors) located in one central research facility (hospital) who would maintain contact with patients remotly. These are only several use cases for the presented one-way configuration of the proposed videoconferencing system. This setup seems to be appropriate, because even though there are some disadvantages such as the fixed head position by one of the interlocutors, the use case still have some benefits.

5 Conclusions

There are some possibilities to enhance telepresence in current videoconferencing systems. The employment of a low-cost eye-gaze tracking system is reasonable and can positively influences the user's immersion in remote environments, making such applications examples of active media. In the paper two configurations of the proposed telepresence system has been proposed: one- and two-way. The one-way configuration has also been developed as a prototype system and its working has been validated in real conditions. It is worth mentioning that the users reported intuitiveness of hands-free controlling the PTZ camera in a remote environment. Our goal is to conduct systematic experiments and to study sense of telepresence for a large group of subjects. Also, some problems related to working of the eye-gaze tracking-based system will be more thoroughly investigated. Since telepresence services can be applicable to various areas such as – remote training, remote education, surgery or interviewing, thus questionnaire forms associated with the applications dedicated to these areas will be proposed to test the correctness of these system configurations. Nowadays, eye-gaze tracking technology is gaining its popularity due to integrating eye trackers with mobile devices (laptops, smartphones) [22], [23], hence eye tracking interfaces are becoming more easily available. Therefore, a novel approach to videoconferencing system presented in the paper indicates a new field of applications of such systems.

Acknowledgements. Research is subsidized partially by the European Commission within FP7 project INDECT, Grant Agreement No. 218086.

References

1. Mortlock, A., Machin, D., McConnell, S., Sheppard, P.: Virtual conferencing. In: Telepresence, pp. 208–226 (1999)
2. E. IST Project IST-1999-10 044, VIRTUE - VIRtual Team User Environment (2003)
3. Redert, P.: Multi-viewpoint systems for 3-d visual communication. Delft Univ. of Technol. (2000)
4. Gibson, J.J., Pick, A.: Perception of Another Person's Looking Behavior. Am. J. Psychol. 76, 386–394 (1963)
5. Cline, M.G.: The Perception of Where a Person Is Looking. Am. J. Psychol. 80, 41–50 (1967)
6. Anstis, S.M., Mayhew, J.W., Morley, T.: The Perception of Where a Face or Television 'Portrait' Is Looking. Am. J. Psychol. 82, 474–489 (1969)
7. Langton, S.R.H.: The mutual influence of gaze and head orientation in the analysis of social attention direction. Q. J. Exp. Psychol. 53, 825–845 (2000)

8. Bayliss, P.T.: Orienting of attention via observed eye-gaze is head-centred. Cognition 94, 1–10 (2004)
9. Murray, N., Roberts, D., Steed, A., Sharkey, P., Dickerson, P., Rae, J.: An assessment of eye-gaze potential within immersive virtual environments. ACM Trans. Multimed. Comput. Commun. Appl. 3(4), 1–17 (2007)
10. Kendon, A.: Some functions of gaze direction in social interaction. Acta Psychol. (Amst). 26, 22–63 (1967)
11. Kleinke, C.L.: Gaze and eye contact: A research review. Psychol. Bull. 100(1), 78–100 (1986)
12. Doherty-Sneddon, G., Anderson, A., O'Malley, C., Langton, S., Garrod, S., Bruce, V.: Face-to-face and video mediated communication: A comparison of dialogue structure and task performance. J. Exp. Psychol. Appl. 3(2), 105–125 (1997)
13. Yoo, D.H., Chung, M.J.: A novel non-intrusive eye gaze estimation using cross-ratio under large head motion. Comput. Vis. Image Underst. 98, 25–51 (2005)
14. Kunka, B., Kostek, B.: Objectivization of audio-visual correlation analysis. Arch. Acoust. 37(1), 63–72 (2012)
15. Kunka, B., Kostek, B.: New Aspects of Virtual Sound Source Localization Research – Impact of Visual Angle and 3-D Video. J. Audio Eng. Soc. 61(5), 280–289 (2013)
16. Kunka, B., Kostek, B.: Exploiting audio-visual correlation by means of gaze tracking. Int. J. Comput. Sci. Appl. 7(3), 104–123 (2010)
17. Kunka, B., Kostek, B., Kulesza, M., Szczuko, P., Czyzewski, A.: Gaze-tracking-based audio-visual correlation analysis employing quality of experience methodology. Intell. Decis. Technol. 4, 217–227 (2010)
18. Kotus, J., Kunka, B., Czyzewski, A., Szczuko, P., Dalka, P., Rybacki, R.: Gaze-tracking and acoustic vector sensors technologies for PTZ camera steering and acoustic event detection. In: 2010 Workshop on Database and Expert Systems Applications (DEXA), pp. 276–280 (2010)
19. Kostek, B., Kunka, B.: Application of Gaze Tracking Technology to Quality of Experience Domain. In: MCSS 2010: IEEE International Conference on Multimedia Communications, Services and Security, pp. 134–139 (2010)
20. Kunka, B., Czyzewski, A., Kwiatkowska, A.: Awareness evaluation of patients in vegetative state employing eye-gaze tracking system. Int. J. Artif. Intell. Tools 21(02), 1–11 (2012)
21. Czyzewski, A., Dalka, P., Kunka, B., Kupryjanow, A., Lech, M., Odya, P.: Multimodal human-computer interfaces based on advanced video and audio analysis. In: The 6th International Conference on Human System Interaction (HSI), pp. 18–25 (2013)
22. Ziegler, C.: Tobii and Lenovo show off prototype eye-controlled laptop (2011), `http://www.engadget.com/2011/03/01/tobii-and-lenovo-show-off-prototype-eye-controlled-laptop-we-go/`
23. Browne, C.: Don't Blink, the Future is Now – Eye-Control Systems for Tablets and Smartphones (2012), `http://nearshore.com/2012/12/dont-blink-the-future-is-now-eye-control-systems-for-tablets-and-smartphones.html`

Using Process-Oriented Interfaces for Solving the Automation Paradox in Highly Automated Navy Vessels

Jurriaan van Diggelen[1], Wilfried Post[1], Marleen Rakhorst[1], Rinus Plasmeijer[2], and Wessel van Staal[3]

[1] TNO, Soesterberg
{jurriaan.vandiggelen,wilfried.post,marleen.rakhorst}@tno.nl
[2] Radboud University, Nijmegen
rinus@cs.ru.nl
[3] Eaglescience, Amsterdam
w.vanstaal@eaglescience.nl

Abstract. This paper describes a coherent engineering method for developing high level human machine interaction within a highly automated environment consisting of sensors, actuators, automatic situation assessors and planning devices. Our approach combines ideas from cognitive work analysis, cognitive engineering, ontology engineering, and task-based prototyping. We describe our experiences with this approach when applying this suite to develop an innovative socio-technical system for fighting the internal battle in navy vessels with a strongly reduced manning.

Keywords: Human-machine interaction, Cognitive engineering, Task-oriented programming.

1 Introduction

The royal Dutch navy expressed its ambition to reduce crew size by 30% to 50% in the next decade. This is because budget cuts force the organization to reduce their personnel costs and because technical personnel is increasingly harder to find (especially in the maritime sector). Although many relevant technologies, such as sensor, network and information processing technologies have evolved tremendously over the last decade, applying more automation does not straightforwardly lead to less workload and a smaller crew size.

This is also known as the automation paradox which states that the more automation is applied, the more crucial the contribution of the human operator becomes, and the more difficult it becomes to implement adequate human-machine interaction [1]. The automation paradox could result in interfaces that keep the human out of the loop which makes it difficult for the human to correct machine errors, or in interfaces which overload the human with too much and too detailed information.

Over the last decades navy vessels have become highly automated with respect to processes such as sensing the environment, assessing the situation, planning a

D. Ślęzak et al. (Eds.): AMT 2014, LNCS 8610, pp. 442–452, 2014.

response, and acting accordingly. Current design methods have led to stove-piped architectures: each automation component is designed individually, with a dedicated human machine interface. This has led to an undesirable situation (i.e. the automation paradox).

A typical example is the alarm flood that occurs in emergency situations such as fire, when each individual smoke detector, heat detector, and potentially damaged subsystems, raises an individual alarm. This typically overloads the users with information that is too detailed and too much focused on the individual (sub)systems. It fails to notify the user of the root cause of failure (i.e. fire), and makes it difficult for the user to prioritize his or her actions to ensure that the critical processes are not endangered.

To properly address these concerns, the system should be regarded as a socio-technical system containing organizational aspects, human factors and system aspects. For the analysis of socio-technical systems, cognitive work analysis (CWA) [8,11] has emerged over the last two decades as a standard way to analyze and represent processes and the way in which systems and humans are taking part in them. Although there has been significant progress in this field, only a few successful applications came out of it (e.g. power plants). One reason that the number of applications remains relatively limited is that CWA still leaves a large gap between analysis and actual system implementation, which makes it difficult to apply the results of CWA. To bridge this gap, the rather abstract and static outcomes of a CWA need to be mapped to an iterative process of software requirements engineering, prototyping and testing (which is common in software engineering methodologies).

This paper describes our attempts at doing this. We will focus on the case of the internal battle (all engineering and damage control activities that ensure proper operation of all ship systems) in highly automated navy vessels with a reduced crew. Based on our experiences, we have extracted some generic methodological guidelines. In particular, we describe our use of ontologies for laying the foundation of the right information representation at the right level of abstraction. Furthermore, we present our task-oriented prototyping platform (based on the iTasks programming language [7]) for designing, refining, and testing socio-technical systems in an early phase of development.

The paper is organized as follows. The next section describes our method for designing complex socio-technical systems, such as fighting the internal battle with a reduced-crew. We describe the use of CWA, the relation to software specification, and the relation to software prototyping. Section 4 describes the case study, followed by a conclusion.

2 Methods

As shown in Fig. 1, we apply three methods: CWA, in which the current situation is explored; Cognitive Engineering, in which the envisioned system is specified; Prototyping & Testing, in which the system is tested. Typically these three phases are performed iteratively. The main outcomes of the different phases are denoted in the blue boxes. Our contribution lies in connecting the main outcomes of the different phases. In the following three subsections, we describe each phase in more detail, paying special attention on how the results of the previous phase can serve as a basis for the current phase.

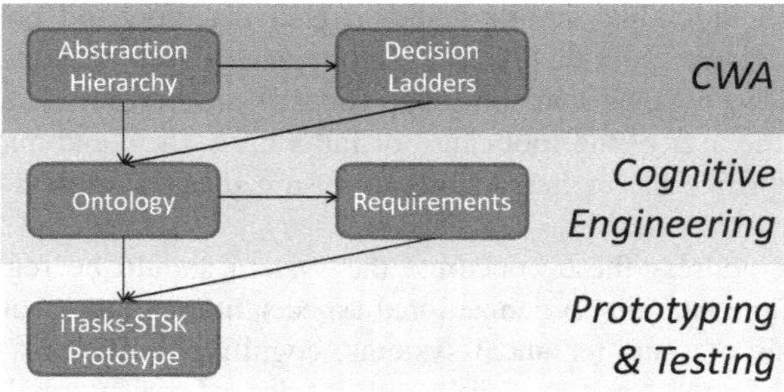

Fig. 1. Different methods and their relations

2.1 Cognitive Work Analysis

CWA provides several useful templates for the different phases of the analysis. We have focused on the Abstraction Hierarchy and Decision Ladders (DL's). The Abstraction Hierarchy is a multi-leveled representation framework which can be used to describe the entire work environment from multiple perspectives (both functional and physical) at different levels of abstraction. One of the layers describes the core functions. For example, in our case of supporting the internal battle for navy vessels, we can identify fire control, flood control, damage control, and personnel control as core functions. Each core function can be analyzed further using a so-called decision ladder (see Fig. 2) which identifies five steps in the process. The low levels in the DL are skill-based and typically suitable for automation. For example, in the case of fire-control, the Sense-step can be done automatically by a smoke-sensor, which automatically triggers an Act-step which can then be performed by a watermist system. In case this skill-based pattern does not work (e.g. due to a malfunctioning system), the process is taken to a higher level (such as Assess or Decide). In fire-control, the decide step could be whether to ignore a fire for the moment because it is not considered a threat to the ship's command aim, and no vital systems are endangered.

The results of a CWA can be applied to design ecological interfaces [12]. The goal of an ecological interface is to present more high-level information (e.g. higher up in the abstraction hierarchy) to the user such that his or her cognitive resources can be devoted to higher cognitive processes such as problem solving and decision making. We follow this philosophy by proposing an interface in which the user has a process-oriented view (which can be visualized using DL's), instead of the traditional system-oriented view (where each system individually communicates with the user which could cause undesirable situations such as alarm floods). The engineering process of such interfaces is described in the next section.

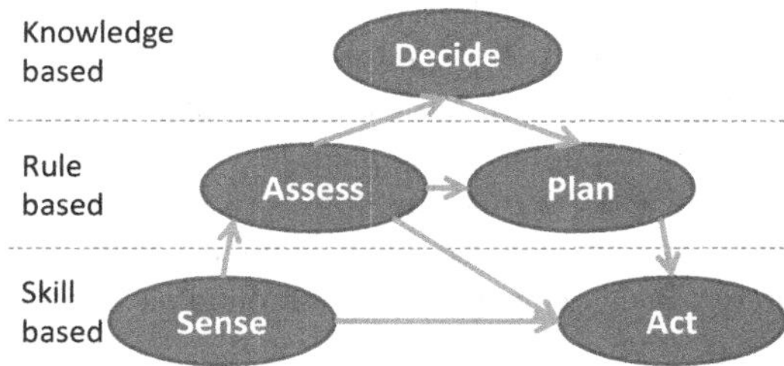

Fig. 2. Decision ladder template

2.2 Cognitive Engineering

To specify the system in terms of use cases, user requirements and claims, we use the situated cognitive engineering framework [3]. We will not introduce the entire framework here, but focus on those aspects that are most important to design the socio-technical system and to ensure that the findings from the CWA find their way through in system design. These aspects are ontologies and user requirements and are further explained below.

In information science, an ontology is defined as a specification of a conceptualization [4]. To specify ontologies, we use the OWL Web Ontology Language [6]. It allows us to formally describe the terms and concepts and relations that are used in our domain. We use ontologies as the main vehicle with which the knowledge gathered in the CWA phase is concretized and used in the cognitive engineering phase. For example, each core function in the abstraction hierarchy is added as a concept in the ontology, given a human-readable definition and enhanced with attributes (for example that fire control is-exercised in some location on the ship). This activity serves two purposes. Firstly, they facilitate communication between stakeholders (domain experts, software engineers, programmers, decision makers) by providing a shared and well-defined vocabulary. For example, they ensure that the user requirements are defined more precisely using a constrained vocabulary. Secondly, the ontologies serve as a data structure that can serve as a basis for prototyping.

User requirements specify what the system is supposed to do from a user perspective. Requirements can be ordered hierarchically. For example, a top-level requirement for our prototype, called IBMt (internal battle management tool), is defined as: "the IBMt shall provide overview, transparency and control on every active internal battle process". By refining this requirement, the cognitive engineer adds sub-requirements that state exactly which device delivers which type of information in which way about which process. For example, "the IBMt shall provide information on temperature development for each Fire-control process". This process is closely related to ontology engineering as the ontology states which types of devices, information and processes are assumed to exist.

Whereas such requirements might seem trivial at first sight, they are currently not fulfilled by state of the art navy ships, despite the advanced forms of automation that have been applied. This is because navy crew members do not access their information at the process-level, but at the component level. Also, the data structures currently used in navy automation (i.e. the ontology) do not allow such information to be represented. Because we derived these requirements from the starting point of the abstraction hierarchy, and we adapted the ontology accordingly, we were able to adopt this fresh perspective.

2.3 Prototyping and Testing

In the explorative phase of tool development, we build prototypes mainly to perform an early evaluation experiment. It is not intended as the final product, and may even be rewritten from scratch before it becomes one. In general, evaluation experiments can differ in *fidelity* (indicating how close the test environment resembles the environment in which the tool is planned to be used) and *realism* (whether the test setup is realized in the real world, a virtual world, or a combination of both) [9]. A cost-effective evaluation entails using an abstract prototype (i.e. low fidelity, low realism). However, it is also important *not* to abstract away from the central concepts of our socio-technical system, because these are the ones we want to test. For our applications, this means that things like multi-user interaction, flexible task allocation to humans and computers, and complex task modelling should be present in the prototype from the beginning. To easily experiment with these aspects of the prototype, while abstracting away from aspects that are not yet relevant, we introduce a prototyping environment which is based on the iTasks framework [7].

iTasks is a task-oriented programming language. This means that *tasks* are used a central concept, unlike the *instruction* which plays a central role in traditional programming languages like Java and C++. Tasks can be allocated to resources (before a run, or at run-time), and tasks can be defined to generate other tasks. What is *not* specified in the program are the user interfaces that establish the collaboration. iTasks automatically generates all required user interfaces and deals with data persistence issues, etc. This greatly reduces programming efforts, allowing more iterations of specification and test to be performed. Furthermore, the concept of a task bridges the gap between an analyst's viewpoint and a developer's viewpoint, providing a shared model that is understandable for both types of experts. This greatly facilitates communication between the two parties. Finally, the iTasks environment enables us to specify tasks at different levels of abstraction. In earlier work by some of the authors, iTasks has been applied to the domain of crisis management [5]. We have built on this work, while realizing the IBMt prototype, and specified an additional within iTasks with reusable common interaction or task execution patterns at an abstract level which can be reused in multiple socio-technical system applications.

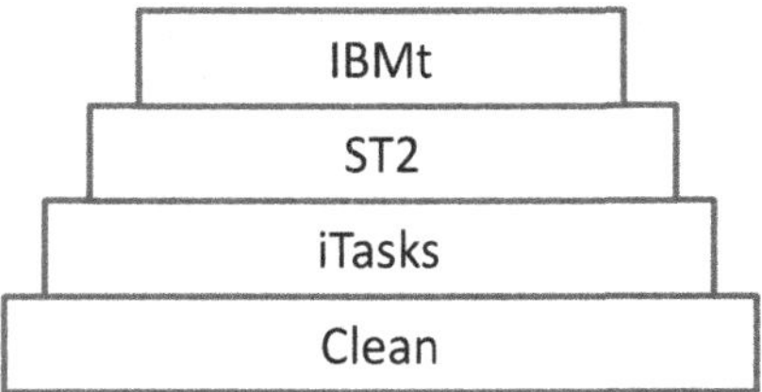

Fig. 3. Layered architecture

We call this layer the socio-technical simulation toolkit (ST2). ST2 is designed to support rapid prototyping of multi-user test applications in the navy domain. We applied ST2 to develop our prototype for internal battle management. The architecture is presented in Fig. 3. The lowest layer concerns the Clean functional programming language [2]. At a higher level of abstraction, the iTasks language is defined to enable task-oriented programming. On top of the iTasks layer, we have designed the ST2 layer as a rapid prototyping environment containing reusable building blocks for socio-technical system simulation. The goal of ST2 is to make all abstract collaboration types available to be able to easily experiment with different configurations. We used ST2 to construct the IBMt (the highest layer), which we will discuss in the next section.

ST2 contains the following features:

- **Scenario control.** The scenario controller can be used to make adjustments to the situation while the experiment is running. For example, a fire or leak can be started, resources can be damaged, etc.
- **Generic decision ladder task structures.** We have developed a generic task structure that follows the stages of a decision ladder (sense, assess, decide, plan, act). This generic task structure can be reused and instantiated to other processes in ontology (such as fire control, damage control, etc.), which can be ultimately traced back to the abstraction hierarchy.
- **Dynamic task allocation.** Tasks can be defined independently from who will execute that task (i.e. the task allocation). This allows us to experiment with different task allocations. We can even set up the system in a way that tasks are allocated at run time by the application itself.
- **Reactive agents for testing.** iTasks is very suitable for multi user prototypes. The user interfaces follows automatically from the task descriptions. Nevertheless, we would also like to be able to test or demonstrate a multi-user application when not all users are available. For this purpose, we have developed reactive agents that can take over the role of an absent human and fill in a value. These agents are only intended to keep the process running, and not to simulate human agents.
- **Intelligent agents for testing.** To be able to perform automated testing, the results that are gathered using reactive agents may not always be realistic.

Sometimes we need simulated humans (taking into account things like reaction time, fatigue, etc.). For this purpose we have enabled the possibility to incorporate intelligent agents in the simulation [10].

3 Internal Battle Management Tool

Using the methodologies (CWA, ontology engineering, user requirements engineering, prototyping), we have developed a prototype of the support tool for the internal battle, i.e. the Internal Battle Management tool (IBMt). This multi user application, programmed in iTasks and ST2, consists of interfaces for the internal battle manager (IBM), and the other persons that are acting in the internal battle (fire fighter, mechanic, and medic). The interaction between these actors is managed within the IBMt.

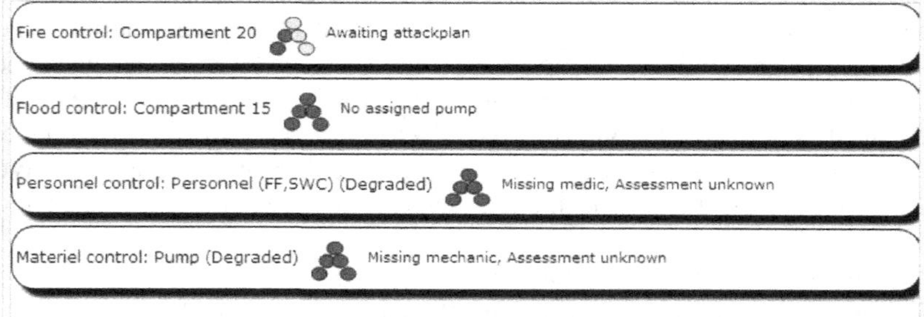

Fig. 4. Process overview window

Fig. 4 shows the process overview window for the IBM, which contains a process-oriented information of the Internal Battle without being too specific (such as presenting each individual system-alarm). Process (represented by the four bars in Fig. 4), become active after the trigger condition of that process is fulfilled. Usually this trigger condition is a detection by a sensor. For example, for fire-control, this can be the detection of smoke by a smoke sensor. If the same process is triggered multiple times, multiple instances of that process are activated. For example, multiple instances of fire control for different fires at different locations may exist simultaneously. The IBMt automatically prioritizes the processes based on the command aim, and shows the processes with highest priority on top of the list.

After a process is activated, the status of the process is assumed to be OK. The status of the process is visualized by a decision ladder icon. Such an icon represents the status of the five steps of the decision ladder by five circles (in the same way as in Fig. 2). A green circle means that the sub process in the decision ladder is running properly; a yellow circle means suboptimal, and a red circle means that it is jammed and requires attention of the IBMt. For example, a decision ladder icon consisting of only green circles, is running OK, and no human intervention is needed. For example, in the case of fire control, a green circle for subprocess *sense* would mean that information about the status of the fire is collected properly, and a green circle for subprocess *act* would mean that the watermist installation is correctly extinguishing the fire. Such a process requires no human intervention. Human intervention is needed when one of the subprocesses are not green. For example, in Fig. 4, the Material control process (shown in the bottommost bar) is running properly all subprocesses, except for *assess* which has status jammed (which means that information about the availability of material is being collected properly, but the system fails to understand the incoming data). This requires human intervention, which can be given by clicking on the corresponding process bar. This opens a new window showing detailed information of the process. An example of this is shown in Fig. 5. The material control process halts at the step Assess (it is red). The resource graph shows the conflict between the resources and the demanding processes. This allows the human to understand what is going on: the mechanic that is necessary to repair the pump is allocated to the fire control process. The IBM needs to decide if the mechanic needs to be reallocated to the material control process, to repair the pump that is necessary for the fire control process.

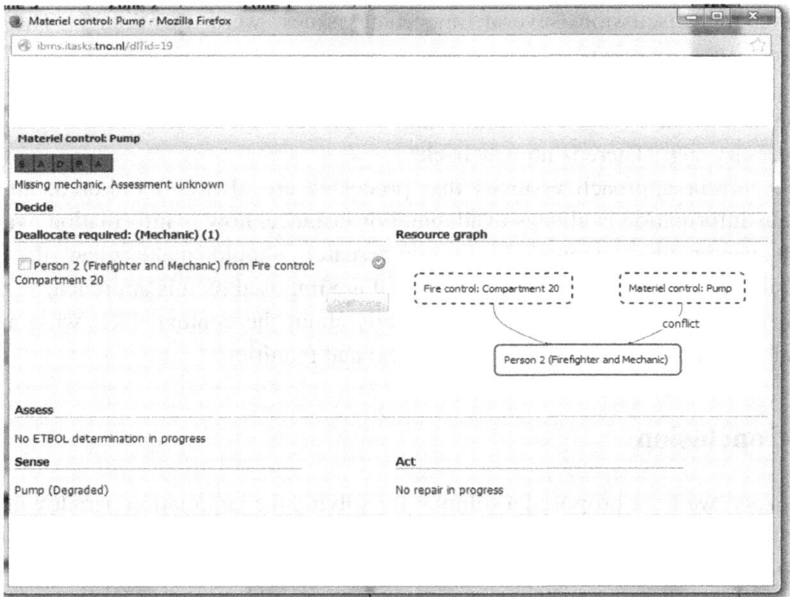

Fig. 5. Detailed window of the material control process

3.1 Cognitive Walkthrough

We have performed a very basic evaluation of the IBMt prototype to collect feedback from navy personnel, and allowed them walk through a basic scenario as if they were an internal battle manager (IBM). The interface consisted of the process overview screen, and the corresponding detail screens. One of the experiment controllers manually created the triggers (such as smoke sensor detection) to activate the processes.

The scenario was as follows:

Suppose the ship has been hit by an RPG (rocket-propelled grenade), causing fire and leakage. The smoke sensor automatically detects fire in Compartment 20, and starts a fire control process. The process automatically allocates the pump to this process which start extinguishing the fire. Because each step in the process is running correctly, the fire control process is represented as fully green in the process view (Fig. 4). Meanwhile, the flood sensor has activated a flood control process. This process requires a pump, which is currently used by the fire control process. This conflict cannot be solved automatically, and hence the process requires intervention by the IBM. The problem ("No assigned pump"), and the jammed decision process is shown in Fig. 4. When the IBM clicks on this process, a detailed view opens (Fig. 5), the user gets information about temperature sensor readings (at the lower left corner), and about the resource conflict (at the upper right corner). The IBM is enabled to reallocate the pump from the fire- to the floodcontrol process.

During the discussion several important issues were mentioned. The most important issues are described below:

- The transition of system monitoring to process monitoring is a logical and necessary step 'There is no way back'.
- The current approach assumes that processes are always monitored, and that status information is always available. For instance, how is information available that personnel is wounded? Future research should make clear the exact requirements on sensor environments that are imposed by this approach.
- Process view should contain more details about the context, like: what are the critical locations of infrastructure (cables) and munition.

4 Conclusion

In this paper, we have proposed a solution for solving the automation paradox in highly automated navy vessels. We believe this type of solution can also be applied to other complex domains where multiple professionals work together in a highly automated environment and where processes can be understood in terms of sensing, situation assessment, decision making, planning, and acting. To make the results applicable in these other domains, we have not only described our solution for navy vessels, but have strongly focused on the method and tools with which we arrived at our solution.

The contribution of this paper is twofold. Firstly, a set of methods has been presented for developing socio-technical systems. Secondly, we have applied that set of methods to develop a concept for an innovative socio-technical system for fighting the internal battle in navy vessels with reduced manning. We present our conclusions on these two topics below.

We presented methodological guidelines to start with a broad and structured work analysis (CWA) and made this more concrete using formal ontologies and user requirements. The step towards ontologies and user requirement engineering is important to bridge the gap between domain exploration (CWA) and system development. Furthermore, it promotes collaboration within larger development teams, which is also necessary given the size and complexity of our innovation projects. To test the envisioned systems, we developed a prototyping environment using the task-oriented programming language iTasks, called ST2. ST2 contains the basic building blocks for socio-technical system design (e.g. process monitoring, task-delegation, agent simulation, etc.). We experienced the ST2 layer in iTasks as a useful tool to create prototypes that are elementary and basic (and therefore do not require much programming effort), and at the same time capture the essence of the envisioned collaboration and task performance.

With respect to our proposed Internal Battle Management Tool, we believe that we have found a practical concept for realizing highly automated navy vessels that can be operated with a reduced crew. The old approach, system monitoring, had reached its limits with respect to scalability, i.e. it was not possible anymore to apply more automation without overloading the user with control tasks. Our proposed new approach, i.e. process monitoring, is better suited for highly automated environments as the human takes the role of a manager over multiple autonomous processes. In terms of the goal manning reduction stated in the introduction, the proposed interface could be used to realize a reduction of five persons to one person for coordinating the internal battle.

We plan to perform future work on the prototyping environment (ST2) and on the prototype for process-monitoring. In this work we have laid out the basic aspects of the ST2 and implemented some standard task structures. We plan to extend ST2 with standard tasks for different ways of communication (e.g. email, computer-mediated, alarm-based, broadcasting, simulated face-to-face communication, etc.). This allows us to experiment with different positioning of actors (which results in different communication structures). With respect to the implemented prototype, we plan to represent more detailed context information in the detailed screens for the different processes. Lastly, we plan to perform tests in a real operational environment, to better understand the differences between the old and the proposed approach.

References

1. Bainbridge, L.: Ironies of automation. Automatica 19(6), 775–779 (1983)
2. Brus, T.H., van Eekelen, M.C.J., Van Leer, M.O., Plasmeijer, M.J.: Clean—a language for functional graph rewriting. In: Kahn, G. (ed.) FPCA 1987. LNCS, vol. 274, pp. 364–384. Springer, Heidelberg (1987)
3. van Diggelen, J., van Drimmelen, K., Heuvelink, A., Kerbusch, P.J., Neerincx, M.A., van Trijp, S., ... van der Vecht, B.: Mutual empowerment in mobile soldier support. Journal of Battlefield Technology 15(1), 11 (2012)
4. Gruber, T.R.: A translation approach to portable ontology specifications. Knowledge Acquisition 5(2), 199–220 (1993)

5. Lijnse, B., Jansen, J.M., Plasmeijer, R.: Incidone: A task-oriented incident coordination tool. In: Proceedings of the 9th International Conference on Information Systems for Crisis Response and Management, ISCRAM, vol. 12 (2012)
6. McGuinness, D.L., Van Harmelen, F.: OWL web ontology language overview. W3C Recommendation 10(2004-03), 10 (2004)
7. Plasmeijer, R., Lijnse, B., Michels, S., Achten, P., Koopman, P.: Task-Oriented Programming in a Pure Functional Language. In: Proceedings of the 2012 ACM SIGPLAN International Conference on Principles and Practice of Declarative Programming, PPDP 2012, pp. 195–206. ACM (September 2012)
8. Rasmussen, J., Pejtersen, A., Goodstein, L.P.: Cognitive systems engineering. Wiley, New York (1994)
9. Smets, N.J.J.M., Bradshaw, J.M., Diggelen van, J., Jonker, C.M., Neerincx, M.A., de Rijk, L.J.V., Senster, P.A.M., Sierhuis, M., ten Thije, J.O.A.: Method and simulation platform to assess human-agent teams for future space missions. Journal Paper Special Issue 'AI Space Odyssey' (IEEE Intelligent systems)
10. van Staal, W.: Agent-based simulation with iTasks for navy patrol vessels, Unpublished Master's thesis (2013)
11. Vicente, K.J.: Cognitive work analysis: Toward safe, productive, and healthy computer-based work. CRC Press (1999)
12. Vicente, K.J., Rasmussen, J.: Ecological interface design: Theoretical foundations. IEEE Transactions on Systems, Man and Cybernetics 22(4), 589–606 (1992)

Multi-agent Solution for Adaptive Data Analysis in Sensor Networks at the Intelligent Hospital Ward

Anton Ivaschenko[1] and Anton Minaev[2]

[1] Samara State Aerospace University, Samara, Russia
anton.ivashenko@gmail.com
[2] Magenta Technology Development Center, Samara, Russia
thebestmauda@gmail.com

Abstract. This paper introduces a multi-agent solution for remote monitoring based on wireless network of sensors that are used to collect and process medical data describing the current patient state. A multi-agent architecture is provided for a sensor network of medical devices, which is able to adaptively react to various events in real time. To implement this solution it is proposed to partially process the data by autonomous medical devices without transmitting it to the server and adapt the sampling intervals on the basis of the non-equidistant time series analysis. The solution is illustrated by simulation results and clinical deployment.

Keywords: Multi-agent technology, intelligent devices, sensor networks, real time sensor processing, smart hospital, telemedicine, computer-aided diagnosis, software architecture.

1 Introduction

One of the essential challenges in computer-aided medical diagnosis is concerned with the implementation of software solutions and architectures capable to process large quantities of data in real time. Widespread medical equipment in hospital wards (e.g. medical monitors) is capable of solving this problem by capturing a number of parameters describing the current status of the patient, but it is quite expensive and requires fixed installation at the medical bed.

Modern mobile devices for diagnostics (like Holter monitors) have their own niche in medicine; some of them are comparatively cheap and portable, which makes them useful for every day patient status monitoring. Still, there are no solutions on the market based on the utilization of these devices for complex monitoring and diagnosis in real time. The reason for this is high complexity of centralized data capturing and analysis.

To cover this gap we propose a multi-agent architecture of a sensor network of medical devices. This network is able to adaptively react to various events in real time. The main difference from the existing approaches is that the process of data flow analysis provided by the proposed solution is distributed between the nodes of intelligent network, formed by multiple mobile devices with autonomous behavior.

D. Ślęzak et al. (Eds.): AMT 2014, LNCS 8610, pp. 453–463, 2014.

This makes it open for possible extension and integration with heterogeneous software and in demand for practical use at a number of hospitals.

2 Motivation

The request for the proposed solution was formed in the process of delivering a number of projects carried out by Magenta Technology [1]. During the last three years we have got extensive experience both in implementing multi-agent technologies for transportation logistics [2] and developing solutions for medical applications [3], which led us to the following conclusions.

Medical applications usually impose high requirements to hardware and software solutions being developed for practical use. This is the reason of their comparatively high costs and difficulty to enter the market. In addition, most providers of medical applications encourage to use separate products of the same brand name that solve different tasks. This helps integrating the pieces of software and hardware and provides high reliability of the whole solution.

From the pragmatic point of view providing the opportunity to integrate heterogeneous devices and associate the software from different providers, which seems to be common for most medical institutions, looks like a reasonable and attractive step. To achieve the desired aim, the IT infrastructure should be developed as an open system using unified protocols for data exchange and providing interoperability for the integrated devices.

It should be mentioned that the data flow in such a web of medical devices will have certain features. First of all it will be multi-directional, which means that each device at a certain node should provide an ability not only to accept and generate certain data, but transmit it as well, supporting the data exchange between other nodes. Next, the process of data exchange will be asynchronous. Time intervals between the messages will vary in time, so the process of data exchange could be described by non-equidistant time series [4]. Finally, the process of data exchange should be highly influenced by the current situation: the density of data exchange will be higher in case of emergency situations, inducing higher traffic in the system.

The combination of these features explains the necessity of some special approach to develop an architecture and technology for data exchange in the open network of medical devices. This process cannot be managed centrally as the time wasted on sending the description about the current situation to the center, analyzing it and providing the solution brings to nothing all the efforts to coordinate it in real time. To overcome this problem the devices themselves should be active and form a complex network of continuously running and co-evolving agents. Such architecture will be close to peer-to-peer (P2P) network [5] that is frequently used to describe and simulate the interaction processes of autonomous agents.

In this case the whole solution can be based on holons paradigm [6] and bio-inspired approach [7]. This paradigm and approach offer a way of designing adaptive systems with decentralization over distributed and autonomous entities organized in hierarchical structures formed by intermediate stable forms. Its implementation in practice requires the development of new methods and tools for supporting fundamental mechanisms of self-organization and evolution, similar to living organisms (colonies of ants, swarms of bees, etc) [8]. The opportunities provided by multi-agent technology in medicine are fully described in [9].

So we came to a challenge to develop a multi-agent solution for the coordination of a network of medical sensors which provides functioning in real time, paying maximum attention to adapting the frequency of data exchange to the rhythm of real processes.

3 State-of-the-Art

Intensive care medicine [10] is a branch of medicine concerned with the diagnosis and management of life threatening conditions, requiring sophisticated organ support and invasive monitoring. One of the most commonly used types of equipment in this area is a fairly wide range of bedside monitors for computer-aided diagnosis. These monitors capture data from a number of sensors, save it and send it to a centralized data storage, and determine emergency situations with visual and audio alarm notification. Bedside monitors can be a part of a distributed system of support for clinical decision-making[11], that has been coined as an active knowledge base, which uses patient data to generate case-specific advice to assist health professionals.

The system architecture for such a solution is illustrated by Fig 1. The algorithms of data analysis and decision-making support, provided by bedside monitors, are usually based on fixed rate sampling that allows consistent analysis of the dynamics of the indicated values, collected by several sensors during a certain period of time.

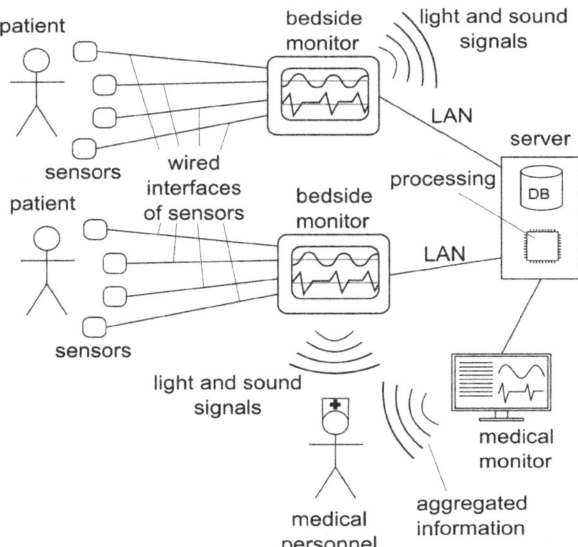

Fig. 1. Architecture of a fixed installation diagnosis system

In order to improve the quality of medical care and personalize it, there is a better solution for remote patient monitoring [12, 13] at general hospital wards. In this solution the vital signs are transmitted via wireless network technologies that provide flexibility and mobility for the patients. This solution is demonstrated in Fig. 2.:

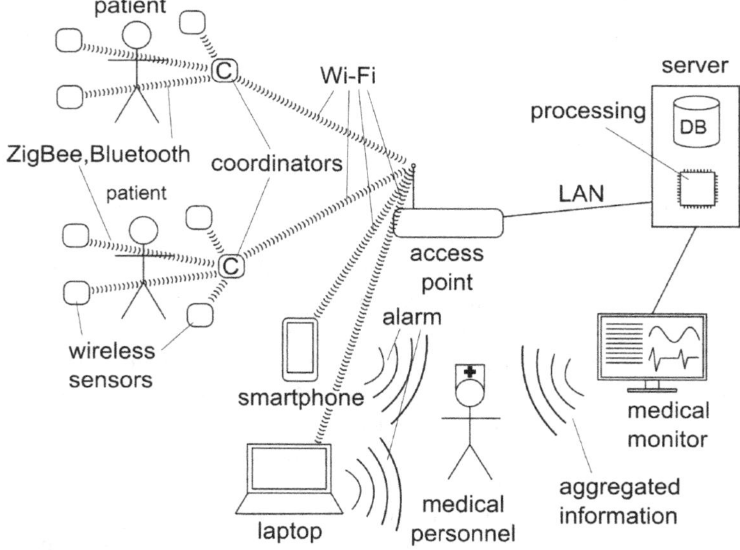

Fig. 2. Remote patient monitoring solution

Here, the sensors and the coordinator interact via ZigBee or Bluetooth protocols. The coordinator collects the data from the sensors and transmits it to the local computer network using Wi-Fi. Besides, the computer network contains the devices of data analysis and visualization in real time, alarm devices and a dedicated server for a specialized knowledge base. So the monitoring process remains centralized: the vital signs are gathered from the patients and sent for processing to a single control unit.

The most considerable benefit of such an approach is its flexibility: each diagnosis sensor is not bound to a particular patient care system, which allows combining a random set of devices to an arbitrary scanning system. The limitations of this approach include the unlicensed radio-frequency range of data transmitting between the sensors and the coordinators at 2.4 GHz, high density of sensors and the limitation of the number of channels supported by Bluetooth and ZigBee. This can cause negative stability of wireless communication. In addition, the data flows of diagnosis information that are translated to local network need to be processed in real time, which gives rise to a new set of requirements to this solution.

Technical capabilities of a medical systems based on sensor networks are also provided by the combined hardware and software platform for medical sensor networks, called CodeBlue [14]. This paper [15] describes a hospital healthcare monitoring system that uses wireless sensor networks: still, its sensors cannot transmit the messages between each other, they support only fixed-rate sampling and cannot be increased in number.

We propose to extend the existing approaches based on wireless technologies by introducing the active behavior of sensors that can be implemented by software agents and interact in P2P mode. Such basic features of P2P models [16, 17] like decentralization, sharing of resources and services, and autonomy make them applicable for a distributed computer-aided diagnosis system for personal medical care.

4 Problem Statement

Let us consider a generalized model where sensors (or devices) s_j, $j = 1..N_s$ are combined into a P2P wireless network to collect the diagnosis data from patients u_i, $i = 1..N_u$. Each measuring operation can be introduced as a Boolean variable: $e_{i,j,k} = e(u_i, s_j, v_{i,j,k}, t_{i,j,k}) \in \{0, 1\}$, where $v_{i,j,k}$ is the value of the parameter of u_i collected by s_j at the time of $t_{i,j,k}$.

Let us select and denote an event that has the best description of the patient's state – enough to make a diagnosis and indicate it as a pattern:

$$\Theta_{i,n} = \{\varepsilon_{i,j,k} = 1, k \in \Omega_{i,n}\} \tag{1}$$

where $\varepsilon_{i,j,k} = \varepsilon(u_i, s_j, v'_{i,j,k}, t'_{i,j,k}) \in \{0, 1\}$ represents the real event of parameter value change in reality (external to our system).

Ideally the adaptive network of devices should perform as follows:

$$\sum_{i=1}^{N_u} \sum_{j=1}^{N_s} \sum_{k=1}^{N_\varepsilon} \varepsilon(u_i, s_j, v'_{i,j,k}, t'_{i,j,k}) \cdot \min_l \left(e(u_i, s_j, v_{i,j,l}, t_{i,j,l}) \cdot t_{i,j,l} - t'_{i,j,k} \right) \to \min \tag{2}$$

This can be calculated as the following:

$$\sum_{i=1}^{N_u} \sum_{j=1}^{N_s} \sum_{k=1}^{N_\varepsilon} \sum_{l=1}^{N_e} \varepsilon_{i,j,k} \cdot e_{i,j,l} \cdot \theta(t_{i,j,l} - t'_{i,j,k}) \cdot \theta(t'_{i,j,k} - t_{i,j,l-1}) \cdot (t_{i,j,l} - t'_{i,j,k}) \to \min \tag{3}$$

where $\theta(x)$ – Heaviside step function [18]: $\theta(x) = \begin{cases} 0, x < 0 \\ 1, x \geq 0 \end{cases}$,

$\forall \varepsilon_{i,j,k}, \exists e_{i,j,l} : t_{i,j,l} > t'_{i,j,k}$, $\forall e_{i,j,l}, l > 1 : t_{i,j,l} > t_{i,j,l-1}$ (the events are prioritized in the order of occurrence).

Statements (2) and (3) define the requirement for the time frames of measurement. The events of data collection should occur just after something happened with the patient. Conversely, in case nothing happens the sensors should not be triggered frequently.

This is a kind of a scheduling problem: the system should construct a plan of $e_{i,j,k}$ and adaptively correct it when needed. So, time is considered to be a continuous parameter, which is sampled in a process of measurements. The measurement plan should adaptively react to the changing situation and be able to collect the amount of data which constitutes a bare minimum to make a decision.

Still, there is another KPI left for consideration – the total number of measurements should be minimized:

$$\sum_{i=1}^{N_u}\sum_{j=1}^{N_s}\sum_{l=1}^{N_e} e\left(u_i, s_j, v_{i,j,l}, t_{i,j,l}\right) \to \min \tag{4}$$

It can be noticed that statements (1) and (3) are in contradiction with each other. This contradiction should be supported by a distributed software solution that will be able to provide the minimum of data broadcast in the network of autonomous devices (which will result in lower network load) and at the same time guarantee adaptive reaction in time.

5 Solution Architecture

To solve this problem there should be proposed an algorithm that will control sampling time interval at some reasonable value and increase it in case of a risk of emergency. To identify such a risk there can be used a simple heuristic like:

a) $v_{i,j,k}$ is outside the time interval $\left(v_{i,j}^{\min}, v_{i,j}^{\max}\right)$,

b) the increment $\dfrac{v_{i,j,k} - v_{i,j,l}}{t_{i,j,k} - t_{i,j,l}} > \Delta v_{i,j}$, $t_{i,j,k} < t_{i,j,l}$.

Otherwise, sampling time interval should adapt to a possible Pattern of the Emergent State (PES) – a hypotheses pattern describing a real situation $\Theta_{i,n}$, which is specific for each patient:

$$\Theta''_{i,n} = \left\{\varepsilon''_{i,j,k''} = 1, k'' \in \Omega''_{i,n}\right\}, \tag{5}$$

where $\varepsilon''_{i,j,k''} = \varepsilon''\left(u_i, s_j, v''_{i,j,k}, t''_{i,j,k}\right) \in \{0, 1\}$.

After identifying the risk, the set of devices responsible for the patient u_i should adapt the sampling intervals of their m\\\\\\easurement to better react to the current situation.

Considering the jitter of real events in a pattern, the time series describing the current patient state should be treated as non-equidistant time series. So the function used for the identification should look like:

$$\rho''\left(u_i, e_{i,j,l_0}, \Theta''_{i,n}, \Delta\tau, \Delta v\right) = \sum_{j=1}^{N_s} \sum_{l=l_0}^{l_0+N''_{i,n}-1} \sum_{k=1}^{N_{\varepsilon''}} e_{i,j,l} \cdot \varepsilon''_{i,j,k} \cdot \theta\left(t_{i,j,l} - t''_{i,j,k}\right) \cdot \theta\left(t''_{i,j,k} - t_{i,j,l-1}\right) \cdot \theta\left(t''_{i,j,k} + \Delta\tau - t_{i,j,l}\right) \cdot \theta\left(v''_{i,j,k} + \Delta v - v_{i,j,l}\right) \cdot \theta\left(v_{i,j,l} + \Delta v - v''_{i,j,k}\right) = N''_{i,n}. \tag{6}$$

This solution was implemented by the intelligent software platform with distributed architecture (see Fig. 3, 4). The data gathered from sensors in real time is partially processed on their side and sent to the server for centralized processing only in case of risk identification. To provide complex analysis, the devices can interact and cooperate by exchanging the messages using ZigBee protocol and coordinating the sampling and accuracy of the measurements.

PES $\Theta''_{i,n}$ patterns (see Fig. 4) are created and stored on a dedicated server and are distributed to the diagnosis devices. According to the current situation each autonomous device determines the risk of emergency using the identification function (6). In case of emergency detected, they initiate data exchange for a deep analysis or an alarm notification sent to the handheld devices of medical personnel.

The benefits of such a solution include flexibility, adaptability to any external events, ability to function in real time and provide diagnosis decision-making support, and capability to process big data. Besides, it allows introducing individual configuration of diagnosis devices (sensors) for each patient without limiting the mobility of the patient.

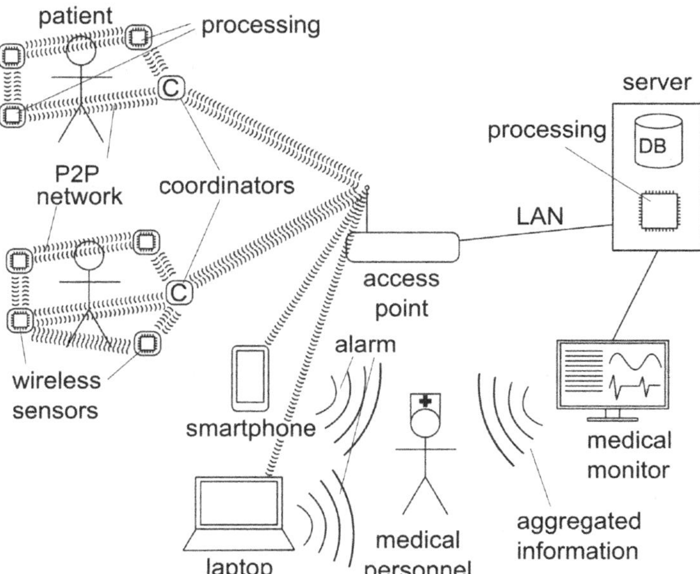

Fig. 3. Solution physical architecture

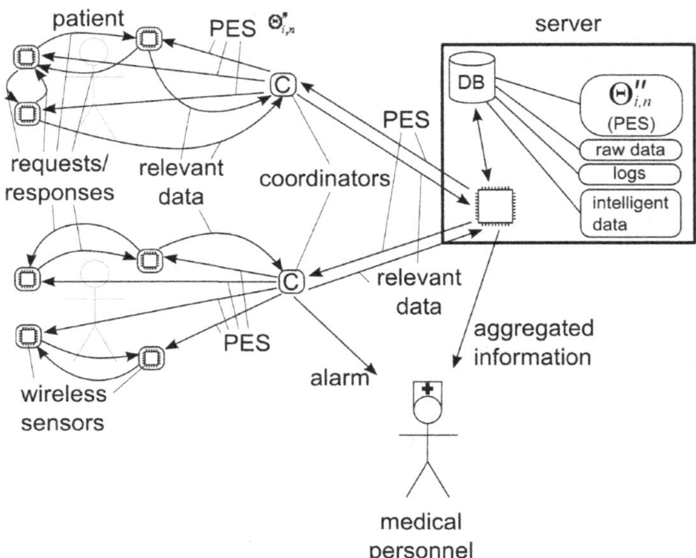

Fig. 4. Logical architecture of the proposed solution

6 Human Aspects of Distributed Medical Monitoring

New performance capabilities of remote medical patients monitoring via sensor networks provide an opportunity to increase the level of medical service and support people in medical, psychological and social aspects of their daily lives. Developments in this area should be first of all oriented on the actual requirements of end users; otherwise even using the most up-to-date technologies will not help getting closer to the expected benefits. The proposed approach for adaptive data analysis addresses this aspect. By reducing the size of the sensors and other medical devices and making them transportable, the users expect to get more comfort with the same reliability. This requirement set a problem of looking for new technologies of data processing, and the proposed approach can give a solution.

Another aspect of human-centric review of the problem of effective medical monitoring is the necessity to function in real time. No technical system can immediately react to the incoming unpredicted events, especially in distributed multi-agent technical environment. At the same time, continuous and permanent monitoring of the patient's medical condition, which can be provided by the sensor networks, allows the diagnostic software to detect changes at early stages.

In this regard, the logic of data collection and analysis should consider the time factor. For a certain period there can be enough time for analysis and decision making, but any moment there can emerge a sudden event that will require immediate reaction. The medical monitoring system should consider this factor and be able to function in adaptive mode in real time. In our opinion, the above-stated approach brings the solution closer.

7 Simulation Results and Implementation

The proposed approach for distributed real time sensor processing in sensor networks was tested and probated by cardio simulation: the results are illustrated in Fig. 5 – 7. A sample electrocardiogram tracking was generated by the simulating engine. In case of low sampling the measured signal can considerably differ from the original one.

Fig. 6 illustrates two possible cases, randomly generated by the simulating engine: the one with the emergency and the one with the normal P-wave. Generating the moments of measurement according to the proposed algorithm, the diagnosis device (see Fig. 7) can adapt its behavior to the real situation and reliably identify all the emergency P-waves. This proves the possibility to partially delegate the functionality of data analysis to the intelligent devices and thus reduce the traffic in P2P network.

The processes of interaction on the basis of the proposed architecture were tested for a couple of devices: a medical monitor and an autonomous infusion drop counter, which was developed in cooperation with the Samara State Medical University, Russia (see Fig. 8). The results of implementation proved the relevance of the proposed approach.

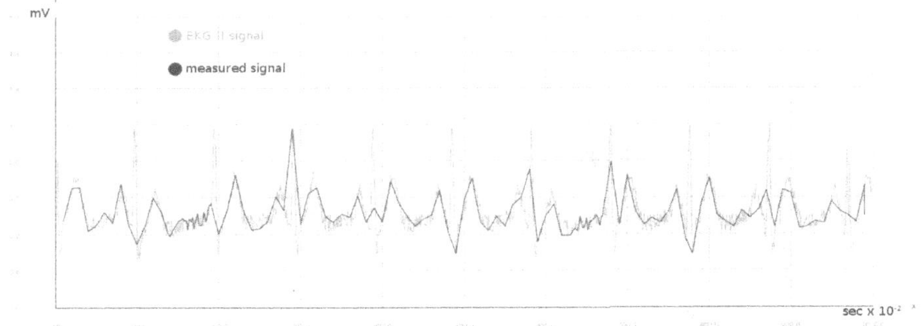

Fig. 5. Electrocardiogram simulation

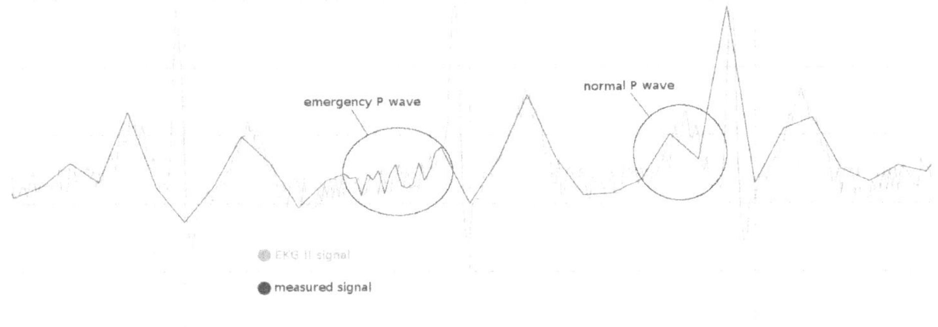

Fig. 6. Fragments of emergency and normal P-waves

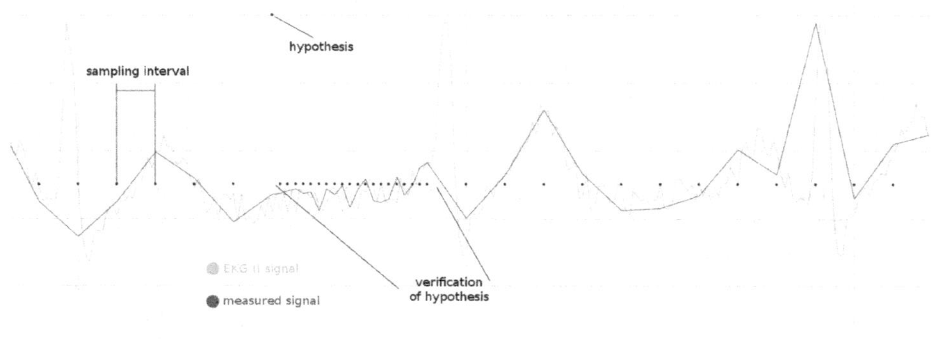

Fig. 7. Sampling times for adaptive analysis

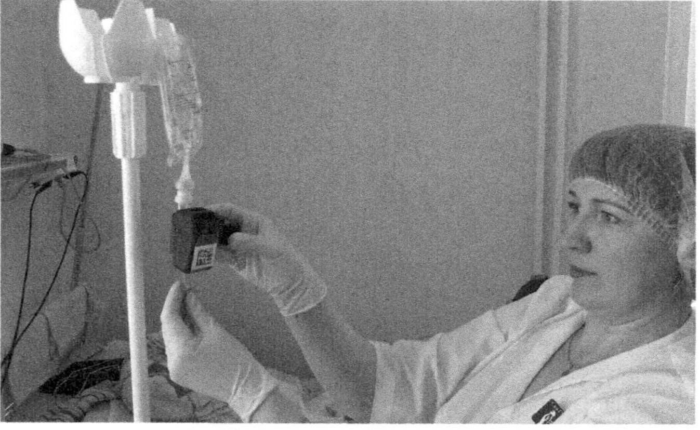

Fig. 8. Intelligent drop counter being deployed in clinic

8 Conclusion

In this paper we have introduced a multi-agent solution for the remote monitoring of medical patients, based on the wireless network of sensors. The process of data flow analysis is distributed between the nodes of intelligent network, formed by multiple mobile devices with autonomous behavior. The benefits of such a solution include flexibility and possibility to function in real time.

One of the first devices developed using to the proposed approach was the autonomous infusion drop counter integrated with a cardio monitor, both able to adapt the sampling intervals and minimize power consumption and network load with no loss of data. First results of its practical use have proven the selected approach, so we plan to extend the number of the supported types of sensors as a further development.

References

1. Andreev, V., Glashchenko, A., Ivaschenko, A., Inozemtsev, S., Rzevski, G., Skobelev, P., Shveykin, P.: Magenta multi-agent systems for dynamic scheduling. In: Proceedings of the 1st International Conference on Agents and Artificial Intelligence (ICAART 2009), pp. 489–496 (2009)
2. Glaschenko, A., Ivaschenko, A., Rzevski, G., Skobelev, P.: Multi-agent real time scheduling system for taxi companies. In: Proceedings of the 8th International Conference on Autonomous Agents and Multiagent Systems (AAMAS 2009), Budapest, Hungary, pp. 29–36 (2009)
3. Ivaschenko, A., Dmitriev, A., Cherepanov, A., Vaisblat, A., Kolsanov, A.: "Virtual Surgeon" training suite for laparoscopy, endovascular and open surgery simulation. In: Proceedings of the 27th European Simulation and Modeling Conference (ESM 2013), pp. 114–118. EUROSIS-ETI, Lancaster University (2013)
4. Prokhorov, S.: Applied analysis of random processes, Samara scientific center of RAS., 582 p. (2007), http://window.edu.ru/resource/665/58665/files/9_pasp.pdf
5. Ivaschenko, A., Lednev, A.: Time-based regulation of auctions in P2P outsourcing. In: Proceedings of the 2013 IEEE/WIC/ACM International Conferences on Web Intelligence (WI) and Intelligent Agent Technology (IAT), Atlanta, Georgia, USA, pp. 75–79 (2013)
6. Leitão, P.: Holonic rationale and self-organization on design of complex evolvable systems. In: Mařík, V., Strasser, T., Zoitl, A. (eds.) HoloMAS 2009. LNCS, vol. 5696, pp. 1–12. Springer, Heidelberg (2009)
7. Kalina, P., Vokřínek, J., Mařík, V.: The art of negotiation: developing efficient agent-based algorithms for solving vehicle routing problem with time windows. In: Mařík, V., Lastra, J.L.M., Skobelev, P. (eds.) HoloMAS 2013. LNCS, vol. 8062, pp. 187–198. Springer, Heidelberg (2013)
8. Gorodetskii, V.I.: Self-organization and multiagent systems: I. Models of multiagent self-organization. Journal of Computer and Systems Sciences International 51(2), 256–281 (2012)
9. Bergenti, F., Poggi, A.: Multi-agent systems for E-health: Recent projects and initiatives. In: 10th International Workshop on Objects and Agents (2009)
10. Reynolds, H.N., Rogove, H., Bander, J., McCambridge, M., Cowboy, E., Niemeier, M.: Telemedicine and e-Health 17(10), 773–783 (2011), doi:10.1089/tmj.2011.0045
11. Decision support systems (February 17, 2009), http://www.openclinical.org/dss.html
12. Sahandi, R., Noroozi, S., Roushanbakhti, G., Heaslip, V., Liu, Y.: Wireless technology in the evolution of patient monitoring on general hospital wards. Journal of Medical Engineering and Technology 34(1), 51–63 (2010)
13. Liu, Y., Sahandi, R.: ZigBee network for remote patient monitoring. In: IEEE 22nd International Symposium on Information, Communication and Automation Technologies, Sarajevo, Bosnia & Herzegovina, October 29-31, pp. 1–7 (2009)
14. Shnayder, V., Chen, B., Lorincz, K., Fulford-Jones, T.R.F., Welsh, M.: Sensor networks for medical care. Technical Report TR-08-05, Division of Engineering and Applied Sciences, Harvard University (2005), http://www.eecs.harvard.edu/mdw/proj/codeblue
15. Aminian, M., Naji, H.R.: A hospital healthcare monitoring system using wireless sensor networks. J. Health Med. Inform. 4, 121 (2013), doi:10.4172/2157-7420.1000121
16. Schoder, D., Fischbach, K.: Peer-to-peer prospects. Communications of the ACM 46(2), 27–29 (2003)
17. Minar, N.: Distributed systems topologies: Part 1 (2001), http://www.openp2p.com/pub/a/p2p/2001/12/14/topologies_one.html
18. Weisstein, E.W.: Heaviside step function. From MathWorld – A Wolfram Web Resource (2013), http://mathworld.wolfram.com/HeavisideStepFunction.html

The Challenges behind Independent Living Support Systems

Giampaolo Bella[1], Pekka Jäppinen[2], and Jussi Laakkonen[2]

[1] Dipartimento di Matematica e Informatica
Università di Catania, Italy
giamp@dmi.unict.it

[2] Software Engineering and Information Management
Lappeenranta University of Technology, Finland
{pekka.jappinen,jussi.laakkonen}@lut.fi

Abstract. Despite their crucial goal of assisting the elderly through their daily routine, Independent Living Support systems still are at their inception. This paper postulates that such systems be designed with a number of requirements in mind, and in particular with safety, security and privacy as fundamental ones. It then correspondingly articulates the three main challenges behind the development of Independent Living Support systems: requirement elicitation, design and correctness analysis. It is found that requirement elicitation will have to cope with a large variety of issues; that design will have to proceed from modularity; and, notably, that correctness analysis will have to be socio-technical. The last finding in particular emphasises that, for a system that prescribes vast interaction with the human, system correctness only makes sense if the system is analysed in combination with the human, rather than in isolation from the human. Building upon previous experience with the socio-technical analysis of Internet browsers, this paper identifies the specific socio-technical challenges that Independent Living Support systems pose, and indicates an approach to succeed in taking them.

1 Introduction

It is widely accepted that ambient intelligence cannot be developed without adequate account on and consideration of the human aspects. Because such intelligence is meant to be expressed in the presence of human beings, the intelligent technology must be conceived in combination with the humans, not in isolation from them. The present paper applies this postulate to a specific form of ambient intelligence, *Independent Living Support* (ILS) systems. ILS systems aim at supporting the elderly or the disabled people through their daily routine. They are sophisticated computer systems typically consisting at least of wearable devices to monitor the user, a robot to carry out various tasks, a computing base and various communications means. With the European population getting statistically older, the motivation to deploy such technology is great, and notable progress has been made lately, for example, within the Mobiserv project [1].

D. Ślęzak et al. (Eds.): AMT 2014, LNCS 8610, pp. 464–474, 2014.

The main contribution of this paper is a well-founded argument that three of the main phases of system development, namely the *requirement elicitation*, the *design* and the *correctness analysis*, become particularly challenging when the system is an ILS one. In particular, among the various requirements that are elicited, the primary roles of safety, then of security and privacy are noted. As for the design, the complexity that should be attained would not be realistic without appealing to modularity.

In terms of correctness analysis, the paper demonstrates why the analysis should be socio-technical by building on past experience with notable systems — a blood infusion pump incident [2], and Internet browsers' displayed icons [3,4]. The paper also identifies and discusses the specific socio-technical issues that ILS systems raise, namely the complexity of the system functionalities, the variety of system users, and the collaborative nature of human interaction with such a system. It then arguments the higher complexity of human interaction with ILS systems than with the notable systems, and concludes by suggesting the use of a specific model.

To the best of the authors' knowledge, this is the first structured attempt to define the main challenges behind ILS systems including the socio-technical ones. Building on the authors' combined experience on systems involving the humans in general, with ILS systems in particular, this paper pinpoints the three main macro-areas where researchers should put most efforts. The presentation is organised accordingly, that is, in sequence: requirement elicitation (§2), design (§3), socio-techical analysis of correctness (§4), and conclusions (§5).

2 Requirement Elicitation

As it can be expected, an effective ILS system should provide a variety of services for the wellbeing of the human user. The Mobiserv project identified several essential functionalities; the main ones were: reminder and encouragement to eat, drink and exercise, a clear control interface for smart home operations, fall detection and direct communication with a care centre, support for social interaction via video communications to family friends and carers. In order to provide these functionalities, several pieces of hardware and software must be deployed. Incorporating these into everyday lives of adults poses a number of requirements, which are summarised below.

The physical embodiments of the system consist of hardware such as cameras for monitoring, central smart home controlling units such as a PC, a variety of smart home sensors and actuators, wearable sensors forming so called smart clothing, and finally a robotic unit. With this amount of hardware and possible distributed intelligence, it is imperative at least that possible failures affecting one part are not going to bring down the entire system. In other words, local failure must not become general. For example, if the robot gets stuck on clothes found on the floor and consequently runs out of battery, it should still be possible to control the smart home through other means; at the same time, the smart clothing should be able to store the data received from its embedded sensors although it cannot transmit them to the robot in real time. In general, expecting error conditions in the system should be part of the system design to avoid single points of failure. Hence the fault tolerance requirement is a primary one.

ILS systems can be acquired by different stakeholders and deployed in a variety of environments. Older adults may acquire the system for their own apartment, while caretaking companies and foundations may acquire them for use in their care homes. The apartments of older adults may vary immensely in terms of layout and materials of the construction. Thick concrete walls can block radio signals and thus more wireless communication routers are required. Older apartments can also prove to be challenging environments for installing ILS system hardware because of limited access to electricity and also of compatibility issues with modern smart home controlling equipment. Furthermore, various levels of home renovation may be needed to enable an unconstrained movement of the robotic unit. For example, too high door thresholds may block wheel-based robots. In short, responding to the entanglements of older apartments may in general raise the cost of deploying the system. A requirement therefore is to build physically robust robots that may cope with virtually all features that a flat may have.

Commonly, an ILS system for a privately owned apartment is funded by the user personally, and therefore it must be possible to cut the costs by removing or at least reducing the non-essential functionalities of the system. In consequence, the robot must function even if surrounded by limited services, for example even if smart home equipment have not been installed in the building. This is an important requirement for flexibility and compositionality of the system.

However, it must be observed that care homes are progressively being designed for facilitating movement, and equipped with a lot of smart home hardware and intelligent home controlling functions. This in turn calls for an additional requirement: the need for interoperability between existing and deployed smart home hardware with the new hardware of a newly-purchased ILS system. In other words, bringing in an ILS system should not require replacement of the existing hardware.

The requirements outlined here form the basis of the research that the Mobiserv project conducted successfully [5]. Having briefly demonstrated their importance above, we would like to stress three (yet more) fundamental requirements: *safety*, *security* and *privacy*. Safety of the user who is supposed to benefit from the ILS system is the most obvious requirement in presence of a robotic unit, hence needs no explanation. However, we shall see below (§4) that, upon the basis of previous experience, meeting it may not be trivial. Also, due to the variety of personal data that ILS systems handle, they raise significant security and privacy requirements that perhaps have been insufficiently treated so far, hence deserve a separate discussion.

2.1 Security and Privacy

Cameras that monitor the user activities within their own home can be extremely intrusive. The same applies to the variety of sensor data that smart clothing gathers and inputs to the system. Therefore, the system impact on user privacy is great, hence demanding effective security and privacy techniques. Without a scrupulous consideration of these fundamental non-functional requirements, it will be hard to build ILS systems and convince people to incorporate them into their lives beneficially.

The need of securing data while they are transmitted through the various components of the system described above is obvious. When wireless communication means are advocated, the system should adopt standard security mechanisms available today, such as strong passkeys for Bluetooth and strong WPA keys for WiFi. Additionally, standard Internet security tools such as HTTPS are mandatory for protecting remote access to the ILS system. However, this care is insufficient to win the security and privacy challenge.

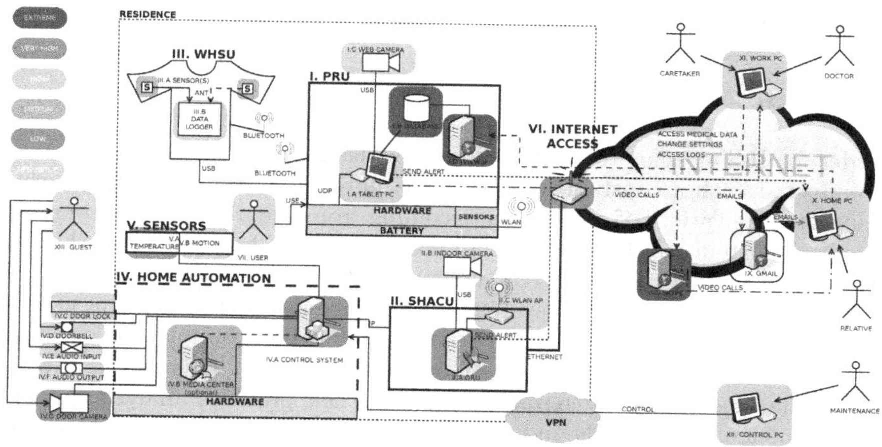

Fig. 1. Risk assessment of an Independent Living Support system

Taking the well-known secure-by-design approach, security and privacy are taken into account ever since the early stages of design. Figure 1 schematises an ILS system in its entirety, hence including the part that lies outside the user residence. A number of components can be appreciated. Since the different capabilities provided by the components as well as their hardware realisation vary from deployment to deployment, the security and privacy preliminary assessment must be carried out independently for each deployment [6]. Unfortunately, this can easily take a significant amount of time. However the component based approach can ease the task, suggesting that the assessment is conducted per component. Thus, when a new deployment is realised, the component based assessment can be used as a basis, and the main focus can be targeted to the information flow between components and the hardware that holds the component. Figure 1 shows an example of security and privacy assessment performed using standard risk assessment. The risk values of software components are combined to hardware components they reside in; different parts of the system are color-coded upon the basis of their risk value. We believe that this helps the analyst pinpoint the potential weak points in terms of security and privacy, where most effort for protective methods should be concentrated.

However, it must be noted that privacy cannot be evaluated in the same fashion as security. Privacy is a multi-disciplinary issue [7] that has no such clear definition as security has [8,9], and there is evidence that the whole concept can be misunderstood or misinterpreted, even by courts [9]. Good security can be achieved with the aforementioned techniques but privacy cannot, and should not, be determined only by the direct damage done, nor should it be perceived as a form of secrecy [9]. Therefore, it is imperative to conduct a separate component-based privacy assessment in order to incorporate the non-technical issues, such as legislation and directives, into the system design. In a system as information intensive as an ILS, privacy of the end-users is a fundamental factor that must be maintained.

3 Design

In order for ILS systems to be successful, the design phase of the system must be conducted with a constant look at the requirements. However, this is challenging due the variety of requirements coming from the different environments where each system will be deployed. In consequence, it becomes clear that the system architecture needs to rely on a component-based modular approach. The capabilities and features of the system are implemented in independent components that are able to communicate with each other. For example, the recording component provides a video image stream to the system. This can in turn be used by the video analysis component, which then provides analysis data for other components to use. The independence of the components means that even if they do not get all the information they need to conduct their task, they will not come to a halt. For example, should the training monitoring component have no access to image data, it could still provide some analysis based on the data it gets from smart clothing sensors. To this extent, the various components accomplish a level of fault-tolerance.

The independence of the software components also means that they should be separated from the hardware and designed as their own if possible. The mapping between software and hardware should be done in the deployment phase only, when the features and potential limitations of the target environment are known. Figure 2 describes an example deployment of components including a smart home, some smart clothing and a robotic unit. More precisely, it features the Physical Robotic unit (PRU), a Wearable Health Supporting Unit (WSHU) and the Smart Home Automation and Communication Unit (SHACU). The various interdependencies between the components are self-explaining. In particular, it can be appreciated that the DataLogger component of the WSHU communicates with the ActivityMonitor component of the PRU, and that the NutritionActivity of the SHACU communicates with the NutritionAgenda of the PRU.

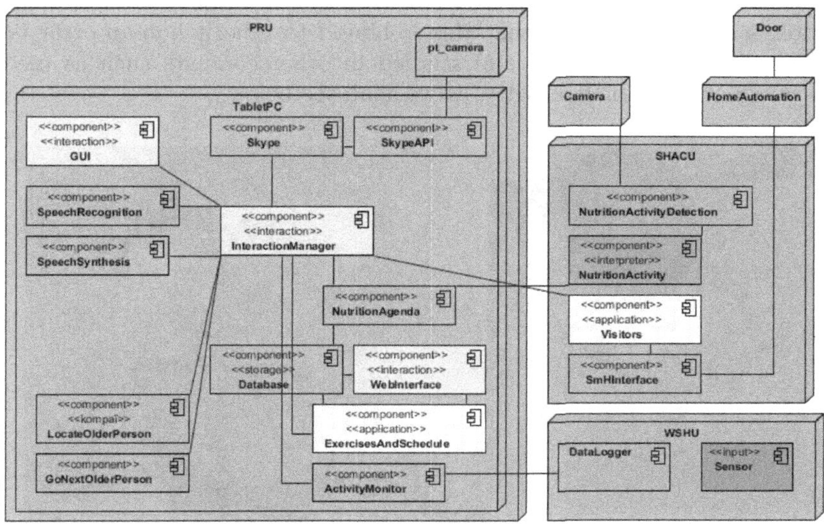

Fig. 2. Deploying software components to hardware

In an application where a robot would be too expensive, several useful components, such as for calling relatives via Skype or for acquiring smart clothing data, could be more cheaply deployed in the SHACU. Also, in order to guarantee interoperability, components that come with open and well documented interfaces stating the data and the format that they can take in must be chosen. Open interfaces enable different companies to build compensatory components for the system, or to create new components that complement or extend the system with new features. Independence of components with open interfaces can also enable dynamic deployment of new capabilites so that the behaviour of the ILS system can be dynamically augmented when necessary or when more resources become available.

4 Socio-Technical Analysis

The first two challenges behind ILS systems have been discussed so far, along with indications on how to face them. One may think that once the requirements are elicited and once it is clear how to design the system to meet them, the biggest effort would be behind. Unfortunately, we are still left with the challenge of analysing if the system works as intended, namely correctly, and in particular that it meets its various requirements, with safety, security and privacy at the forefront.

This Section demonstrates that the analysis of correctness of an ILS system, no matter if conducted over the design or over a working prototype, is very challenging indeed. Figure 3 shows a relative of an elderly user of an ILS system while she is configuring the daily tasks of the Physical Robotic Unit through its user interface. We emphasise that this is bound to inherit human error issues that have already been found and studied in other contexts such as medical device safety and Internet security, as we shall see below.

Fig. 3. User interface to a Physical Robotic Unit [5]

Let us consider a digital system that is supposed to meet certain requirements, such as safety, security, privacy and usability. The main hypothesis of socio-technical analysis is that, if a system prescribes interaction with the human, a *purely technical* analysis that the system meets its requirements is insufficient. For example, the security of a cryptographic scheme and the correctness of programming code traditionally stem from purely technical analysis. These are typically carried out by means of standard mathematical techniques and tools, such as probability theory and static/runtime analysis, either by pen and paper or with some computer assistance.

Insufficiency of purely technical analysis of systems that prescribe human interaction is due to the fact that technical analysis lacks an appropriate treatment of the human node in the system, which, by contrast, socio-technical analysis puts at its core. The challenge is great because humans cannot be treated like deterministic automata, as they are subject to very complex decision making processes. These are influenced by factors such as education, culture and social dimensions in general, which form such a large variety of paths of practice that it is challenging to enumerate it completely. Users may at least forget instructions, make errors, or do incomplete string matching when they read from a user interface, for example exchanging 'greenhouse' with 'green hose'. In short, users cannot be expected to read a system manual that engineers carefully wrote down, and fully comply with it in their everyday lives.

4.1 Motivation

The motivation for socio-technical analysis can be easily derived from the literature. For the sake of illustration, this Section outlines two past experiences where socio-technical analysis could have helped thwart significant problems. One, about medical device safety, concerns a drug, the other one, about Internet security, concerns mobile browsers; we shall see that both experiences are rooted in the socio-technical aspects of human interaction with technology.

The Fluorouracil drug incident killed a patient during chemotherapy [2]. The patient had been prescribed a continuous blood-infusion treatment with the drug, to take place over four days. Due to the obvious impracticality of manual infusion, the drug was instilled using a blood-infusion pump. It would seem that the pump was set to instill the prescribed dose of Fluorouracil over four hours rather than over four days, and this setting was entered through the simple user interface consisting of a standard liquid crystal display and plastic buttons illustrated in Figure 4.

The OperaMini and the Nokia browsers' padlocks gave their users a false sense of security [3,4]. OperaMini used to be one of the most common browsers worldwide because it is computationally lightweight, hence can run also on low-end or old smartphones. When the user requests an URL that responds via HTTPS, OperaMini displays a padlock. The padlock is the traditional means to indicate that the connection to the server corresponding to the URL is secure. However, it was found that, when the padlock is displayed, there in fact is no end-to-end security between the browser and the server because an Opera server proxies cleartext traffic between the two ends. The same was found of the Nokia browser, also fairly popular.

Both experiences described above demonstrate issues on human interaction with the technology. However, they bear subtle differences. The blood infusion pump of the Fluorouracil incident seems to have shown hours rather than days *correctly*, hence the root cause of the incident could only be found among error or decision (with the official investigation establishing the former) of the human user. By contrast, the padlock of the two mobile browsers mentioned above was displayed *incorrectly* because the corresponding transaction in fact did not enforce end-to-end security. Still, it must be observed that, although the positive meaning in terms of security of the padlock seems universal, we are not aware of socio-technical studies confirming that the padlock enjoys the universal human interpretation (in terms of security) that time measurement units do (in terms of duration). This observation of course hinders the universality of the claim that the padlock was displayed incorrectly. Only assuming the socio-technical result that a padlock universally means end-to-end security, would the root cause become a purely technical one — namely that the browser displayed a padlock when it should not have done so. This would in turn imply either error (leading to a buggy browser) or decision (purposely leading to an incorrect browser) of the human team that designed and developed the browser.

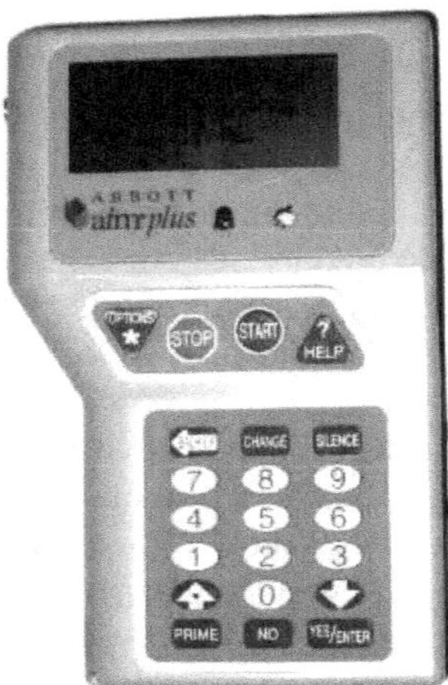

Fig. 4. User interface to the blood infusion pump used in the Fluorouracil incident [2]

The challenge of socio-technical analysis therefore is to ensure that user interfaces with the technology minimise the risks of error (or misinterpretation) of the users. Not only does this concern the interface of a technology hence the end user of the technology, but also the interface adopted to develop the technology hence the developer (e.g. the designers and the coders) of the technology.

4.2 ILS System Specificities

ILS systems raise specific socio-technical issues. The three main ones are:

The complexity of the functionalities that an ILS system offers raise the risks against its correct programming, as well as the risks against its correct, intended use. Life support functionalities, as seen above, are arguably more varied and heterogeneous than just infusing a liquid, hence in particular the user interface of the PRU is expected to be more complicated than that of the blood infusion pump. This raised complexity can be only partially appreciated by comparing Figure 3 to Figure 4. If we consider the number of life-saving functionalities that the PRU interface can manage, the risks that led to the Fluorouracil incident appear to multiply dramatically.

The variety of users of an ILS system also raise the risks against its correct use. By showing the various points of contact between human beings and technology, Figure 1 indicates that, in addition the the target elderly who is indicated as USER, there also are other roles: DOCTOR and CARETAKER, who setup the medical programme; RELATIVE, who inspects and corrects the system to ensure that everything works well; MAINTENANCE, who does the routine technical operations; GUEST, who also interacts with the system from time to time. All of these roles except the last one may perform system operations that are sensitive in terms of safety, or security or privacy. Also, these roles may be played by principals with varying skill in terms of technology interaction, exacerbating the socio-technical risks.

The collaborative nature of human interaction with an ILS system complicates its design and programming. The roles that interact with the system are not only varied, as noted above, but are also supposed to work collaboratively towards a common aim, the wellbeing of the elderly user. For example, the RELATIVE may raise a health concern with the DOCTOR, who may prescribe certain medicines, or MAINTENANCE may temporarily suppress certain services and instruct the CARETAKER of a temporary workaround. Such a collaborative, multi-user system raises standard task control challenges to establish who can set or reset what operations, and access control challenges to establish who can see (portions of) what data. Facing these challenges appropriately also requires a socio-technical approach because the computer security literature shows that users often strive to circumvent the limitations that the system imposes — neglecting the fact that they might have been aimed at safety, security and privacy.

4.3 A Useful Model

The socio-technical problem is currently far from being solved. Having seen above why it also spans over ILS systems, we argue that existing approaches to tame this problem should also be adopted over ILS systems. In particular, the multi-layer model of Bella and Coles-Kemp [10] is an example of transdisciplinary effort that can effectively help to pan out the problem and then address it. It classifies the socio-technical system into five layers, covering technology as well as user expression of personas in front of it, and under the influence of society. The significance of the model is that technical guarantees that the technology works as intended should traverse the layers up to the user, who is also influenced by societal stimuli, not just by technology. Those guarantees would then become profitable to the users beyond the risk of human error.

5 Conclusion

This paper presented the challenges behind Independent Living Support systems, and provided an architectural approach to take them. In particular, it observed that safety, security and privacy are primary requirements though far from trivial to meet. It explained that design should be based on modular components with well documented interfaces in order for the system architecture to become widely usable in various environments. It finally discussed the necessity and the difficulty of carrying out a correctness analysis in socio-technical terms. The worldwide capacity to respond to the need of ILS systems is increasing steadily.

Acknowledgements. The authors would like to thank the members of Mobiserv project for useful discussions.

References

1. Heinilä, P. (ed.): Deliverable 3.1: Technical Specifications, Test and Implementations Scenarios. Technical report, Mobiserv project report (2010)
2. URL: Fluorouracil Incident Root Cause Analysis (2007), `http://www.ismp-canada.org/download/reports/FluorouracilIncidentMay2007.pdf`
3. Radke, K., Boyd, C., Gonzalez Nieto, J., Brereton, M.: Ceremony Analysis: Strengths and Weaknesses. In: Camenisch, J., Fischer-Hübner, S., Murayama, Y., Portmann, A., Rieder, C. (eds.) SEC 2011. IFIP AICT, vol. 354, pp. 104–115. Springer, Heidelberg (2011)
4. URL: How the Nokia Browser Decrypts SSL traffic: "A Man in the Client" (2013), `https://freedom-to-tinker.com/blog/sjs/how-the-nokia-browser-decrypts-ssl-traffic-a-man-in-the-client/`
5. URL: Mobiserv project website (2014), `http://www.mobiserv.info/`
6. Laakkonen, J., Annala, S., Jäppinen, P.: Abstracted architecture for Smart Grid privacy analysis. In: 2013 ASE/IEEE International Conference on Privacy, Security, Risk and Trust (2013)
7. Hui, K.L., Png, I.: The Economics of Privacy. Industrial Organization 0505007, EconWPA (2005)
8. Ward, M.R.: The economics of online retail markets (2001)
9. Solove, D.: 'I've Got Nothing to Hide' and Other Misunderstandings of Privacy. Social Science Research Network Working Paper Series (2007)
10. Bella, G., Coles-Kemp, L.: Layered analysis of security ceremonies. In: Gritzalis, D., Furnell, S., Theoharidou, M. (eds.) SEC 2012. IFIP AICT, vol. 376, pp. 273–286. Springer, Heidelberg (2012)

Exploring Patterns as a Framework for Embedding Consent Mechanisms in Human-Agent Collectives

Stuart Moran, Ewa Luger, and Tom Rodden

Mixed Reality Lab,
University of Nottingham, UK
{firstname.lastname}@nottingham.ac.uk

Abstract. With ever increasing developments in computing technology, approaches to attaining informed consent are becoming outdated. In light of this ongoing change, researchers have begun to propose several new mechanisms to meet the emerging challenges of consent in pervasive settings. Unfortunately a particular problem arises when considering consent in the context of Human-Agent Collectives (HACs). These large-scale heterogeneous networks, of multiple co-operating humans and agents are particularly complex and it is difficult to know *what*, *where* and *how* to introduce these new mechanisms. In this paper we explore the potential of patterns of interactional arrangement as a framework for embedding consent mechanisms in HACs and other ubiquitous systems.

1 Introduction

The process of embedding agency within computing systems raises unprecedented challenges to human privacy. The requirement for consent, as the means by which we agree to invasion of our private selves, is an ethical principle enshrined within our contemporary social expectations. Simply put, we expect to have a choice over who does or does not have access to certain aspects of our lives and, increasingly, our digital selves. Our interaction with the digital has occurred principally through tangible interfaces, which have to some extent been informed by social expectations and familiarity. An example of this is the re-interpretation of physically signing a paper document [2] as a digital checkbox [3] as a means of giving consent. However, the emergence of embedded computing brings with it a new set of challenges on how we deal with consent, as "not in a single one of these dimensions is the experience of [pervasive computing] anything like that of personal computing" [1]. When these ubiquitous computing systems are implicated within our daily interactions, it will likely break social expectations, becoming less clear how, when and where consent should occur. As such, not only the mechanisms, but also the concept, of attaining consent must evolve and adapt [4]. This motivated researchers to develop a number of new techniques and modes of interaction for attaining consent in ubiquitous computing. However, even with these new consent mechanisms, in complex systems it is difficult to effectively know which to use and when. Human-Agent Collectives [36] (HACs) are one example of particularly complex ubiquitous systems, as they consist of large-scale real-time networks of multiple teams of humans and software agents working in

D. Ślęzak et al. (Eds.): AMT 2014, LNCS 8610, pp. 475–486, 2014.

collaboration, with different degrees of autonomy. In this paper, we draw upon the notion of patterns [35] as an analytical lens through which to formalize the recurrent interactional themes in HACs. These themes contribute toward an emerging typology, which could simplify the selection and implementation of consent mechanisms in ubiquitous computing systems.

2 Consent in Pervasive Computing

With the ongoing advancements in mobile and embedded devices, we are rapidly approaching an era of computing where technology literally pervades every aspect of our lives [5]. Pervasive computing sees a fundamental change to the way humans perceive and interact with computers. No longer will a system be limited to one device per user, but with several, even hundreds of, devices serving each person throughout their daily lives. Interaction with these devices will vary significantly. Some will work in the background, hidden from view, while others will require more direct interaction. This new technology, with its wide range of configurations, has the potential to create highly novel services and social environments [6]. However, the current lack of real world deployments mean that the social considerations of this technology are not yet fully understood [34]. Furthermore, this computational trend stands to challenge many of the well-established practices associated with current technologies. For example, the number and frequency of user interactions with pervasive technology are likely to increase in comparison to existing technology; but more disconcerting is that so too will the numbers of implicit/unknown interactions. This is especially problematic for current approaches to attaining consent, with dynamic systems rendering the idea of consent at a single point redundant [7].

Subsequently, the complexity of the infrastructure, interaction and the available choices (or lack thereof) requires a rethinking of the idea of consent in pervasive computing [7]. Consent, in its most theoretical form, must be (a) voluntary; the user must be free to give consent, it should not be coerced or the result of fraud, (b) competent; the user should be capable of giving consent; for example, they should be an adult and should not be otherwise vulnerable, (c) informed, in that the user should be meaningfully and fairly furnished with sufficient information of the conditions of the agreement, and (d) comprehending, in that the user should fully understand those conditions in order that there is a shared understanding between the consent-giver and consent-seeker [8]. Only at the point that all these conditions are met should a signal of assent be secured, for example the signature or mark of agreement as currently represented by the check box. Whilst consent in the context of pervasive systems has received only limited recent attention, some studies have sought to deal with designing for consent explicitly in the context of more traditional online interactions.

2.1 Designing for Consent

With particular reference to informed consent, Friedman et al. [9] offer a series of value sensitive design principles to shape the development of informed consent within online systems; disclosure, comprehension, voluntariness (non-coercion), competence, agreement and minimal distraction. This latter point is considered important as undue distraction might undermine the act of consenting due to (a) the desensitizing effects and impact upon attention that could occur if a user is constantly notified to consent to minor issues, and (b) if the notification becomes too intrusive, the user may choose to bypass the consent process by ignoring or moving past the distraction altogether. Having made explicit the criteria of informed consent, the authors focus upon web-based interactions and seek to apply these criteria to the proposed redesign of both cookie notifications [10] and browsers in the belief that "informed consent provides a critical protection for privacy, and supports other human values such as autonomy and trust" [11 p.1]. Extending a value-driven view, the Principled Electronic Consent Management (ECM) is an approach suggested by Bonnici and Coles-Kemp [12]. The authors argue for a framework (consent theory, ECM norms, and manifestation of those norms) on the basis that (a) principled ECM addresses consent before and after the consent decision, (b) it considers a broad range of contributory factors such as both organizational and software processes, and (c) it builds upon theory in order to enhance consistency at the point of application [12].

Further studies also make suggestions in relation to how consent might be better designed within existing systems; though this is dominated by the field of bioethics, health and health data. Prasad and Kotz [13] make suggestions for solutions to the binary nature of consent management mechanisms through 'privacy management interfaces', which might incorporate the use of (a) privacy icons to support greater comprehension of risks to personal privacy (b) clearer 'interface documentation' designed to detail trade-offs/benefits related to each act of sharing, (c) the system could present beneficial recommendations based upon peer-behaviour, and (d) delegation of the decision to a specialist (e.g. in the case the doctor) [13].

In terms of the proposition informing consent, research has explored how the SMOG measure of text complexity can be used to assess whether terms and conditions can be understood by their readers [14]. A similar approach has also been developed to visualize such documents [15]. Whilst interesting, such approaches are not customizable to user preferences. More interestingly, it may be possible to automatically parse the consent-supporting text by machine and customize the output to suit the consent-giver (e.g. [16]). Given that consent is contextual [17], this could also be tailored to support differentiated consent scenarios, using the system to adapt the content and means of attaining consent. Hence, rather than undermining consent, system agency could be employed to support user agency in the consent process.

With much potential for adaptive technology, the question arises: what if a machine were able to act on our behalf and give consent? If we conceive of a raft of future systems, all interacting in a variety of different ways with different information, it is hard to imagine that humans will be able to process this information unsupported [18]. Hence it will become necessary to distribute some aspects of control and autonomy to intelligent software agents. Whilst seemingly dystopic, this is merely an extension of current practices of distributed cognition, such as using a computer for complex numeric processing or for setting reminders. The emerging field of human-agent collectives (HACs) aims to explore this type of interactional arrangement in large-scale pervasive networks of multiple co-operating software agents and humans; which brings its own complexity to consent attainment.

3 Human-Agent Collectives

The idea of software agents has been around for a long time, and their relationship to humans has been explored in the field of HCI [19]. However, to date, there has been a tendency to focus on a limited scale one-to-one relationship between humans and agents. This is due to a focus on specific tasks and problems where agents can most effectively support humans. As the potential of the technology becomes increasingly evident, we can begin to explore new and more complex challenges that require us to move beyond the one-to-one relationship. Subsequently, a body of research has grown within the study of large groups of humans and software agents, in a variety of configurations, acting in what has been termed 'Human-Agent Collectives' (HACs) [36]. It is intended that these configurations will help solve, support and manage problems that humans cannot easily complete alone, such as extremely dynamic, cognitively demanding, and timed constrained tasks [37]. Disaster response is one example of a complex socio-technical problem where agents could assist humans in data collection within inaccessible areas (e.g. scouting aerial drones), and decision-making (e.g. knowledge of the 'bigger' picture that is incomprehensible to humans) [20]. HACs are envisaged as large, complex networks of real time interactions. These networks might consist of professionally trained personnel carrying out a task alongside members of the public who volunteer to help during a difficult time. Unlike current technologies, in HACs there is an intriguing inter-play between different configurations of professionals, members of the public and software agents and the distribution of agency amongst these actors [23]. Conceptually, consent is also predicated upon the idea of agency (the power to act) in the decision making process. So, *if HACs are to reconfigure the locus of agency, what does this mean for consent?* With such complex adaptable and dynamic networks of interaction, the difficulty and need to explore how consent to participation or exchange of information is attained or promoted becomes imperative; particularly so if agents are to act on our behalf.

3.1 Spectrum of Control, Visibility and Complexity

With intelligent, adaptable and dynamic software agents come different degrees of autonomy; some agents may mimic the contribution of a human, and take control of a task entirely. The main point for consideration is that there is a spectrum of human-to-agent control during completion of a task. What makes this relationship particularly interesting is that the level of autonomy and agency may change during the course of an interaction, with humans handing over and taking control as and when is needed. This has significant implications for attaining and maintaining consent, as any change in levels of autonomy and agency need to be appreciated and understood by the human; particularly if a machine is autonomously taking control. These different exchanges can also lead to agents invisibly performing processes in the background, while a human completes a task, unaware of that processing. Equally, an agent could be completely transparent about its actions, explicitly showing the user its actions as they occur. Software systems may act on a user's behalf in ways that they may not fully be aware of at the point of action. For this reason, it is necessary to consider the ways users will be kept fully informed about the system they are a part of, in addition to ensuring that they understand the choices they make and the actions they are offloading to machines. The ideas of a flexible spectrum of autonomy, and visibility of the processes involved, potentially add even further complexity to the system [36].

There is a clear need to understand and explore how individuals (and groups) might give consent to participate or exchange information in a HAC and also how they might use an agent to do this on their behalf. The difficulty in examining these questions is to know how best to make sense of the complex interactions, and the implications of the propagation of their effects. We suggest that what is needed is a framework with which to effectively capture and analyze the points of interaction and changes in autonomy. This framework must be scalable in order to deal with the different configurations of humans and agents. As such, in this paper, we propose to operationalize HACs into a series of discrete core configurations, or patterns, to allow consent-based mechanisms to be embedded in the interaction.

4 Patterns of Consent

Many real-life applications can be modeled as a system of interacting actors. By using the idea of patterns [35] as a theoretical lens, examination of any such system could allow us to identify discrete interaction points; and thus highlighting instances within which consent might be embedded. Furthermore, breaking down and modeling the system in this way could also potentially allow us to visualize information trajectories (in the form of interactions and exchanges of information). These trajectories could then be translated into 'consent trajectories' that model the consequences of giving or withholding consent throughout the entire HAC. To begin to utilize such a framework, we must examine the most fundamental patterns of human-agent interaction. In order to identify core-interactional arrangements between software agents and humans, we examine a number of existing systems and their patterns of interaction, and consider what primary class of entity makes up a pattern.

In Fig.1. circles represent Humans, which are the users, the people who interact with the software/embodied agent or the core system. Triangles represent an Agent, which is an often semi-autonomous software/embodied agent which can act standalone or as an interface to the core system. Finally, squares represent the Core System, which is the technology that serves as the foundation of (or direct solution to) the task at hand. This could be in the form of storage, processing, operating, routing or even the main tool for the task (e.g. a plane). This is one aspect of the entire system

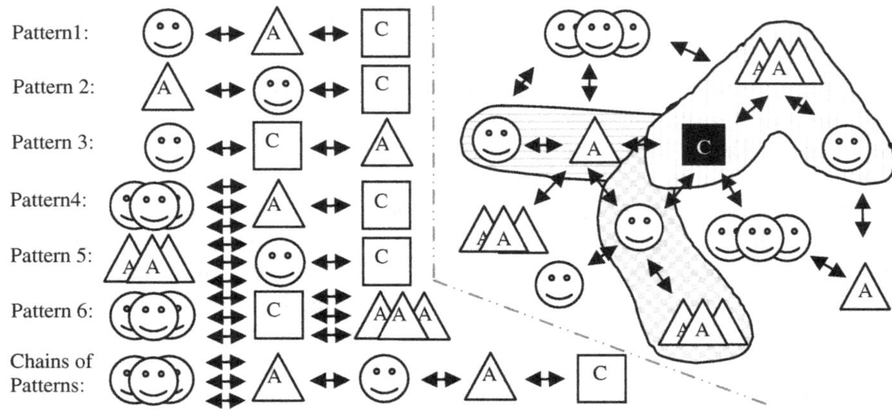

Fig. 1. Patterns of interactional arrangement (left), and example human agent collective (right)

which would include all of the above components. Initially we explore agent-interactions that are typically found in computing systems developed in recent years, focusing on an individual user and individual agent.

4.1 Pattern 1: Intermediary/Proxy Interaction

An intermediary interaction is the most basic pattern (see Fig. 1.) where a human interacts with the core system through a software agent acting as an intermediary. This agent is essentially an intelligent user interface, and could take on a variety of forms, ranging from a purely virtual agent that supports tasks on a computer, to a physically embodied agent (e.g. a robot) through which information is exchanged. The point of interaction with this core system is via the agent, and in some ways the agent will be a representation and spokesperson for the human. The interaction could take different forms, including (a) the exchange and propagation of instructions to the core system, (b) the agent observing/monitoring the activities of the human, or (c) after an initial interaction the agent acts on the human's behalf. Consider a first responder during a disaster interacting with an unmanned ground vehicle (UGV) (e.g. [21]). This UGV relays richer information between HQ and the first responder than that which is possible using current approaches. Some other existing examples of this type of pattern/configuration include human-robot interaction [20], e-commerce systems [22] and mixed reality games [23]. The implications of this pattern in terms of consent centre on whether the agent supports or undermines the agency of the user. For example, the user may choose not to interact with the agent, but simply to perceive the agent as a standalone technology, rather than a gateway. In this way, user knowledge of the core system is undermined, raising a number of questions (related to awareness) that need to be addressed when users are exposed to this pattern of interaction: (a) *Does the user have control over whether the agent acts on their behalf?* (b) *Do users have an appreciation of the trajectories of their information?* (c) *Are users aware of the wider-consequences of their consent-based choices in 'the-moment'?*

4.2 Pattern 2: Supervised/Advisory Interaction

A supervisory interaction, while similar to intermediary, places the user in direct contact with the core system, while a software agent supervises/observes (see Fig. 1.). The interesting aspect of this pattern is the (possible) disconnect between the agent and the core system, with the agent becoming an almost independent entity. The agent may act as a prompter giving advice to the user as they complete their work. Equally, it may take on a more proactive role acting as a quality assurance. In some cases, the agent may even be a safety backup system that is programmed to act when the situation is deemed unsafe for the user. Consider a user being advised by a SatNav on directions to their specified destination. The SatNav is collecting data about the user's actions (i.e. current location) and making recommendations for future actions to achieve the user's goal. The user interacts directly with the core system (the car) which the SatNav has no control over (in this instance). Other examples might include agents which take control of planes [24], manufacturing systems [25] and intelligent surveillance systems that asses suspicious behaviors [26]. With the agent being a more independent entity this will require a degree of user approval prior to/during interaction, in addition to interaction with the core system. Where the agent intervenes in critical situations, one could imagine a model of consent where the user implies or infer consent as a default but inbuilt explicit moments of revocation? This unique arrangement highlights a number of key points for the process of consent in respect of voluntariness and agency: (a) *At what point does a user consent to the core system taking control? Can users take back control of the core system, after an agent has taken it?* (b) *Where does the accountability lie when an agent is instructing a human?* (c) *Do users have a choice whether they are advised?*

4.3 Pattern 3: Remote/Tele-interaction

A tele-interaction explores how a user can observe the actions and assume control of an agent (see Fig. 1.). The human uses the core system to remotely operate and receive information from an agent in the field, or at a significant distance. This type of pattern is typical in reconnaissance or in environments that are dangerous or impossible for humans to reach. The user may not necessarily directly control the agent, and may simply observe and analyze the information it transmits. The point of interest here is that the interaction occurs at a distance, with a greater disconnected experience with the agent. Consider two soldiers tele-operating a UAV drone. They have a variety of different information streams, and make decisions in real-time. This serves to allow soldiers to reach targets in difficult terrain, but also promote the safety of the lives of the soldiers themselves. Other examples of this type of system include: telerobotics such as NASA mars landers and probes [27] and submersible mapping systems [28]. With a significantly greater disconnect between human and agent (rather than agent/core system), a series of important considerations should be made for consent. These relate most directly to the issue of competence, autonomy and accountability: (a) *When should a user take control of an agent (or agent become autonomous)?* (b) *Who is accountable for the actions of the agent in situ; both during agent and human control?* (c) *How might humans in situ approve/engage with interaction with the agent?* (d) *What affect does tele-operation have on human judgment; during both automation and control?* Following on from a single user and agent, we now extend the patterns to include multiple agents and multiple humans. Applications that involve these numerous agents include crowd sourcing, crowd management, and manufacturing/assembly line computing systems.

4.4 Pattern 4: Consolidated Intermediary Interaction

Consolidated intermediary interaction expands the intermediary pattern to include additional humans, which provide a larger corpus of information to an agent than an individual would (see Fig. 1.). Unlike the intermediary pattern, this agent has additional processing requirements in order to distinguish between the humans it interacts with, and the range of data types it receives. The main difference in this pattern is the unique exchange that may occur between the humans prior to interaction with the agent. Consider a busy pedestrian cross roads: the traffic light systems used to control thoroughfare during peak and off peak times are the same agent interacted with by both drivers and pedestrians (each who will act slightly differently, but generally conform to a series of rules/laws). The coordinating core system communicates information to different embodied agents, which then relay that information in real time. Users can interact with the core system through embodied agents that take the form of road crossing buttons. This pattern can also take other forms such as crowd sourcing [29], intelligent crowd control [30] or scheduling of shipping operations [31]. The same considerations as intermediary are applicable here (Pattern 1), but in addition we must also consider the factors brought about by the multiple human interactions, such as: (a) *How might a user exercise control of how their data is shared with others?* (b) *What information is exchanged between users?*

4.5 Pattern 5: Consolidated Advisory Interaction

Consolidated advisory interaction adds multiple agents to the advisory pattern, which can provide different sources of information or even different interpretations of data (see Fig. 1.). The agents may take different physical and virtual forms, or interact in different ways (via audio, actuation or visually). In this pattern, the user has potentially an additional cognitive load added by the increased number of software agents observing/providing them with information. Consider a CCTV operator receiving multiple streams of information from different agents (cameras). This information is then used to interact with the core system, intended to relay important and timely information on to the appropriate persons. Equally, multiple agents may work to take control of different parts of the core system – e.g. an aeroplane may have multiple agents controlling multiple facets. Many of the same considerations apply from advisory interaction (Pattern 2), but the following should also be considered in respect of consent: (a) *Is the user aware of the information exchanged between agents?* (b) *Does the user need to give consent to each agent (local vs global consent)* (c) *Can the user make sense of the information exchanged and produced from multiple agents?*

4.6 Pattern 6: Common Pool

With the core patterns identified, we now consider the final more complex pattern, which is the common pool: multi-human multi-agent collectives working together via a core system (see Fig. 1.). This could take on a variety of different levels of autonomy and control, with humans and agents working independently, for or via each other. Consider multiple humans and multiple embodied agents working together to complete a task, such as in a car manufacturing plant, where there is a need for both humans and machines; they work together via the core system (the factory line). Some of the machines are autonomous, while others will be supervised/controlled by humans. Other examples of this pattern might include industrial tasks supported by robots [32], astronauts [33] and the envisaged applications of HACS in disaster response. This pattern is on the edge of succumbing to complexity. All the teleoperation considerations are important (Pattern 3), but also the social relationships between the people like in the advisory interaction (Pattern 2).

4.7 Chains of Patterns

Having explored some of the possible fundamental patterns, we can see how these can join together in chains of patterns (see Fig. 1.). Patterns may repeat, or may connect via a core system, agent or human as the networks of interaction grow. Consider a pattern that joins remote telecommunication with intermediary interaction. A field responder might interact with a software agent that is controlled by a human, this combines these two patterns. It may be possible that within a pattern, the nature of the interaction can change, and indeed the pattern as well (depending on the role of the agent and its situation). For example, one moment a UGV could be relaying information between HQ and the user, the next it could make advisory points about the situation independent of the core system, then a human could control it.

5 Using Patterns

With a series of fundamental, but not necessarily exclusive, patterns identified we now have a means to explore the operationalization of complex large-scale HACs in order to embed consent mechanisms in the network. For example, a complex HAC, which consists of multiple humans and agents interacting, centered around a core system presents considerable challenges (see Fig. 1.). To explore aspects of consent in these collectives we must first break the HAC down into as many 'major' patterns as possible. There are likely to be individual and consolidated versions of intermediary and advisor patterns, all existing within a larger common pool pattern, including: Intermediary/Proxy Interaction (Horizontal Line), Agent-Consolidated Intermediary/Proxy Interaction (Vertical Lines), and Agent-Consolidated Advisory Interaction (Checkered) (see Fig. 1.).

5.1 Incorporating Consent Mechanisms

Having identified a number of patterns in the HAC, we examine each in more detail and attempt to resolve some of the considerations outlined above.

Intermediary/Proxy Interaction (Horizontal Line): The key things for consideration in this pattern are whether or not the human understands what information is collected and how it is used; in addition to whether or not a user has any control over the agent. If the core system is part of a planned interaction that occurs regularly, then ensuring the human is informed about and understands the purpose of the agent and the data it collects is adequate. However, if the interaction is momentary or ad hoc, then one recommendation would be to have the agent express its purpose and intent (in an accessible way), and to proceed only when a human has given approval.

Agent-Consolidated Intermediary/Proxy Interaction (Vertical Lines): In this pattern we see multiple consolidated agents acting as an intermediary to the main system. Again, consideration should be made about how aware the user is of the agents and whether or not the interaction is fleeting. Unlike the example given above this sees a single user interacting with multiple agents. With agents for different purposes it will likely be difficult for a user to fully comprehend the implications of their actions. For this reason, it would be beneficial to instigate a 'representer-agent', such that a user has a single point of control/interaction. This would reduce the cognitive burden on the user, and also create a more manageable situation. The agent may have periodic exchanges with the user about changes to the wider system, or a need for permission from one of the many agents in the pattern.

Agent-Consolidated Advisory Interaction (Checkered): This pattern sees multiple agents giving advice/information to a single user. It is questionable whether a user is able to fully comprehend the way that the agents exchange information amongst themselves, or the impact of providing consent to a single agent amongst others. Again, there is a clear need for an interface agent, one that can consolidate the advice from multiple agents and manage the exchange between users. There are also a number of questions with respect to agents taking control of the situation, with perhaps each agent taking control of different parts of the system. The way this information is expressed is critical, as is whether or not a user is able to take control back.

6 Conclusions

Consent is an important part of everyday life. With the ever increasing power of mobile computing systems the traditional approaches to attaining consent are no longer applicable. This has motivated researchers to explore new more appropriate consent-based mechanisms for the technology. However, complex large-scale heterogeneous networks of interacting software agents and humans pose new sets of problems. As such, we have proposed a nascent, exploratory use of patterns as a possible framework for breaking down complex systems into more easily examinable parts. Even our rudimentary overviews begin to offer some, key insights into the challenges arising. Examining each pattern more closely can help select the most appropriate approach to attaining consent, but also highlight many of the other socio-technical issues that arise. The next steps for this work are to continue to explore and refine the core patterns of interactional arrangement, to develop a pattern language for HACs in order to cross-compare systems, and to empirically test the framework by introducing consent mechanisms to a real world deployed HAC.

References

1. Greenfield, A.: Everyware: The dawning age of ubiquitous computing. New Riders, Berkeley (2006)
2. Wendler, D., Rackoff, J.E.: Informed consent and respecting autonomy: What's a signature got to do with it? Ethics & Human Research 23(3), 1–4 (2001)
3. Averitt, J.: Legal Ethics and the Internet: Defining a Lawyer's Professional Responsibility in a New Frontier. J. Legal Prof. 29, 171 (2004)
4. Luger, E.: Consent reconsidered; reframing consent for ubiquitous computing systems. In: Proceedings of the 2012 ACM Conference on Ubiquitous Computing, pp. 564–567 (2012)
5. Weiser, M.: The Computer for the 21st Century. Scientific American 265(3), 94–104 (1991)
6. Callaghan, V., Clarke, G., Chin, J.: Some socio-technical aspects of intelligent buildings and pervasive computing research. Intelligent Buildings International (1), 56–74 (2009)
7. Luger, E., Rodden, T.: Terms of Agreement: Rethinking Consent for Pervasive Computing. Interacting with Computers 25(3), 229–241 (2012)
8. Faden, R., Beauchamp, T.: A History and Theory of Informed Consent. Oxford University Press (1986)
9. Friedman, B., Lin, P., Miller, J.K.: Informed Consent by Design. In: Cranor, L.F., Garfinkel, S. (eds.) Security and Usability, pp. 503–529. O'Reilly Media Inc. (2005)
10. Millett, L.I., Friedman, B., Felten, E.: Cookies and Web browser design: Toward realizing informed consent online. In: Proc. CHI 2001, pp. 46–52. ACM Press (2001)
11. Friedman, B., Howe, D.C., Felten, E.: Informed Consent in the Mozilla Browser: Implementing Value Sensitive Design. In: Proc. HICSS 2002, vol. 8, pp. 247–257. IEEE Computer Society, Washington, DC (2002)
12. Bonnici, C.J., Coles-Kemp, L.: Principled Electronic Consent Management: A Research Framework. In: Proc. 2010 International Conference on Emerging Security Technologies, pp. 119–123. IEEE (2010)
13. Prasad, Kotz, D.: Can I access your data? Privacy management in mHealth. In: Proc, USENIX Workshop on Health Security and Privacy. USENIX Association (2010)
14. Luger, E., Moran, S., Rodden, T.: Consent for All: Revealing the Hidden Complexity of Terms and Conditions. In: Proceedings of the SIGCHI Conference on Human Factors in Computing Systems, pp. 2687–2696 (2013)
15. Kay, M., Terry, M.: Textured Agreements: Re-envisioning Electronic Consent. In: Proc. SOUPS 2010, pp. 1–13 (2010)
16. Carreira, R., Crato, J.M., Gonçalves, D., Jorge, J.A.: Evaluating adaptive user profiles for news classification. In: Proceedings of the 9th International Conference on Intelligent User Interfaces (2004)
17. Luger, E., Rodden, T.: An Informed View on Consent for Ubicomp. In: Proc. Ubicomp 2013. ACM (2013)
18. Gonçalves, D.J., Jorge, J.A.: Ubiquitous access to documents: Using storytelling to alleviate cognitive problems. In: Proceedings of the Tenth International Conference on Human-Computer Interaction, pp. 374–378 (2003)

19. Green, S., Hurst, L., Nangle, B., Cunningham, P.: Software agents: A review (1997)
20. Fischer, J.E., Jiang, W., Moran, S.: AtomicOrchid: A Mixed Reality Game to Investigate Coordination in Disaster Response. In: Mobile Gaming Workshop (MOGA) 2012 as a part of Proceedings of the 11th International Conference on Entertainment Computing (ICEC 2012), pp. 572–577 (2012)
21. Murphy, R.R.: Human-robot interaction in rescue robotics. IEEE Transactions on Systems, Man, and Cybernetics, Part C: Applications and Reviews 34(2), 138–153 (2004)
22. Wang, N.D., Shen, X., Georganas, X.: A fuzzy logic based intelligent negotiation agent (FINA) in ecommerce. In: Electrical and Computer Engineering CCECE 2006, pp. 276–279. Canadian (2006)
23. Moran, S., Pantidi, N., Bachour, K., Fischer, J., Flintham, M., Rodden, T.: Team reactions to voiced agent instructions in a pervasive game. In: International Conference on Intelligent User Interfaces, pp. 371–382 (2013)
24. Vaščák, P., Kováčik, J., Betka, P., Sinčák, F.: Design of a Fuzzy Adaptive Autopilot. In: The State of the Art in Computational Intelligence, pp. 276–281 (2000)
25. Shen, W., Hao, Q., Yoon, H.J., Norrie, D.H.: Applications of agent-based systems in intelligent manufacturing: An updated review. Advanced Engineering INFORMATICS 20(4)
26. Hengstler, H., Prashanth, S., Fong, D., Aghajan, S.: MeshEye: a hybrid-resolution smart camera mote for applications in distributed intelligent surveillance. In: 6th International Symposium on Information Processing in Sensor Networks, p. 360 (2007)
27. Fong, T., Thorpe, C.: Vehicle teleoperation interfaces. Autonomous Robots 11(1), 9–18 (2001)
28. Williams, H., Newman, S.B., Dissanayake, P., Durrant-Whyte, G.: Autonomous underwater simultaneous localisation and map building. In: Conference on Robotics and Automation, vol. 2, pp. 1793–1798 (2000)
29. Savage, N.: Gaining wisdom from crowds. Communications of the ACM 55(3), 13–15 (2012)
30. Jin, J., Xu, X., Wang, J., Huang, C.C., Zhang, S.: Interactive control of large-crowd navigation in virtual environments using vector fields. Computer Graphics and Applications 28(6), 37–46 (2008)
31. Steenken, D., Voß, S., Stahlbock, R.: Container terminal operation and operations research-a classification and literature review. OR Spectrum 26(1), 3–49 (2004)
32. Heyer, C.: Human-robot interaction and future industrial robotics applications. In: IEEE/RSJ International Conference on Intelligent Robots and Systems (IROS), pp. 4749–4754 (2010)
33. Carignan, D.L., Akin, C.R.: Using robots for astronaut training. IEEE Control Systems 23(2), 46–59 (2003)
34. Moran, S., de Vallejo, I.L.: Introduction to the special section on the social implications of embedded systems. Interacting with Computers 25(3) (2012)
35. Alexander, C., Ishikawa, S., Silverstein, M.: A pattern language. Ch Alexander 60 (2006)
36. Jennings, N.R., Moreau, L., Nicholson, D., Ramchurn, S.D., Roberts, S., Rodden, T., Rogers, A.: On human-agent collectives. Communications of the ACM (in press, 2014)
37. Fischer, J.E., Jiang, W., Kerne, A., Greenhalgh, C., Ramchurn, S.D., Reece, S., ... Rodden, T.: Supporting Team Coordination on the Ground: Requirements from a Mixed-Reality Game. In: Proc Intl. Conf. on Design of Cooperative Systems, COOP (2014)

Increasing Physical and Social Activity through Virtual Coaching in an Ambient Environment

Arjen Brandenburgh, Ward van Breda, Wim van der Ham, Michel Klein, Lisette Moeskops, and Peter Roelofsma

Ambient Assisted Living Group, Department of Artificial Intelligence, Faculty of Sciences, Vrije Universiteit, Amsterdam, The Netherlands
{a.h.brandenburgh,w.r.j.van.breda,w.f.j.vander.ham, michel.klein,e.j.p.c.moeskops,p.h.m.p.roelofsma}@vu.nl

Abstract. This paper describes the development and the validation of an ambient system (AAL-VU) that empowers its users in self-management of daily activities and social connectedness. The system combines state-of-the-art psychological knowledge on elderly user requirements for sustained behavior change with modern ICT technology in suggesting an adaptive ambient solution for the prevention and management of chronic diseases, inactivity and loneliness, thus resulting in a higher quality of life. Specifically, the AAL-VU system stimulates beneficial levels of activity in elderly as well as social connectedness. The focus on physical and social activity was chosen as this is recognized as a crucial element for the prevention, cure, and management of many chronic illnesses.

Keywords: virtual coaching, physical activity, social connectedness, self-management, ambient assisted living.

1 Introduction

It is widely known that the overall activity of older people is generally decreasing, partly because of reduced physical abilities, and partly because of fewer societal responsibilities and a smaller social network. The use of ambient solutions in these interventions is a new research area. In approaches from this area the focus is either on social connectedness or on physical activity. This ignores the relatedness of both aspects for elderly persons.

In the AAL-VU project an advanced technical system is developed to help elderly persons change their lifestyles on both aspects, taking this interrelatedness into account. More specifically, a personalized 3D virtual coach motivates the elderly to break a sedentary lifestyle and increase their overall level of physical and social activity. Although the breaking of the sedentary lifestyle and increasing activity are the focus domain in the project, the underlying theoretical notions are also applicable for changing lifestyles that are related to other domains, like unhealthy eating habits, addiction, and procrastination. In this paper, we describe the AAL-VU system and report about the first experiences with using this system in practice. In the next

D. Ślęzak et al. (Eds.): AMT 2014, LNCS 8610, pp. 487–500, 2014.

section, we first motivate the use of an ambient assisted living system based on virtual coaching for elderly persons. Next, section 3 provides a theoretical framework for behavioral interventions based on virtual coaching. The system and its various components are described in detail in section 4. Section 5 presents two feasibility studies that have been performed with the system. The studies provide insight in the experiences of users that were coached by the system and give a first indication of the effect on their behavior. Finally, in section 6 we conclude the paper with directions for future work.

2 Motivation

2.1 Increasing Physical Activity

There is a worldwide acceptance among medical authorities that activity constitutes a fundamental element of healthy living [13]. Activity has positive effects in several health domains [26], e.g. cardiovascular disease [3], diabetes type II [37] and multiple sclerosis [23] Nowadays problems related to inactivity are one of the major behavioral risk factors to health in modern society [17]. Several reports with activity recommendations have been published [25], [24]. However, research also demonstrates that more than 70% of adults fail to meet current activity recommendations [17], [33]. Astrup [4] mentions that recently issues concerning healthy living and physical activity have been increasingly prevalent, especially in developed countries.

2.2 Focus on Elderly People

Of all age groups, elderly persons are the most inactive and have relatively greater difficulty to comply with the recommendations [20]. Research shows that activity leads to a significant health increase in older adults: a reduced risk in occurrence of disabilities [18], [22] in the number of falls in elderly persons [9], an increase in self-esteem and self-efficacy [21], an increase in cognitive functioning [26], a decrease in dementia[26], and lower anxiety and depression levels in older age [31]. The conclusion, that people in general and elderly persons in particular have to be more active, seems to be a given fact.

A variety of intervention types have been made for this purpose. This has led to a variety of intervention program recommendations [12]. The overall conclusion of the evaluation of these intervention programs is that intervention programs can lead to significant increases in the amount of physical activity, but that the effects are in general only moderate and they often do not lead to reaching the recommended target for daily activity [11]. This demonstrates the need for research on alternative approaches and strategies for intervention programs.

2.3 Virtual Coaching

Recently a growing amount of studies has been performed on different types of 'serving digital agents', virtual agents designed to help people in daily task activities [34]. Virtual characters or agents act as a new medium to interact with system information and system users. The use of virtual characters can be stimulating and improve human-computer interaction [8]. The characters have high-quality graphics and make numerous types of interfaces possible. Another benefit is that they can be helpful for a variety of trainings: for example technical trainings (e.g. nurses, medical students, etc.) and social skills trainings. In addition, the gender, age, race, physical attractiveness, apparel, or roles of a virtual character can be changed to match the needs at hand [38]. Since elderly persons have relatively greater inability in dealing with ICT [32], the use of virtual coaching can be a promising venture to provide smart support in caregiving.

Research by Lin et al. [19] has confirmed that virtual agents can be effective in several domains, including motivating exercise. Frost, Boukris and Roelofsma have also demonstrated that virtual coaches can be effective in exercise [15]. Their argument is that in the interaction between humans and virtual characters, humans will feel more personally addressed as compared to when in interaction with standard ICT devices. As a consequence, a so-called para-social relationship will develop between the human being and the virtual agent [7], [10]. Frost, Boukris and Roelofsma demonstrated that through the development of such a para-social relation, users' intrinsic motivation will increase for task performance [15].

The AAL-VU system concentrates on increasing activity by mediated communication through virtual coaching. Change in physical activity will be explained by persuasive communication performed by a virtual character [6], [5]. In most studies, theoretical frameworks regarding the influence of virtual coaching on human behavior are absent. Such a framework should provide prescriptions on how changes should be achieved [27].

3 Theoretical Basis

A common phenomenon in behavioral change is that people with the best intentions often fail to commit to their decisions [1], [14], [29]. Sniehotta et al. label this phenomenon the intention-behavior gap [35]. They often end up not being able to resist the temptation of a variety of vices that may come up in between.

The intrapersonal dilemma of resisting upcoming vices while working toward the realization of virtuous goals is represented in Figure 1 [from [8]. In this figure the x-axis represents time, passing from left (T_0) to right (T_3). The y-axis represents the attractiveness, or subjective utility, of an event or an outcome. The unbroken line represents the subjective utility over time for a virtuous goal, for example: 'reaching a reduction in BMI from 30 to 20 at T_3 through a twelve month physical activity program starting at T_0'. As the figure shows, the subjective value of the virtue increases gradually over time. It is highest at the moment of realization of the goal at T_3 and lowest at T_0 when it is still only a promise that waits for realization. The subjective utility of the vice is represented in the dashed line, for example: 'sitting on the couch, eating chips and drinking coke, while watching a soap in the evening at T_2'.

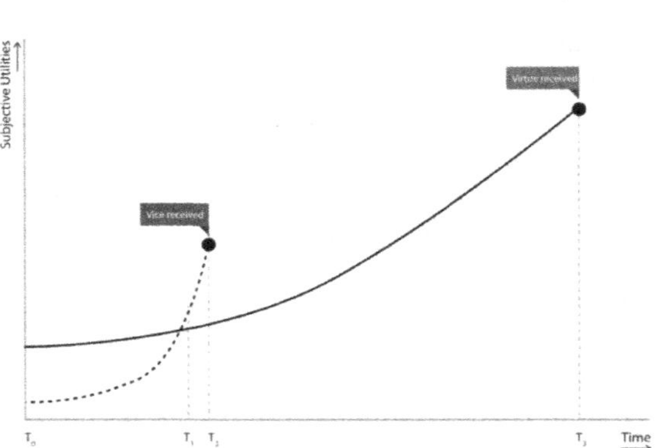

Fig. 1. Subjective utilities for vice and virtue

At T_0 the subject may be presented with the following decision problem presented by a caregiver: 'Break your sedentary lifestyle now and follow a twelve month activity program to reduce your BMI by registering now' (virtue). In Figure 1 at T_0 the value of the remote virtuous prospect is perceived to be higher than the value of the imminent prospect of sitting on the couch. At such a moment people will decide to register and pay for the physical activity program. However, as time goes by during the day the value of the vice will rapidly increase, more rapidly than the value of the virtue. At that moment the option to stay at home for the rest of the day may seem misleadingly attractive. Accordingly, the subject falls for this imminent temptation. This is represented in the dashed curved line crossing the unbroken line at T_1. More specifically, the period between T_1 and T_2 in the figure represents the moments of susceptibility to temptations for a variety of environmental vices.

The period between T_2 and T_3 represents the period where the subject has just given into temptation. By staying at home the moment of realization of the virtue has been delayed. The investments made so far have been (partly) lost and the reward is further away in the future. The subject may feel regret and become out of control.

The crossover point and the timing of the vice divide the figure into three time periods. In each time period the subject will have different goals and intentions. The subject's decision frame will be different in each period. In the time period A, the first period, the subject will choose the virtue because that has a higher present value. The moment of performing behavior is still far away. Over time, the competing goal of relaxing on the couch becomes increasingly attractive. Once the crossover point has been passed and period B is present, we can say that the vice is dominant and the consideration of that being imminent now leads the decision maker to prefer it. The decision frame of the subject has changed into either ignoring the lure of imminent temptation or giving in to it. The attractiveness of the vicious goal becomes increasingly larger than that of the virtuous goal. At last, once the vice has passed, there is a third period in which the decision maker regrets his impulsive choice or feels relief over the conflict resolution of self-control, and that period is named C.

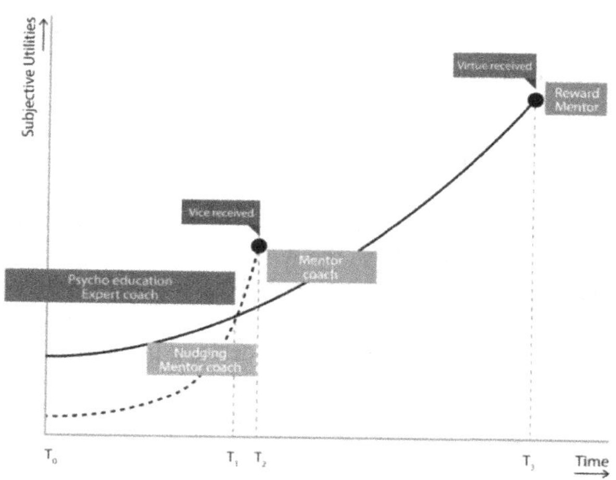

Fig. 2. The different roles for the coach in each period

Figure 2 can be used as a representation of different intervention components for behavioral change in physical activity. The first moment for intervention is the period before T0. This is an adequate period for intervention components that focus on changing intentions and plans. A second moment for intervention is in period A. Here relative utilities for virtues and vice change dynamically. The subject still wishes to reach the virtue; but the subject does not activate behavior yet. In both these periods intentional change programs based on cognitive behavioral approaches can be used, based, for example, on the theory of planned behavior [2] or on other intentional motivational theories like Self-determination theory [42]. More specifically, the virtual coach can provide persuasive expert information, like why it is healthy to move. In addition, the subjective norms or perceived behavioral control can be influenced by such persuasive communication by the virtual coach.

A third moment for intervention is the period before and after T_1. Here the subjective utility of the vice becomes higher and a change in the decision evaluation occurs. A fourth moment for intervention is in period B. In both of these periods, the subject will be most susceptible for motivational and action driven stimulation. This can be achieved by prompts and nudges that remind the subject of their plans. Nudging cues are used as implicit motivational strategies that guide people to the right behavior [6]. These are also adequate moments for the coach to act as a motivator providing support to strengthen the subject's belief not to give up.

Yet another period for intervention is in period C. Subjects have either fallen for temptation or been successful in self-control. This is an adequate moment for mentoring and counseling. After period C there is a final moment for intervention with a focus on specific rewards by the virtual coach for consolidation of the behavioral change. The combination of the expert, motivator and mentoring role is a promising venture [5], [36]. The AAL-VU system uses the framework illustrated in Figures 1 and 2 to apply the strategies during the process of behavioral change. Our hypothesis is that this combined approach of prompts, nudging and cognitive behavioral strategies will result in increased overall activity.

4 System Description

The AAL-VU system consists of several hard- and software components that are connected to each other. Together they form the AAL-VU system in the home of the user. The most central part of the system is the Google Nexus 7 tablet with Android 4.0. On this tablet the AAL-VU app is installed, which is the main interaction hub between the user and the system. The tablet can control a desktop pc through the use of a wireless internet connection. For the internet connection a Vodafone mobile Wi-Fi router R206 was used, because not all participants had access to internet connections. On the desktop pc the home software runs, which shows a living room environment with a virtual coach. The movements and speech of this coach are controlled by the intelligence present in the tablet. As part of the program, the users of the system carry an activity sensor during the day. This sensor can be plugged into the desktop pc, which reads the data and sends the data to the server.

The server contains an administrator panel, so new users can be added easily and test groups with specific users can be formed. This panel also allows a review of the sensor data and the possibility to export these from the database. On the server the activity data and the usage data of the tablet are combined. Through the use of an API, the tablet downloads these data in order to analyze them and to give personalized feedback to the user, based on this analysis.

The AAL-VU app on the tablet is the most important software component of the system [28]. This application provides a framework for the user to make appointments, monitor themselves and their social environment, Skype with friends, conduct friendship lessons, and interact with the coach. The calendar has basic functionalities like adding, editing and removing events.

The events are used to induce the intentions of the user and are compared to the activity data retrieved from the sensor [30]. Based on the total activity planned for a day, and the total activity measured with the sensor for that day, feedback is given to the user. There are three moments during the day on which a different kind of feedback is provided. In the morning the need for physical activity is emphasized. In the evening feedback on the current progress towards the activity goal of the day is provided. In the afternoon the user is complimented if the goal was achieved, or encouraged to do better next time if it was not achieved.

Furthermore, a list of contacts is accessible and editable by the user. Each contact contains some personal information and a possibility to Skype with the person. The tablet contains a course about friendship that is divided into eight lessons. The lessons are based upon a face-to-face group lesson but have been rewritten so they can be used on a tablet with a virtual coach. Lessons are presented in the form of text and pictures on the tablet and voice and bodily expressions by the virtual coach on the desktop pc.

The tablet also contains two questionnaires that can be filled out by the user. One is about how the user experiences the use of the various functionalities of the tablet and one is about how the user is currently feeling. All the answers of these questionnaires and all the data about the use of the tablet are stored on the server.

The home platform software has been installed on the desktop pc. This software shows a graphical representation of a living room with a virtual coach in it. The coach is able to speak to the user and to show some basic expressions or movements. The expressions and movements are, a.o. smiling, winking, thumbs up.

The activity sensor measures three-dimensional positioning during short time intervals. The data are read from the sensor when the sensor is plugged into the desktop pc. A specific program converts the data to levels of activity for specific time intervals. The level and duration of the activity are sent to the server for further analysis.

The last software module of the system consists of a web portal, specifically designed for the caregivers and relatives of the user. For each user a username and password are generated, with which the related caregivers and relatives can login into a web portal. Within this web portal the caregivers can view specific activity-related information in real-time: general tablet activity, calendar activity, Skype activity, and physical activity. General activity represents the general use of the AAL-VU system in terms of number of screens viewed each day. Calendar activity represents the number of appointments for each week. Skype activity represents the number of outgoing Skype calls for each day. Physical activity represents physical sensor activity in terms of percentage of activity versus inactivity for each hour, from current time to 48 hours back. Based on this information, the system gives a few recommendations to the caregiver to stimulate certain behavior or to indicate that something might be wrong. For instance, when a user has low physical activity, the system suggests that the caregiver tries to support the user in going out more. Something similar is done for calendar activity and general activity. By viewing the data and recommendations, the caregivers and relatives have insight in (global) information regarding the elderly user's physical and social behavior, which enables possible proactive decision making beneficiary for the elderly user.

5 Feasibility Study

To evaluate the user experience and assess the effect of this system, two small scale living lab studies were performed. This means studies were performed at the participants' homes, in close cooperation with the participants, who served as important informants for alterations to the system. The first study served as a small scale test of the technical setup and the logistics of the system installation and use, besides having the goal of stimulating the participants to increase their amount of physical activity. The second study was meant to scale up the implementation of the system for use by six to twelve users at once, also allowing for a longer duration of the program, and had the goal of reducing loneliness in the participants.

5.1 Study I

Setting. A field study was performed in the Amsterdam ABCC Living Lab [30]. Two subjects were selected based on an interview about openness for new technology and the selection user criteria for the project (e.g. 65+, living independently, single). The subjects received a personal system instruction. They were allowed to get acquainted with the system for a few days before the intervention. Subjects received a dinner voucher of 50 euros for their participation. The intervention took four to six days and consisted of several elements. First, persuasive information that focused on (intentional) attitude change towards physical activity was presented by the coach in the morning. This was the cognitive component. Here, the virtual coach took the role of the expert, providing information on why physical activity is important and good for subjects' well-being. Second, specific nudges were given and presented as prompts. With each subject a personalized activity agenda had been set. Nudges were given for the separate activities in the agenda. Finally, in the afternoon motivational behavioral feedback was given and in the evening the coach took the mentor role and provided mentoring feedback.

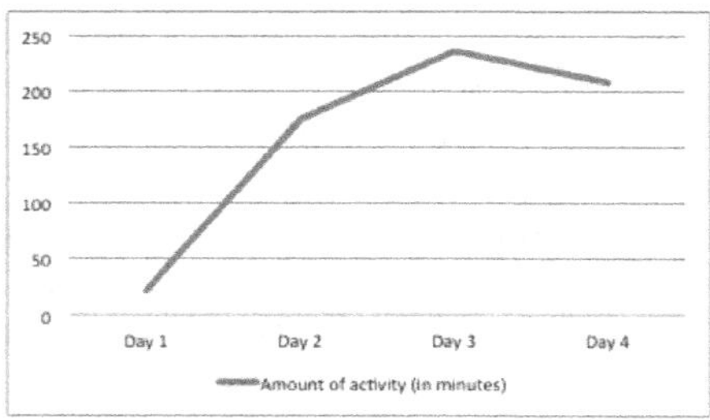

Fig. 3. Physical activity of participant 1

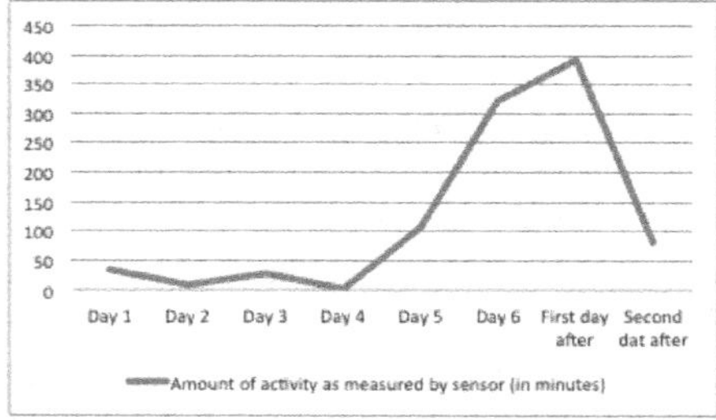

Fig. 4. Physical activity of participant 2

Results / experiences. The study with two subjects will be treated as two instances of a case study. We will present the results of the studies below, see Figures 3 and 4. *Case 1:* In the concluding interview the user mentioned that 'she amused herself with the system'. She mentioned that she found it 'surprising' and a very luxurious system'. 'I liked the exercises'. The advice that was given by the coach was 'good'. 'The advice was good and stimulating, but I like to have more types of exercises every day'. 'I did use the sensor quite often. Sometimes I forgot it. [...] But the tips from the coach on physical exercise were good and stimulating.' She also mentioned that it was a complex system and not fully finished yet and there were still some usability issues. 'Sometimes I got lost in all the options. But I think the system can help'.

Case 2: The user mentioned that she in particular liked 'the sensor part'. 'But I like to have more concrete exercises. I also have the impression that the sensor did not register all my activities. The coach wanted me to move more, while I thought I had moved enough already! He was a bit impolite, I think. That can be improved.' 'The content of the advice on physical activity was good, but I felt no contact with the coach! The mobile component was easy to use and to move. I found the information clear and understandable. In general I do not like too much advice. [...] I liked the fact that I could easily oversee the system. [...] I used the sensor a lot. But I forgot it a few times. [...] For people who are stuck at home the system is usable. Also, for measuring blood pressure, memos, or for alarm when you need someone, or calling a doctor or nurse. I think more functionality is needed. '

5.2 Study II

Setting. Seven participants (three women and four men), aged 64 - 77 years old, were recruited to participate in a friendship program [28]. They were screened on loneliness and openness to technology. All of them went through the same procedure. They were first, in their own homes, provided with the user part of the system as described in the previous section. Two-hour instruction sessions were carried out at their homes in order to teach the participants how to use the tablet and the app on it, and also how to use the home platform and the accelerometer, independently. They were informed that they were supposed to use the system on a daily basis for at least six weeks, but that they could miss one or a few days in this period. In a plenary startup session they received information and instruction on the general course of the program.

Throughout the following weeks, i.e. the study period, the users worked with the system. They received prompts (a pop-up would appear on the tablet, and be read aloud by the virtual coach) to go for a walk with someone each morning and to fill out questionnaires on how they were feeling every evening, and they also received prompts to follow a lesson and make the accompanying exercises about every five days. The researchers communicated with the participants on a regular basis, through Skype, phone calls and house visits. Three additional plenary meetings took place during this period, in which participants could address the researchers with all their questions, and in which some of the exercises of the course were discussed.

Results / Experiences

Loneliness data: To establish the loneliness scores the six-item De Jong Gierveld scale [16] was completed before and after the intervention by all participants. The items are:

- Q1 There are plenty of people I can rely on when I have problems.
- Q2 There are many people I can trust completely.
- Q3 There are enough people I feel close to.
- Q4 I miss having people around.
- Q5 I experience a general sense of emptiness.
- Q6 I often feel rejected.

There has been a decrease in the overall loneliness score among six participants. One participant reported almost no change whatsoever. Three participants had a significant decrease in the final loneliness score. Since this was a case study in the end and since we are describing our participants, this can be seen as a significant achievement.

Figure 5 shows the mean scores on each of the items. It is clear that the score on every single item decreased over the intervention. The system did have an effect on the lives of these participants. This is also giving us hope for further studies on a larger scale.

The participants reported their experience with the avatar as a need for improvement from the visual to the actual activity of the avatar. Overall, it was clear that most of the participants liked the virtual coach. Most of them wished he should have a nice and a comfortable home, it can be said that they cared about him. It seems plausible that para-social relations were created between six participants and the avatar.

There is a tendency in the decrease of the loneliness scores; however, a follow-up is needed. Nevertheless, our seven participants did use the system and provided us with their valuable insights as potential users.

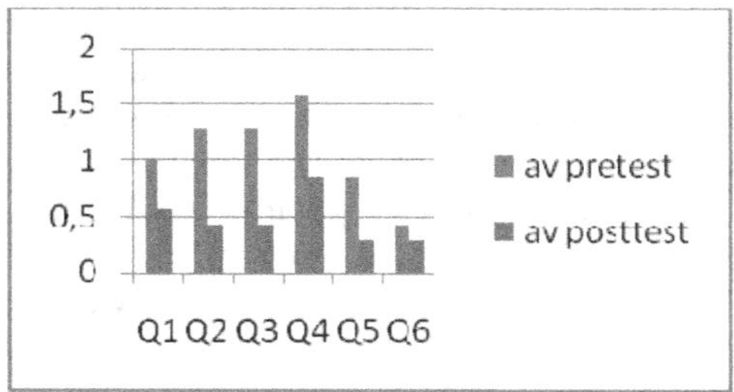

Fig. 5. Means of pre- and posttest scores

Fig. 6. Aggregated and normalized interaction

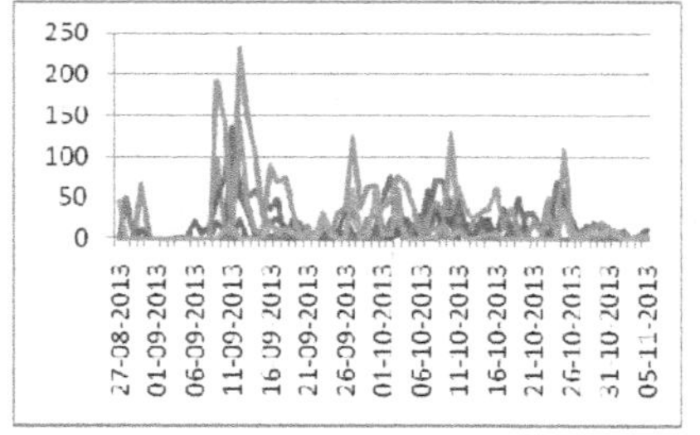

Fig. 7. Aggregated and normalized activity

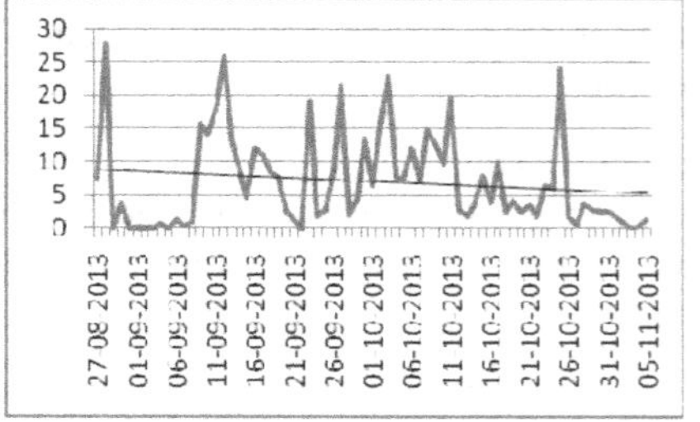

Fig. 8. Raw system interaction data (# per day)

For the results to be more conclusive, a larger sample over a longer period of time is necessary.

Physical activity: Figure 6 provides an overview of the normalized physical activity of all participants over time. For each participant the average physical activity over the total time of the study was calculated. The daily activity was normalized by dividing it by the personal average. The graph shows the average of the seven normalized daily scores. This measure gives some insight in the development of the physical activity over time. It can be seen that the trend is increasing for the first month, but after that it is decreasing.

Interaction with the system: Figure 7 provides a similar overview for the interaction with the tablet. This graph shows that there is a decrease in interaction. However, this is skewed by the fact that two participants stopped using the system altogether during the last 8 and 11 days respectively. As can be seen in Figure 8 the other persons kept on using the system until the end of the study.

6 Discussion and Conclusions

The study showed that in this project we were able to reach, at least for some of the participants, a habitual use of the system. After the intervention all participants but one had lower loneliness scores than before. This one participant was the person who did not use the system much and was the only participant that did not seem to have developed a para-social relationship with the virtual coach. It can be concluded that the overall concept of all the features is successful and effective for those participants in the study that used the system. There was an actual effect of the program in decreasing the loneliness scores among six participants.

In relation to the tablet data, two participants had decreased activity during the assessment process that lasted for eight weeks. The other five participants were most active in using friendship lessons and least active in creating events and story-sharing. It appears that the content was used almost on daily basis by the participants. The story-sharing function seemed to be more problematic and needs improvement.

The AAL-VU approach integrates a variety of management strategies that are tailored towards the elderly individual by taking relevant personal, social, and environmental factors into consideration. One of the main challenges in common AAL solutions is the acceptance of the systems by the users. To emphasize the benefit and convenience for the user, the AAL-VU system introduces a virtual 3D coach that mediates person input and system output and presents itself in a personalized, credible, easy to handle, and enjoyable way. Contrary to many available ICT solutions that failed to attract a sustained end user acceptance, the system applies advanced motivational models and involves the user in the design process. It thus turns traditional AAL from head to toe by giving control back to the user and by ascertaining deliberate and enjoyable use of the technology and interaction between all the actors in the social-health care chain.

The suggested framework of interventional support is promising and needs further research. Lifestyle change, like becoming more physically active, is more a matter of implicit motivation than of explicit motivation. For future developments a fruitful direction might be to focus more on how and not just on what behavior should be changed. Mediated communication by virtual characters which directly address the user verbally and non-verbally are the future of AAL systems.

Acknowledgments. This paper was based on funding received from: ZonMw, AAL, Agentschap NL and the city council of Amsterdam for the following projects: A2E2, V2me, V2VU, Healthlab and ABCC. We thank all consortium partners in these projects for their contribution.

References

[1] Agerström, J., Björklund, F.: Moral concerns are greater for temporally distant events and are moderated by value strength. Social Cognition 27(2), 261–282 (2009)

[2] Ajzen, I.: Models of human social behavior and their application to health psychology. Psychology and Health 13(4), 735–739 (1998)

[3] Archer, E., Blair, S.N.: Physical activity and the prevention of cardiovascular disease: From evolution to epidemiology. Progress in Cardiovascular Diseases 53(6), 387–396 (2011)

[4] Astrup, A.: Healthy lifestyles in europe: prevention of obesity and type ii diabetes by diet and physical activity. Public Health Nutrition 4(2b), 499–515 (2001)

[5] Baylor, A.L.: Promoting motivation with virtual agents and avatars: Role of visual presence and appearance. Philosophical Transactions of the Royal Society B: Biological Sciences 364(1535), 3559–3565 (2009)

[6] Baylor, A.L., et al.: The impact of three pedagogical agent roles. In: Proceedings of the Second International Joint Conference on Autonomous Agents and Multiagent Systems, pp. 928–929. ACM (2003)

[7] Bickmore, T.W., Picard, R.W.: Establishing and maintaining long-term human-computer relationships. ACM Transactions on Computer-Human Interaction (TOCHI) 12(2), 293–327 (2005)

[8] Blanson Henkemans, O.: epartner for self-care: How to enhance ehealth with personal computer assistants (2009)

[9] Carter, N.D., Kannus, P.: Exercise in the prevention of falls in older people. Sports Medicine 31(6), 427–438 (2001)

[10] Chorianopoulos, K.: Animated character likeability revisited: The case of interactive tv. Journal of Usability Studies (4), 171–184 (2006)

[11] Conn, V.S., Hafdahl, A.R., Mehr, D.R.: Interventions to increase physical activity among healthy adults: Meta-analysis of outcomes. American Journal of Public Health 101(4), 751–758 (2011)

[12] Cress, M.E., Buchner, D.M., Prohaska, T., Rimmer, J., Brown, M., Macera, C., Dipietro, L., Chodzko-Zajko, W.: Best practices for physical activity programs and behavior counseling in older adult populations. Journal of Aging and Physical Activity 13(1), 61–74 (2005)

[13] Edmunds, J., Ntoumanis, N., Duda, J.L.: Adherence and well-being in overweight and obese patients referred to an exercise on prescription scheme: A self-determination theory perspective. Psychology of Sport and Exercise 8(5), 722–740 (2007)

[14] Frederick, S., Loewenstein, G., O'donoghue, T.: Time discounting and time preference: A critical review. Journal of Economic Literature 40(2), 351–401 (2002)

[15] Frost, J., Boukris, N., Roelofsma, P.: We like to move it move it!: Motivation and parasocial interaction. In: Proceedings of the 2012 ACM Annual Conference Extended Abstracts on Human Factors in Computing Systems Extended Abstracts, pp. 2465–2470. ACM (2012)

[16] Gierveld, J.D.J., Van Tilburg, T.: A 6-item scale for overall, emotional, and social loneliness confirmatory tests on survey data. Research on Aging 28(5), 582–598 (2006)
[17] Health, D.: Choosing health? choosing activity: A consultation on how to increase physical activity. Department of Health/Department of Culture Media and Sport (2004)
[18] Keysor, J.J.: Does late-life physical activity or exercise prevent or minimize disablement?: A critical review of the scientific evidence. American Journal of Preventive Medicine 25(3), 129–136 (2003)
[19] Lin, J.J., Mamykina, L., Lindtner, S., Delajoux, G., Strub, H.B.: Fish'n'steps: Encouraging physical activity with an interactive computer game. In: Dourish, P., Friday, A. (eds.) UbiComp 2006. LNCS, vol. 4206, pp. 261–278. Springer, Heidelberg (2006)
[20] Martinez-Gonzalez, M.A., Varo, J.J., Santos, J.L., De Irala, J., Gibney, M., Kearney, J., Martinez, J.A.: Prevalence of physical activity during leisure time in the EU (2001)
[21] McAuley, E., Elavsky, S., Motl, R.W., Konopack, J.F., Hu, L., Marquez, D.X.: Physical activity, self-efficacy, and self-esteem: Longitudinal relationships in older adults. The Journals of Gerontology Series B: Psychological Sciences and Social Sciences 60(5), P268–P275 (2005)
[22] Miller, M.E., Rejeski, W.J., Reboussin, B.A., Ten Have, T.R., Ettinger, W.H.: Physical activity, functional limitations, and disability in older adults. Journal of the American Geriatrics Society (2000)
[23] Motl, R.W., McAuley, E., Snook, E.M.: Physical activity and multiple sclerosis: A meta-analysis. Multiple Sclerosis 11(4), 459–463 (2005)
[24] Nelson, M.E., Rejeski, W.J., Blair, S.N., Duncan, P.W., Judge, J.O., King, A.C., Macera, C.A., Castaneda-Sceppa, C., et al.: Physical activity and public health in older adults: recommendation from the american college of sports medicine and the american heart association. Medicine and Science in Sports and Exercise 39(8), 1435 (2007)
[25] Oja, P., Bull, F., Fogelholm, M., Martin, B.: Physical activity recommendations for health: What should europe do? BMC Public Health 10(1), 10 (2010)
[26] Proctor, D.N., Singh, M.A.F., Salem, G.J., Skinner, J.S.: Position stand (2009)
[27] Read, D., Roelofsma, P.: Hard choices and weak wills: The theory of intrapersonal dilemmas. Philosophical Psychology 12(3), 341–356 (1999)
[28] Reljic, G., Roelofsma, P.H.: V2me final report on the effectiveness study. Tech. rep., University of Luxemburg and Vrije Universiteit Amsterdam (2013)
[29] Roelofsma, P.H.M.P.: Modelling intertemporal choices: An anomaly approach. Acta Psychologica 93(1), 5–22 (1996)
[30] Roelofsma, P.H.M.P.: A2e2 evaluation report - phase 2. Tech. rep., Vrije Universiteit Amsterdam (2013)
[31] Ross, C.E., Hayes, D.: Exercise and psychologic well-being in the community. American Journal of Epidemiology 127(4), 762–771 (1988)
[32] Saunders, E.J.: Maximizing computer use among the elderly in rural senior centers. Educational Gerontology 30(7), 573–585 (2004)
[33] Services, H.: Healthy people 2010. Government Printing Office (2000)
[34] Skalski, P., Tamborini, R.: The role of social presence in interactive agent-based persuasion. Media Psychology 10(3), 385–413 (2007)
[35] Sniehotta, F.F., Scholz, U., Schwarzer, R.: Bridging the intention–behaviour gap: Planning, self-efficacy, and action control in the adoption and maintenance of physical exercise. Psychology & Health 20(2), 143–160 (2005)
[36] Thaler, R.H., Sunstein, C.R.: Nudge: Improving decisions about health, wealth, and happiness. Yale University Press (2008)

[37] Uusitupa, M., Louheranta, A., Lindström, J., Valle, T., Sundvall, J., Eriksson, J., Tuomilehto, J.: The finnish diabetes prevention study. British Journal of Nutrition 83(S1), S137–S142 (2000)

[38] Zanbaka, C., Goolkasian, P., Hodges, L.: Can a virtual cat persuade you?: The role of gen-der and realism in speaker persuasiveness. In: Proceedings of the SIGCHI Conference on Human Factors in Computing Systems, pp. 1153–1162. ACM (2006)

Enhancing Communication through Distributed Mixed Reality

Divesh Lala, Christian Nitschke, and Toyoaki Nishida

Kyoto University, Graduate School of Informatics, Kyoto University, Kyoto, Japan

Abstract. A navigable mixed reality system where humans and agents can communicate and interact with each other in a virtual environment can be an appropriate tool for analyzing multi-human and multi-agent communication. We propose a prototype of our system, FCWorld, which has been developed to meet these requirements. FCWorld integrates various technologies with a focus on allowing natural human communication. In this paper we discuss the requirements for FCWorld, the technical issues which it must address, and our proposed solutions. We intend it to become a novel tool for a variety of communication tasks such as real-time analysis and facilitation.

1 Introduction

Technology has removed the need for people to be in the same physical space to communicate, but still has not adapted to completely replace or even accurately simulate real-world communication. There is a lack of environments which are both navigable and allow the user to interact naturally inside them. Scenarios in these environments produce a rich source of interactions, group dynamics and discourse. Our long-term goal is to create a distributed navigable virtual environment which can support multi-modal communication between combinations of humans and agents. Additionally, this environment should preserve a user's communicative behavior, such as facial expressions and gestures. Achieving this goal allows us to use the environment in many application domains, such as conversational analysis and management.

We consider the following example scenario as a motivation for this work. A group of users are playing a game of virtual basketball. Their world is based on a real location in Kyoto city. Users shoot and pass a virtual ball between each other and communicate using facial expressions, gesture and verbal utterances. One user feels tired and says they wish to take a break and walk around Kyoto. They leave the group and navigate throughout a virtual city. The goal of this paper is to implement this type of scenario with the following functionalities:

- **TaskEnv:** A synchronized task environment based on a real location
- **UserRep:** A representation of the user based on their physical self
- **Scale:** Easily scalable number of physical environments
- **Navig:** The ability to freely move around the world

D. Ślęzak et al. (Eds.): AMT 2014, LNCS 8610, pp. 501–512, 2014.

Fig. 1. The top images show a human user interacting inside an immersive display system through a Kinect sensor. They can communicate naturally with others inside the virtual environment, which is based on a real location. We can see that the actual user is displayed as opposed to an avatar. In the bottom images, the world is completely virtual, with the basketball user represented by an avatar. In this environment, virtual objects can be manipulated through algorithms which recognize passing and shooting.

Fig. 2. The target FCWorld mixed reality system which combines elements from each of our partially implemented systems

- **Agent:** The ability to integrate intelligent agents and interactive virtual objects
- **UI:** A non-intrusive, uncalibrated natural user interface

The next section provides related research on current systems and their limitations. Section 3 presents our preliminary implementation of FCWorld. Section 4 describes an analysis of the major problems to be solved and how we plan to address them using FCWorld.

2 Related Research

A system which can support these requirements has not yet been realized. However, there are various systems which can function as partial implementations. For example, there already exists a virtual basketball environment[1] in which users can play the game in a completely virtual world. We assess the limitations of other systems through the perspective of the system requirements stated in Section 1. Table 1 outlines this analysis.

Table 1. Topology of multi-user communication systems. △ indicates that this feature may be possible in the system.

System	**TaskEnv**	**UserRep**	**Scale**	**Navig**	**Agent**	**UI**
Second Life meeting[2][3]	✓	×	✓	✓	△	×
Telehuman[4]	×	✓	✓	×	×	✓
MirageTable[5]	✓	✓	✓	×	△	✓
t-Room[6]	△	✓	×	×	×	✓
Group telepresence[7]	△	✓	×	✓	△	△
Multiplayer HMD[8]	✓	△	△	✓	△	×
Virtual basketball[1]	✓	×	✓	✓	✓	✓

A world can be virtual but not necessarily be remotely distributed, such as a multi-player video game where multiple participants are sharing the same screen. We only consider virtual worlds in which more than one remote user can participate in at the same time and changes are observed by all the participants. Second Life[2][3] is a virtual world in which human users can interact with each other. In this world users are represented by avatars, so unlike a real user representation cannot fully express communicative behavior. Telehuman[4] is a cylindrical display which displays a representation of a remote user, but does not act in a virtual environment. MirageTable[5] allows users to interact with digital objects in a task space which is a real environment. Like Telehuman, the environment is restricted.

In t-Room[6] users interact with each other from remote locations through video telepresence, but the setup is complex, users cannot navigate and virtual characters cannot be integrated. The group telepresence system[7] creates a virtual world in which users interact using their real bodies. However it involves very complex engineering and specialized equipment, so is extremely difficult to scale.

Virtual worlds in which users interact using head-mounted displays[8] are another common system, but these require equipment which needs to be calibrated. There is also the issue of limited navigation—the user is restricted to travel in a physical space. Our virtual basketball system[1] meets almost all the requirements but uses avatars rather than real user representation. Additionally the environment is limited to a basketball scenario, whereas the concept of FCWorld is to accommodate multiple scenarios. We conclude that current systems do not address all these requirements. This can be partly explained by the limited scope of the applications.

This is not an exhaustive list, as there are countless other types of virtual environment systems[9–11]. These are largely in the same domain as the examples in Table 1, and as such have the same limitations. So far we have not encountered any system which is able to meet all our above requirements. Therefore, we propose that there is a research gap which we intend to fill through the construction of FCWorld.

3 FCWorld

This section describes our target system, named FCWorld (Flexible Communication World), in which we aim to realize the integration of real artifacts with virtual environments. Currently, we have implemented two existing systems which are shown in Figure 1. These are an immersive world based on Google Street View and a completely virtual basketball environment. Our next step is to realize our target system in Figure 2 by integrating features of the two systems. Users in different physical environments share a common immersive task environment, and can communicate with real representations of other participants.

To frame FCWorld, we distinguish between two types of environment. The first is the physical environment (PE), which describes the real world. In all remote communication tools, PE is separated as participants are not in the same physical space. The second environment is the task environment (TE), which describes the environment of the task itself. The TE can differ depending on the application, but we would like the TE to be based on a real location and inhabited by virtual agents. Figure 3 shows this concept. Supporting distributed PE and shared TE brings about many key questions which we must address.

In Figure 2, two humans are playing a game of basketball with another agent player. Humans interact in an immersive display system, where they can see a virtual environment displayed around them. This environment is based on a real location in Kyoto. The humans see each other as representations of their real selves. Kinect sensors track their body movements and provide gesture recognition so that they may pass and shoot the ball. Agents also recognize their gestures and respond to them. This scenario is an ideal motivation for us, because it contains elements of natural human communication such as pointing, human-agent interaction, and virtual object manipulation—making it an ideal testbed for communication analysis.

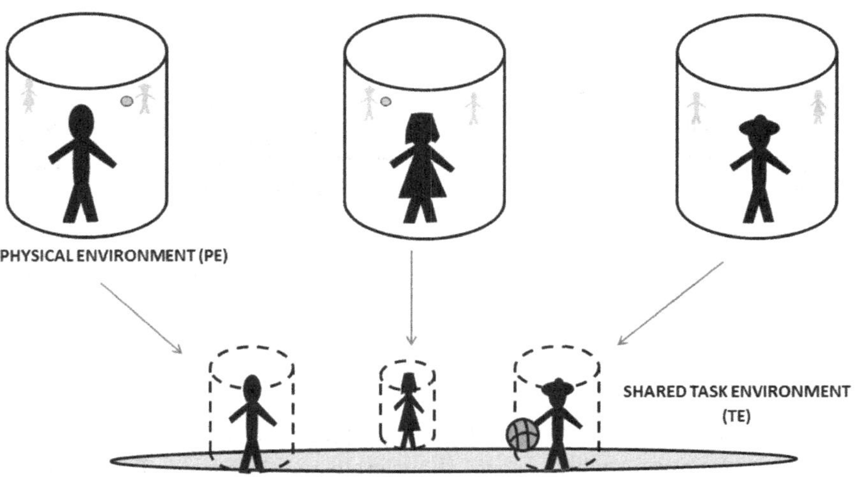

Fig. 3. Overview of FCWorld. Participants are located in remote physical spaces but share the same virtual space. The representation of all participants is the same as their physical form.

4 Design Issues

We now identify several design issues which FCWorld must overcome in order to become a fully fledged system. Some of these issues have been solved in our current implementation, while others are to be addressed in future prototypes.

4.1 Key Hardware Features

First we briefly describe the system itself. FCWorld is based on the VISIE system[12] used to create the virtual basketball system[1]. VISIE is not an application in itself, but it serves a framework to create various virtual environments. The hardware consists of a surrounding tile-screen immersive display which allows the user to see the world around them in a 360-degree view. The user stands in front of a Kinect sensor, which uses a depth camera to capture their body movements. In order to navigate, the user walks on a foot pressure pad, facing the direction they wish to walk. Figure 4 shows a conceptual and real implementation of the system.

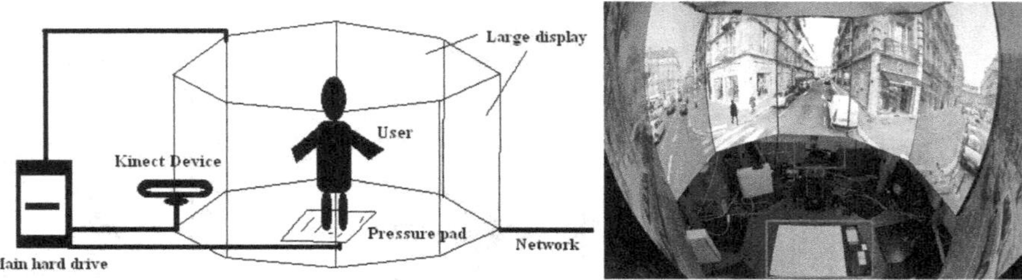

Fig. 4. Conceptual diagram of a FCWorld terminal and the actual implementation seen through a fish-eye lens. Red and green boxes indicate the location of the Kinect sensor and foot pressure sensor respectively.

4.2 Technical Goals

There are several technical goals which must be achieved in order to make FC-World a viable system which meets the system requirements. The major goals can be described as follows:

- Creating a virtual environment based on a real location that is seen by all users
- A method of natural navigation which can be used by independent users
- Creating a user representation which preserves communication modalities
- Allowing the ability to accurately refer to a spatial object and have it recognized by others, such as pointing
- Integrating gesture recognition to allow users to communicate with agents and manipulate virtual objects

To achieve each of these goals, there are various difficulties which must be overcome. We discuss each of these in turn and present our proposed solutions.

4.3 Interaction Environment

The creation and display of an environment in virtual space based on a real location presents four major difficulties. We must first obtain the data representing the environment from somewhere, display it appropriately, and then synchronize the environment between all users. The final, and most crucial, difficulty is providing this environment with collision and occlusion, so the user feels like they are located inside a world with the same physical properties as reality.

To solve these problems we can make use of existing resources. To obtain the data, we query Google Street View[13] online to extract images of any surrounding area which has been captured. To display the data, we create a full panorama image which can be shown around the user. Synchronization is achieved by using the server to specify each client's individual scene. We can then preserve the relative locations of each participant in the game. If a user is standing in front of a building, he will be seen that way by everyone else.

While these problems are solved, the major difficulty of collision and occlusion remains. Our prototype has not yet addressed this, but it is possible to use techniques to construct ground and object planes by using additional depth map information from Google[1] to generate ground planes and reconstruct objects. This will allow us to give buildings depth and to have users walk on a virtual surface.

4.4 Navigation

Several difficulties are involved in creating a navigable environment. Firstly, we must ensure that users do not need peripherals to navigate. Secondly, we have to create a method which gives users the sense of moving through the virtual environment as opposed to discrete scene changes.

To allow users to move independently, we provide them with a foot pressure pad, using an algorithm which detects the direction in which the user is walking[14]. The user can simply use a natural walking motion to navigate. Their upper body becomes free to interact with no need to hold any peripherals. Currently, FCWorld only uses scene transitions when navigating, so the second challenge still remains. Our proposed strategy is to make these scene transitions continuous rather than discrete. This feature can potentially be achieved through the use of standard image-based rendering algorithms[15].

4.5 User Representation

Two major challenges of creating an appropriate user representation exist. The first is that we require a technique which allows users to exhibit natural communication. That is, modalities such as facial expression, posture and gesture should be preserved as much as possible. Avatars do not allow this flexibility, so this introduces an additional constraint. The second is providing each participant with different perspectives of each other.

Our solution to the first challenge is to generate a real image of the user which can be displayed in real-time. In the current prototype, we make use of a Kinect sensor and a "Green Screen" application from the Kinect SDK toolkit[2] to segment the user image and stream the data to all participants. Figure 5 shows this process.

Consequently, this solution is at odds with the second challenge, because we use a 2-d video plane. However, by using the same information plus additional depth image data, we can solve both issues by creating a point cloud model of the user in real-time and sending it to all participants. The benefit of this approach is that the model can move and rotate around the virtual world, creating any number of perspectives. Recent progress has been made which shows the ability to implement this using multiple Kinect sensors[16].

---

[1] Google Maps Javascript API.
`https://developers.google.com/maps/documentation/javascript` (2014)

[2] Kinect for Windows SDK version 1.6. Microsoft, 2013.

Fig. 5. The user is segmented from the background and then displayed in the program. The Kinect image is mirrored, so we reverse this to preserve the correct image.

4.6 Spatial Representation

One main difficulty of achieving this goal is that while the environment is synchronized, the perspective of the users are not. This means that we need a way to display this synchronized perspective to all users. Take the case of pointing. In a real environment, pointing direction can be reasonably inferred. For a distributed environment, a pointing gesture made by user A is seen by user B. But because A and B have separate perspectives, the referred object will be in slightly different localized positions. This can cause a breakdown in communication, and must be solved by FCWorld.

Our solution to this issue is the same as for the user representation. If we construct a 3-d model of the user, then their individual perspectives can be adjusted in the displayed environment. Therefore, it can be manipulated (i.e. rotated) to produce a correct displayed image.

4.7 Gesture Recognition

The final goal is to allow users to both communicate with agents and manipulate virtual objects, such as in our basketball scenario. The difficulty is deciding which recognition model should be used and how gestures can be recognized. We must ensure that these gestures are useful for human and agent interaction.

We have already seen that a solution to this problem can be found in partially implemented systems, such as virtual basketball[1]. It is possible to integrate these models into our system. However, rather than basketball-related motions such as shooting, it is more beneficial to cater for more generic gestures. We are considering to use simple gestures for throwing an object or "handing" an object to another user. For communication with an agent, recognition of gestures such as pointing and waving should also be considered. Recognizing these is also beneficial for automatic communication analysis.

5 Discussion

We have tested initial prototypes of FCWorld in both immersive display and desktop environments, and conducted virtual meetings with participants at remote locations. These informal meetings can be constructed in one of three locations with immersive displays which we maintain in Kyoto. Videos demonstrating our system and such meetings can be found online[3,4].

The overall impression of FCWorld from users is that it is a novel way to communicate in a common workspace. Users seemed particularly impressed with the ability for every participant in the virtual meeting to have a shared environmental context. We can successfully use natural commands such as "Look at this temple" or "Let's go this way" while pointing in the general direction of their target. We now discuss potential applications for FCWorld. The flexibility of FCWorld to serve a wide range of task domains is a major benefit.

5.1 Addressing the System Requirements

At the beginning of this paper, we identified several key requirements for FCWorld. We return to these and evaluate how our system fares.

TaskEnv. FCWorld uses panoramic image data as the basis of the TE. The advantage of this is that we can create a wide navigable environment without having to model and create a whole new world. We can even use our own panoramas to implement any real-world environment we wish, which is crucial for creating a wide range of applications. The TE is synchronized through the use of a server-client architecture. Importantly, the relative spatial poses of different users are preserved so that each individual has their own perspective.

UserRep. We take body data directly from the Kinect sensor for a user representation. This means that all communication modalities can be preserved in interactions. One issue is that the current representation is only displayed using information from one Kinect sensor, giving the impression of a flat plane projection. We can improve this by using multiple Kinect sensors to create a 3-d model of the participants. The main task is not retrieving the information itself but fusing them together to create a representative point cloud or mesh model.

Scale. FCWorld exploits the advantages of the immersive displays, but it can just as easily utilize a desktop implementation. Of course, some features become redundant, such as pressure pad navigation. However, the ability to join a virtual

[3] Remote Meeting and Interaction in Immersive Shared Environment. `http://www.youtube.com/watch?v=GD4X1H_nOyo`. (2014)

[4] Google Street View Navigation on Immersive 360° Display. `http://www.youtube.com/watch?v=V-9SKpcMrzk`. (2014)

space from any terminal gives FCWorld an advantage over other systems which utilize many complex technologies. Currently we have connected three users in our VISIE installations without noticeable delays or lag. We aim to analyze the limits of the number of users which can function in the system.

Navigation. Currently, FCWorld supports independent world exploration the clients. The avatar of the user can move throughout the environment and be seen by all participants. As we do not have a full 3-d body model, the next step will be to ensure that the correct perspective of the user is preserved. For example, if another user is seen walking away, their back should be displayed. Additionally, the construction of a ground plane will enable avatars to appear as if they are moving along the ground.

Agent. While we have not yet integrated virtual agents into FCWorld, these can be easily implemented. The advantage of our particular scene (a surrounding of textured surfaces) is that agents can be rendered with consistent spatial information, allowing them to move freely around the world. We can simply make the surrounding scene transparent. Through depth map information, we can introduce occlusion properties to further create the perception of an agent inside a real location.

UI. FCWorld ensures natural communication by using non-intrusive devices, such as the Kinect sensor and pressure pad. The user can simply enter the environment interact without any need for calibration. We feel this accessibility is a major benefit of FCWorld, particularly for users who are not so proficient with computers.

5.2 Current Implementation and Future Work

Figure 6 shows an example of demonstrations of FCWorld with outside participants. We successfully allowed participants to share attention in the virtual space and navigate around the environment. Feedback was positive towards FC-World, so the next stage is to implement a more thorough evaluation. Our current prototype of FCWorld shows promise, but we have identified some areas for improvement. From our problem analysis, the following appear to be the most crucial tasks:

- Creation of ground and object planes for occlusion and collision purposes
- Implementation of scene blending to evoke the feeling of a walking motion
- Change the user representation from a 2-d image to a 3-d model

Techniques for accomplishing these tasks already exist. The integration of them into FCWorld will be the focus of our next prototype system. We have already successfully implemented a version of FCWorld which uses point cloud avatars from a single Kinect sensor (bottom image in Figure 6). Additional Kinect sensors should provide us with the ability to create full body models. The problems which we have identified can be solved by an adequate completion of the above tasks.

Fig. 6. Images taken from demonstrations of FCWorld. The top two images show participants in different locations (Mt. Fuji and Prague) focusing on a shared attention point. The bottom images display participants interacting with a remote user who is represented by a point cloud avatar.

6 Conclusion

In this paper we presented FCWorld, an environment in which distributed participants share a common virtual environment and navigate through it. This type of system is the culmination of integrating several technologies into an immersive virtual world. We described the technologies FCWorld uses to represent participants as well as the methods which allow them to navigate in a virtual environment based on a real location. Initial demonstrations of the FCWorld prototype were met with positive responses from outside participants. We identified several challenges which we must address in future prototypes and our solutions to these challenges.

References

1. Lala, D., Nishida, T.: Joint activity theory as a framework for natural body expression in autonomous agents. In: Proc. of the 1st Intl. Workshop on Multimodal Learning Analytics, MLA 2012, pp. 2:1–2:8 (2012)
2. da Silva, C., Garcia, A.: A collaborative working environment for small group meetings in second life. Springer Plus 2(1), 1–14 (2013)
3. Bredl, K., Groß, A., Hünniger, J., Fleischer, J.: The avatar as a knowledge worker? How immersive 3d virtual environments may foster knowledge acquisition. Electronic J. of Knowl. Mgmt. 10(1), 15–25 (2012)
4. Kim, K., Bolton, J., Girouard, A., Cooperstock, J., Vertegaal, R.: Telehuman: Effects of 3d perspective on gaze and pose estimation with a life-size cylindrical telepresence pod. In: Proc. of the SIGCHI Conf. on Human Factors in Comp. Sys., CHI 2012, pp. 2531–2540 (2012)
5. Benko, H., Jota, R., Wilson, A.: Miragetable: Freehand interaction on a projected augmented reality tabletop. In: Proc. of the SIGCHI Conf. on Human Factors in Comp. Sys., CHI 2012, pp. 199–208 (2012)
6. Hirata, K., Harada, Y., Takada, T., Aoyagi, S., Shirai, Y., Yamashita, N., Kaji, K., Yamato, J., Nakazawa, K.: t-room: Next generation video communication system. In: Glob. Telecom. Conf., pp. 1–4. IEEE GLOBECOM (2008)
7. Beck, S., Kunert, A., Kulik, A., Froehlich, B.: Immersive group-to-group telepresence. IEEE Trans. on Visualization and Comp. Graph. 19(4), 616–625 (2013)
8. Zhou, Z., Tedjokusumo, J., Winkler, S., Ni, B.: User studies of a multiplayer first person shooting game with tangible and physical interaction. In: Shumaker, R. (ed.) Virtual Reality, HCII 2007. LNCS, vol. 4563, pp. 738–747. Springer, Heidelberg (2007)
9. Misawa, K., Ishiguro, Y., Rekimoto, J.: Ma petite chérie: What are you looking at?: A small telepresence system to support remote collaborative work for intimate communication. In: Proc. of the 3rd Augmented Human Intl. Conf. AH 2012, pp. 17:1–17:5 (2012)
10. Demeulemeester, A., Kilpi, K., Elprama, S.A., Lievens, S., Hollemeersch, C.-F., Jacobs, A., Lambert, P., Van de Walle, R.: The ICOCOON virtual meeting room: A virtual environment as a support tool for multipoint teleconference systems. In: Herrlich, M., Malaka, R., Masuch, M. (eds.) ICEC 2012. LNCS, vol. 7522, pp. 158–171. Springer, Heidelberg (2012)
11. Cassola, F., Morgado, L., de Carvalho, F., Paredes, H., Fonseca, B., Martins, P.: Online-gym: A 3d virtual gymnasium using Kinect interaction. Procedia Technology 13, 130–138 (2014); SLACTIONS 2013: Research conference on virtual worlds Learning with simulations
12. Lala, D., Nishida, T.: VISIE: A spatially immersive interaction environment using real-time human measurement. In: 2011 IEEE Intl. Conf. on Granular Computing (GrC), pp. 363–368 (2011)
13. Google: Google Street View Image API (2014). `https://developers.google.com/maps/documentation/streetview/` (Online; accessed February 17, 2014)
14. Lala, D.: VISIE: A spatially immersive environment for capturing and analyzing body expression in virtual worlds. Masters thesis, Kyoto University (2012)
15. Shum, H., Kang, S.B.: Review of image-based rendering techniques. In: Proc. of SPIE, vol. 4067, pp. 2–13 (2000)
16. Alexiadis, D., Zarpalas, D., Daras, P.: Real-time, full 3-d reconstruction of moving foreground objects from multiple consumer depth cameras. IEEE Trans. on Multimedia 15(2), 339–358 (2013)

A New Graphical Representation of the Chatterbots' Knowledge Base

Maria das Graças Bruno Marietto, Rafael Varago de Aguiar, Wagner Tanaka Botelho, and Robson dos Santos França

Universidade Federal do ABC, Brazil
{graca.marietto,wagner.tanaka}@ufabc.edu.br,
rafael.varago@aluno.ufabc.edu.br,
robsonsfranca@gmail.com

Abstract. This paper discusses chatterbots that use the Pattern Recognition technique and the Artificial Intelligence Markup Language (AIML) regarding their design and implementation. Usually, chatterbot's Knowledge Base (KB) is written with the AIML language, and it is based on the stimulus-response type block approach. Considering that the chatterbots' KB might be a large collection of question-answer pairs, there is an increasing need of visual design proposals for the chatterbots' dialogues. However, there is a lack of projects related to the chatterbots' KB display. In order to fill this gap, this paper introduces a new graphical model called Dialogue Conceptual Diagram (DCD). It is a unified theoretical framework to describe the chatterbot's knowledge in a graphical, intuitive and compact way. Finally, a subset of the ALICE's KB was modeled using the DCDs, showing that DCD is a suitable framework to graphically represent the human-machine dialogues.

Keywords: Artificial Intelligence, Chatterbot, Pattern Recognition, AIML Language, Knowledge and Data Engineering, Graphical Representation.

1 Introduction

According to [1] the purpose of Ambient Intelligence (AmI) is "*...to broaden the interaction between human beings and digital information technology through the usage of ubiquitous computing devices*". Into the ubiquitous computing area researches have been performed for developing the "natural interfaces". Such interfaces are designed to ease the communication process by applying human-like interaction features such as natural language, gestures and vision. Among these interfaces this paper highlights the chatterbots, computational systems designed to interact with humans through natural language. To be more specific, this paper analyzes the chatterbots that use the Pattern Recognition technique and the Artificial Intelligence Markup Language (AIML) [2]. The AIML allows the modeling of natural language dialogues between chatterbots and humans through a stimulus-response approach. For each possible sentence from the user (stimulus) there are pre-programmed replies available in the chatterbot's Knowledge Base

D. Ślęzak et al. (Eds.): AMT 2014, LNCS 8610, pp. 513–524, 2014.

(KB). The ALICE chatterbot was the first bot to use the AIML language and its interpreter.

The chatterbots' KBs builded over the Pattern Recognition framework and the AIML language are becoming a massive collection of data that requires proper methodologies for their processing, considering three operations: storage, retrieval and viewing. However, despite the relevance of the organization and portrayal of chatterbots' KBs, it is unusual to find very few references in the literature about these topics. This observation becomes even more surprising since one can find a considerable number of of chatterbots available to the research community. Therefore, to help filling this gap, this study presents a new graphical model called Dialogue Conceptual Diagram (DCD). It is a theoretical frame of reference to visually model the sequence of the human/chatterbot dialogues.

The formal structure of the DCD's graphical representation - the one proposed by this paper - is used to build the KBs because: (i) it highlights which dialogues the chatterbot is ready to keep with users (the bot's knowledge); (ii) it allows a shared view of the dialogues' structure; (iii) it works as a guide for a critical analysis of the bot's KB; (iv) it eases the recycling and sharing of KBs; (v) the DCDs can be used by a multidisciplinary team, which some members are responsible to build the KB while others are going to deal with the computational matters; (vi) by working with DCDs, the designer can focus on the chatterbot's KB modeling instead of the implementation details; (vii) the abstraction provided by the DCDs allows that a single DCD can be implemented in different ways, working as a blue print for the chatterbot's KB.

This paper is organized as follows. Section 2 describes the syntactical and semantical structure of the Dialogue Conceptual Diagrams. An application of DCDs for a graphical portrayal of the ALICE chatterbot is provided in the Section 3. Lastly, Section 4 outlines the conclusions of this paper.

2 Dialogue Conceptual Diagram

This section introduces the DCD as a frame of reference to graphical representation of dialogues among humans and chatterbots. It is visually depicted as a horizontal flowchart, which the dialogue sequence must be read from left to right. Also, the DCD's structure is composed of predefined elements that are graphically represented by symbols. The dialogue flow is established by the interaction of three components called actors and defined as follows: (i) User: it is the user's data input; (ii) Inference Machine: it is the computational process performed by the chatterbot; (iii) Chatterbot: it is the chatterbot output that is shown to the user.

Figure 1 presents the DCD's basic structure with three actors and their interactions established during the dialogue flow. The flow starts at ① with the user's input as a sentence. After that, such sentence is simplified, normalized and transferred to chatterbot's inference machine at ②. In that stage the Pattern Recognition described in [2] occurs. When the most appropriate answer is found,

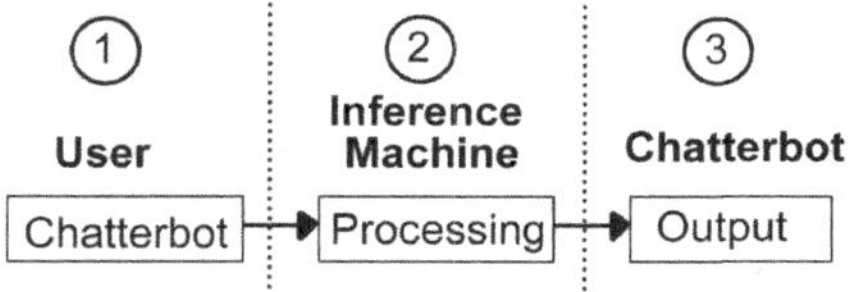

Fig. 1. Dialogue Conceptual Diagram Structure

the inference machine sends the message to be displayed for the user through the output device at ③.

As previously stated, DCDs are built using graphical elements. For each element it is established a syntax and a set of layout rules, as well as semantics related to its meaning. This paper proposes and describes in the next sections ten graphical elements for the DCDs.

2.1 Dialogue Conceptual Diagram's Graphical Elements

This section presents and describes each DCD's graphical element proposed in this work. It is based on its syntactical and semantical structure.

User's Data Input Element. The User's Data Input Element expresses the text informed by the user during the dialogue with the chatterbot. Figure 2(a) shows this element with the text inside as the user's input sentence. This diagrams refers to AIML's <PATTERN> tag.

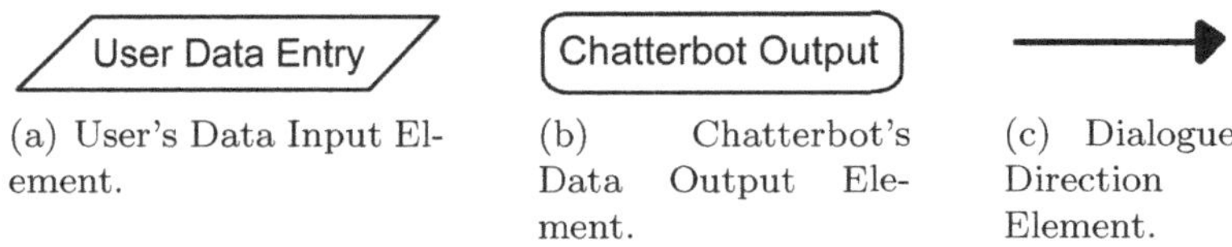

(a) User's Data Input Element. (b) Chatterbot's Data Output Element. (c) Dialogue Direction Element.

Fig. 2. DCD's Elements: User's Data Input, Chatterbot's Data Output and Dialogue Direction

Chatterbot's Data Output Element. The Chatterbot's Data Output element shown in Figure 2(b) is applied to represent chatterbot's reply to the user. Also, in the AIML language it is implemented using the <TEMPLATE> tag.

Dialogue Direction Element. Figure 2(c) illustrates the Dialogue Direction element, which is used to designate DCD's dialogue flow. It is depicted by a directed line segment with an arrow to indicate the dialogue's flow.

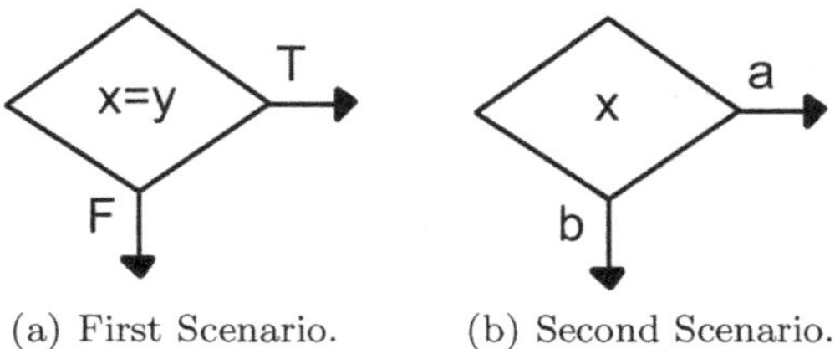

(a) First Scenario. (b) Second Scenario.

Fig. 3. Bivalent Decision Element

Bivalent Decision Element. The Bivalent Decision element is related to the conditional structures, also known as decision or selection structures, which are common in programming languages. This element dictates the running of certain code fragments according to a logical test performed in a control variable.

Into the DCDs' context, the Bivalent Decision element allows the distinction of two scenarios. The first happens when the control variable is compared with a certain value, and such comparison is either true or false. Figure 3(a) shows the scenario with the X control variable being compared to the Y value. If it is true (symbol T), the dialogue is directed to a certain path. Otherwise, the symbol F leads the dialogue to another path.

The second scenario defined in Figure 3(b) happens when the control variable is able to have two values. In the figure, the X control variable can take A or B as a value. There are two ways to program the Bivalent Decision element in AIML: by the <CONDITION> and <LI> tags, or by the <IF> and <ELSE> tags.

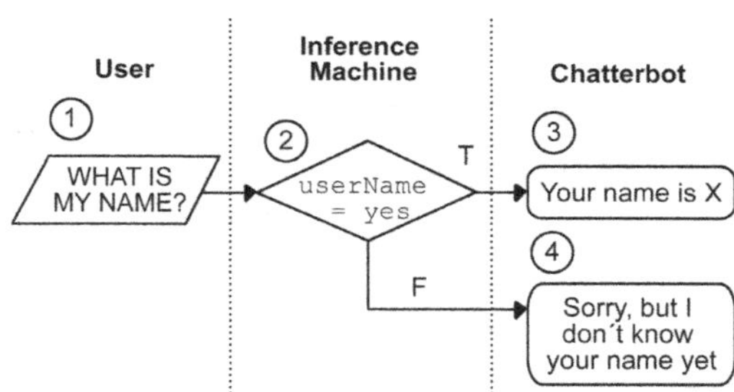

Fig. 4. Dialogue Conceptual Diagram with the Bivalent Decision Element

In order to provide an example of Bivalent Decision element in a dialogue context, Figure 4 shows a DCD that, at first in ① the user enters with the sentence "WHAT IS MY NAME?". After that, the inference machine, using the Bivalent Decision element in ② evaluates the conditional test upon the USERNAME control variable. This variable states whether the chatterbot has the user name in its memory or not. If this variable contains the YES value - implying that the user has already informed his/her name - the outcome from the comparison is the true value. Thus, the reply from the chatterbot is "Your name is X" in ③, as X being the user name stored in the chatterbot's memory. However, if the comparison ends up in a false value, the reply becomes "Sorry, but I don't know your name yet" in ④.

Listing 1.1 presents the AIML code that matches with the DCD shown in Figure 4.

Listing 1.1. AIML Code for the DCD in Figure 4

```
<?xml version="1.0" encoding="UTF-8"?>
<aiml version="1.0">
<category>
   <pattern>WHAT IS MY NAME?</pattern>
   <template>
      <condition name = "userName">
         <li value = "yes"> Your name is <get name = "
             userName"> </li>
         <li> Sorry, but I don't know your name yet </li>
      </condition>
   </template>
</category>
</aiml>
```

Multivalent Decision Element. Just like the Bivalent Decision element, the Multivalent Decision element is a conditional decision structure. However, the control variable can take from 3 to n distinct values, where $n \in N | n \geq 3$. Therefore, these values define the direction of the commands running flow.

Figure 5(a) shows the element that represents the multivalent decision commands. For each X control available values, depicted in the element's edges, there is a specific dialogue sequence. This element can be implemented using the AIML in two ways: using the <CONDITION> and <LI> tags or the <IF> and <ELSE> tags.

Figure 5(b) provides a DCD as an application of the Multivalent Decision element. The user in ① starts with the sentence "I LIKE ICE CREAM". This message is forward to the inference machine by ②, and later the decision command is applied. The inference machine uses the information found in the FAVORITEIC control variable to direct the dialogue flow.This variable stores the user's choice for favorite ice cream, and it can take the values: CHOCOLATE, STRAWBERRY or PINEAPPLE. With these values, the chatterbot replies as follows: (i) If FAVORITEIC is CHOCOLATE the chatterbot replies in ③ with "Chocolate ice cream is yummy"; (ii) If FAVORITEIC is STRAWBERRY the chatterbot replies in ④ with "I like strawberry ice cream too"; (iii) If FAVORITEIC is PINEAPPLE the chatterbot replies in ⑤ with "I would like a pineapple ice cream too".

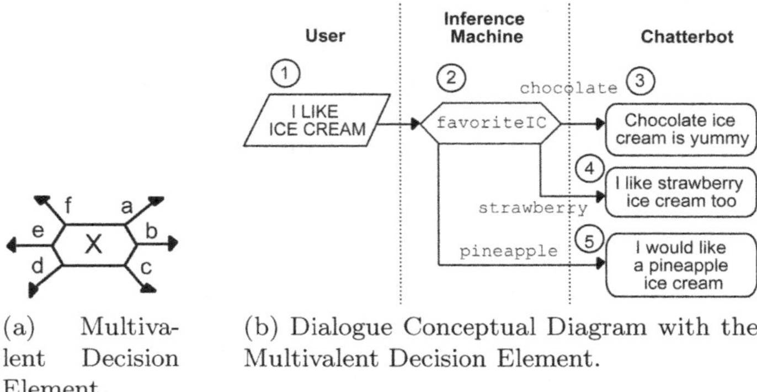

(a) Multivalent Decision Element.

(b) Dialogue Conceptual Diagram with the Multivalent Decision Element.

Fig. 5. Multivalent Decision Element

By lack of space the AIML code related to this element and the followings will not be showed.

Category Call Element. The Category Call element directs the flow of distinct user inputs to a single chatterbot's reply. Thus, the bot identifies whether two or more user sentences are similar (with the same meaning), even when they are typed in different ways. It is also shown in Figure 6(a), where X is the pattern that defines the category to be called. When the category X is called into the context of another category Y (for instance), the category X's instructions are performed. When that processing of the category X is over, the chatterbot's running flow returns to the category Y, and the dialogue flow continues.

The DCD shown in Figure 6(b) in ① depicts the situation when the user types "I LIKE *", which the * wildcard marks any character sequence that the user typed. This DCD is ready to identify the following user inputs with the * wildcard: "I LIKE HARDWARE" and "I LIKE SOFTWARE". Therefore, if the user types any of these sentences, the inference machine directs the dialogue flow by the Bivalent Decision element defined in ②. It compares the value stored in * with HARDWARE and SOFTWARE values. Based on the outcome of such comparison, two dialogue flow directions are available: (i) If * is HARDWARE, the Category Call element calls the category with the "HARDWARE" pattern, as shown in ③. Thus, the chatterbot's reply related to ⑤ and ⑥ will be "Hardware is a collection of physical elements that constitutes a computer system"; (ii) If * is SOFTWARE, the Category Call element calls the category with the "SOFTWARE" pattern, as shown in ④. Based on such choice, the chatterbot reply is defined by ⑦ and ⑧ items as "Software is a program that enables a computer to perform a specific task".

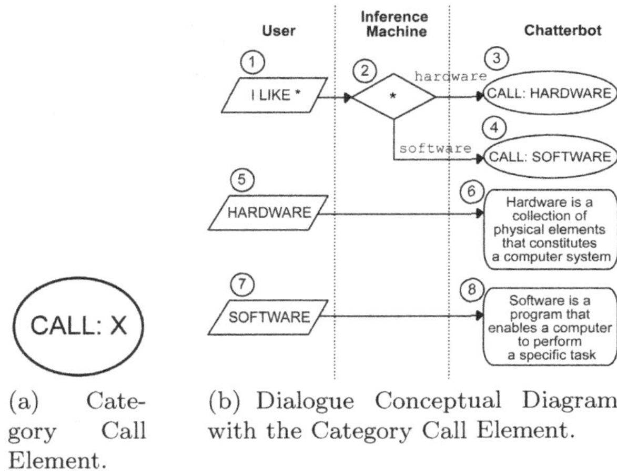

(a) Category Call Element.

(b) Dialogue Conceptual Diagram with the Category Call Element.

Fig. 6. Category Call Element

Variable and Temporary Files Storage Element. The Variable and Temporary Files Storage element depicts the files storage and/or variables that will be solely available to the chatterbot when the program is running. Then, if the program ends the files or variables are erased from the memory. This element is shown in Figure 7(a), where the X variable or temporary file has the "a" value. AIML uses the <SET> tag to perform the temporary variable storage.

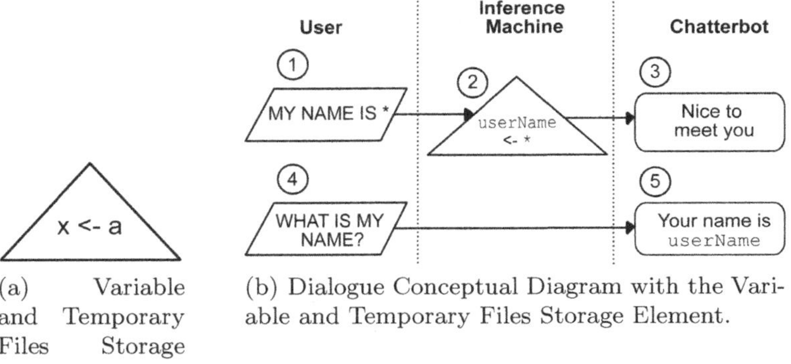

(a) Variable and Temporary Files Storage Element.

(b) Dialogue Conceptual Diagram with the Variable and Temporary Files Storage Element.

Fig. 7. Variable and Temporary Files Storage Element

Figure 7(b) shows an application of DCD's Variable and Temporary Files Storage Element. In ①, the user enters with the sentence "MY NAME IS *", where the * wildcard represents any character sequence typed by the user. In this particular instance, the goal is to grab the user's name with *. After that, this message goes to the inference machine, as seen in ②, where the value found in * is stored in the USERNAME variable. In the next step ③, the chatterbot presents to the user the message "Nice to meet you". In the dialogue ④, the user asks "WHAT IS MY NAME?". To answer this question, the chatterbot uses the previously stored information found in the USERNAME variable, and replies to the user "Your name is USERNAME", as shown in ⑤.

Permanent File Storage Element. The Permanent File Storage Element allows the chatterbot's data persistence. Figure 8(a) shows this element details, and the "b" value is the information that the chatterbots needs to store.

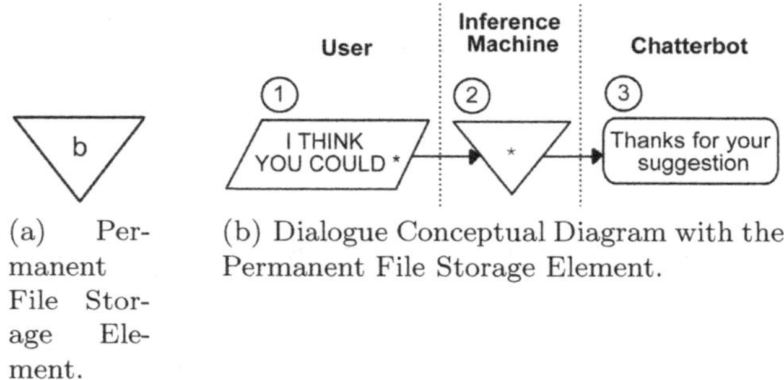

(a) Permanent File Storage Element.

(b) Dialogue Conceptual Diagram with the Permanent File Storage Element.

Fig. 8. Permanent File Storage Element

The Permanent File Storage element is implemented using the AIML's <GOSSIP> tag. For example, Figure 8(b) shows a DCD that uses this element. The user starts to type in ① "I THINK YOU COULD*". The proposal for this message is to grab a user's suggestion to the chatterbot with the * wildcard. After that, the inference machine, through the Permanent File Storage element, saves the information found in *, as shown in ②. Finally, the chatterbot replies with the text "Thanks for your suggestion" (DCD Item ③).

Topic Change Element. A topic is a theme, a subject that the chatterbot is able to talk. For instance, a bot can be able to talk about topics/subjects like gastronomy, ioga, philosophy, and so on. A topic groups the categories that deal with the same subject, which allows the chatterbot to simulate an important feature of human dialogues: to direct the dialogue to a specific subject, to talk about this subject and later identify when there is a subject change during the dialogue (adding the ability for the bot to shift topics during chat).

In Figure 9(a), the Topic Change element is used when the subject is changed. It occurs when the TOPIC variable is set with the new subject, indicated by the NEWTOPIC value. Finally, it is describe in AIML with the <TOPIC> tag.

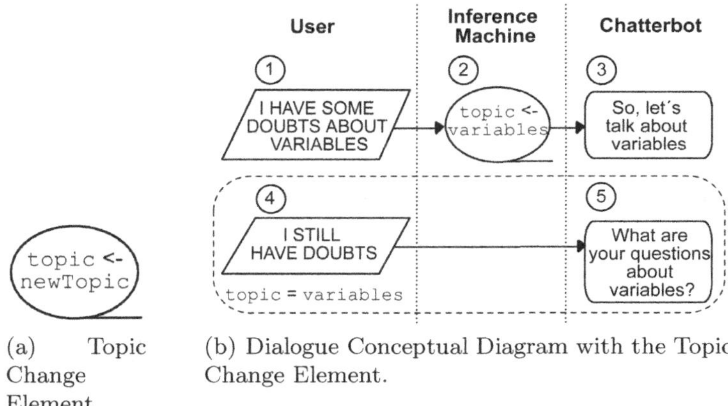

(a) Topic Change Element.

(b) Dialogue Conceptual Diagram with the Topic Change Element.

Fig. 9. Topic Change Element

Figure 9(b) shows an example of the VARIABLES topic definition that is used to group the categories related to the "variable" concept in the programming logic. In this figure, the topic and its categories are wrapped by dashed lines. In ① the user says "I HAVE SOME DOUBTS ABOUT VARIABLES". After that, in ② the inference machine sets the TOPIC variable with the VARIABLES value, pointing out the dialogue's theme change. From now on, the next user inputs detected by the chatterbot are sought initially in the VARIABLES topic category. Next, the chatterbot shows in ③ the sentence "So, let's talk about variables".

In the dialogue's following moment at ④, the user types the sentence "I STILL HAVE DOUBTS". Since the topic variable was set as TOPIC = VARIABLES, the chatterbot performs the pattern matching first in the VARIABLES' topic categories. Thus, the inference machine acknowledges that the user's intention is to talk about programming languages, and the chatterbot shows in ⑤ the reply with the question: "What are your questions about variables?".

Internal Processing Element. The Internal Processing element, depicted in Figure 10(a), represents the processing activities performed by the chatterbot that should not be seen by the users. In the AIML language there is a relationship between the Internal Processing element and the <THINK> tag.

Figure 10(b) shows an example of the Internal Processing element application. In this case, the chatterbot stores a certain user preference in a variable, but it does not display the data related to that action. At first, the user types the sentence "I LIKE TO PLAY SOCCER" in ①. After the recognition of that input pattern, the inference machine puts the SOCCER value into the SPORTUSER variable. Since this action is not supposed to be displayed to the user, the elements Variable and Temporary Files Storage is placed in the Internal Processing scope in ②. In ③, the chatterbot shows the sentence "I also like to play soccer".

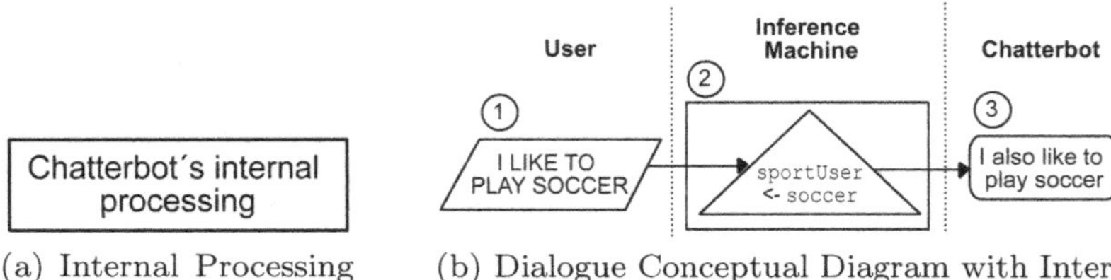

(a) Internal Processing Element.

(b) Dialogue Conceptual Diagram with Internal Processing Element.

Fig. 10. Internal Processing Element

3 Dialogue Conceptual Diagram Graphically Displaying ALICE Chatterbot's Knowledge Base: A Case Study

In order to verify if DCDs are proper frames of reference for the graphical representation of chatterbots' dialogues, this section presents a DCDs usage's case study in the graphical representation of an ALICE [2] chatterbot KB's snippet. The ALICE chatterbot was chosen due to its acceptance as a model for chatterbots' developers. Also, it is an open source system with very specific implementation instructions and source codes. The code reuse allows the design of systems with no need to build them from scratch.

Table 1. Number of Occurrences for AIML Tags in the ALICE Chatterbot's KB

Tag	Occurrences	Tag	Occurrences
CATEGORY	93,607	THAT	1,542
PATTERN	93,607	CONDITION	49
TEMPLATE	93,607	TOPIC	10
SRAI	25,894	GET	8,265
SET	8,265	GOSSIP	0
THINK	3,071	IF	0
STAR	2,325	ELSE	0

The ALICE current version has 45.000 categories arranged in sixty-four (64) AIML files. Among these files it is possible to find the following: AI.AIML, ALICE.AIML, ASTROLOGY.AIML, GEOGRAPHY.AIML, POLITICS.AIML, RELIGION.AIML and WALLACE.AIML. ALICE KB's graphical representation using the DCDs is defined in two steps. First, a survey of how many times each AIML tag was used in the ALICE's KB was performed. Table 1 displays the outcome of this survey.

Table 1 gives an overview of which DCDs are the most common in the ALICE's KB. For instance, the Bivalent Decision and Multivalent Decision elements can be implemented with the CONDITION tag. Considering that this tag appears 49 times in the ALICE's KB, then this bot has dialogues that can be graphically represented by these elements. In a second step, all the 64 ALICE KB's related

Listing 1.2. AIML Code of the ALICE's KB Related to the Category Call Element.

```
<category>
<pattern>WHAT IS A CHAT ROBOT</pattern>
   <template>
      A chat robot is a program that attempts to simulate
          the conversation or "chat" of a human being. The
          Chat robot "Eliza" was a well-known early attempt
          at creating programs that could at least
          temporarily fool a real human being into thinking
          they were talking to another person.
   </template>
</category>
<category><
   pattern>WHAT IS A CHATTERBOT</pattern>
   <template>
      <srai>WHAT IS A CHAT ROBOT</srai>
   </template>
</category>
```

files were analyzed, while dialogues were chosen according to the tags shown in Table 1 and modeled with the DCD elements.

Hereafter the application of the Category Call element for the graphical representation of ALICE's KB is presented. Such element can be found in the file AI.AIML of the ALICE's KB, as shown in Listing 1.2. In this example, the user wants to ask what is the meaning of chatterbots. Figure 11 shows the DCD related to the code in the Listing 1.2.

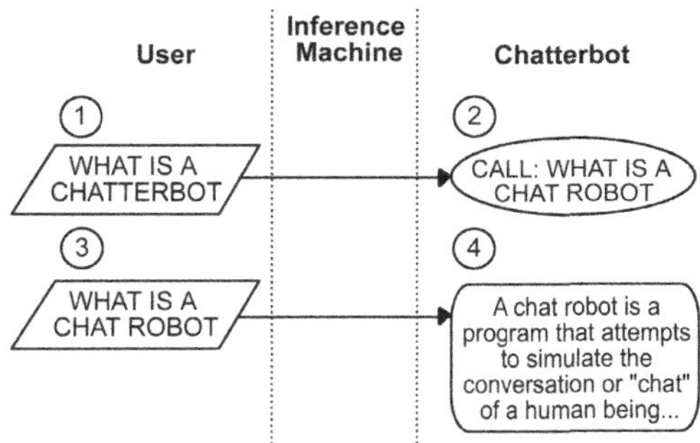

Fig. 11. Dialogue Conceptual Diagram Related to Listing 1.2

In ① the user enters the text "WHAT IS A CHATTERBOT". After that, the dialogue flow changes with the Category Call element for the category that corresponds to the pattern "WHAT IS A CHAT ROBOT" in ②. The category ③ has the chatterbot reply in ④.

4 Conclusions

The DCDs proposed by this work provide graphical common languages that can be used by domain specialists and computing researchers/developers with no extra knowledge. Domain experts' main concern is modeling the chatterbot's KB, while the programmers and computing experts handle the implementation details. Thanks to the DCDs, both parties can see the big picture and have a better control of the entire process using such graphical portrayal as a guideline.

In order to present the DCD as a tool and how it works, this paper started showing two related works and the AIML language with its main tags. Also, the correlation between dialogue's visual elements and AIML tags was considered. Another challenge this work surpassed was the presentation of a practical use case to see the usage of DCDs in an ordinary, real-life situation. A subset of the ALICE's KB was modeled using DCDs in order to portray how robust and viable they are as devices to graphically display bots' KBs.

The studies shown in this paper is not the end of the DCD, it is hunt the beginning. The purpose of this paper is to invite researchers, domains specialist from varied fields and artificial intelligence designers to try out a new, graphical and complete way of describing the knowledge for natural language processing. Thus, the next DCD's challenge will be the test of time and real-life applications.

References

1. Alcaniz, M., Rey, B.: New Technologies for Ambient Intelligence. In: Riva, G., Vatalaro, F., Davide, G., Alcaniz, M. (eds.) Ambient Intelligence: The Evolution of Technology, Communication and Cognition Towards the Future of Human-Computer Interaction, pp. 3–15 (2005)
2. Wallace, R.: The Elements of AIML Style. ALICE A.I. Foundation (2003)

Adaptive System for Intelligent Traffic Management in Smart Cities*

Paweł Gora and Piotr Wasilewski

Faculty of Mathematics, Informatics and Mechanics
University of Warsaw
Banacha 2, 02-097 Warsaw, Poland
{p.gora,piotr}@mimuw.edu.pl

Abstract. In the paper we propose an adaptive system for intelligent traffic management in smart cities. We argue why the traffic is difficult and complex phenomenon, why such traffic management systems are necessary for the smart city, what is the state of the art in the traffic science and traffic management and how to improve existing solutions using methods that we develop, based on the Perception Based Computing paradigm.

Keywords: urban traffic, transport management system, adaptive systems, smart cities, Perception Based Computing.

1 Introduction

Large vehicular traffic congestion in cities became a serious civilizational problem, causing travel delays, stress of drivers, environmental pollution, increase of fuel consumption, more car accidents, problems with organizing public transport and designing road network, significant noise level. According to [8], drivers in 7 largest polish cities wasted in 2013 over 3.5 billion PLN because of traffic jams. Similarly, according to [48], large traffic congestion caused urban drivers in USA to travel over 5.5 additional billion hours and to purchase over 2.9 billion gallons of fuel in 2011. Many other negative factors, such as noise and air pollution, were not taken into account in those reports.

The following issues became very important in the urban traffic research:

- How to manage and optimize vehicular traffic (standard or after closing parts of the road network, e.g. because of road works or car accident)?
- How to prevent occurrence of traffic jams and other undesired traffic states?

In the paper we propose the adaptive system for managing vehicular traffic in cities, which may help solving or reducing effects of mentioned problems

* We would like to express our gratitude to Professor Andrzej Skowron for his valuable comments and advices. The research has been supported by the research grant 2011/01/D/ST6/06981 from the Polish National Science Centre.

D. Ślęzak et al. (Eds.): AMT 2014, LNCS 8610, pp. 525–536, 2014.

and give answers for the raised issues. The main feature of this system is intelligence, i.e. ability of learning from experience and knowledge of experts and self-improvement. The adaptive system proposed in our work employs results of the latest development of the Perception Based Computing theory, aiming to create a general framework for dealing with spatio-temporal data which come from complex systems, such as vehicular traffic.

The rest of the paper is organized as follows: in section 2 we introduce the concept *smart city* and argue while vehicular traffic is a complex concept in cities, which is important and difficult for scientific investigation. Section 3 describes theoretical background of the paradigm that we employ in our approach, Perception Based Computing. Section 4 presents existing approaches to design traffic management systems. Section 5 presents details of the adaptive system that we propose, section 6 concludes the paper.

2 Smart Cities and Vehicular Traffic

The *smart city* is a relatively novel and still vague concept, different scientific works try to define what a smart city is. According to [5] the city is smart when it *"utilizes information and communication technologies with the aim to increase the life quality of their inhabitants while providing sustainable development"*. [7] defines that the city is smart when *"investments in human and social capital and traditional (transportation) and modern (ICT-based) infrastructure fuel sustainable economic growth and a high quality of life, with a wise management of natural resources, through participatory government"*.

Both popular definitions include the indispensable utilization of ICT (Information and Communication Technologies) as a key requirement for a city to be smart. Indeed, nowadays the communication, especially using information technology, is the major medium of interactions, and solutions from the area of ICT, which integrate many communication and computation devices and applications, are necessary to ensure required quality and range of communication.

Scientists conduct extensive studies on how to organize and manage life and communication in the city using new technologies ([10], [15], [41]). Such studies must have an interdisciplinary level and employ not only solutions from the ICT area, but also from mathematics, physics, psychology, sociology and other sciences. The study of complex systems and complexity science is especially important, since when we look at contemporary cities, we inherently see that they are composed of many agents (citizens), which interact with each other (in 2 possible ways: by communication or direct, physical actions) and with the environment (buildings, roads), forming an open system (people may travel between cities) which is nonlinear (e.g. at some point the city becomes too crowded and couldn't function properly) and exhibits some emerging properties (e.g. crowd, communication network overload) which are difficult to be analyzed and predicted. Thus, the large city itself possesses all features of a complex system, according to characteristics of complex systems in [20], [22]. Understanding the nature and consequences of interactions that lead to situations which are difficult to be predicted, is crucial to design appropriate tools to manage the complexity in cities and making the city smart, but this is still a difficult task.

The complexity of phenomena that occur in the city as a result of interactions between citizens is just the one difficult problem for scientists. The second problem, important especially from the ICT standpoint, is the large amount of data that have to be collected from different devices. The data may come from many different sources, such as mobile phones, personal computers, servers, GPS navigation devices, sensors. They may have different format, all must be transmitted through the network, interpreted, stored in large databases, analyzed in large computational centers to give an understanding of the state of the process in a city, and provided in a user-friendly way for different people and organizations. It may be difficult to manage such large data which have different formats and it is reasonable to classify such data as Big Data, which is a common name for data that have to be processed quickly, but are difficult to be processed using traditional data management tools, because of their size and structure, and require parallel software which runs on thousands of servers.

Vehicular traffic could be also seen as a complex system. Drivers use their cars to achieve some goals (usually reaching from point A to point B safely and as soon as possible, using existing infrastructure and driving according to some rules), they interact with each other and with the infrastructure, they are also exposed to the changing environmental conditions, which are difficult to be predicted (e.g. the weather). The interaction between agents may lead to emerging, undesired phenomena, such as car accidents or traffic jams formation. Such phenomena may have significant, negative impact on people and the city as a whole. Hence, it is necessary to design and implement systems, that will be able to predict and prevent such undesired situations.

The problem with vehicular traffic in cities and other complex processes is that they are extremely difficult to be modeled, predicted, analyzed and controlled. Even the traffic on highways is a complex phenomenon, however there are many mathematical models, ranging from macroscopic models ([32]), which describe dependency between congestion, density and speed (e.g. based on analogies to fluid dynamics ([23, 24]), kinetic gas theory ([16], [17], [35]), through mesoscopic models which are based on traffic flow theory, but propagate individual vehicles or packets ([6]), to microscopic models based on computer simulation (e.g. models based on probabilistic cellular automata [26, 27]). It is difficult to say which models are best (i.e. which models reproduce the traffic with the best accuracy and realism), it may depend on the road structure or region/country (e.g. the model of drive in Poland is different than model of drive in Germany). Most of popular models can reproduce some traffic phenomena such as spontaneous traffic jam formation ([26]) or a shockwave ([24]) and classify traffic states with a satisfactory accuracy ([18], [21]).

The case of urban traffic is more difficult, since it is necessary to include different types of roads, different crossroads and traffic lights which influence traffic dynamics in cities. The problem of understanding the phenomena of vehicular traffic in cities, its dynamics and properties, is an important scientific field of study in the transport area and there exist scientific projects and laboratories which work on this problem ([9], [19]). However, we still can't say that we fully understand the traffic dynamics in cities. There also exist traffic simulation tools such as VISSIM ([49]) or PARAMICS ([33]), which implement different urban traffic models and those tools are used in the traffic engineering industry for transport planning purposes. Also, there are traffic simulation tools which are used for scientific purposes, such as TSF ([11], [13]), which we use to conduct our research on adaptive optimization of vehicular traffic and acquiring traffic-related domain knowledge from experts. In the past it was very difficult or even impossible to conduct such research, because of lack of real traffic data. Nowadays, there are methods to collect such data, e.g. using inductive loops, mobile phones or GPS navigation systems, but such data are not always shared by their owners because of the law or commercial reasons. Also, there are still important issues related to store, process and analyze such large datasets quickly.

The next difficulty with the urban traffic is that it has properties which emerge from interactions between agents (cars with drivers), depend on many factors, so are hardly predictable. For example, random events such as car accidents or rainy weather may dramatically influence traffic properties. Existing traffic management methods don't cope with such unpredictable events properly.

3 Analysis of Smart Cities Data: Perception Based Computing Approach

Poggio and Smale claim in ([34]) that the problem of learning is the basis of for understanding intelligence in brains and machines and for discovering how the human brain works. These two challenges are roads to making intelligent machines that, as children, could learn from experience and improve their competencies ([34]). Designing and intelligent managing of smart cities is a good example for this claim. Smart cities as complex systems generate huge amount of data, which will be referred as smart city data or in short: SC data. These data have to be collected, proceed, analyzed and some knowledge and concepts should be learn from them.

Data collected from a smart city are good example for the idea of Big Data. Processing, analysis, knowledge discovery and concept learning in smart city data are really challenging. Concerning quantitative complexity (wrt the size of data) those tasks can not be achieved by humans in reasonable time and with reasonable resources usage and so they should be somehow automatized and supported using AI methods and tools taken from machine learning, data mining and knowledge discovery. Even more, concerning quantitative complexity of smart city data as well as its qualitative complexity (SC data includes data of various types) data mining, knowledge discovery, and especially complex concept learning hardly can be made using classical universal AI methods and algorithms ([34], [50]). If we assume that intelligent managing in smart cities should be made at least partially on line, this calls for fast and frugal processing of SC data to face the problems of data mining, complex concept learning and knowledge discovery in databases with dynamically evolving complex data (e.g. stream data or sensory data from traffic). These challenges can be met by knowledge discovery systems dialogizing with experts or users during the discovery process or by domain oriented algorithms designing ([50]). Such challenges can also be met by adaptive learning systems changing themselves during the learning process as the response to evolving data. All of challenges connected with smart cities data can be dealt using Perception Based Computing methods (see e.g. [45–47]).

Perception Based Computing (PBC) paradigm, as classical, hard artificial intelligence, is not intended to model human or animal perception. PBC is aimed at following principles that govern perception processes taking place in humans and animals in order to optimize the perception function [45, 46] i.e. to optimize action and perception results. Perception can be treated as understanding of sensory information (see [45], [47]): on the input of perception process are sensations or sensory measurements and on the output are some pieces of knowledge needed for planning and control of action performance. The main principle of PBC is that perception is action oriented [2], [29]. It reflects the fact that perception is needed in systems aimed at performing some actions. In the context of adaptive learning it states that perception evaluation depends on successfulness of actions made on the basis of perception results. According to contemporary psychology and cognitive science perception is hierarchical information processing i.e. it is separated in some levels organized hierarchically stating from sensual/sensory level where pieces of information from higher level are constructed on the basis of information [25]. Perception is also not purely passive and knowledge is involved in perception on some of its levels [25].

In Perception Based Computing perception is a process of computation performed in a hierarchical way on objects called granules in which basic granules are pieces of sensory information, usually vectors of sensory measurements results (see e.g. [44, 45], [47]). Granules form higher levels are constructed on the basis of granules from lower levels, e.g. granules from a given levels can be generalizations of granules from its antecedent levels or a coalition of granules from a lover level can be a granule on the next level of hierarchy. In PBC both granules and levels can be represented by means of interactive information systems [44, 45], [47]. As an example of PBC computations can serve a hierarchical approximation of complex concepts [44]. Because of hierarchical information processing, PBC paradigm can be very useful in modeling smart city, including smart city traffic, as a complex systems with a hierarchical nature. PBC paradigm also admits interactive computations on granules from the same level as well as from different levels of hierarchy [43–45], [47]. It also makes PBC very useful in modeling of smart cities since they are interactive systems where many processes interact or interfere each other. Smart city traffic is a good example for this phenomenon: traffic jam can be seen as a result of interactions between cars and traffic lights. Because of a hierarchical organization and interactivity of smart city traffic, its intelligent management should also use hierarchical and interactive methods and algorithms [4] what makes PBC paradigm very needed also in smart city traffic management.

4 Existing Approaches to Manage Urban Vehicular Traffic

Managing vehicular traffic is cities is a subject of extensive research from many years. There are few possible approaches to the traffic control, such as managing configurations of traffic lights at crossroads, proposing routes for cars, adapting speed limits, managing parking places, road pricing. There are even professional implementations of those approaches deployed in many cities.

In case of controlling traffic lights, which is mostly considered in our research, the first such system was SCATS ([39]) introduced in Sydney in early 70's. SCOOT ([40]) was started in Great Britain few years later, other popular systems are RHODES [38] and OPAC [30]. Such systems implement adaptive strategies (i.e. they change their state in response to the changing traffic and road conditions) and optimize different parameters, e.g. cycles, offsets, splits.

The next interesting method for optimizing the traffic is an *intelligent parking system*. It is assessed that large part of the whole traffic in the city center is generated by people driving around searching for a parking place [1]. The solution may be a "smart" system integrated with mobile devices, on-board computers in cars and the parking infrastructure in the city, that enable reserving parking places remotely and navigating drivers to such places. Some companies (e.g Streetline [42]) already provide such "smart parking" solutions.

Another approach to manage and reduce the traffic is the road pricing: charging fees for drive in the city center and making drivers to choose another routes. This approach is introduced e.g. in London, Stockholm, Singapore.

There are much more approaches to manage the traffic, e.g. modifiable speed limits, building new streets, detours, extending public transport. Also, the traffic is affected by many factors, every new street, detour, bus line, new buildings, housings may change traffic properties significantly. Thus, it is important to have access to comprehensive data from many sources, in order to manage the traffic and plan development of the city. All sources should be integrated and taken into account in calibrating traffic models and traffic management systems.

The existing traffic management solutions are quite effective in case of standard traffic, that could be investigated by long observations, or in cases that could be planned beforehand, such as building new housings, detours, road works, changes in public transport etc. However, in case of such a complex system as the urban traffic, the small changes in the initial conditions (e.g. larger than usual number of cars, car accidents, weather hazard) may cause significant, nonlinear changes in the whole traffic. The effective traffic management system must respond to such changes properly, but the existing traffic management tools don't cope with such situations effectively. The reason is that existing solutions base on quantitative changes in traffic parameters that could be measured (often the number of data is even insufficient) and not on understanding the traffic and its dynamics. For example, existing engineering approaches often define the concept "Level of Service" to differentiate traffic states, but definitions use only approximations based on observations and are not inferred from sensory data.

We argue, that the next progress in traffic management should be made by designing intelligent traffic management tools, where intelligence would mean that the systems will understand the traffic, e.g. will be able to approximate (from sensory data) complex, vague, traffic-related concepts, such as *traffic jam on a crossroad*, *large traffic congestion*, and to give information which are not obvious from sensory data. Also, the intelligence should mean that the system will learn from experience and self-improve to cope with new, unprecedented situations. Such properties may be obtained by applying machine learning and artificial intelligence methods, e.g. the Perception Based Computing 3.

5 Adaptive System for Managing Traffic in Smart Cities

The proposed system should be composed of the following elements:

- sources of traffic data: client applications in mobile devices or on-board computers, traffic management centers, inductive loops,
- servers (located in different city districts) to collect traffic data with databases to store traffic data,
- computational centers to analyze and interpret data and making decision regarding complex actions required to optimize the traffic,
- traffic lights (and potentially other tools to manage the traffic).

Figure 1 presents overview of system's architecture (without traffic signals) and the following subsections describe components of the system.

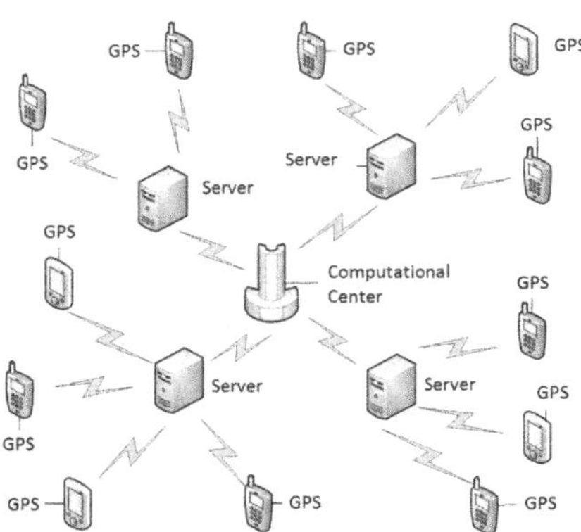

Fig. 1. Architecture overview

5.1 Sources of Traffic Data: Client Applications in Mobile Devices or On-Board Computers

Real traffic data are indispensable for understanding the traffic and should be collected online to give the up-to-dated information about the traffic, and stored in databases for further analysis. Currently, traffic data provided to traffic engineers originate mostly from detectors, such as inductive loops. In the proposed solution traffic data would come mostly from mobile devices which may be localized either by GPS navigation system or localization provided by the network operators. Currently, many road users (drivers and passengers) travel equipped with a mobile phones or other portable devices. Some of them uses GPS navigation applications which already collect information such as position and velocity and send them to computational centers of GPS navigation producers in order to propose the best routes for other drivers ([3], [28]). This approach would be extended if data were collected from all mobile phones and on-board computers.

Besides the traffic data, the system would need to collect additional information, such as information about the weather, car accidents, special events or other atypical situations, which may be useful for better understanding the current traffic situation and to predict the future states. Those data must be collected from different sources integrated with the system (e.g. taxi drivers, traffic management centers, inductive loop) and will be complex and difficult to analyze not only because of large size, but also because of different formats.

5.2 Servers to Collect Smart City Traffic Data

According to the proposed architecture, servers with databases should be located in different districts of the city, the number of such servers should depend on the size of the city and size of data that will be transferred through servers and stored in databases. The grounds for correctness of such architecture is the fact, that the traffic on a given region of the city, its dynamics, future state and decisions regarding its short-term management, depend mostly on the current and past state of the traffic in that area. It is possible that the traffic in a given region may be influenced by the traffic in neighboring regions, thus the servers should also communicate with servers from neighboring areas, exchanging some information inferred from traffic data, but there will be relatively few interactions between a given server and servers in distant areas of the city.

The amount of traffic data that should be collected online by servers is huge. In a large city we have about $10^5 - 10^6$ cars driving through the network simultaneously and we would like to collect information about move of every car, e.g. its speed and location. Such data should be transferred with an adequate frequency (e.g. once per $1 - 5$ seconds). In order to achieve even greater knowledge about the traffic and to make more accurate prediction and better decisions, it may be necessary to collect much more data, as was stated in the section 5.1.

Servers will be responsible not only for collecting and storing data, but also for doing initial analysis of those data, which, together with information about traffic lights configuration, could be used to assess velocities, delays, sizes of queues on crossroads. Also, since data come from mobile devices, it could be used to determine the mean of transport (e.g. car, bus) to give a better knowledge about the traffic, e.g. using methods from [37].

The quality of traffic analysis will depend on the penetration rate of data (haw many traffic participants sends their data). There are works which study how to cope with low penetration rates of data ([36]) and we expect much higher penetration rates in the future, thanks to using on-board computers in cars and increasing popularity of mobile devices and GPS navigation systems.

The next step in analyzing data would be approximating complex, high-level, spatio-temporal traffic concepts (such as *formation of traffic jam on a crossroad, large traffic congestion*) from acquired low-level, sensory data. The step would be implemented by constructing hierarchy of classifiers, using methods described in the paper [14]. The method for constructing classifiers, described in [14], uses knowledge acquired by interaction with domain experts, so the system would also learn from that knowledge while constructing required classifiers approximating complex, spatio-temporal concepts from collected sensory data.

Such high-level concepts, approximated from sensory data, constitutes knowledge which may be exchanged (together with some sensory data) between servers to improve their classification, and to the computational center to improve its traffic prediction and optimization algorithms. This part will be a crucial property of the proposed system, marking its intelligence, i.e. ability to learn from experience and self-improve.

5.3 Computational Centers to Analyze Data and Makings Decision Regarding Complex Actions Required to Optimize the Traffic

Computational center would be responsible for modeling and predicting the global traffic, making decisions regarding traffic optimization and initializing their realization. The center would also collect preprocessed traffic data and induced knowledge from servers. Since collected traffic data are low-level and concern speeds and positions of every car, it would be possible to construct a microscopic traffic model, e.g. it may be a large-scale agent-based traffic simulation model being developed in the software Traffic Simulation Framework ([11], [13]) or similar models implemented in tools such as VISSIM ([49]) or Paramics ([33]). The model will be calibrated using real traffic data (positions and speeds of cars) and will be used to perform fast, short-term traffic simulation (say, $10 - 15$ minutes ahead, the proper time should be based on simulation's accuracy and needs of traffic prediction and traffic optimization methods).

While the simulation is calibrated to predict the traffic with a satisfactory accuracy, it may be used to test different traffic optimization techniques, e.g. it is possible to test many different traffic lights configurations and choose the one that will be best according to some metrics, such as the total travel time, delay, fuel consumption etc. The question is how to find the best configuration. In general the problem is difficult, since the set of possible solutions is extremely large. In addition, we can not be sure that the solution is certainly best, since solutions will be verified using traffic simulation (and, possibly, traffic data from the past), while the real traffic may be in the unstable state, with different possible evolution paths, in addition we cannot predict car accidents and other unexpected situations (we can model it in the traffic simulation, but we can't compute all possible situations). Indeed, the problem is difficult and existing solutions mentioned in the section 4 do not solve the problem, there is still a lot of work to do and it must be a subject of further extensive research.

However, it is not necessary to find the best solution, since it may be sufficient to find suboptimal configurations, that will be relatively good, e.g. there will be high chances, that the traffic would be much better then in most other cases, and not much worse than in the best possible case, which is unknown. Methods that brought promising results in that research are based on a genetic algorithms [12] an other metaheuristics, but their usefulness is still limited because of large computational complexity (this is a common issue of most similar methods).

In addition, the best approach to cope with car accidents, seems to be just detecting that state from sensory data as soon as possible (or acquiring such information from the traffic management center or other sources), so the system will *understand* that such incident has occurred, and run traffic simulation and optimization algorithms assuming that some parts of the road network are closed. Also, there are some approaches, which study how the traffic dynamics propagates in the network in case of such situations [31].

6 Conclusions

In the paper we proposed an adaptive system for intelligent traffic management in smart cities. The most important feature of such system, which may potentially make it successful and more efficient than existing solutions, is the ability to learn from experience, self-improvement and understanding traffic dynamics using methods based on the Perception Based Computing paradigm. Details of the system, e.g. methods for approximating complex, spatio-temporal traffic concepts from sensory data, and methods for adaptive optimization of traffic parameters are still a subject of our research, but results of initial experiments are promising.

The development of our solution may lead to elaborating a new generation of intelligent traffic management systems. In section 5 we presented overview of the proposed architecture. We suppose that such approach may be deployed in every normal city. Detailed requirements could be a subject of a larger research, for example, one of the main requirements is the possibility of data collection, which in some countries may be restricted by the law.

Also, some submodules of the proposed solution may be integrated into existing, comprehensive traffic management systems (e.g. [39], [40], [30]).

References

1. Anderson, S.P., de Palma, A.: The economics of pricing parking. Journal of Urban Economics 55(1), 1–20 (2004)
2. Arbib, M.A.: The Metaphorical Brain 2: Neural Networks and Beyond. Willey & Sons (1989)
3. AutoMapa, `http://www.automapa.pl` (last access: April 13, 2014)

4. Bazan, J.: Hierarchical classifiers for complex spatio-temporal concepts. In: Peters, J.F., Skowron, A., Rybiński, H. (eds.) Transactions on Rough Sets IX. LNCS, vol. 5390, pp. 474–750. Springer, Heidelberg (2008)
5. Bakici, T., Almirall, E., Wareham, J.: A Smart City Initiative: The Case of Barcelona. Journal of the Knowledge Economy 4(2), 135–148 (2013)
6. Bliemer, M., Raadsen, M., de Romph, E., Smits, E.S.: Requirements for Traffic Assignment Models for Transport Planning: A Critical Assessment. In: Australasian Transport Research Forum 2013 Proceedings (2013)
7. Caragliu, A., Bo, C.D., Nijkamp, P.: Smart cities in Europe. Business Administration and Econometrics, Series: Serie Research Memoranda number 0048 (2009)
8. Deloitte, Targeo.pl. Traffic jams report in 7 largest polish cities., `http://korkometr.targeo.pl/Raport_Korki_2013.pdf`
9. Urban Transport Systems Laboratory at Ecole Polytechnique Federal de Lausanne, `http://luts.epfl.ch` (last access: April 08, 2014)
10. Future ICT project, `http://www.futurict.eu` (Last access: April 13, 2014)
11. Gora, P.: Traffic Simulation Framework - a cellular automaton-based tool for simulating and investigating real city traffic. In: Recent Advances in Intelligent Information Systems, pp. 641–653 (2009)
12. Gora, P.: A genetic algorithm approach to optimization of vehicular traffic in cities by means of configuring traffic lights. In: Emergent Intelligent Technologies in the Industry, pp. 1–10 (2011)
13. Gora, P.: Traffic Simulation Framework. In: 14th International Conference on Modelling and Simulation, pp. 345–349 (2012)
14. Gora, P., Wasilewski, P.: Inducing Models of Vehicular Traffic Complex Vague Concepts by Interaction with Domain Experts. In: The Fifth International Conference on Advanced Cognitive Technologies and Applications, COGNITIVE 2013, pp. 120–125 (2013)
15. Helbing, D.: FuturICT-New science and technology to manage our complex, strongly connected world. arXiv:1108.6131 (2011)
16. Helbing, D.: From microscopic to macroscopic traffic models. A Perspective Look at Nonlinear Media. From Physics to Biology and Social Sciences 503, 122–139 (1998)
17. Helbing, D., Hennecke, A., Shvetsov, V., Treiber, M.: MASTER: Macroscopic Traffic Simulation Based on A Gas-Kinetic, Non-Local Traffic Model. Transportation Research Part B: Methodological 35(2), 183–211 (2001)
18. Helbing, D., Treiber, M., Kesting, A., Schonhof, M.: Theoretical vs. Empirical Classification and Prediction of Congested Traffic States. The European Physical Journal B 69(4), 583–598 (2009)
19. IBM Research Dublin, Smarter Urban Dynamics, `http://researcher.watson.ibm.com/researcher/view_group.php?id=2522` (last access: April 13, 2014)
20. Johnson, N.F.: Simply complexity: A clear guide to complexity theory. Oneworld Publications (2009)
21. Kerner, B.S.: Introduction to Modern Traffic Flow Theory and Control. The Long Road to Three-Phase Traffic Theory (2009)
22. Ladyman, J., Lambert, J., Wiesner, K.: What the complex system is? European Journal for Philosophy of Science 3(1), 33–67 (2013)
23. Lustri, C.: Continuum Modelling of Traffic Flow (2010)
24. Lighthill, M.J., Whitham, G.B.: On kinematic waves. II. A theory of traffic flow on long crowded roads. Proceedings of the Royal Society A 229, 317–345 (1955)
25. Marr, D.: Vision: A Computational Investigation into the Human Representation and Processing of Visual Information. Freeman (1982)

26. Nagel, K., Schreckenberg, M.: A cellular automaton model for freeway traffic. Journal de Physique, 2221–2229 (1992)
27. Nagel, K., Wolf, D.E., Wagner, P., Simon, P.: Two-lane traffic rules for cellular automata: A systematic approach. Physical Review E (Statistical Physics, Plasmas, Fluids, and Related Interdisciplinary Topics) 58(2), 1425–1437 (1998)
28. NaviExpert, `http://www.naviexpert.pl` (last access: April 13, 2014)
29. Noë, A.: Action in Perception. MIT Press (2004)
30. Gartner, N.H., Pooran, F.J., Andrews, C.M.: Implementation of the OPAC adaptive control strategy in a traffic signal network. In: Proceedings of Intelligent Transportation Systems, pp. 195–200 (2001)
31. Pan, B., Demiryurek, U., Shahabi, C., Gupta, C.: Forecasting Spatiotemporal Impact of Traffic Incidents on Road Networks. In: International Confderence on Data Mining 2013, pp. 587–596 (2013)
32. Papageorgiou, M.: Some remarks on macroscopic traffic flow modeling 32(5), 323–329 (1998)
33. Paramics software, `http://www.paramics-online.com`. (last access: April 13, 2014)
34. Poggio, T., Smale, S.: The mathematics of learning: Dealing with data. Notices of the AMS 50(5), 537–544 (2003)
35. Prigogine, I., Herman, R.: Kinetic Theory of Vehicular Traffic. Elsevier (1971)
36. Quang, T.M., Kamioka, E.: Adaptive Approaches in Mobile Phone Based Traffic State Estimation with Low Penetration Rate. Journal of Information Processing 20(1), 297–307 (2012)
37. Quang, T.M., Kamioka, E.: Granular Quantifying Traffic States Using Mobile Probes. In: 2010 IEEE 72nd Vehicular Technology Conference Fall, pp. 1–6 (2010)
38. Mirchandani, P., Head, L.: RHODES: A real-time traffic signal control system: architecture, algorithms and analysis. Transportation Research. Part C: Emerging Technologies 9(6) (2001)
39. SCATS, `http://www.scats.com.au` (last access: April 13, 2014)
40. SCOOT, `http://www.scats.com.au` (last access: April 13, 2014)
41. Smart City Lab, `http://smartcity.csr.unibo.it` (last access: April 13, 2014)
42. Streetline, `http://www.streetline.com` (last access: April 13, 2014)
43. Skowron, A., Wasilewski, P.: An introduction to perception based computing. In: Kim, T.-h., Lee, Y.-h., Kang, B.-H., Ślęzak, D. (eds.) FGIT 2010. LNCS, vol. 6485, pp. 12–25. Springer, Heidelberg (2010)
44. Skowron, A., Wasilewski, P.: Information Systems in Modeling Interactive Computations on Granules. Theoretical Computer Science 412, 5939–5959 (2011)
45. Skowron, A., Wasilewski, P.: Interactive information systems: Toward perception based computing. Theoretical Computer Science 454, 240–260 (2012)
46. Skowron, A., Wasilewski, P.: Interactive Grammars: Toward Perception Based Computing. In: Watada, J., Watanabe, T., Phillips-Wren, G., Howlett, R.J., Jain, L.C. (eds.) Intelligent Decision Technologies. Smart Innovation, Systems and Technologies, vol. 16, pp. 391–402. Springer (2012)
47. Skowron, A., Wasilewski, P.: Introduction to Percpetion Based Computing. In: Ramanna, S., Jain, L.C. (eds.) Emerging Paradigms in Machine Learning, pp. 249–275. Springer (2013)
48. Texas Transportation Institute. 2012 Urban Mobility Report (2012), `http://d2dtl5nnlpfr0r.cloudfront.net/tti.tamu.edu/documents/mobility-report-2012.pdf`. (last access: April 13, 2014)
49. PTV VISSIM, `http://vision-traffic.ptvgroup.com/en-us/products/ptv-vissim` (last access: April 13, 2014)
50. Vapnik, V.: Learning Has Just Started. An interview with Prof. Vladimir Vapnik by R. Gilad-Bachrach (2008), `http://learningtheory.org` (last access: 2012)

Implementing a Holistic Approach for the Smart City

Roberto Requena, Antonio Agudo, Alba Baron,
Maria Campos, Carlos Guijarro, Jose Puche, David Villa,
Felix Villanueva, and Juan Carlos Lopez

School of Computer Science, University of Castilla-La Mancha,
Ciudad Real, Spain
{roberto.requena,jose.puche,alba.baron,
antonio.agudo,maria.campos,carlos.guijarro}@alu.uclm.es,
{david.villa,felix.villanueva,juancarlos.lopez}@uclm.es

Abstract. Extending the services offered by the city requires, most of the times, reimplementation efforts. This paper presents our on-going efforts to develop a platform for the Smart City that focuses in providing the appropriate solutions for an easy integration of new services and devices. This endeavor is accomplished by abstracting communication issues using a middleware platform and by standardizing the way services are instantiating.

1 Introduction

The Smart City paradigm is gaining credit as technology improvements are enabling new and more ubiquitous ways of interaction. In the Spanish context, several cities are leading the way towards the Smart City such as Malaga, Barcelona, or Santander. However, most of these cities make the emphasis in deploying *black boxes* of sensors and services. On the contrary, the Smart City paradigm envisions more versatile and flexible solutions than just a set of services attached to a specific set of sensors. A Smart City should tackle a broader objective as it is providing a basic platform where new devices and services could be easily deployed and integrated.

Ideally, new devices could be integrated in the Smart City by simply implementing a *plug-and-play* approach. Similarly, new services could be implemented, that would make use of the data gathered by those devices, in a standardized manner without having to be aware of the low-level details of every different device available in the Smart City.

However, this is not a trivial matter since several issues need to be addressed, such as a mechanism for abstracting communication between different protocols and technologies or a syntactic and semantic standardization to support an orthogonal instantiation of services.

This paper presents a holistic approach to the Smart City, called Civitas, that implements a bottom-up approach. The proposed solution consists in a layered

D. Ślęzak et al. (Eds.): AMT 2014, LNCS 8610, pp. 537–548, 2014.

framework, in which different functionalities are organized by levels of complexity. In this sense, the lowermost level deals with device-integration aspects. Then, an intermediary layer is taking care of device-communication issues, abstracting device-like particularities. Finally, the uppermost layer provides a set of services built upon the previous layers. Figure 1 depicts the different layers along with the different devices and services supported by each of them. Examples of every layer are provided underneath.

This paper is organized as follows. First, the state of the art for Smart City framework is studied, summarizing not only specific solutions for the Smart City but also comprehensive frameworks upon which devices and services can be deployed. This section will pose the most relevant challenges that need to be addressed from the point of view of the Smart City paradigm, as well describes the specific details of how standardization is accomplished, providing concrete examples of how this standardization is implemented in the different layers of the proposed architecture. Finally, the most relevant conclusions drawn from the proposed framework will be presented in Section 5.

2 Smart Cities Platforms

Lately, a lot of services are appearing devoted to provide smart functionality in smart city field. There is still a lack of standards or platforms that we could take as facto standards in service development and specification.

One of the main efforts in smart city application field is SmartSantander project [7] where one of the most large testbed has been deployed in Santander (Spain). This project can be seen as an city-scale experimental facility which enables research in smart city field identifying key technologies, services and socioeconomic factors involved in smart city ecosystem.

One of the main issues of smart city is "information island" problem which appear if we develop services as standalone applications. Examples of this type of services are traffic management [8], human dynamics [9], bicycle routes [10], mobile crowdsensing [11], etc.

We can clasify the approach followed by these works since middleware point of view in the following solutions:

- A complete design solution from scratch focusing in the functionality of the service designed without middleware considerations. They do not usually deal with common problems in a real deployment (e.g scale, basic security mechanism, etc.)
- A middleware designed from scratch ignoring existent solutions.
- The use of specific protocols as a middleware, for example XMPP (Extensible Messaging and Presence Protocol) in [11], without any other consideration.

Our approach is different to any other, we take an efficient, robust, well-proved generic object-oriented distributed middleware and we add specific smart-city related layers. Currently we are working with Internet Communication Engine (ICE) [6], an object-oriented distributed middleware which provide us with basic security mechanisms, scalability mechanisms, event broker, etc.

With this starting point, we add what we consider essential common services for future smart city services (e.g. reasoning layer or visualization layer) and we integrate other protocols in order to provide a flexible framework for service development (e.g MQTT).

3 The Civitas Framework

Figure 1 shows an overview of Civitas framework, the lowest layer is a sensor/actuator layer since a lot of process in the city has to be monitorized and controlled by mean sensors and actuators. As we will see in the next subsection by mean a pollution service monitoring example, according to the type of service, a set of appropriate sensors/actuators are deployed in the city. The information from sensors/actuators, together with other type of information (e.g. from citizens), is gathered using an Information and Communications Technologies (ICT) Infrastructure.

We strongly believe that, similarly to services which are provided using common infrastructures (electric/water supply network, transport networks, etc.), an smart city should provide with an ICT layer to help service and device deployment. At logical level, Civitas backbone layer has similar aim that ICT layer does at physical level. In the Civitas backbone, information is gathered and distributed according to each application field (Homeland Security, Waste management, social media, etc.). These application fields, in turn, are divided into specific services (semaphore control, pollution monitoring, traffic monitoring, parking management, etc.)

Over these application field Civitas offers a reasoning and analytics layer in order to deal with information extraction from raw data. Again, this layer is specialized by application field and the information is provided to different entities (Government, corporations and citizens) which also provide with information.

Finally, an advanced visualization and control layer enables human advanced interaction in order to monitor and to control different process in the city.

Every layer and service in Civitas exposes one or several object-oriented interfaces in order to interact with its functionality. We describe these interfaces by mean an Interface Description Language (IDL) which provide us with an excellent tool to specify a “contract” between the developer who provides the service and the developer who uses the service.

Additionally, as we will see later and in order to easily integrate sensors from different partners, in the sensor/actuator layer we also support other protocols.

About the infrastructure of the smart city we are deploying a set of Service nodes with four interfaces each one(figure 2):

- 802.15.4 [12] at 868.6 MHz for information gathering coming from sensors. We use this technology for sensor/actuator wireless network.
- 802.11 at 2.4 GHz for citizen public service enabling them to connect to the smart city infrastructure at specific places in order to access directly to the services.

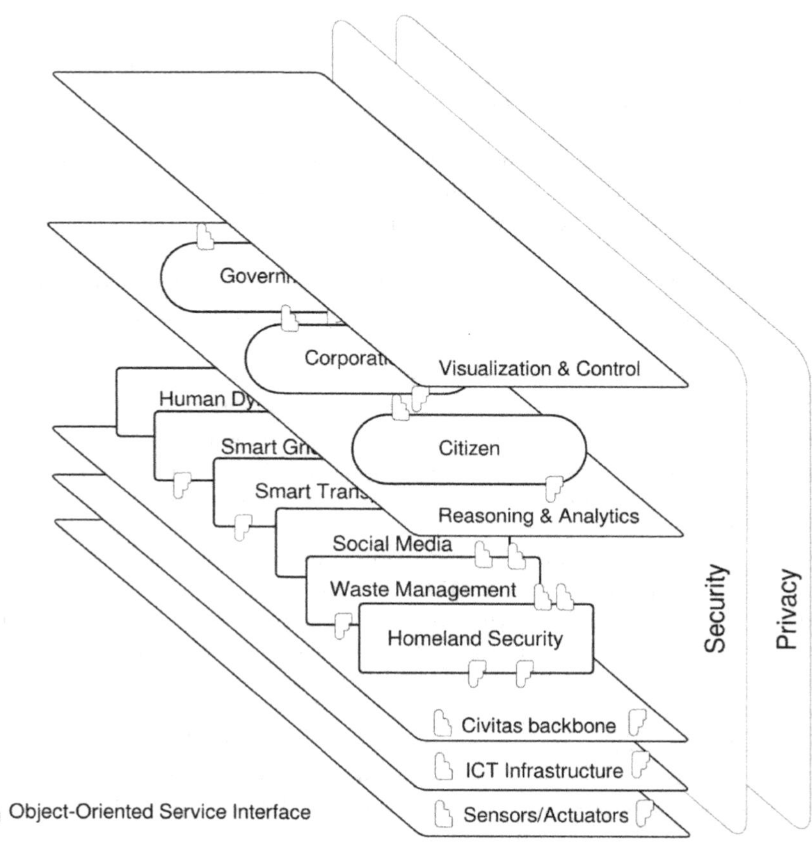

Fig. 1. Civitas framework overview

- 802.11 at 5.8 GHz for connecting the service node between them in a mesh network if there is no Ethernet connection (e.g ADSL modem) available.
- Ethernet interface for connecting the service node to the rest of the infrastructure.

The best way to understand Civitas is through the current implementation of services and layers described in the following subsections.

3.1 Sensor Integration: Pollution Monitoring

The lowermost layer of the framework integrates different devices, in order to create a "*wireless sensor/actuator network*" (WSN). In the case of monitoring air pollution, the aim of this WSN is to gather and monitor the values of some pollution parameters like O3 or CO2. This activity is justified because there are several worrying information about pollution, and it is known that big cities generate approximately 70% of the gases that cause the greenhouse effect [2].

Fig. 2. Civitas pollution monitoring node (left) and service node (right) deployed in the school of computer science

This layer describes how to integrate different technologies into this WSN, and how to offer the pollution service as a Civitas service. One of the devices integrating this WSN is "*Waspmote Plug&Sense Smart Environment*"[1], in charge of periodically gathering air pollution values using different sensors(Figure 2). Then, the gathered data are sent to a Civitas service node[2], that receives frames with the values and some other extra information, like battery level or "waspmote" coordinates. Finally, a Civitas node device with a "Node.js" web server consults a MySQL database that contains sensor values sent by the gateways.

Gathering pollution values would not have sense if these values could not be seen. Consequently, an implementation of the visualization layer and accessible way to observe and monitor them is needed. In our case, we choose as implementation of visualization layer a Node.js web server which is responsible for serving html5 pages, containing updated polluting values, to the users (Figure 3). Citizens can therefore use laptops, smartphones or tablets to visualize them in a web browser. Node.js is a platform built on Chrome's JavaScript run-time that is highly scalable. This is very important to satisfy all user's requests. Moreover,

[1] `http://www.libelium.com/development/plug-sense/`

[2] `http://www.libelium.com/development/meshlium`

a library of OpenStreetMaps [13] project is used in the html5 page to show a map with the real location of the Waspmotes and their values, providing to the citizen a more integrated experience with the city environment. Additionally, we expose the collected values by mean an object-oriented interface specified in the mentioned interface definition language.

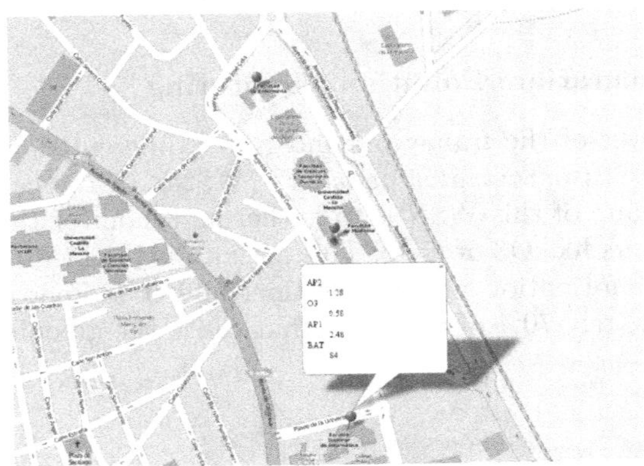

Fig. 3. Civitas visualization layer example

The proposed interface for this service should look:

```
module Pollution {
    struct Coord {
        double latitude;
        double longitude;
        double altitude;
    };
    struct Gas {
        long sensorID;
        long timestamp;
        Coord point;
        double CO;
        double CO2;

    };
    interface PollutionListener {
        void report(Gas m);
    };
};
```

3.2 Protocol Integration: MQTT

From the pollution monitoring service we extracted a set of requirements for Civitas. The Civitas framework, as a middleware for the Smart Cities, needs to be able to adapt to new technologies in order to easily adopt new ways of collecting/distributing data. The Sensor/Actuator and the backbone layer of the Civitas framework addresses the integration of an event-based infrastructure which uses the deployed WSN in order to publish information about the environment. This information will be disseminated to those parties subscribed to the communication channel (e.g emergency teams, security people, etc.). The previous subsection has introduced different types of devices comprising a WSN. This section focuses in the dissemination infrastructure and more specifically, in the use of MQTT-SN and MQTT protocols. MQTT [3] is a royalty free, very lightweight, publish/subscribe, asynchronous messaging transport protocol designed by IBM company. Its design principles makes MQTT ideal for emerging paradigms of connected devices like M2M or IoT and also for mobile applications and sensor networks. MQTT is undergoing a standardization process at OASIS[3] and the protocol specification has been openly published with a royalty-free license. It is also important to take into account the more than possible increased use of MQTT due to Eclipse Paho project which is currently developing some greats MQTT clients. But WSNs usually do not have TCP/IP as transport layer. They have their own protocol stack such ZigBee on top of IEEE 802.15.4 layer. Thus, MQTT which is based on TCP/IP cannot be directly run on WSNs. For this reason MQTT-SN was developed, which is an extension of MQTT for WSNs.

MQTT-SN is designed to be as close as possible to MQTT, but it is adapted to the peculiarities of a wireless communication environment such as low bandwidth, high link failures, short message length, etc. Clients are WSN nodes, which communicate via a gateway to a broker on IP network [3].

According to the infrastructure described at the beginning of this section, Figure 4 describes the communication architecture. As it can be seen the Waspmote Plug&Sense devices (recall section 3.1) forms a WSN with Meshlium gateways using MQTT-SN over 802.15.4 layer like the protocol communication. The gateway is the responsible for communicate the WSN with a TCP/IP network. So in the Civitas Service node, a protocol gateway translation between MQTT-SN and MQTT is needed. With this process of translation a point-to-point communication with the MQTT broker can be established.

All of this involves an infrastructure which allows WSN devices to publish environment data to a server in a TCP/IP network. The only task that it needed to make is translate this information for Civitas platform format.

3.3 Service Instantiation: Action Recognition

Civitas is here considered not only as a middleware framework where services and devices can be deployed, but also as an urban area consisting of infrastructures, smart platforms and networks (Civitas backbone and ICT Infrastructure). In this scenario, citizens play an essential role, increasingly connected one to each other, allowing themselves and their mobile phones to work as sensors.

[3] `https://www.oasis-open.org/committees/tc_home.php?wg_abbrev=mqtt`

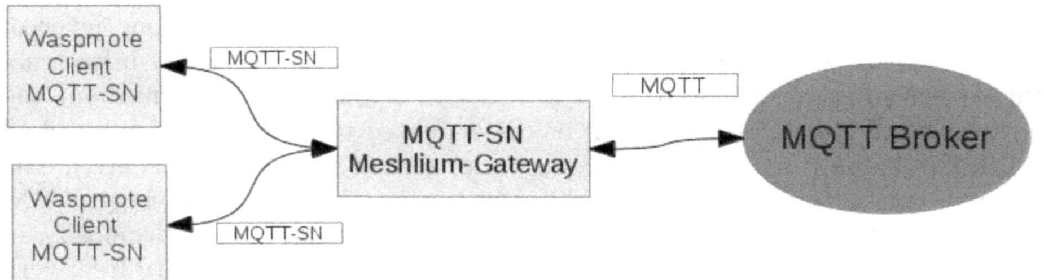

Fig. 4. Communication architecture

Making the most out of this circumstance, Civitas has devised a way of integrating mobile phones as though they were traditional sensors. These sensors provide context-aware services since they accompany their owners all day long. In order to prove the capability of such devices, Civitas provides a service that identifies the output of a given set of actions regarding the tasks being performed by the phone holders, we use this implementation as example of reasoning & analytics layer. Particular attention is paid at those tasks that involved transportation.

A machine learning approach is followed in this service, based in a learning by example technique. First of all, in order to state different means of transport, an application has been develop in order to collect values from sensors available in a smart phone. This application has been developed considering a Samsung Galaxy S4 and therefore the sensors this phone provides (gestural sensor, proximity sensor, gyroscope, accelerometer, geomagnetic sensor, temperature and humidity sensor, barometer, RGB light sensor, GPS and positioning). The application is in charge of automatically collecting these sensor values and characterize the signal through a feature extraction process. The resulting feature vectors will be used first for a training process and once the system has been trained, and a model for each means of transport has been computerized, feature vectors will be used for the recognition process. Please note that the training process is carried out just once, offline. Once the service is deployed in Civitas, sensor values are used for recognizing means of transportation. The result of the recognition process is communicated to the middleware framework using a Publish/Subscribe approach, so that other services interested in such information can subscribe to these channels and therefore use this information for more elaborated services.

Examples of services that could make use of this basic means of transport recognition are services analyzing traffic congestion in a city, public safety, or even private services. Regardless, the result of the recognition tasks should be provided in a standardized way to the Smart City management framework, so these can be automatically used by other specialized services.

This service is therefore useful not only for the Smart City but also for the Ambient Assisted Living (AAL) [4] and Ambient Intelligence (AmI) [5]. The main objective of the AAL is to provide support services to elder people, it uses technological devices that are present in the environment naturally. This is where comes the second paradigm, which promotes an environment where technology is always available and fully integrated, almost imperceptible way for its users.

If we also take into account that human attention is a limited resource, these technologies can help overcoming this limitation by providing daily assistance based on these ubiquitous technologies such as mobile phones.

The role that smart phones can play in enabling contexts conceived by AAL or AmI is critical because they often come with us in our daily lives and their capabilities are constantly increasing.

We could recognize the activity the person is performing at each moment in time and the location in which it is being carried out, so that any other system can provide more sophisticated functionalities based on these data, adjusting to the peculiarities of each user. Some examples might be:

- Make a call to the emergency services if it detects that a person takes too many hours without moving and may be seriously ill.
- Notify parents or guardians of a child if caught on his mobile phone that is involved in a fight.
- Submit a attendance unit looking for an Alzheimer's patient that has been lost and walking away from home.

4 A Standardization Mechanism for Services and Devices in the Smart City

The Smart City concept presents a challenge in the current conception of cities through the most efficient creation or renovation of infrastructures, intelligent and respectful with the environment. By means of a combination of technologies the Smart City tries to offer a better quality of life to the citizen and a reduction of the environmental impact. All these changes do not consist simply of a technical procedure, but also it is necessary to bear in mind the change produced in, for example, a company in general.

The proposed Smart City paradigm is not solely based on the employment of new technological devices, but it is also based on information that these devices can provide and how to treat such information. Therefore, it is possible to define the Smart City as a knowledge system, whose sources of information are all devices or sensors deployed throughout the city. However, it is important to highlight that the majority of the collected information from the Smart City do not follow a concrete standard, showing formats difficult to handle and originating information islands.

To be able to work with the information generated by the Smart City, it is necessary to initially identify the concepts that compose the area of the city, so that later on, it might be possible to relate and provide them with the appropriate semantics and a common vocabulary. It is therefore necessary to count on an ontology for the Smart City, so that it could carry out the standardization of the different services that the city offers, obtaining this way a deeper knowledge and a precise specification of the concepts that these include.

An ontology defines a common vocabulary to describe the concepts of a certain context, in this case the domain of the city, through a set of basic terms and relations between them. The use of an ontology in the Smart City will provide better communication, integration and sharing between the different services it could offer. Across the ontology one tries to reach the semantic interoperability of the smart cities, carrying out a homogenization of the diverse services they offer. Implementing this approach a global solution is obtained that can be apply to any city.

However, the employment of an ontology is not sufficient. Once the information offered by the city is collected, it is necessary to be later on processed in order to support intelligent decisions-making, giving the impression that it is the city itself the one reasoning, similarly as a person would do. To this end, the approach presented here proposes the use of a common-sense reasoning[1] as a way of enhancing the semantic held by the ontological concepts and providing means to infer new knowledge out the information present in the knowledge base.

4.1 Service Composition: A Service Based on Other Services

As we mention before, the Smart City paradigm heavily depends on how to obtain information about the city and how to process it, and in this endeavour, Civitas can play an essential role by providing a great number of ways of integrating and providing devices and services. Previous sections have proved how different technologies, protocols, and services could be integrated into Civitas. This section is intended to prove how citizens can use all that information held in Civitas.

The interaction between the Smart City and citizens is one of the most interesting application conceived by the Smart City paradigm. Citizens are a continuous source of information, as they give life to the city while they enjoy it. So, the data they produce is a big valuable resource that the Smart City should collect. However, at the same time, the Smart City cannot be conceived merely as an information consumer, but on the contrary it has to provide citizens with useful services. This can easily be achieved in Civitas by developing applications built upon the functionality provided by previously studied layers.

This subsection shows a Civitas application in which interaction between the citizen and the Smart City is leveraged. Potentially, thousands of applications can be deployed into the Smart City framework provided by Civitas. The one presented here is a prototype that can be used as a model of how this interaction is supported.

First of all, the application must provide a service to the user. In this case, this application proposes a geolocalization service and an interaction application. This means, using all the services deployed over the Smart City (restaurants, bus stops, parkings, etc.), this application prototype should be able to detect and geolocalize them. Note that, when the application works for one service –for example, the restaurant service–, it can be widely scalable for the rest of them. The more services the Smart City presents, the more interactions and geolocalizations the application could implement.

Regarding the technologies available for developing this application, many available ones could be employed, however, the following ones have been proposed to this end:

- Bluetooth technology is chosen for service discovery and communication support between them and the smartphone. Bluetooth is a very strong and stable specification that allows data transmission between many different devices. Moreover, it can be easily implemented in a smartphone.
- For the geolocalization task, the already mentioned OpenStreetMap project comes into scene. It is an open-source project which provides free geographic data for all over the world.

The use of these two technologies support geolocalization and communication between Smart City services. This information can be exchanged with citizens which while providing citizens with these two services can collect user data. Using this approach, both communication sides obtain useful data from the interaction. The user information obtained and stored by the Smart City can be sold to external companies interested in which are citizens' customs, habits or preferences. Moreover, these data increase the knowledge held by the Smart City.

5 Conclusions

This paper presents a description of how to implement different granularity services for Civitas, a Smart City framework. To this end, different technologies and communication protocols have been employed, therefore demonstrating how new services and devices can be easily integrated into the framework.

Starting from an existent middleware we are more close to a final solution than other approaches. We can also focus in the research of new middleware services as reasoning layers or protocol integration. This ongoing work has to be validated in real testbeds and provide with examples of integration of final services and devices.

Acknowledgments. This research was supported by the Spanish Ministry of Science and Innovation through project DREAMS (TEC2011-28666-C04-03).

References

1. Fahlman, S.E.: Marker-Passing Inference in the Scone Knowledge-Base System. In: Lang, J., Lin, F., Wang, J. (eds.) KSEM 2006. LNCS (LNAI), vol. 4092, pp. 114–126. Springer, Heidelberg (2006)

2. Trenberth, K.E., et al.: Observations: Surface and atmospheric climate change. Climate Change 2007: The Physical Science Basis. In: Contribution of Working Group I to the Fourth Assessment Report of the Intergovernmental Panel on Climate Change. Cambridge University, Cambridge
3. Stanford-Clark et al.: MQTT for Sensor Networks (MQTTs), pp. 235–236 (October 17, 2007)
4. Pieper, M., Antona, M., Cortes, U.: Editorial: Ambient Assisted Living. ERCIM News, Special Theme: Ambient Assisted Living 87, 18–19 (2011)
5. Ducatel, K., Bogdanowicz, M., Scapolo, F., Leijten, J., Burgelman, J.C.: Scenarios for ambient intelligence in 2010. ISTAG, Brussels (2010)
6. Henning, M., Spruiell, M.: Distributed Programming with Ice Revision 3.4. ZeroC Inc. (June 2010)
7. Theodoridis, E., Mylonas, G., Chatzigiannakis, I.: Developing an IoT Smart City framework. In: 2013 Fourth International Conference on Information, Intelligence, Systems and Applications (IISA) (July 2013)
8. Khekare, G.S., Sakhare, A.V.: A smart city framework for intelligent traffic system using VANET. In: 2013 International Multi-Conference on Automation, Computing, Communication, Control and Compressed Sensing (iMac4s), pp. 302–305 (March 2013)
9. Vakali, A., Angelis, L., Giatsoglou, M.: Sensors talk and humans sense Towards a reciprocal collective awareness smart city framework. In: 2013 IEEE International Conference on Communications Workshops (ICC), June 9-13, pp. 189–193 (2013)
10. Cosido, O., Loucera, C., Iglesias, A.: Automatic calculation of bicycle routes by combining meta-heuristics and GIS techniques within the framework of smart cities. In: New Concepts in Smart Cities: Fostering Public and Private Alliances, SmartMILE (2013)
11. Szabo, R., et al.: Framework for smart city applications based on participatory sensing. In: 2013 IEEE 4th International Conference on Cognitive Infocommunications (CogInfoCom), December 2-5, pp. 295–300 (2013)
12. IEEE 802.15.4-2006 standard, http://www.ieee802.org/15/pub/TG4.html
13. OpenStreetMap Wiki 2010, http://wiki.openstreetmap.org/wiki/Main_Page (cit. March 15, 2014)

Elimination of Moving Shadow Based on Vibe and Chromaticity from Surveillance Videos

Huaxiang Zhao[1], Min Wook Kang[2], Kyoung Yeon Kim[2], and Yoo-Sung Kim[2]

[1] Chongqing Key Laboratory of Computational Intelligence,
Chongqing University of Posts and Telecommunications, Chongqing, 40065, China
[2] Knowledge Management System Laboratory, Inha University, Incheon, 402-751, Korea
yskim@inha.ac.kr

Abstract. Shadow removal is one of the most important parts of moving object recognition in the field of intelligent video surveillance since the shadow definitely affects the recognition performance. This is caused from that shadows share the same movement patterns and similar magnitude of intensity to those of the foreground objects. Therefore, in this paper, to effectively remove moving shadows from video, a new approach based on chromaticity and a well-known universal background subtraction named as Vibe was proposed. Experimental results prove that moving shadows can be removed effectively by the proposed approach than the other ones.

Keywords: Moving shadow, Background subtraction, Object recognition, Vibe, Chromaticity.

1 Introduction

In recent years, with the rapid development of image processing technologies, object recognition and tracking is more and more widely used in intelligent surveillance systems. In this context, a reliable detection of moving objects is the most critical requirement for intelligent surveillance systems. However, current detection approaches have a typical drawback: moving shadows are usually misclassified as part of the moving object, resulting in inaccurate object recognition and tracking. This happens because shadows share the same movement patterns and have a similar magnitude of intensity change as those of the foreground objects. The incorrect classification of shadow as foreground will result in inaccurate detection and decrease tracking performance.

For this reason shadow removal is imperative for robust and stable intelligent surveillance systems. The shadow classification approaches could be categorized into two folds: (i) Model-based approach which needs scenario, object, illumination, and priori knowledge, therefore the applicability of this category might be limited. (ii) Feature-based approach which is on the basis of the assumption that shadow is generally has low luminance, but identical chromaticity and texture to the background. Moreover, according to the proposed approaches during the last decade, the features

D. Ślęzak et al. (Eds.): AMT 2014, LNCS 8610, pp. 549–558, 2014.

which are used for shadow detection are also categorized into the following four subgroups[3]: (i) Chromaticity-based approach uses the assumption that regions under shadow become darker but retain their chromaticity[1], (ii) Physical approach in which the updated Gaussian Mixture Model (GMM) of the 3D color features with candidate pixels, posterior probabilities are used to classify each pixel in the foreground as object or shadow[4], (iii) Geometry-based approach uses orientation, size and even shape of the shadows with proper knowledge of the illumination source to split the shadow from object[5], (iv) Texture-based approach uses color features to first create large candidate shadow regions, and then discriminate shadow from objects using gradient-based texture correlation[6].

Over the years, many background subtraction techniques have been proposed. The most classical one is the Gaussian Mixture Model (GMM). Background model is able to cope with the multimodal nature of many practical situations and lead to good performance when repetitive background motions, such as tree leaves or branches, are encounter [2]. However, GMM needs a large computational time, and GMM-based approach cannot extract the foreground effectively when the update rate of the GMM background model is so fast, since moving objects involved in the background will be misclassified as background. This will result in GMM misclassify moving objects as a background, making moving objects inside remain holes, finally leading to tracking errors.

Even if the approaches above can remove shadow precisely in some cases, much of them have to face a challenging problem that some parts of objects which are similar with shadow region in color space may be detected and removed as shadow, and this will make some holes in the objects region. To handle this problem, in this paper, Vibe is firstly adopted to extract a relatively complete foreground, and next take advantage of chromaticity-based approach to remove shadow precisely. Finally, morphological optimization is used to fill the possible holes which remained in vehicle after shadow removing.

The remainder of this paper is as follow. In Section 2, some related works will be discussed. In Section 3, our novel approach will be described in detail. In Section 4, some of these algorithms is selected which use GMM combined with chromaticity and texture to be implemented to make a comparison with the proposed approach. Section 5 concludes this paper.

2 Related Work

For removing shadow from video, the background identification is an important step. The problem tackled by background subtraction is to divide the image of video sequence into foreground which contains the objects of interest and background that does not contain the objects of interest. One of most famous background subtraction algorithms is GMM first presented in [7]. This model consists of modeling the distribution of the values observed over time at each pixel by a weighted mixture of Gaussians. The GMM value of a particular pixel is represented as a mixture of Gaussian.

This is different from some other algorithms modeling explicitly the values of all the pixels as one particular type of distribution. However, the drawbacks of GMM algorithm are resulted from its strong assumptions that the background is more frequently visible than the foreground and that its variance is significantly low [2].

Most of feature-based shadow removal approaches are adopted widely because of their easy implementation and high efficiency. Physical approach mentioned in Section 1 updates a GMM of the 3D color features with candidate pixel and posterior probabilities of the model are used to classify each pixel in the foreground as object or shadow. However, since they are still limited to spectral properties, their main disadvantage involves dealing with objects having similar chromaticity to those of the background. Geometry-based approach in [5] uses orientation, size and even shape of the shadows since these can be predicted with proper knowledge of the illumination source to split the shadow from object. However, current geometry-based approaches are not designed to deal with objects having multiple shadows or multiple objects detected as a single foreground blob. Texture-based approach in [6] uses color features to first create large candidate shadow regions, and then discriminates from objects using gradient-based texture correlation. However, texture-based approach tends to be slow because it makes a comparison between one or several neighbor pixels. Performances of these approaches are not as good as expected.

3 A Novel Approach for Moving Shadow Removal

In this section, the brief block diagram of the proposed approach to remove moving shadow from video is depicted in Fig. 1. First, our proposed approach normalizes the frame and adopts Vibe to store a set of values taken previous frames at the same location or in the neighborhood of each pixel and then compares this set to the current pixel value in order to determine whether that pixel belongs to the background, and modifies the model by choosing randomly which values to substitute from the background model. Second, we adopt a chromaticity approach based on HSV color space to remove shadow. Finally, morphological optimization is adopted to fill the remained holes to get an effective result.

3.1 Background Subtraction

In this paper, a more effective approach is adopted which is proposed by Olivier Barnich et al [2] to improve above-mentioned drawbacks of GMM, which models each background pixel with a set of samples instead of with an explicit pixel model which is different from other algorithms. Follow the steps outlined below. Vibe algorithm is generally divided into two parts: segmentation part and update part, as described in [2].

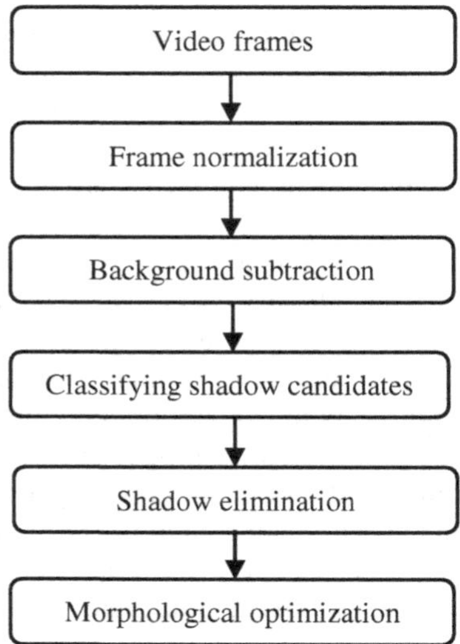

Fig. 1. Workflow for moving shadow elimination

a) Segmentation part: the pixel value $v(x)$ is classified according to M (x) of Eq (1).

$$M\left(x\right) = \left\{v_1, v_2, \ldots, v_n\right\} \tag{1}$$

This is a collection of n background sample values taken in previous frames, and each background pixel x is modeled by $M(x)$. Where $v(x)$ represents the pixel value in a given Euclidean color space located at x in the image, v_i represents a background sample value with an index i.

Firstly compare a pixel value $v(x)$ to the closest values within the set of samples whose scope is a circle of radius R with $v(x)$ as the center of the circle $S_R(v(x))$, then next the pixel is classified as background if the cardinality, denote *#min*, of the set intersection of the circle and $M(x)$ is larger than or equal to a given threshold *#min*, that pixel is classified as background if the following condition of Eq (2) is satisfied:

$$\#\left\{S_R\left(v\left(x\right)\right)\cap\left\{v_1, v_2, \ldots, v_n\right\}\right\} \geq \#min \tag{2}$$

The process for each pixel can be halted as soon as the condition is satisfied. In this paper, we adopt appropriate values 20 for radius and 2 for cardinality *#min* from the experiments in [2]. Next, we will introduce the mechanism of how to initialize the model from one single frame.

In this paper, a novel approach in [2] is adopted to initialize the model from a single frame, which means a new model is initialized in every frame and is proved to be robust while sudden light changes. Neighboring pixels is assumed to share a similar temporal distribution so that we fill the pixel model with values randomly in the 8-connected neighborhoods of each pixel in the first frame.

To be specific, the samples are selected according to the following method in Eq (3),

$$M^0(x) = \{v^0(y \mid y \in N_G(x))\} \tag{3}$$

where $t=0$ indexes the first frame and $N_G(x)$ is a spatial neighborhood of the pixel and location x, and y represent location in the first frame, respectively.

As described above, we have detailed the segmentation part. So in the following part, how to update the model is explained in update part.

b) Update part: update is composed of three components: (i) A *memoryless* update policy, which ensures a smooth decaying lifespan for the samples stored in background pixel model by offering an exponential monotonic decay. This approach improves the time relevance of estimation by allowing a few old samples to remain in the pixel model. The probability of expected remaining lifespan of any sample value of model decays exponentially is equal to Eq (4).

$$P(t, t+dt) = e^{-\ln\left(\frac{N}{N-1}\right)dt} \tag{4}$$

$p(t,\ t+dt)$ represents the probability for the interval $(t,\ t+dt)$, and the probability is assumed to be included in the model before time t, which means probability is independent of t and past has no effect on the future. We called this property memoryless property. (ii) A random time subsampling, in order to further extend the size of time windows covered by the fixed size pixel model, is introduced in this situation. In practice, there is no need to update each background pixel model for each new frame. In consideration of the use of fixed subsampling, intervals might prevent the background model from properly adapting to these motions due to the presence of periodic or pseudo-periodic background motions, a random subsampling is introduced. When a pixel value has been classified as background pixel, a random process determines whether this value is used to update the corresponding pixel model. (iii) Spatial consistency through background sample propagation, a mechanism that propagates background pixel samples spatially to ensure spatial consistency and to allow the adaptation of the background pixel models that are masked by the foreground. According to a policy which neighboring background pixels have similar temporal distribution, background models hidden by foreground will be updated with background samples from neighboring pixel locations from time to time. In this paper, a famous test video is chosen in the field of shadow removal to verify our approach.

Fig. 2. Current frame

Fig. 3. Result of background subtraction

The Fig. 2 is the current frame with the vehicle passing, Fig. 3 show the white region is the result processed by Vibe algorithm we used in this paper. In subsection 3.2, how to detect and remove the shadow will be described, and morphological processing is adopted to get a more accurate vehicle region in subsection 3.3.

3.2 Shadow Detection and Removal

Most shadow detection approaches based on spectral feature use color information. On the basis of the assumption that regions under shadow become darker but still retain their chromaticity. HSV color space proposed by [1] is adopted in this paper since HSV is sensitive to the difference between chromaticity and illumination, which is proved to be robust and fast for shadow detection.

This shadow detection is also widely used in the intelligent surveillance system. We realize the color space conversion from RGB to HSV. According to [1], pixels in shadow usually have a lower value v by comparison to pixels in background, hue of shadow is changed within a certain limit and saturation is lower in shadow. Therefore, a pixel p is classified to be a part of shadow if the following three conditions in Eq (5) which can be obtained from [3] are satisfied:

$$\begin{aligned} &\beta_1 \le \left(F_p^v / B_p^v\right) \le \beta_2 \\ &\left(F_p^s - B_p^s\right) \le \tau_s \\ &| F_p^h - B_p^h | \le \tau_h \end{aligned} \tag{5}$$

where F_p^v, F_p^s, F_p^h, B_p^v, B_p^s and B_p^h represent the values of the pixel position p in the frame F and background image B, β_1, β_2 , τ_s and τ_h represent thresholds that are optimized empirically.

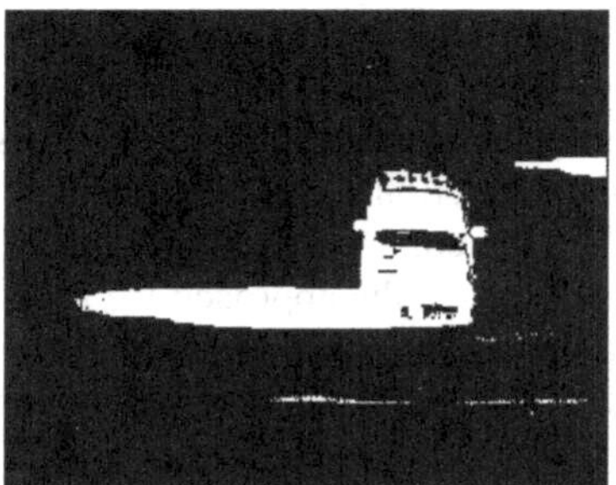

Fig. 4. Vehicle region

Fig. 5. Vehicle without shadow

As shown in Fig. 5, the shadow is removed by using the HSV color space. However, there are some holes remained in the vehicle in Fig. 5. Detection and Tracking performance are affected by these remained holes [8], which cause inaccurate tracking and even cut the vehicle into two vehicles. For this reason, we will describe how to fill the holes and get a more accurate vehicle region in subsection 3.3.

3.3 Morphological Optimization

It is necessary to fill the holes remained in the vehicle region for accurate tracking [8]. Many hole-filling algorithms are widely used, such as topology-based approach and morphology-based approach, the most common one is morphological operation: dilation and erosion. The effect of dilation operation on a binary image is to gradually enlarge the boundaries of regions of foreground pixels. Thus areas of foreground pixels grow in size while holes within those regions become smaller. In the same way, the effect of erosion operation is to erode away the boundaries of regions of foreground pixels. Thus areas of foreground pixels shrink in size, and holes within those areas become larger. Therefore, we integrate the two approaches to realize hole-filling. According to the experimental observation, we choose optimized factors of dilation and erosion operations for accurate tracking. The result showed in Fig. 7 shows that the holes in vehicle region are filled mostly.

Fig. 6. Vehicle after shadow removing

Fig. 7. Vehicle after hole-filling Experiment Evaluation

In this section, four experiments are presented to compare the performance of the approaches we mentioned in Section 1 on our platform (64-bit 3.40 GHz CPU, Intel(R) Core i5-3570K, 8 GB of RAM, C implementation). Four experimental results are shown by respectively using GMM foreground extraction combining with texture-based approach and chromaticity-based approach, Vibe combining with large texture-based approach and chromaticity-based approach to prove the approach we proposed outperforms others, all of which include morphological optimization processing. And in order to prove that shadow detection we proposed outperforms others, four experiments are also presented firstly in the following part.

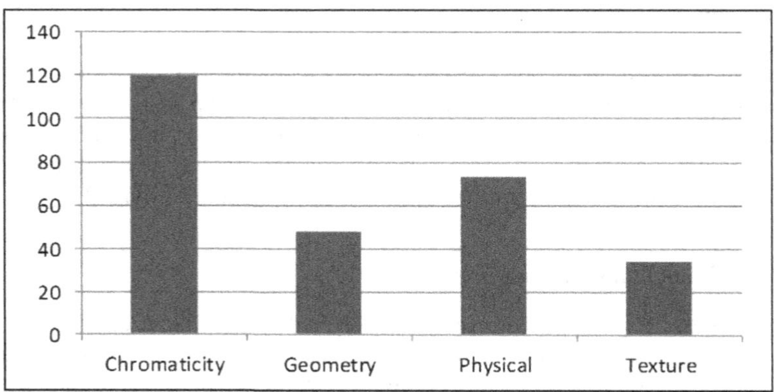

Fig. 8. The number of processed frames per second

As shown in the Fig. 8, we compared the computation times of four shadow detection approaches we introduced in Section 1, and expressed the number of processed frames per second of these approaches. We see that chromaticity clearly outperforms the three other approaches: its processing speed is as high as a frame rate of 120 FPS.

The first column in Fig. 9 shows an original frame of video sequence we used; the remaining columns show the observed results for each approach. Not only the effect of foreground extraction by using Vibe, but also the computation time is also superior to GMM. Both the texture-based approaches are affected by road texture, and fail when the road texture is similar with the vehicle. Visually, the result of Vibe combined with chromaticity looks better and is also the closest to expected result. Fig. 10 shows processing speed, the number of processed frames per second of these approaches. Each bar represents the average of the FPS on different approaches. We also see that Vibe with chromaticity approach clearly outperforms the three other approaches: its processing speed is as high as a frame rate of 8 FPS. According to the above experimental results, it comes to a conclusion that this novel approach we adopted is quite effective. Compare with other approaches, the approach proposed in this paper can accurately and effectively remove the shadow.

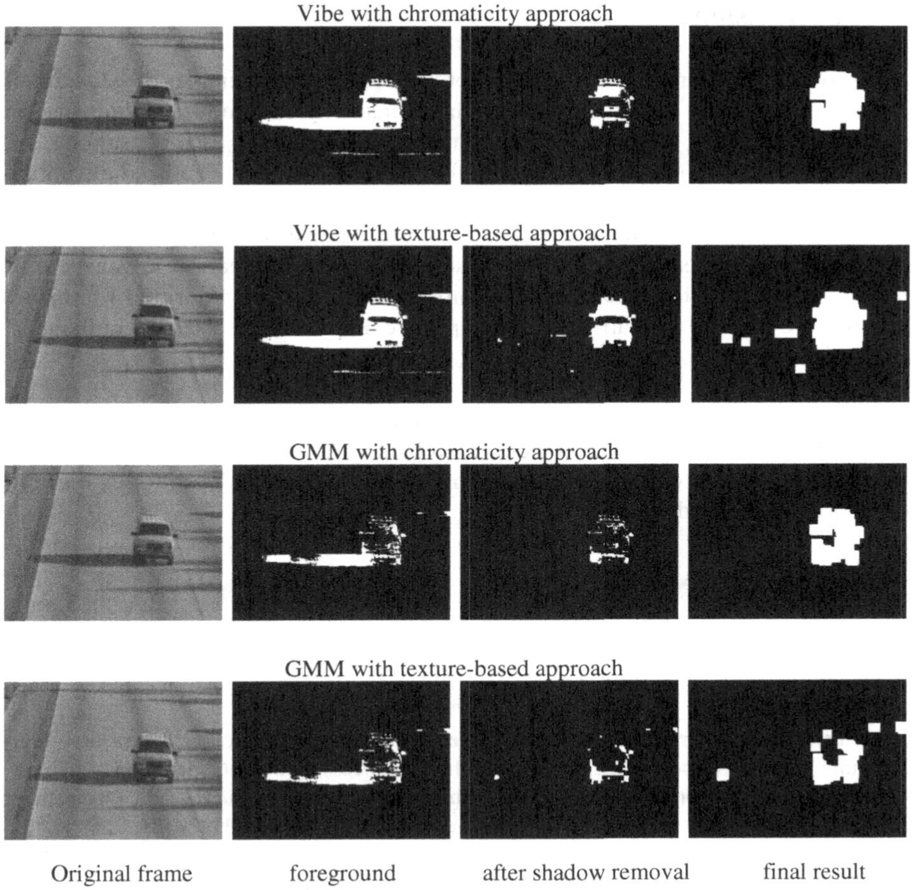

Vibe with chromaticity approach

Vibe with texture-based approach

GMM with chromaticity approach

GMM with texture-based approach

Original frame foreground after shadow removal final result

Fig. 9. Four shadow removal results

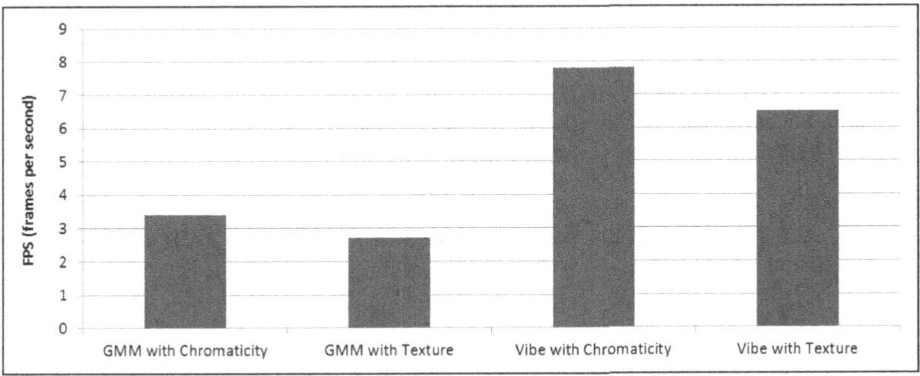

Fig. 10. The number of processed frames per second

4 Conclusion

In this paper, we introduced a novel approach combining a universal background subtraction algorithm named as Vibe with chromaticity to remove moving shadow from video. Usually, GMM has considerable computational time-consuming and GMM-based approach cannott extract the foreground effectively. To handle with this problem, we firstly adopted Vibe to extract a relatively complete foreground, and next take advantage of chromaticity-based approach to remove shadow. Finally, morphological optimization is used to fill the possible holes which remained in vehicle after shadow removing. The experimental results indicate that the algorithm of shadow removing we adopted is relatively fast and precise, improves the accuracy of the moving object detection.

Acknowledgements. This research was funded by the MSIP(Ministry of Science, ICT & Future Planning), Korea in the ICT R&D Program 2014. Authors would like to express many thanks to Mingwei Sun and his advisor, Dr. Yong Yang for their efforts on the previous related works to this paper.

References

1. Cucchiara, R., Grana, C., Piccardi, M., Prati, A.: Detecting moving objects, ghosts, and shadows in video streams. IEEE Trans. on Pattern Analysis and Machine Intelligence 25(10) (October 2003)
2. Barnich, O., Van Droogenbroeck, M.: Vibe: A Universal Background Subtraction Algorithm for Video Sequences. IEEE Trans. on Image Processing 20(6) (June 2011)
3. Sanin, A., Sanderson, C., Lovell, B.C.: Shadow Detection: A Survey and Comparative Evaluation of Recent Methods. Pattern Recognition 45(4), 1684–1695 (2012)
4. Nadimi, S., Bhanu, B.: Physical models for moving shadow and object detection. IEEE Trans. on Pattern Analysis and Machine Intelligence 26(8) (August 2004)
5. Chang, C.-J., Hu, W.-F., Hsieh, J.-W., Chen, Y.-S.: Shadow elimination for effective moving object detection with Gaussian models. In: Proc. of the 16th Int'l Conf. on Pattern Recognition, pp. 540–543 (2002)
6. Sanin, A., Sanderson, C., Lovell, B.C.: Improved Shadow Removal for Robust Person Tracking in Surveillance Scenarios. In: Proc. of the 20th Int'l Conf. on Pattern Recognition, pp. 141–144 (August 2010)
7. Stauffer, C., Grimson, E.: Adaptive background mixture models for real-time tracking. In: Proc. of IEEE Int. Conf. on Computer Vision and Pattern Recognition, pp. 246–252 (June 1999)
8. Zhu, Z., Lu, X.: An Accurate Shadow Removal Method for Vehicle Tracking. In: Proc. of Int'l Conf. on Artificial Intelligence and Computational Intelligence (AICI), pp. 59–62 (October 2010)

A Resemblance Based Approach for Recognition of Risks at a Fire Ground*

Łukasz Sosnowski[1,2], Andrzej Pietruszka[3,1,2], Adam Krasuski[4], and Andrzej Janusz[3]

[1] Dituel Sp. z o.o.
ul. Ostrobramska 101 lok. 206, 04-041 Warsaw, Poland
[2] Systems Research Institute, Polish Academy of Sciences
ul. Newelska 6, 01-447 Warsaw, Poland
[3] Institute of Mathematics, University of Warsaw
ul. Banacha 2, 02-097 Warsaw, Poland
[4] Section of Computer Science, The Main School of Fire Service
ul. Słowackiego 52/54, 01-629 Warsaw, Poland
{l.sosnowski,a.pietruszka}@dituel.pl,
krasuski@inf.sgsp.edu.pl, janusza@mimuw.edu.pl

Abstract. This article focuses on a problem of a comparison between fire & rescue actions for a decision support at the fire ground. In our research, we split the actions into a set of frames which compose a timeline of a firefighting process. In our approach, the frames are represented as compound objects. We extract a set of features in order to represent these objects and we apply a comparator framework for the evaluation of similarities between the processes. The similarity constrains allow us to recognize the risks that appear during the actions. We justify our approach by showing results of a series of experiments which are based on reports describing real-life incidents.

Keywords: Fire service, process mining, networks of comparators, compound objects resemblance.

1 Introduction

A fire & rescue (F&R) action is considered as a one of the most challenging environment for modeling in decision support systems. There were only a few attempts so far to at least partially automate the decision making process in this area [1,2]. One such attempt is the R&D project called ICRA[1]. The main goal of ICRA is to build a modern AI-based, risk-informed decision support system

* This work was supported by National Centre for Research and Development (NCBiR) grant No. O ROB/0010/03/001 in the frame of Defence and Security Programmes and Projects: "Modern engineering tools for decision support for commanders of the State Fire Service of Poland during Fire &Rescue operations in the buildings" and by Polish National Science Centre (NCN) grants DEC-2011/01/B/-ST6/03867 and DEC-2012/05/B/ST6/03215.

[1] http://www.icra-project.org

D. Ślęzak et al. (Eds.): AMT 2014, LNCS 8610, pp. 559–570, 2014.

for the Incident Commander (IC), which improves the safety of firefighters and extends a situational awareness of the IC during F&R actions. The developed system is ought to analyze sensory data with regard to the threats and vulnerabilities that may occur at the fire ground. It should also evaluate a risk level for particular stages of the ongoing action.

Within the framework of the project a F&R action is perceived as a process. The process is a sequence of states and transitions, ordered in time [3]. A state represents the situation at the fire ground at a given time point, resembling a frame in a movie. Each of the frames is a complex object which is composed of other, possibly complex objects, such as buildings, equipment, rescuers, occupants and others. The objects within the frame are in mutual relations. For example: *an occupant is in a room; a firefighter extinguishes the fire.* The process changes its state/frame when one of the objects within the frame changes its parameter or a relation to other objects. The transitions of frames during a F&R action are reflected in a radio communication which is recorded in a control room. Then for a subset of actions the transcripts are created defining an event log [4]. An event log of F&R is originally written in a natural language. It allows to identify complex objects and vague concepts, however, a special framework is needed in order to facilitate modeling of such complex processes [2].

The representation of a F&R action in a form of the transcript is currently reserved only for serious or peculiar actions. In a future, it may be done automatically by processing a sensory data using a system such as the one we are developing in the ICRA project. Currently, however, its availability is very limited. In this paper we would like to present the problem of handling a *cold start*. We show that our method can be used in order to assess incidents described by reports from the EWID[2] system. Such data can be used in ICRA before we can obtain better suited real-life data. This issue decomposes into several smaller problems such as the representation of a rescue action object, finding similar objects and foreseeing the risks.

The remainder of the paper is as follows: in Section 2 we present the F&R context and some details about the ICRA project. Section 3 introduces a method that we use for designation of the similarity of rescue actions and depicts our solution which is based on networks of compound objects comparators. In Section 4 we describe our experiments which we conduct to validate the approach and we present the results. Consecutive section presents a discussion and the interpretation of the experimental results. The article is concluded with final evaluation of presented method and a perspective for future work.

2 Fire Rescue Actions Context

The IC is responsible for all the activities performed by firefighters at the fire ground. IC operates under uncertainty as her/his situational awareness is based on the reports received orally from firefighters located in danger zone. The uncertainty, mental and time pressure make the decision making challenging.

[2] Polish Incident Data Reporting System used by Polish State Fire Service.

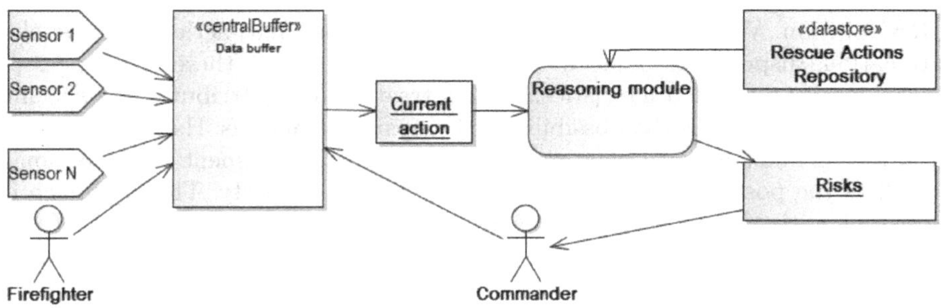

Fig. 1. An overview of the presented context of a part of the designed ICRA system. A list of sensors as a source of atomic data. A data buffer as a module to present a chunk of data that is ready for further processing and interpretation. A current action object built from available data in the data buffer. Reasoning module which finds similar historical actions and retrieves risks from them to warn the commander.

To facilitate the process of decision making and reliable risks monitoring we created a decision support system with a central data repository for F&R actions. The system gathers information from the fire ground using sensors, as well as notifications obtained from firefighters.

The gathered information covers more details of a F&R action and is more reliable. However, the IC in unable to comprehend such amount of low level data. Therefore we introduced a hierarchical data processing approach in order to aggregate the information into human-readable concepts. The process of concept approximation begins in the *data buffer* which is the first level of the sensor data aggregation. This module of our system is responsible for making quick overviews and presenting only newest information in the defined time window.

The IC interacts with the system, acquires the needed information and asks for any details in order to make better decisions. The highest level concepts which are used to describe the situation at the fire ground originate from the risks management approach [5].

One certain element of the schema presented in Figure 1 is a repository of historical, well described F&R actions. It stores serialized action processes which are ready for the retrieval of knowledge about risks and recommendations for the prevention measures. The actions are stored in the repository in a form that is suitable for describing its consecutive stages. The beginning of the stage is defined by sequences of actions consisting of notification and confirmation. Furthermore, the course of an action is divided into the following phases: a preliminary phase, a phase of the emergency, a phase of the evacuation and a phase of firefighting efforts. As the ending stage we consider the sequences which are executed after finishing the firefighting activities. In the case of ICRA we consider a process as a series of snapshots with some features. In the final solution there will be a couple of snapshots after each phase.

In this paper we present four sub-processes which can be described by the four corresponding snapshots. This is due to the fact that in EWID system we have less detailed information than the assumed data available for the final ICRA solution. We list the four sub-processes of the main action process: the notification, disposal, recognition and actions. For each of these sub-processes we can specify some characteristics. We can treat them as attributes of a specific F&R action. It gives us the possibility to compare the actions. Having historical data, we can find similar actions and try to predict a development of new actions, as well as the possible threats to human health and property. This approach is consistent with the general CBR cycle [6]. Its application in the case of our system is shown in Figure 1. The further discussion will focus on the problem of a proper representation of the actions and finding similar action objects to the currently investigated one.

In the described case we use a compound object as a representation of the process. The objects, in general, can be divided into two groups: compound and simple objects. The simple objects are elements of the real world, which have their representation capable of being expressed by the adopted ontology. In addition, the following properties arise from their ontological representation:

1. An object always belongs to a certain class or a fixed number of classes in the ontology. A single object may belong to several classes.
2. An object has some properties within its class. Its features may vary with a class.
3. An object may be in relation to other objects in the same ontology.

A compound object is composed of other objects. It can be defined, using ontologies (connects them), and creates a new entity. The compound object has its specific structure, relations and connections between sub-objects.

In our approach, each characteristic point of the process will correspond to an attribute in the representation of the object. This enables us to use attribute values to make comparisons between compound objects. We consider two types of comparisons: by sets of features and by values of features. In order to determine the characteristic points, we analyzed historical data of F&R actions within the particular parts that constitute the sub-processes.

Every F&R action has an unique course. However, we can state that the actions share a common set of events, decisions, risks, etc. Our goal is to foresee the potential events, risks, etc., for an ongoing action. Our idea is based on an assumption that similar actions should be associated with similar decisions or risks. The repository discussed in this section contains a set of actions described by quantitative attributes. Additionally, those actions were assessed by domain experts who assigned them risks defined by the *threat matrix* [7]. In our approach, for current ongoing action, we find the most similar action in our repository and label it with the corresponding risks.

A procedure for defining attributes of objects is quite complex. A balance should be kept between a detailed description and the usability of the representation for processing in a computer system. The representation of an action that is too specific will be impractical in terms of matching with other objects and the assessment of the similarity. On the other hand, too general description will cause finding many falsely similar objects in spite of the lack of information. Therefore, we use a kind of granulation of available historical data to create clusters of processes. The list of the utilized attributes (features), which are processed in particular stages of comparing the sub-processes can be found on Figure 2 in form of comparators. This is a result of our research on the representation of a fire rescue action, and the available data in the EWID system.

3 Proposed Solution and Methods

Our approach is based on the analysis of similarity of F&R actions. We find the most similar action for the currently ongoing one and if the similarity is good enough, then we retrieve the labeled risks. Similarity expresses a *degree of identity*, which can be expressed by a question: in what degree an object a is alike an object b? Values of the similarity degree are often put into an interval [0,1], where *1* means that two objects are indiscernible. Properties of a similarity function are dependent on a domain of objects which are compared. Its values often depend on a context, and thus the similarity is difficult to model using standard methods [8, 9].

We use a framework of compound object comparators [10] for modeling resemblance of F&R actions. This is a new approach to solve the described problem. The earlier research presented in [7] was based on a different modeling techniques for calculating the resemblances.

A network of comparators is a tool for building complex (composite) structures based on a captured specification of compound objects. The network is composed of layers. The subsequent layers can take into account results obtained at the previous layers. The network structure can be derived either from a context associated with a group of features (treating an object as a single entity) or from the analysis of a structure of objects and their relations [11].

In this case, we use a network with two layers. The first layer is composed of four comparators responsible for considering a similarity of particular sub-processes: notifications, disposals, recognitions, actions. Each comparator examines selected features and returns results independently from the others.

Compound object comparators are multi-layered structures consisting of several closely interlinked components for determining the similarity between objects. A comparator can be formally described as a function $C_B : A \to 2^{B\times[0,1]}$, where A is a set of input objects and B is a set of reference objects. Comparator outcomes take a form of weighted subsets of reference objects $C_B(a) = F(\{(b, g(\mu(a, b)) : b \in B\})$, where F is a function responsible for filtering partial results, e.g. *min*, *max*, *top*. Furthermore, $\mu(a, b)$ is a membership function of the fuzzy relation [12], which returns a similarity degree between $a \in A$ and $b \in B$, and $g(x)$ is an activation function which filters out too weak results. We put

$$g(x) = \begin{cases} 0 : x < p, \\ x : x \geq p \end{cases} \tag{1}$$

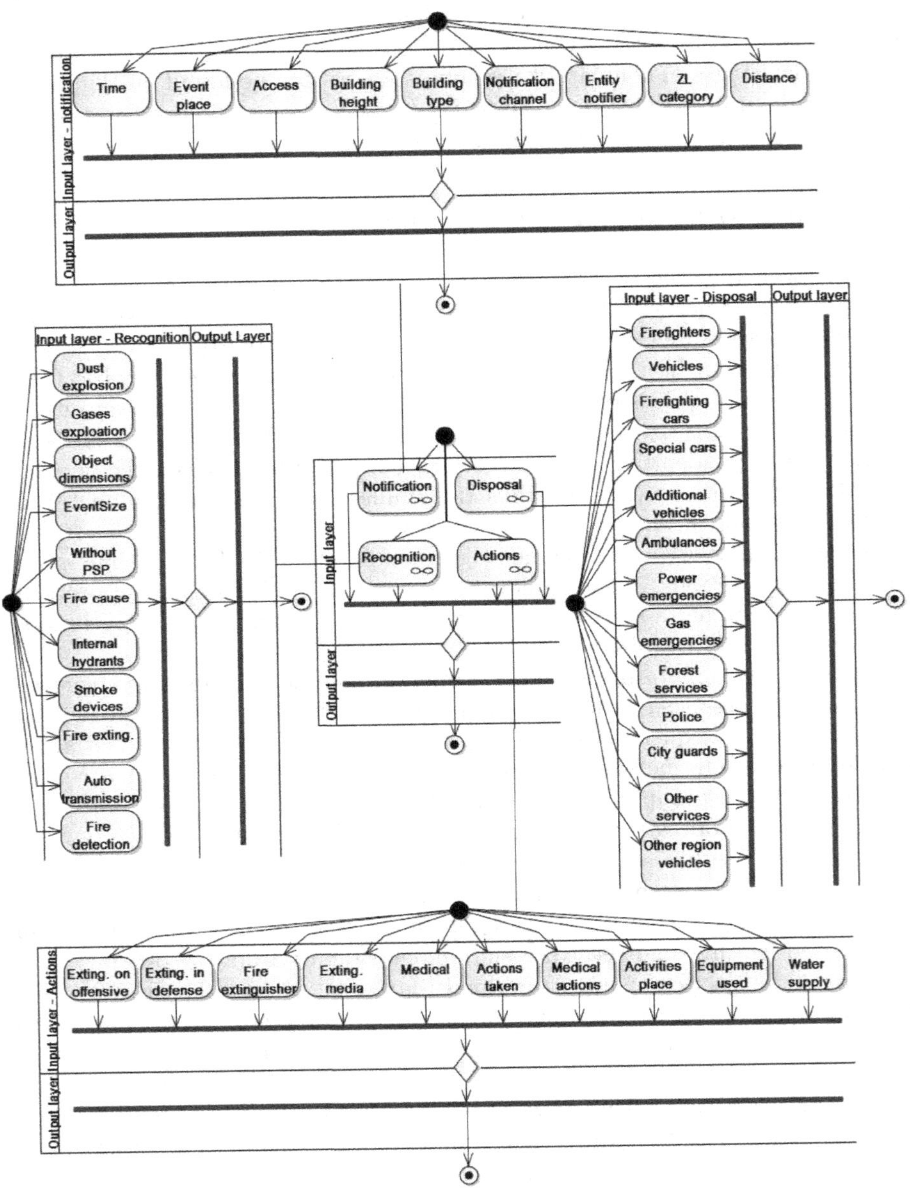

Fig. 2. Four local networks embedded into one heterogeneous network of comparators for finding similar F&R actions. Long black lines are the aggregators, oval boxes - comparators of particular features, diamonds - the translators of results. Abbreviations used: PSP - State fire brigade, ZL category - a threat to humans category

where p denotes the lowest acceptable similarity. One may also introduce some constraints which make $\mu(a, b) = 0$ based on so called exception rules [13].

In the F&R case we have composite comparators which means that they are independent subnetworks of comparators. This in turn means that they have their own layers and comparators. Figure 2 shows details of the constructed network and subnetworks. It indicates all of the proposed comparators for examining particular features. In our solution we use default values of the parameter p=0.5, but in future its value can be optimized. We use different types of the similarity measures (implemented as the functions μ). First type is based on a similarity matrix. This is an expert-driven knowledge, limited to the considered cases. For example, for the purpose of comparing building heights, one can use the following similarity matrix:

$$\mu[v_a][v_b] = \begin{bmatrix} 1 & 0.7 & 0.5 & 0.3 \\ 0.7 & 1 & 0.7 & 0.5 \\ 0.5 & 0.7 & 1 & 0.7 \\ 0.3 & 0.5 & 0.7 & 1 \end{bmatrix}, \quad (2)$$

where v_a and v_b are clustered values of the building height attribute. The optimization of similarity matrices, e.g. using genetic algorithms, has already been studied in the CBR literature [14]. Although the employment of this approach seems to be a promising direction for the future research, in this study no adjustment of the values given by the experts was performed.

One can also use binary measures expressing an identity and a dissimilarity. The next type of measures is used for calculating a similarity of numerical attributes. It is in a form:

$$\mu(v_1, v_2) = \begin{cases} 1 - \frac{|v_1 - v_2|}{\max(v_1, v_2)}, & v_1 \neq 0 \vee \; v_2 \neq 0 \\ 1, & v_1 = 0 \wedge v_2 = 0 \end{cases}, \quad (3)$$

where v_1, v_2 are the numerical non-negatives values of a given attribute. The next one is like a token based measure. It counts compatibility of tokens and finds the most similar pair. The last type implements a hierarchy of investigated features. It makes to find a similarity between objects by considering assignments to a class or a super-class of objects. Additionally, the domain knowledge [15] can be easily applied. We have implemented a couple of domain knowledge standards, such as a building category, a fire category, etc. All the measures have a big impact on the obtained results. For all the comparators used in this study there were no exception rules applied.

After the calculation of the resemblance on a local level we aggregate them into a global similarity for each of the processed pairs of objects. This approach is in accordance with the methods for computing similarity using hierarchical ontologies [16]. We aggregate the results from particular comparators in each input layer of all four subnetworks. We use a default method proposed by the framework. This is a mean value for particular pairs of objects, calculated based on the value returned by each comparator in the layer. The global aggregator

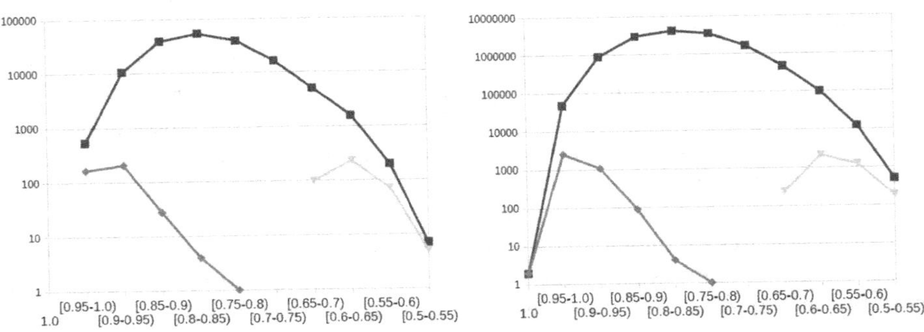

Fig. 3. Similarity distribution for: all, best and worst object pairs, when using leave-one-out. Left chart corresponds to a data set with 406 F&R actions and the right one correspond to the data that describes 3736 F&R interventions. Both charts use logarithmic scale. Y axis shows the number of pairs and the X axis shows the clustered similarity value.

on the subnetwork level do not change the data. It only passes the results to an upper level which is the main network.

The next step is the global aggregation of the main network. It is realized in the output layer. Analogically as in the case of subnetworks, the default aggregation method is the mean. In general, we use two types of the aggregation. The default one and the weighted average. It leads to an interesting optimization problem of finding the best values of weights. There are a lot of other methods which can be used to aggregate the partial results. For calculating the weights we use an evolutionary algorithm. The result of the aggregation is a set of pairs in a form of $\{(b, similarity\_value) : b \in B\}$.

After finding the most similar pairs of objects, for a given action we retrieve the risks labels from the most similar reference object, e.g. we use the 1-NN approach [17]. We treat those risks as a prediction with an assigned chance of materialization.

4 Experiments and the Results

We conducted two experiments. One was performed on a small data set which came from EWID – it consisted of 406 reports from F&R actions. We utilized this set in order to compare our results with the previous experiments [7]. The second experiment was performed on a bigger set of the EWID data. It consisted of 3736 reports. For both experiments we used the leave-one-out method. In total, the data sets contained 164430 and 13953960 pairs of objects, respectively.

This is not a full Cartesian product because some of the pairs were eliminated by the activation functions. The main statistical characteristics for the both data

Table 1. Statistical characteristics of the results set, * - results for 406 objects, ** - results for 3736 objects. All - all F&R pairs, Best - pairs with maximum similarity value taken for each F&R action, Worst - pairs with minimum similarity value taken for each F&R action.

Characteristic	*-all	*-best	*-worst	**-all	**-best	**-worst
max	0.985	0.985	0.690	1.000	1.000	0.676
min	0.530	0.791	0.530	0.477	0.797	0.477
avg	0.815	0.940	0.627	0.811	0.956	0.606
median	0.820	0.946	0.630	0.824	0.961	0.608
standard deviation	0.061	0.028	0.031	0.062	0.023	0.030
variance	0.004	0.001	0.001	0.004	0.001	0.001

Table 2. A performance comparison between several risk classification methods for F&R actions. Abbreviations: ESA - Explicit Semantic Analysis, k-NN - k nearest neighborhood, NoC - Network of Comparators, NoC WA - Network of Comparators with Weighted Average aggregation.

Method	Precision	Recall	F1-score
Naive Bayes	0.68	0.64	0.61
ESA	0.48	0.70	0.54
k-NN canberra	0.74	0.74	0.69
NoC	0.73	0.70	0.66
NoC WA	**0.79**	**0.75**	**0.71**

sets are shown in Table 1. The distribution of the similarity values for the results is presented in an analogical way and is illustrated in Figure 3.

The described results have been achieved with the default parameters of the network such as the activation function value p - 0.5 and global aggregation method - *mean*, with $\frac{1}{n}$ weights for NoC $\frac{1}{n}$. On the other hand, the NoC WA results were obtained using the weighted average global aggregator and the evolutionary algorithm for learning the weights. The winning weight values are: $\frac{2}{89}, \frac{60}{89}, \frac{1}{89}, \frac{26}{89}$ for notification, disposals, recognition and actions comparators, respectively. The weights were learned using a training set consisting of 136 out of 406 F&R actions (33%). The procedure was repeated ten times. The final weights are the ones with the maximum evaluations. The evaluation function took into account the number of recognized risks, as well as the overall prediction quality.

All parameters of the model can be further tuned. The p value can limit the quantity of processed objects, but on the other hand the weights can control the importance of the feature selection. The final results are checked by the risks similarity layer. We compared the already found pairs with their risks. The risks come from the threat matrix described in [7] which is a form of a higher level risks communication to the IC in ICRA.

The final analysis has been divided into two steps. Firstly we have calculated the performance using Precision, Recall and F1-score separately for each of the

Table 3. Comparison (using F1-score) of the classification methods relative to the risks of F&R. The risks are defined as a Cartesian product of threats and objects from the Threats Matrix [2]. Abbreviations of algorithm names: ESA - Explicit Semantic Analysis, kNN - k Nearest Neighborhood, NoC $\frac{1}{n}$ - Network of comparators with the arithmetic mean aggregator, NoC WA - Network of comparators with weighted average aggregator, * - results for 406 F&R actions, ** - results for 3736 F&R actions.

Risk	Naive Bayes*	ESA*	kNN Canberra*	NoC $\frac{1}{n}$*	NoC WA*	NoC $\frac{1}{n}$ **
A1_MA	0.38	0.45	0.34	0.36	**0.39**	0.47
A1_ME	0.86	0.82	**0.91**	**0.91**	0.90	0.91
A1_T	–	0.07	0.09	0.12	**0.16**	0.14
A2_MA	0.81	0.84	**0.89**	0.85	0.88	0.70
A2_ME	0.83	0.84	**0.90**	0.89	0.89	0.84
A2_S	**0.29**	0.22	0.09	0.17	0.1	0.20
A2_T	0.05	0.14	0.09	0.13	**0.17**	0.17
A2_U	0.39	0.30	0.44	0.34	**0.45**	0.38
A4_G	–	0.08	0.21	0.11	**0.25**	0.14
A4_MA	0.30	0.22	0.35	0.24	**0.47**	0.34
A4_ME	0.27	0.17	**0.41**	0.16	0.36	0.33
A4_S	–	–	**0.40**	0.23	0.30	0.34
A4_T	–	0.13	–	–	**0.67**	0.06
E1_MA	–	0.11	**0.48**	0.22	0.42	0.40
E1_ME	–	–	**0.22**	–	–	0.21
E2_MA	0.11	**0.31**	0.17	0.07	0.24	0.30
E2_ME	–	**0.24**	0.20	0.09	0.14	0.24
E2_S	–	**0.15**	0.13	–	0.12	0.03
E3_G	–	0.12	**0.40**	–	–	0.08
E3_MA	–	**0.50**	0.16	0.13	0.17	0.28
E3_ME	–	–	0.12	**0.28**	0.13	0.42
E4_MA	–	–	–	–	**0.14**	0.13
E4_ME	–	–	–	–	–	0.14
E4_S	–	–	–	–	–	0.12

F&R action, and then we averaged the results. Table 2 contains obtained values and compares them with the ones which have been achieved in the earlier research performed on the same set of data. Secondly, we have done the analysis of the performance for particular risks came from *threat matrix*. Table 3 contains a comparison of the results which have been obtained for various methods.

5 Discussion of the Results

As we can see in Table 2 and Table 3, the best results for the considering problem were achieved by the NoC WA method. In the first experiment we evaluated methods using three known measures from the information retrieval domain [18]: Precision, Recall and F1-score. For all these measures our approach showed the best performance. The average Precision was 0.79 while the score of the second best classification algorithm was about 0.05 worse. Analogical results were obtained for the other two measures. These results mean that NoC WA method correctly classified the biggest number of the risks and that it had the smallest number of false classifications.

Further investigation of the results revealed, that even when we look at them from the particular risks point of view, our approach gives the highest results. When we analyze the detailed results shown in Table 3, we can see that the comparison of each pair of presented methods gives the first place in the ranking to NoC WA. We use two evaluations method: first is the number of classified risks and the second is the number of risk with a greater F1-score. In the first case our method has the first place ex aequo with k-NN using the Canberra similarity measure. Both methods classified 20 risks and in two cases these methods were the only ones. One of them is E1_ME (k-NN Canberra) and the second is E4_MA (NoC WA). The second evaluation method compares classifiers in pairs. For each pair and each risk we evaluate which score is better. In this paper we focus on the comparison of the NoC method and the remaining methods. In this competition the NoC WA wins as well. The second best is k-NN Canberra. In this pair NoC WA wins in 12 risk categories and k-NN Canberra wins in 8.

6 Conclusions

We presented an approach to the problem of risks assignment, which is based on the network of comparators. The implementation of a network of comparators is quite easy and intuitive. A domain knowledge, dedicated for a given feature or for the whole processing can be easily injected into the presented solution. Conducted experiments and obtained results show that this is an efficient method for this kind of problems.

The results obtained in this study can be further optimized by checking an impact of different aggregation methods on the global level, as well as the local aggregators for the NoC method.

The problem of the approximation of vague concepts, such as the risks at a fire ground, attracts many researchers from the AI domain. Our approach is a contribution in this field. One of the shortcomings of our approach is a quite laborious process of the definition of the features for the NoC and the qualitative relations between the features (measures of the similarity). However, the effort is rewarded by the good performance of the classification. Moreover, the general trend of defining the ontology in every domain may facilitate the process of the definition of the input and the structure of the comparators.

The future work will focus on a development of complementary expert knowledge base about F&R actions. It can be used to derive new features of the processes representing the actions. Such features will enrich the set of attributes processed within the ICRA project and may contribute to even better performance of our model.

References

1. Han, L., et al.: Firegrid: An e-infrastructure for next-generation emergency response support. Journal of Parallel and Distributed Computing 70(11), 1128–1141 (2010)
2. Krasuski, A., Jankowski, A., Skowron, A., Ślęzak, D.: From sensory data to decision making: A perspective on supporting a fire commander. In: Web Intelligence/IAT Workshops, pp. 229–236 (2013)
3. Bazan, J.G.: Hierarchical classifiers for complex spatio-temporal concepts. In: Peters, J.F., Skowron, A., Rybiński, H. (eds.) Transactions on Rough Sets IX. LNCS, vol. 5390, pp. 474–750. Springer, Heidelberg (2008)
4. van der Aalst, W., Adriansyah, A., van Dongen, B.: Replaying history on process models for conformance checking and performance analysis. Wiley Interdisciplinary Reviews: Data Mining and Knowledge Discovery 2(2), 182–192 (2012)
5. ISO 31000 - Risk management (2009)
6. Aamodt, A., Plaza, E.: Case-based reasoning: Foundational issues, methodological variations, and system approaches. Artificial Intelligence Communications 7(1), 39–59 (1994)
7. Krasuski, A., Janusz, A.: Semantic tagging of heterogeneous data: Labeling fire & rescue incidents with threats. In: FedCSIS, pp. 77–82 (2013)
8. Tversky, A., Shafir, E.: Preference, Belief, and Similarity: Selected Writings. Bradford books. MIT Press (2004)
9. Janusz, A.: Algorithms for similarity relation learning from high dimensional data. In: Peters, J.F., Skowron, A. (eds.) Transactions on Rough Sets XVII. LNCS, vol. 8375, pp. 174–292. Springer, Heidelberg (2014)
10. Sosnowski, Ł., Ślęzak, D.: How to design a network of comparators. In: Brain and Health Informatics, pp. 389–398 (2013)
11. Nguyen, S.H., Bazan, J., Skowron, A., Nguyen, H.S.: Layered learning for concept synthesis. In: Peters, J.F., Skowron, A., Grzymała-Busse, J.W., Kostek, B.z., Swiniarski, R.W., Szczuka, M.S. (eds.) Transactions on Rough Sets I. LNCS, vol. 3100, pp. 187–208. Springer, Heidelberg (2004)
12. Kacprzyk, J.: Multistage Fuzzy Control: A Model-based Approach to Fuzzy Control and Decision Making. John Wiley & Sons, Limited (2012)
13. Sosnowski, Ł., Ślęzak, D.: Networks of compound object comparators. In: FUZZ-IEEE, pp. 1–8 (2013)
14. Stahl, A., Gabel, T.: Using evolution programs to learn local similarity measures. In: Ashley, K.D., Bridge, D.G. (eds.) ICCBR 2003. LNCS (LNAI), vol. 2689, pp. 537–551. Springer, Heidelberg (2003)
15. Szczuka, M.S., Sosnowski, Ł., Krasuski, A., Krenski, K.: Using domain knowledge in initial stages of kdd: Optimization of compound object processing. Fundam. Inform. 129(4), 341–364 (2014)
16. Schickel-Zuber, V., Faltings, B.: Oss: A semantic similarity function based on hierarchical ontologies. In: Proceedings of the 20th International Joint Conference on Artifical Intelligence, IJCAI 2007, pp. 551–556. Morgan Kaufmann Publishers Inc., San Francisco (2007)
17. Mitchell, T.M.: Machine Learning. McGraw Hill series in computer science. McGraw-Hill (1997)
18. Rinaldi, A.M.: An ontology-driven approach for semantic information retrieval on the web. ACM Transactions on Internet Technology 9(10), 10:1–10:24 (2009)

Author Index